Reproductive Strategies in Insects

Reproductive Strategies in Insects

Edited by
Omkar and Geetanjali Mishra

CRC Press
Taylor & Francis Group
Boca Raton London New York

CRC Press is an imprint of the
Taylor & Francis Group, an **informa** business

Cover photo courtesy of Ms. Shriza Rai, Dept of Zoology, University of Lucknow.

First edition published 2022
by CRC Press
6000 Broken Sound Parkway NW, Suite 300, Boca Raton, FL 33487-2742

and by CRC Press
2 Park Square, Milton Park, Abingdon, Oxon, OX14 4RN

Library of Congress Cataloging-in-Publication Data
Names: Omkar, editor.
Title: Reproductive strategies in insects / edited by Omkar and Geetanjali Mishra.
Description: First edition. | Boca Raton : CRC Press, 2022. |
Includes bibliographical references and index.
Identifiers: LCCN 2021041336 | ISBN 9780367488574 (hardback) | ISBN
9781032191959 (paperback) | ISBN 9781003043195 (ebook)
Subjects: LCSH: Insects–Sexual behavior. | Insects--Behavior. |
Insects--Reproduction. | Sexual behavior in animals.
Classification: LCC QL496 .R39 2022 | DDC 595.715--dc23/eng/20211018
LC record available at https://lccn.loc.gov/2021041336

ISBN: 978-0-367-48857-4 (hbk)
ISBN: 978-1-032-19195-9 (pbk)
ISBN: 978-1-003-04319-5 (ebk)

DOI: 10.1201/9781003043195

Typeset in Times
by MPS Limited, Dehradun

Contents

Preface

'Let us investigate more closely this property common to animal and plant, this power of producing its likeness, this chain of successive existences of individuals, which constitutes the real existence of the species'. – *Comte Georges-Louis Leclerc de Buffon*

Reproduction is one of the most primitive, yet essential function that is fundamentally vital to the very existence of life and furtherance of generations. It takes myriad forms and mechanisms depending on the taxa and the complexity of the living being, ranging from asexual to sexual, each of which comes in numerous variants, but it is always there; a sustained mechanism that forms the primal basis of existence and sustenance of all organisms. Or as Louis Schwartzberg said, 'Nature has invented reproduction as a mechanism for life to move forward. As a life force that passes right through us and makes us a link in the evolution of life'.

Reproduction is so primitive and fundamental that the mechanism involved is highly complex and not yet clearly understood. An overview of all taxa from prokaryotes to eukaryotes, plants to animals, the simplest to the most complex, provides an astonishing range of mechanisms that have evolved to ensure the successful propagation of each living organism. These mechanisms have evolved to enhance the way organisms mate, produce offspring or raise them, thereby increasing their direct or indirect fitness levels. However, if there is a single group of organisms that encompasses the most diversity in reproductive mechanisms and strategies, that would be insects.

Insects form one of the most varied and eccentric taxon across the entire range of organisms. Not only are they immensely diverse in shape, size, habitat, and ecosystem services; they employ an equally diverse range of reproductive strategies; encompassing almost all possible modifications across the animal world. Reproduction in insects ranges from purely sexual to parthenogenetic, which itself comes in a wide number of variants.

The search for mates, number of mates, display of mate quality, assessment of mate quality, acceptance of mate, rejection of mates, forced copulations, the fight for paternity pre, during, and post copula, the modulation of paternity, ovipositional strategies, and parental care together form the basis of reproductive strategies that are employed to successfully transmit the genome to future generations. As these myriad steps interconnect for successful reproduction of individuals, it is essential to understand that at each of these steps, the male and female usually do not 'hold hands' but are rather at antagonistic warring ends with each other, a process scientifically known as sexual conflict. In fact it is sexual conflict that plays a major role in the evolution of reproductive strategies.

Further to add to the complexity of the entire situation, most of these behaviours are not fixed in nature but are modulated by environmental factors, both abiotic and biotic. For example, a darker or more melanic individual would turn out to be an exemplary mate in colder climates as it would provide its offspring the warming cloak of darkness. However, if the weather turned warm, the selection of a melanic mate would be a folly. Furthermore to limit oneself to one mate when multiple mates are easily accessible would be as big a folly as scrambling to find more mates, when they are too widely dispersed.

While there are multiple books on animal behaviour, there are relatively fewer on reproductive strategies. Those that concentrate on reproduction are usually detailed volumes on any one aspect such as mating systems, sperm competition, sexual selection, or parental care. This book deals with the morphological, physiological, and ethological aspects of the entire gambit of reproductive strategies, and also deals with their ecological and evolutionary aspects. Further being inspired and focused on insects make this a niche book for ethologists, biologists studying behavioural evolution, and entomologists. Also because of being concentrated on insects, it presents a wide ranging diversity of reproductive strategies in one place.

In this endeavour we were supported by a whole bevy of people. We are really grateful to our research team members; Dr. Shashwat Singh, Dr. Swati Saxena, Ms. Apoorva Shandilya, Ms. Priya

Singh, Ms. Chandni Verma, Ms. Tripti Yadav, Ms. Shriza Rai, Ms. Gaurvanvita Singh, Ms. Lata Verma, and Ms. Dipali Gupta for their constant support throughout the planning and execution of this project. I (Geetanjali) am especially indebted to my parents, Mrs. Munni Mishra and Lt Col. Ramesh Chandra for their unwavering faith and support. We are also thankful to our spouses, Ms. Kusum Upadhyay and Dr. Ravi Shanker Verma for constant encouragements and sacrifice for sparing me for this work. We are also thankful to Dr. Renu Upadhyay and her entire team from CRC press for their constant encouragement and follow-up that helped us complete the work in time.

Editors

Editors

Prof. Omkar, FNASc, has been associated with teaching for about 35 years and research for more than 42 years. He is Former Head, Department of Zoology, University of Lucknow, Lucknow-226007, Former Coordinator, UGC-SAP (DRS-II), DST-FIST, DST-PURSE, Centre of Excellence, Govt. of UP, programs. He has worked on more than 10 projects of state/central funding agencies. He specializes in Environmental Toxicology and Entomology, with particular reference to the Insect Pest Management. He is also an Associate Editor, *International Journal of Tropical Insect Science*, Springer Nature and Chief Editor, *Journal of Applied Bioscience*, besides being President of International Society of Applied Biology.

He is a recipient of several awards, such as Saraswati Samman by the Department of Higher Education, Govt. of UP; Prof. T N Ananthakrishnan Foundation Award, Rescholar Award of Excellence in Agricultural Entomology by Association of Entomologists, Prof. G. S. Shukla Gold Medal by The Academy of Environmental Biology, Dr. S. Pradhan Memorial Lecture Award by The Division of Entomology, IARI & The Entomological Society of India, ACCLAIM and Uddeepan awards from University of Lucknow. He is a Fellow of The Entomological Society of India, Zoological Society of India, Society for Biocontrol Advancement, and Fellow of The National Academy of Sciences India (FNASc), besides many others. He has 16 books to his credit, such as Pesticides, Man & Biosphere, Ecofriendly Pest Management for Food Security, Elsevier, Industrial Entomology, Springer Nature, Pests & Their Management, Springer Nature, Sucking Pests of Crops, Springer Nature and Polyphagous Pests of Crops, Reproductive Strategies of Insect Pest Management, etc. He has published 235 research papers, 10 reviews, 40 reviews as book chapters, and 17 popular science articles. Prof. Omkar has guided more than *two* dozen PhD students. He has visited the School of Biological Sciences, University of East Anglia (2006) for 3 months and Department of Zoology, University of South Bohemia and Czech Academy of Sciences, Ceske Budejovice, Czech Republic (May 2016).

Dr. Geetanjali Mishra, has been associated with teaching for about 14 years and research for more than 21 years. She is a passionate behavioural ecologist, evolutionary biologist and an ardent entomologist. She has been working on insect behaviour and biology for the past 21 years using ladybird beetles as her model of choice. She is Member, Executive Council, Indian Society of Evolutionary Biology. She is also an Associate Editor, *Journal of Applied Bioscience*, and reviews extensively for journals of repute in the field of entomology, behaviour, and ecology.

She has been a Commonwealth Academic Fellow at the Centre for Ecology & Conservation, University of Exeter, Penryn Campus, UK. She has also visited California State University, Long Beach, USA as a part of a select team of teachers to brainstorm over teacher education quality. She has been a university Gold Medalist during her post graduation in Zoology and has also been awarded Prof. T N Ananthakrishnan Young Scientist Award and Prof. B.V. David Women Scientist Award. She has published 1 proceeding, 1 technical bulletin, 85 research papers, 12 reviews as book chapters, and 5 popular science articles. She is also an avid popular science speaker and panelist in radio and TV shows. Dr. Mishra has guided 5 PhD students.

Contributors

Biswas, Payel
School of Biotechnology, Kalinga Institute of Industrial Technology (KIIT)
Bhubaneswar, India

Boopathi, T
ICAR- Indian Institute of Oil seeds Research
Hyderabad, India

Chattopadhyay, Aradhya
School of Biotechnology, Kalinga Institute of Industrial Technology (KIIT)
Bhubaneswar, India

Chaudhary, Desh Deepak
Indira Gandhi National tribal University Amarkantak
Madhya Pradesh, India

Cordero-Rivera, Adolfo
ECOEVO Lab, E.E. Forestal
Universidade de Vigo
Campus A Xunqueira
Pontevedra, Spain

Dubey, Ankita
Ladybird Research Laboratory, Department of Zoology
University of Lucknow
Lucknow, India

Evans, Richard
USDA-ARS National Biological Control Laboratory
Stoneville MS, USA

Ghosh, Shampa M.
School of Biotechnology, Kalinga Institute of Industrial Technology (KIIT)
Bhubaneswar, India

Grodowitz, Michael J.
USDA-ARS National Biological Control Laboratory
Stoneville, MS, USA

Guru Pirasanna Pandi G.
ICAR- National Rice Research Institute
Cuttack, India

Jalali, Sushil K.
ICAR-National Bureau of Agricultural Insect Resources
Karnataka, India

Körner, Maximilian
Institute of Evolutionary Animal Ecology
University of Bayreuth
Bayreuth, Germany

Kramer, Jos
Department of Quantitative Biomedicine
University of Zürich
Zürich, Switzerland

Kumar Bhupendra
Department of Zoology
Banaras Hindu University
Varanasi, India

Meunier, Joël
Institut de Recherche sur la Biologie de l'Insecte (IRBI)
University of Tours
Tours, France

Mishra, Geetanjali
Ladybird Research Laboratory,
Department of Zoology
University of Lucknow
Lucknow India

Navik, Omprakash
ICAR-National Bureau of Agricultural Insect Resources
H. A. Farm Post
Hebbal, Bangalore
Karnataka, India

Nebapure, Suresh
ICAR- Indian Agricultural Research Institute
New Delhi, India

Omkar
Ladybird Research Laboratory,
Department of Zoology
University of Lucknow
Lucknow, India

Pervez, Ahmad
Biocontrol Laboratory, Department of Zoology, Radhey Hari Government P.G. College
Kashipur, Udham Singh Nagar
Uttarakhand, India

Prasannakumar, N.R.
ICAR-Indian Institute of Horticulture Research
Bengaluru, India

Rai, Shriza
Ladybird Research Laboratory, Department of Zoology
University of Lucknow
Lucknow, India

Ribeiro, José Ricardo Inacio
Laboratório de Estudos da Biodiversidade do Pampa (LEBIP)
Universidade Federal do Pampa
Campus São Gabriel, Avenida Antônio Trilha
São Gabriel RS, Brazil

Rodriguero, Marcela S.
Departamento de Ecología, Genética y Evolución, Facultad de Ciencias Exactas y Naturales
Universidad de Buenos Aires
Av Intendente Güiraldes
Pabellón II, Ciudad Universitaria
Ciudad Autónoma de Buenos Aires, Argentina

Sagar D
ICAR- Indian Agricultural Research Institute
New Delhi, India

Saxena, Swati
Ladybird Research Laboratory, Department of Zoology,
University of Lucknow
Lucknow, India

Singh, Garima
Department of Zoology, University of Rajasthan
Jaipur, India

Singh, Priya
Ladybird Research Laboratory, Department of Zoology
University of Lucknow
Lucknow, India

Singh, Rajendra
Department of Zoology
Deendayal Upadhyay Gorakhpur University
Gorakhpur, India

Singh, Shashwat
Institute of Plant Protection, Department of Entomology, Agricultural Research Organization, the Volcani Center
Rishon LeTsiyon, Israel

Soumia P.S.
ICAR- Directorate of Onion and Garlic Research
Pune, India

Stefanello, Fabiano
Laboratório de Biologia Comparada e Abelhas (LBCA), Departamento de Biologia, Faculdade de Filosofia, Ciências e Letras deRibeirão Preto (FFCLRP)
Universidade deSão Paulo Ribeirão
Preto, Brazil

Subramanian S
ICAR- Indian Agricultural Research Institute
New Delhi, India

Varshney, Richa
ICAR-National Bureau of Agricultural Insect Resources
Bangalore, Karnataka, India

Yele, Yogesh
ICAR- National Institute of Biotic Stress Management
Raipur, India

1

Various Modes of Reproduction

S. Subramanian[1], T. Boopathi[2], and D. Sagar[1]
[1]*ICAR-Indian Agricultural Research Institute, New Delhi, India*
[2]*ICAR-Indian Institute of Oil seeds Research, Hyderabad, India*

CONTENTS

DOI: 10.1201/9781003043195-1

1.1 Introduction

Reproduction is a fundamental biological process for propagation of species. Asexual and sexual are two different forms of reproduction. In asexual reproduction, an organism can reproduce without involvement of another organism, creating a genetically similar or identical copy of itself. In sexual reproduction, two reproductive cells (male and female gametes) are created by meiosis and by fertilization of a female gamete with a male gamete, which leads to formation of a zygote. The zygote develops into an embryo to give rise to offspring, whose genetic make-up is derived from both of their parental organisms.

Insects are one of dominant life forms on earth; a state that can partly be attributed to their diverse reproductive strategies. They are equipped with high reproductive rates and adopt a number of behavioural and physiological adaptations for survival in diverse ecological habitats.

Sometimes, eggs are retained by females after fertilization and eggs continue to grow before they are laid. If this internal period of retention of eggs is longer, eggs develop into larvae and female gives birth to larvae (Viviparous). Viviparity assumes different forms, such as ovoviviparity, pesudoplacental, haemocoelous, and adenotrophic viviparity. Sometimes, more individuals develop from a single egg (polyembryony). Certain species of insects mature precociously and start producing young ones even when they are still at larval or pupal stage (paedogenesis). At times, whole of their life cycle or in alternate generations, some insects undergo asexual mode of reproduction, as eggs develop without being fertilized (parthenogenesis).

Different modes of reproduction in insects have earlier been reviewed by several workers: Blackburn (1999) on viviparity and oviparity, Hagan (1951) on viviparity, Suomalainen (1962) and White (1964) on parthenogenesis, Ivanova-Kasas (1972) on polyembryony, and Kerr (1962) on sex determination. These studies provide an overview of reproductive and functional significance, role of reproductive symbionts on manipulation of sex of insects, and as such on mode and physiology of reproduction.

1.2 Modes of Reproduction in Insects

1.2.1 Oviparity

Most insects are oviparous in nature, i.e. female insects lay eggs and embryonic development occurs after oviposition by utilizing yolk in deposited eggs. The process of egg laying consists of ovulation followed by fertilization and oviposition. Females usually deposit their eggs either on or near food source through posterior abdominal segment modified as ovipositor or an egg-laying organ (Figure 1.1). In Lepidoptera, Coleoptera, and Diptera, posterior abdominal segments of female abdomen itself function as ovipositor, as they can be protracted into a telescopic tube in which opening of egg passage is close to the distal end.

1.2.2 Viviparity

Some species are viviparous in that the egg is retained by mother after fertilization and hatching of eggs and development of young ones takes place within mother. The nutrition required for development of embryos is directly received from the parent. Generally, viviparous species produce fewer offspring than oviparous species and this may be due to lesser number of ovarioles.

There are four distinct types of viviparity found in different insect groups.

1.2.2.1 Ovoviviparity

Eggs are incubated inside the reproductive tract of a female and hatching of eggs occurs just prior to or soon after oviposition; young ones are released immediately after hatching. As eggs contain yolk for nourishment, no special nutritional structures are present. Ovoviviparity occurs across several insect orders, such as Ephemeroptera, Dictyoptera, Psocoptera, Hemipera, Coleoptera, etc. Fecundity is

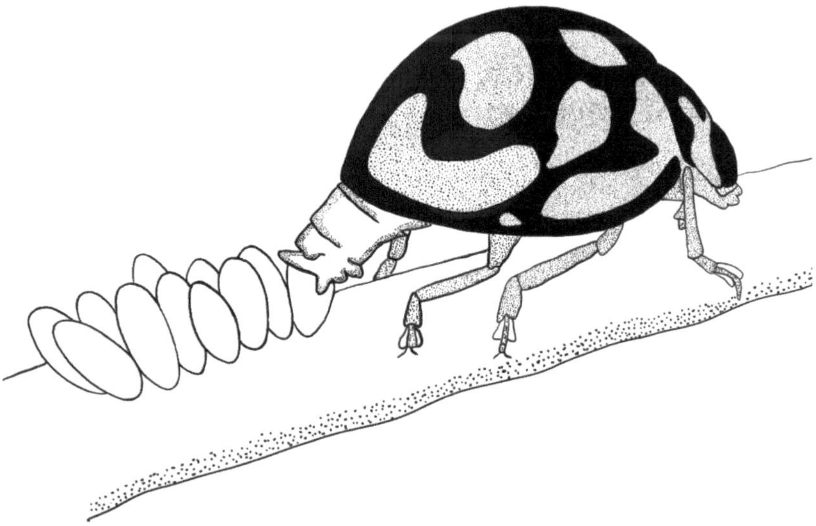

FIGURE 1.1 Oviposition by South African ladybird beetle (Oviparous type of reproduction).

relatively low when compared to oviparity as greater protection is afforded to eggs carried by females. Size of eggs is also relatively larger as eggs contain more nutrients to support embryo development beyond normal hatching and larvae are born in late stage of development.

1.2.2.2 Pseudoplacental Viviparity

The eggs contain little or no yolk. Yolk-deficient eggs develop inside the genital tract of a female. Nutrients are transferred to developing embryos through special maternal structures called pseudoplacentae. It does not involve any oral feeding and larvae are laid upon hatching. Oocyte is accompanied by a single nurse cell and enclosed by a follicular epithelium, which differentiates into a form maternal pseudoplacenta. The embryo develops in pseudoplacental cavity and is connected with pseudoplacentae through cytoplasmic processes extending from cells of amnion and serosa of egg. It is presumed that nutrients pass from pseudoplacentae to fluid surrounding the embryo, thus allowing developing young ones to obtain nutrients (aphids, earwigs, psocids, and polytenid bugs).

1.2.2.3 Haemocoelous Viviparity

Embryos develop freely in a female's haemolymph with nutrient uptake taking place by osmosis. This occurs in Strepsiptera and some gall midges. Haemocoelous viviparity differs from other forms of viviparity, as development of young ones takes place in haemocoel of a female adult. In Strepsiptera, no oviducts are present and mature oocytes are released into haemocoel by rupture of ovarian cells. Sperm enters through genital canal present on ventral midline of female adults and fertilize eggs in haemocoel. Development of young ones occurs in haemocoel with nourishment from haemolymph. The eggs hatch and larvae find their way out through genital canals.

1.2.2.4 Adenotrophic Viviparity

When a poorly developed larva hatches from an egg, it feeds orally on accessory (milk) gland secretion within the uterus of mother's reproductive system. The embryonic development follows as in ovoviviparity, but when larva hatches, it remains in uterus and receives nutrition from special maternal glands. The full-grown larva is deposited and pupation occurs immediately. In tsetse flies, oocytes are fertilized in uterus and embryonic development occurs within. The larva hatches inside uterus, on the ventral wall of which there is a small pad of glandular cells with a cushion of musculature.

This structure is known as choriothete and is responsible for removing chorion of eggs. The larvae feed in uterus on secretion from specialized accessory glands known as milk glands; nutrients to which are derived from blood meal taken by parent fly.

1.2.3 Parthenogenesis

Parthenogenesis is common in animal kingdom. Charles Bonnet was the first to discover the parthenogenesis phenomenon in aphids in 1740. During the next hundred years, further experiments with aphids, bagworm moths, drone bees, and silkworm moths verified this phenomenon.

There are different types of parthenogenetic forms based on occurrence, such as facultative (bees), obligate (sticky insects), and cyclic or sporadic (aphids). Based on sex produced by parthenogenesis, it can be classified into three forms, such as Arrhenotoky (bee), Thelytoky (aphids), and Amphitoky or deuterotoky (cynipid wasp). Based on cytological basis, it can be categorized as Apomictic (no meiosis) or Automictic (meiosis occurs but diploidy is maintained). Detailed description of parthenogenesis is available in later part of this book.

1.2.3.1 Parthenogenetic Forms Based on Sex

1.2.3.1.1 Thelytoky

There are two distinct forms of thelytokous parthenogenesis in insects: facultative and obligate. Hymenopteran insects (ants, bees, and wasps) are characterized by a haplodiploid sex-determining system, wherein females arise from unfertilized eggs. However, in ant, *Cataglyphis cursor,* unmated females may produce diploid daughters from unfertilized eggs through thelytokous parthenogenesis (Pearcy *et al.*, 2006). Thelytoky allows females to pass along their successful genotypes to all of their offspring so as to produce only daughters, thereby maximizing rate of increase and eliminate need for finding or attracting a mate.

Facultative or obligate thelytoky occurs sporadically in over 80 families of super class Hexapoda. Thelytoky is present in Thysanoptera, Psocoptera, Hemiptera (especially suborder Sternorrhyncha) and Phasmatodea, while it is of low occurrence in orders Lepidoptera, Diptera, and Coleoptera. However, thelytoky occurs at high rate in weevils (Coleoptera: Curculionidae), bagworm moths (Lepidoptera: Psychidae), and chironomid midges (Diptera: Chironomidae). In Hymenoptera, it has been documented in some taxa, including Cynipidae, Tenthredinidae, Aphelinidae, Ichneumonidae, Apidae, and Formicidae.

1.2.3.1.2 Arrhenotoky

Arrhenotoky is often referred to as "haplodiploidy" with males developing from unfertilized haploid eggs, while females develop from fertilized diploid eggs. The entire order Hymenoptera and many species in Homoptera, Thysanoptera, and Coleoptera are arrhenotokous. It also occurs in some scale insects (Hemiptera: Margarodidae), whiteflies (Hemiptera: Aleyrodidae), some bark beetles, and *Micromalthus* sp. of monotypic family Micromalthidae.

1.2.3.1.3 Deuterotoky

In deuterotoky, both males and females are developed parthenogenetically. In some species that usually reproduce by thelytoky, a very small percentage of individuals develop into males, which may or may not be able to mate. The term deuterotoky is also sometimes applied to cyclic and facultative parthenogenesis, which is seen in aphids (Simon *et al.*, 2002). Deuterotoky involves development of unfertilized eggs into either males or females, and at least one insect, a mayfly, is reported to exhibit facultative deuterotoky.

1.2.3.2 Parthenogenetic Forms Based on Cytology

In parthenogenetic forms, eggs develop into new organism without fertilization and this absence of fertilization causes alterations cytologically and genetically. Parthenogenetic forms are often accompanied by lack of chromosome conjugation or reduction, when compared to normal ones having undergone fertilization. The genotype of gametes in parthenogenetic forms is not haploid as in sexual reproduction but it is diploid set of chromosomes as in somatic cells. In some parthenogenetic forms, diploidy is also restored albeit in an abnormal manner. Hence, an understanding of cytological basis of parthenogenetic forms gives insight into genetic nature of parthenogenesis. Parthenogenesis can be divided into three main types on cytological basis: apomictic, automictic, and generative or haploid parthenogenesis.

1.2.3.2.1 Apomictic

In apomictic parthenogenesis, meiotic features are lacking fully or in part. Only one maturation division takes place in eggs and division is an equatorial one. No chromosome reduction takes place during maturation division, and zygoid chromosome number and whole genome is thus maintained (Suomalainen, 1962). Apomictic parthenogenesis is most common in Homoptera (Suomalainen, 1962), Diptera (White, 1964), and Hymenoptera (Peacock & Sanderson, 1939).

It is the simplest type of parthenogenesis. Meiosis is totally absent and, consequently, genetic recombination is not possible; offspring retains exact genetic constitution as mother. Genotypically new forms may arise only through mutations.

1.2.3.2.2 Automictic

Early stages of meiosis in egg are quite normal in this type of parthenogenetic forms, i.e. it is similar to those of forms with fertilization. The chromosomes pair at zygotene, crossing over occurs between them, and they form bivalents. As a result of chromosome reduction during meiosis, gametes are haploid in nature. However, it is restored soon to diploid status via fusion of two zygotic nuclei. The fusion of nuclei occurs due to some kind of endomitosis. The developed animals always have a diploid soma and are homozygous. This mode of parthenogenesis has been recorded in Phasmidae (Nakano *et al.*, 2019), Homoptera (Nur, 1971), Thysanoptera, Diptera, and Hymenoptera (Suomalainen *et al.*, 1976).

1.2.3.2.3 Generative or Haploid Parthenogenesis

Regular chromosome conjugation and reduction take place in eggs in this type of parthenogenetic forms. In spite of this, eggs may develop either through fertilization and give rise to a diploid female or without it resulting in production of haploid male. The eggs of a single female may thus develop into both females and males; latter being of parthenogenetic origin. Generative parthenogenesis is, therefore, always facultative and arrhenotokous (male-producing). This type of parthenogenesis has been reported in Homoptera, Thysanoptera, and Hymenoptera (Whiting, 1945).

1.2.3.2.4 Cyclical Parthenogenesis

It is characterized by a heterogeneous alternation of generations, one or usually several parthenogenetic generations alternating with a bisexual generation. This kind of alternation of generations is favourable for organism, as species is able to reproduce quickly through thelytoky. The speed of reproduction is also accelerated in certain forms of paedogenesis (*Micromalthus*) and by viviparity (many aphids). In addition, favourable gene combinations may be widely spread in animals with cyclical parthenogenesis.

1.2.4 Paedogenesis

Sometimes immature insects that mature precociously are able to reproduce by a phenomenon known as paedogenesis. Life cycles are reduced by loss of adult and pupal stages. In this precocious stage, gonads develop and give birth to young ones by parthenogenesis or by viviparity.

Larval paedogeneis or larval giving birth to larvae or even oviposition is more common in *Miaster* (Phasmidae) and *Micromalthus* sp. (Coleoptera). Young larvae are set free in body cavity of paedogenetic larvae and they feed on maternal tissues and eventually emerge out from body wall of parent. In *Micromalthus* sp., besides adult male and female, there are larvae producing males or females or both. The form emerging from egg is called a triungulin, which moults into apodous larvae that can develop into pupae and into an adult female; or to male-producing paedogenetic larvae or female-producing paedogenetic larvae.

Pupal paedogenesis sporadically occurs in gall midges. The embryos are formed in haemocoel of paedogenetic mother pupa termed Hemipupa, which differs from a normal pupa. A brood of larvae is released into haemocoel of parent larvae and they survive in Hemipupa for up to 18 months, from whereafter they escape.

1.2.4.1 Paedogenesis and Neoteny

Paedogenesis appears to have evolved to allow maximum use of locally abundant but ephemeral larval habitats, such as a mushroom fruiting body. When a gravid female detects an oviposition site, eggs are deposited, and larval population builds up rapidly through paedogenetic development. Adults are developed only in response to conditions adverse to larvae, such as food depletion and over-crowding. Adults may be female only, or males may occur in some species under specific conditions.

There are no reproductive adaptations beyond precocious egg development in true paedogenetic taxa. In contrast, in neoteny, a non-terminal instar develops with reproductive features of adult, including ability to locate a mate, copulate, and deposit eggs (or larvae) in a conventional manner. For example, scale insects (Hemiptera: Coccoidea) have neotenous females. In coccids, final nymphal instars moult into a winged adult male; whereas one or more instars are omitted in development of female that is sedentary nymph-like or larviform with a vulva and developing eggs. Neoteny also occurs in Strepsiptera with female development ceasing at puparium stage.

1.2.5 Polyembryony

This form of asexual reproduction involves production of two or more embryos from one egg by subdivision, mostly observed in parasitic insects (*Platygaster*). Nutrition for a large number of developing embryos cannot be supplied by original egg and is acquired from host's haemolymph through a specialized enveloping membrane called trophamnion.

Many insects are terrestrial organisms that expose their eggs to atmosphere. Adaptations to develop under these conditions require presence of a strong chorion that protects embryo against desiccation and an abundant yolk supply to provide nutrients required for growth. For this type of egg morphology, most insects undergo syncytial cleavage, resulting in several thousand nuclei residing in a specific cytoplasm before cellularization. Then these nuclei migrate to egg surface, creating a cellular blastoderm. This blastoderm forms a single-layered epithelium above yolk, forming primordium for future embryonic development. Future epithelial invaginations may form insect's mesodermal and endodermal derivatives. This mechanism is surprisingly conserved in insects, although size of initial embryonic primordium from blastoderm can differ greatly in individual species (Sander, 1976).

Embryogenesis of polyembryonic wasps greatly varies from that of other insects, as thousands of progenies would come from a single egg to clonally develop. Unlike other insects, for example, polyembryonic species are subject to holoblastic cleavage such that they are embryogenized from an early stage in cellularized environments (Grbic *et al.*, 1996). Another interesting characteristic associated with insect polyembryony is that during its development embryonic mass increases in size (Ivanova-Kasas, 1972). This developmental characteristic is related to mammals (Davidson, 1990; Gurdon, 1992) and is considered unusual in other metazoans. Because insect polyembryony emerges solely from endoparasitic taxa, developmental changes in endoparasitic lifestyles are likely to lead to conditions that favour development of polyembryony (Strand & Grbic, 1997).

Marchal (1898) reported occurrence of polyembryony in insects (1898) and since then it has been described in Hymenoptera (bees, wasps, and ants) and Strepsiptera (Ivanova-Kasas, 1972). In Hymenoptera, most extreme examples of obligatory polyembryony are those of some species that contain more than 2,000 embryos from each of their laid eggs (Strand & Grbic, 1997). Since polyembryony in strepsipterans is so little known, most discussions are focused on polyembryony in wasps (Noskiewicz & Poluszynski, 1935).

Polyembryony was initially identified in parasitic wasps using light microscopy in early 20th century (Baehrecke & Strand, 1990). *Copidosoma floridanum*, polyembryonic wasp, is perhaps the most extreme form of polyembryony that has been recorded in literature. Usually, up to 2000 embryos develop from a single egg (Baehrecke *et al.*, 1993). *C. floridanum* oviposits its egg into *Trichoplusia ni* egg. The wasp egg is tiny (60 µm), without yolks, and is cleaved into a single embryo (Grbic *et al.*, 1996). After host egg hatches, *C. floridanum* embryo extends to thousands of embryos in host larva. When host larva enters its final (fourth) position, these embryos begin morphogenesis and ultimately grow into adult wasps. It remains unclear how embryonic polarity in each of 2000 clonal progenies is formed following such large cell division. A gene that controls late stage embryonic patterns in cellular environment in *C. floridanum* is expressed as a preserved periodic pattern in relation to *Drosophila* (French, 1996). The mechanism by which this regulation occurs within individual embryos remains unknown.

Polyembryony likely has spread over Hymenoptera four independent times (approximately 40 million years is divergence time) and polyembryonic species are fairly closely related in each case to monoembryonic species (Strand & Grbic, 1996). This repeated and very rapid occurrence of this extreme development shift indicates that improvements required for transition to polyembyrony can be found in comparing morphogenesis and patterning mechanisms of different hymenopteran species. A full course of polyembryonic development from current literature cannot be reconstructed. For an understanding of embryonic patterning processes in polyembryonic wasps and developmental changes leading to polyembryony, precise knowledge of cellular events during embryogenesis needs to be collected.

1.2.5.1 Polyembryony in Hymenoptera

In a comparative review (Strand & Obrycki, 1996), an analysis of basic life history characters revealed that polyembryonic wasps in all families have similar life history. All polyembryonic wasps lay small, yolkless eggs, which appear holoblastic. Polar bodies form an enveloping membrane, proliferation occurs in an increasing number of embryos, when extraembryonic membrane partitions are rounded, loosely aggregated embryonic cells. As in *C. floridanum*, germ-band extension is generated by an extension of embryonic primordium in polyembryonic platygasterids, emphasizing that despite apparent lack of any syncytial stage, at least a few other polyembryonic wasps are likely to be long germ-band insects. This conservation of polyembryonic wasps in their external characteristics is remarkable, although there are differences in early development of these species (Anderson, 1972). Such a punctuated alteration in early development reveals a shift in history of life in polyembryony in Hymenoptera. It is thus instructive to look at what this change may have been and how embryological processes that regulate polyembryony may have come from their closest ancestors.

1.2.5.2 Evolution of Polyembryony

When considering evolution of polyembryony, we must first evaluate direction of development of Hymenopteran history. The majority of hymenopterans are of the sub-order Apocrita, which consists of free living and parasitic species. The free-living apocritans, like honeybees and ants, are confined to a single group (Aculeata) while rest are parasitic. Parasitic wasps exhibit two different strategies for development (Strand, 1986). Some species are ectoparasitoids, which deposit their eggs on outside a host and feed their larvae by rasping a hole through cuticle of host. Other species develop as endoparasitoids by injecting their eggs into haemocoel of host, which feeds blood and tissue of their progeny. The ectoparastic Orussoidea is considered as sister group to Apocrita (Whitfield, 1992; Dowton & Austin, 1994) based on molecular and morphological data. This is consistent with an observation of ectoparasite basal groups in most apocritan superfamilies. It also indicates that: (1) all apocritans and free living species have probably evolved from an ectoparasitic ancestor, and (2) endoparasitism has evolved independently from different ectoparasitic ancestors at least eight times.

1.2.6 Hermaphroditism

The occurrence of male and female gonads in same individual is called hermaphoditism. It may be a functional hermaphroditism as in case of cottony cushiony scale, *Icerya purchasi* (Hemiptera: Margarodidae). Several species of *Icerya,* that have been studied cytologically are gynomonoecious hermaphrodites, as they are female-like but possess an ovo-testis (a gonad that is a part testis and a part ovary). In these species, occasional males arise from unfertilized eggs and are apparently functional, but normally self-fertilization is assured by production of male gametes prior to female gametes in body of one individual (protandry of hermaphrodite). Non-functional hermaphroditism is noticed in case of stonefly, *Perla marginata.* In this species, testes are fused anteriorly and as soon as testicular follicles are differentiated, oogenesis does not proceed further. Oogenesis is not completed and partially grown eggs degenerate.

1.3 Alternation of Generations

Some insects are reproduced by alternating sexual and asexual generations. Females typically produce males and females, but at some point, females cease producing males and only produce females parthenogenetically. Subsequent generations of males start development again, allowing sexual reproduction before parthenogenetic process begins again. This occurs most commonly in Hymenoptera and Hemiptera.

1.3.1 Aphids (Hemiptera: Aphididae)

Aphids are remarkable as different female morphs can reproduce parthenogenetically and become virginopara/vivipara without mating as males do not exist. At other times, females may sexually reproduce and be oviparous. The development of female embryo depends on whether it is meant to become a vivipara (live birth) or ovipara (lays eggs) and is influenced by genetic and environmental factors. The aphid life cycle is discussed in much greater detail in Chapter 13.

1.3.2 Gall Wasps (Hymenoptera: Cynipidae)

Many cynipids have a dynamic life cycle that requires heterogeny or alternation of generations. As a result, a bisexual (both male and females) generation alternates with a unisexual generation (all females). By parthenogenesis, female generation reproduces and unfertilized eggs become sexual offspring. Due to haplodiploid genetics of wasps, all males will develop from unfertilized eggs. Females are, therefore, product of chromosome replication in unfertilized nucleus of cell. The wasps from two generations also look different morphologically and target same or different plant structures and create very different galls. As a result, insects of both generations were sometimes misrepresented as separate species.

1.4 Castration

The destruction or alteration of gonad tissues is called castration. It is one of the strategies used by parasites to modify host tissues and is also called parasitic castration. Parasitic castration is an infectious strategy that requires eventual intensity-independent elimination of host reproduction as primary means of acquiring energy (Lafferty & Kuris, 2002). The parasite responsible for reduction or suspension of fecundity is termed as parasitic castrator. In parasitic castration, there will be modification of host secondary sexual characteristics and may also have physiological and behavioural effects. Parasitic castration is one of the reproductive disturbances induced by parasites of insects, where there is widespread occurrence of parasite induced reduction in host fecundity via manipulation of hosts by influence of parasite genes. Suppression of host reproduction by parasitic castration falls into one of the three following categories: inhibition of host mating due to morphological or behavioural changes, destruction of host reproductive tissue, and reduction of egg output due to malfunction of ovarian yolk sequestration. Parasitic castrators involve some strespiterans and *Rodolia* sp.; they consume adult hosts of braconids and Diptera (Kuris, 1974). In doing so, they consume reproductive tissues early, a cautious step in lengthy and precise process of fully consuming an adult host (Hall *et al.*, 2007). Apart from insects, fungi (Microspora), flat worms (Dilepididae and *Acanthocephala*), and horse hair worms (Nematomorpha) also act as parasitic castrators (Lafferty & Kuris, 2009).

1.5 Genetic Systems of Diverse Reproductive Forms

Most insects are diploid (2n) in their somatic cells and haploid (n) in their gametes. Other systems can be found; some insect groups are parthenogenetic and may be polyploid, including species in Orthoptera (Blaberidae, Tettigoniidae), Homoptera (Coccidae Delphacidae), Embioptera (Oligotomidae), Lepidoptera (Psychidae), Diptera (Chamaemyiidae, Chironomidae, Psychodidae, Simuliidae), Coleoptera (Ptinidae, Chrysomelidae, Curculionidae), and Hymenoptera.

Thelytokous insect species have females only. Thelytoky may be the only mode of reproduction in a species, or it may alternate with sexual reproduction in regular manner (cyclical thelytoky) in some cases of thelytoky; eggs only develop after penetration by a sperm (pseudogamy or gynogenesis). In Homoptera, both arrhenotoky and thelytoky occur but even more complex genetic systems can be found. For example, in some mealybugs (Pseudococcidae), both males and females develop from fertilized eggs, but in embryos that develop into males, paternally derived chromosomes become heterochromatic, genetically inactive, and are not transmitted to male progeny. This genetic system has been called parahaploidy.

1.5.1 Ploidy in Insects

The discussion of ploidy is confusing because, in most insects, some of the somatic tissues exhibit high levels of endopolyploidy. Haploid male honey bees have about same amount of DNA, as females in some of their somatic tissues because nuclei of male undergo compensatory endomitosis so that equal amounts of DNA are present. Polyploid insects usually are 3n or 4n, but exceptions include curculionid weevil species that are 5n and 6n. Owing to lack of fertilization, there is no gene recombination and subsequent segregation, and if conjugation and reduction are also absent, new gene combinations cannot rise even by crossing over.

1.6 Role of Reproductive Symbionts

1.6.1 Direct Manipulation of Host Sex Determination

Sex determination can be affected by inherited bacterial endosymbionts (See Table 1.1. For bacterial endosymbionts associated with sex manipulation in insects). Disrupting mode of sex determination of their hosts could be advantageous for endosymbionts because they are predominantly transmitted vertically through female egg cytoplasm, and not via male sperm. Thus, males represent dead ends for such microorganisms. Consequently, any effect of endosymbiont that distorts host sex ratio towards females will be selectively advantageous for endosymbiont. Recent surveys showed that >30% of sampled arthropods is infected by most common reproductive parasites. Among them, *Wolbachia* is not only the most common bacterial endosymbiont involved in reproductive parasitism but is also the only one known till date to induce all four commonly recognized types of reproductive manipulations:

i. Cytoplasmic incompatibility: Sperm–egg incompatibility leading to post-zygotic sterility between infected males and females that are uninfected or infected with a different endosymbiont strain.

TABLE 1.1

Bacterial endosymbionts associated with sex manipulation in insects

Endosymbiont	Infected Arthropod Host Groups	Manipulation Phenotypes
Wolbachia	Insects, crustaceans, mites, spiders	F, PI, CI, MK
Cardinium	Insects, mites, spiders	F, PI, CI
Rickettsia	Insects, spiders	PI, MK
Spiroplasma	Insects	MK
Flavobacteria	Insects	MK
Arsenophonus	Insects	MK

F: feminization of genetic males; PI: parthenogenesis induction; CI: cytoplasmic incompatibility; MK: male killing.

ii. Male killing (MK): Sex-ratio distortion towards females through targeted death of male progey.
iii. Feminization of genetic males: Sex-ratio distortion towards females through conversion of genetic males into functional females.
iv. Parthenogenesis induction: Sex-ratio distortion towards females through induction of asexual daughter development.

1.6.2 Cytoplasmic Incompatibility

It favours endosymbiont transmission without host sex-ratio distortion. Among different reproductive manipulations induced by bacterial endosymbionts, cytoplasmic incompatibility (CI) is most widespread phenotype: it has been described in mites, isopods, and insects and is caused by *Wolbachia* and *Cardinium*. Unlike feminization, parthenogenesis induction and MK, CI does not induce host sex-ratio distortion. Instead, CI favours infected female reproduction and, thereby, transmission of maternally inherited endosymbionts. CI has been particularly studied in *Wolbachia*.

CI *Wolbachia* is thought to cause sperm modification in infected males. Although modified sperm can be rescued by infected oocytes and lead to viable progeny, uninfected females fail to rescue modified sperm and abort. CI-induced sterilization of uninfected females, thus leads to spread of females carrying *Wolbachia* in uninfected populations.

Bidirectional CI is also observed when both sexes are infected by different *Wolbachia* strains, which are not able to rescue sperm modification induced by other strain. Several studies have shown that bidirectional CI between diverging populations could promote speciation that can be reinforced by premating isolation.

Although CI molecular mechanisms remain largely unknown, cytological studies in different insects and isopod species have described asynchrony in development of male and female pronuclei, leading to defects of first mitotic division of embryo. Delay of histone H 3.3 phosphorylation in male pronucleus, which is required for initiation of mitosis, induces late male chromosome condensation during first metaphase and exclusion during following anaphase, leading to embryonic lethality. If both male and female insects are infected with *Wolbachia* – progeny will be infected. If female is infected and male is not infected, progeny will all be infected. If female is not infected and male is infected, there will not be any progeny (See Figure 1.2 for schematic illustration of *Wolbachia*-induced cytoplsmic incompatibility).

1.6.3 Male Killing: Sex-Ratio Distortion without Manipulation of Host Sex Determination

MK is an adaptation to maternal transmission used by several bacterial endosymbionts, such as *Wolbachia, Rickettsia*,,, and *Flavobacteria* that are found in five insect orders and acari. Embryonic male killers (i.e. early male killers) are found in host species with environmental, XO, XY, ZW, and haplodiploid sex-determination systems. Contrary to feminization and parthenogenesis induction, which

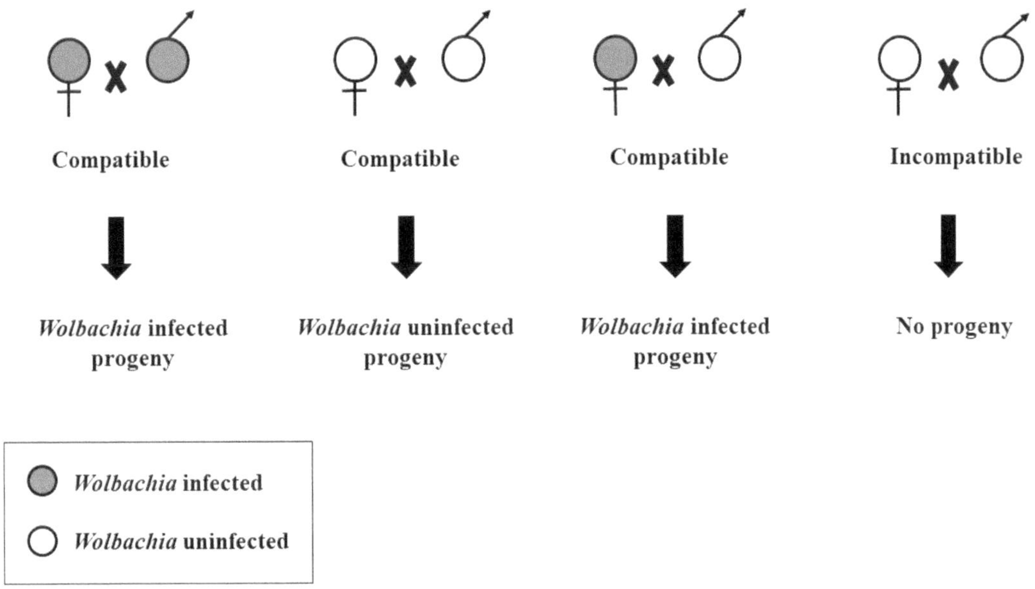

FIGURE 1.2 Cytoplasmic incompatibility influenced by reproductive parasite *Wolbachia* in insects.

both consist of converting non-transmitting males into transmitting females, MK involves death of sex that does not vertically transmit endosymbionts (i.e. males). In ladybird beetles, which are 'hotspots' for male killers, MK-induced death of males benefits their infected sisters by sibling egg consumption, decreased intensity of antagonistic interactions between siblings, and reduced levels of inbreeding. As a result of this fitness compensation, infected females produce daughters with a higher probability of survival than uninfected ones, allowing endosymbionts to spread in population. However, fitness compensation is generally imperfect (death of male progeny commonly increases sister host survival probability by 10% or less). Selection for an MK endosymbiont is, therefore, much smaller than that for a feminizing endosymbiont, such that male killer prevalence in majority of hosts is lower than 40%, although high-prevalence infections do occur. The low drive also increases sensitivity of male killer prevalence to environmental conditions.

The mechanism by which sex specificity of virulence is achieved has been demonstrated in association between *Spiroplasma poulsonii* and *Drosophila melanogaster*. A functional dosage-compensation complex, a major component of sex-determination pathway in *Drosophila,* is required for MK by *S. poulsonii*. Endosymbionts failed to kill males lacking any of five protein components of dosage-compensation complex. This result can be exploited to yield further insights into MK mechanism, which still remains unknown.

1.6.4 Feminization

Sex determination in insects is a cellular process. Consequently, endosymbionts have to infect all host cells and interact with genetic control of sex determination in all somatic cells for feminization to be effective. This view changed, however, with discovery of *Brevipalpus phoenicis* mites feminized by bacterial endosymbiont *Cardinium*. Cases of feminization in insects have also been discovered recently, mediated by *Wolbachia* in butterfly *Eurema hecabe* and leafhopper *Zyginidia pullula* and by *Cardinium* in wasp *Encarsia hispida*.

In *E. hecabe* and *Z. pullula*, *Wolbachia*-infected females produce all-female broods in laboratory conditions and antibiotic treatment leads to male-biased progenies, which suggested that these phenotypic females possess a male genotype. In *E. hecabe*, genetic sex determination follows female heterogamety (ZZ males and ZW females) and cytological observations have confirmed that *Wolbachia*-infected females are in fact ZZ genetic males inverted into phenotypic females.

1.6.4.1 Mechanisms of Feminization

In insects, *Wolbachia* could interact with key genes that control somatic sex determination, such as *Drososophila melanogaster doublesex* or transformer homologs. *Doublesex* (a switch gene at bottom of sex-determining cascade) and transformer (responsible for female-specific splicing of *doublesex*) have been identified in several insect orders. In moth *Ostrinia scapulalis*, *Wolbachia* endosymbionts typically induce MK but, at lower density, they also have a feminizing effect on genetic males. It was recently shown that *Wolbachia* can manipulate *O. scapulalis* sex determination by interfering with sex-specific splicing of *doublesex* homolog or with an upstream gene in sex-determination cascade. In addition, *doublesex* is not regulated at level of pre-mRNA splicing but it rather exhibits different expression levels between males and females. Alternatively, it has been hypothesized that *Wolbachia* could modulate host sexual phenotypes by interacting with hormonal pathways involving ecdysteroids. It is suspected that bacterial endosymbionts are able to interact with several different molecular pathways to achieve feminization of their arthropod hosts.

Endosymbiont-mediated reproductive manipulations directly impact host sex determination:

a. Feminization of genetic males, characterized by sex-ratio distortion towards females through conversion of genetic ZZ males into phenotypic ZZ females.
b. Parthenogenesis induction characterized by sex-ratio distortion towards females through conversion of genetic (haploid) males into genetic (diploid) females. Black (white) coloration: individual carries (does not carry) endosymbionts. ZZ/ZW, homo/heterogametic status of individual. n/2n, haploid/diploid status of individual. To simplify, endosymbiont transmission rate from mother to offspring is assumed to be 100% (See Figure 1.3 for schematic illustration of endosymbiont mediated feminization in insects).

1.6.5 Induction of Parthenogenesis in Insects

Feminizing endosymbionts drive female development to enhance their vertical transmission. However, successful endosymbiont transmission not only requires production of males and females but also mating. From an endosymbiont perspective, ultimate manipulation would consist in driving female development while making males superfluous. This strategy is used by at least three endosymbionts: *Wolbachia*, *Cardinium*, and *Rickettsia*. In all cases, these endosymbionts induce parthenogenesis in haplodiploid insects (hymenopterans and thrips) and acari. In these taxa, sex is normally regulated by ploidy of embryo: males develop from unfertilized haploid eggs and females develop from fertilized diploid eggs.

Parthenogenesis-inducing (PI) endosymbionts are able to convert non-transmitting males into transmitting females by enabling unfertilized eggs to develop as females. This is achieved through doubling of chromosome number in unfertilized eggs, rendering them diploid. As a result, infected parthenogenetic females are in turn able to produce endosymbiont-transmitting female progeny without need for sexual reproduction, egg fertilization and, thus, males as for feminization; precise molecular mechanisms underlying parthenogenesis induction are unknown. However, cytogenetic observations have shown that parthenogenesis is induced by endosymbionts in at least three different ways (See Figure 1.4 for an overview of endosymbiont mediated parthenogenetic induction in insects).

In hymenopterans, endosymbiont-mediated diploidization is caused by disruption of cell cycle during early embryonic development in two different ways:

i. In *Trichogramma* wasp species, two haploid sets of chromosomes do not separate during anaphase of first mitotic division, resulting in one diploid nucleus with two identical sets of haploid chromosomes instead of two haploid nuclei.
ii. In wasp *Muscidifurax uniraptor*, first mitotic division is normal, leading to two cells with haploid nuclei, and diploidy is restored by fusion of two cell nuclei after completion of first mitotic division.
iii. A third mechanism occurs in *Bryobia* mites, in which *Wolbachia* induces parthenogenesis by meiotic modification in infected eggs, resulting in diploid gametes.

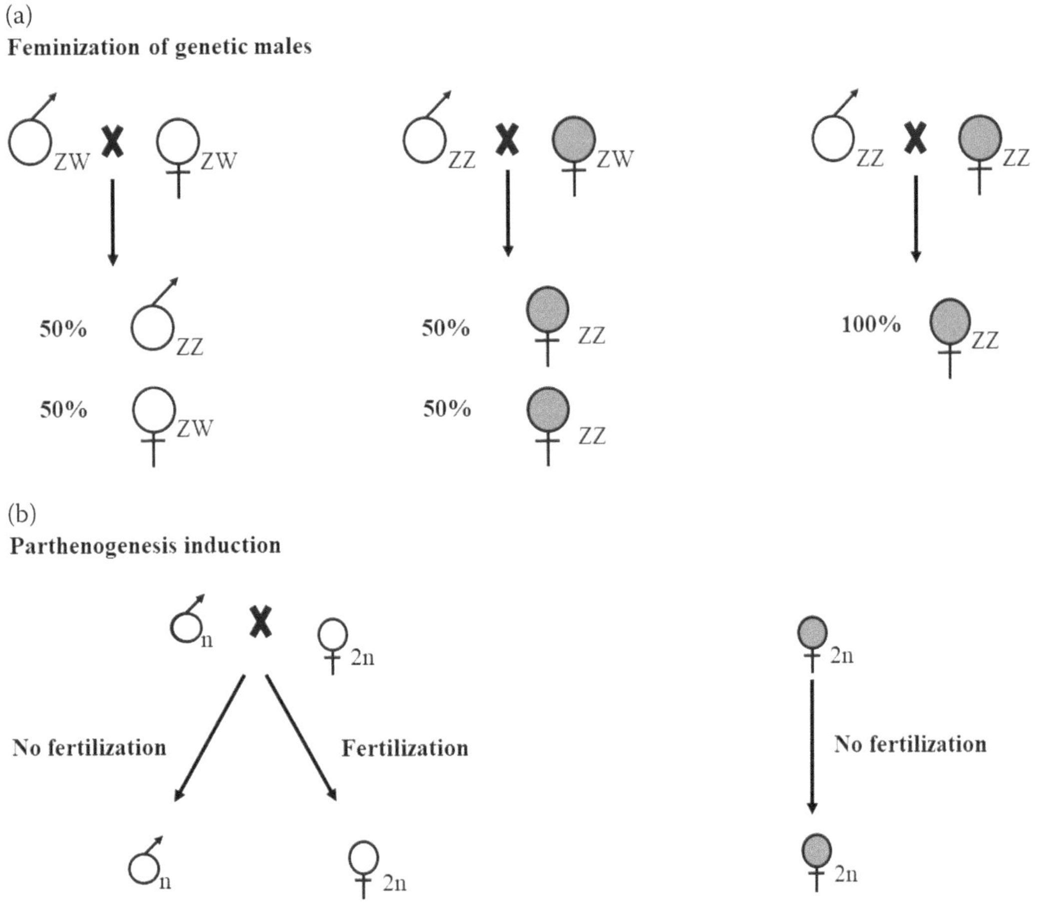

FIGURE 1.3 Endosymbiont-mediated feminization in insects: (a) Feminization of genetic males by sex-ratio distortion. (b) Feminization of induction of parthenogenesis induction accomplished through conversion of genetic (haploid) males into genetic (diploid) females.

Similar to feminizing endosymbionts, PI endosymbionts sometimes produce intersex phenotypes when some tissues become diploid, whereas others remain haploid during embryogenesis. This is influenced by rearing temperature of mothers, which is thought to affect density of endosymbionts (which are thermosensitive).

For example, in *Trichogramma* species infected by *Wolbachia*, all-female progenies are produced at temperatures below 26.8°C because endosymbiont density is high enough to diploidize unfertilized (haploid) eggs. By contrast, higher temperatures (>30.8°C) eliminate *Wolbachia* endosymbionts. As a result, unfertilized (haploid) eggs are not diploidized, which leads to production of all-male progenies. At intermediate temperatures, *Wolbachia* titer is moderate and many infected females produce males, females, and gynandromorphs (i.e. individuals in which some tissues are male whereas others are female) from unfertilized (haploid) eggs. In gynandromorphs, gender of tissue is determined by level of ploidy of cell from which tissue is derived. In these individuals, *Wolbachia*-mediated diploidization could be somewhat repressed, and this does not take place during first mitotic division but at a later stage in a subset of cells.

Overall, outcomes of parthenogenesis induction and feminization of genetic males are fairly similar in many respects. They both convert males into females and occasionally produce intersexes. The main difference is that feminizing endosymbionts invert genetic males into phenotypic females, whereas PI endosymbionts invert genetic males into genetic females.

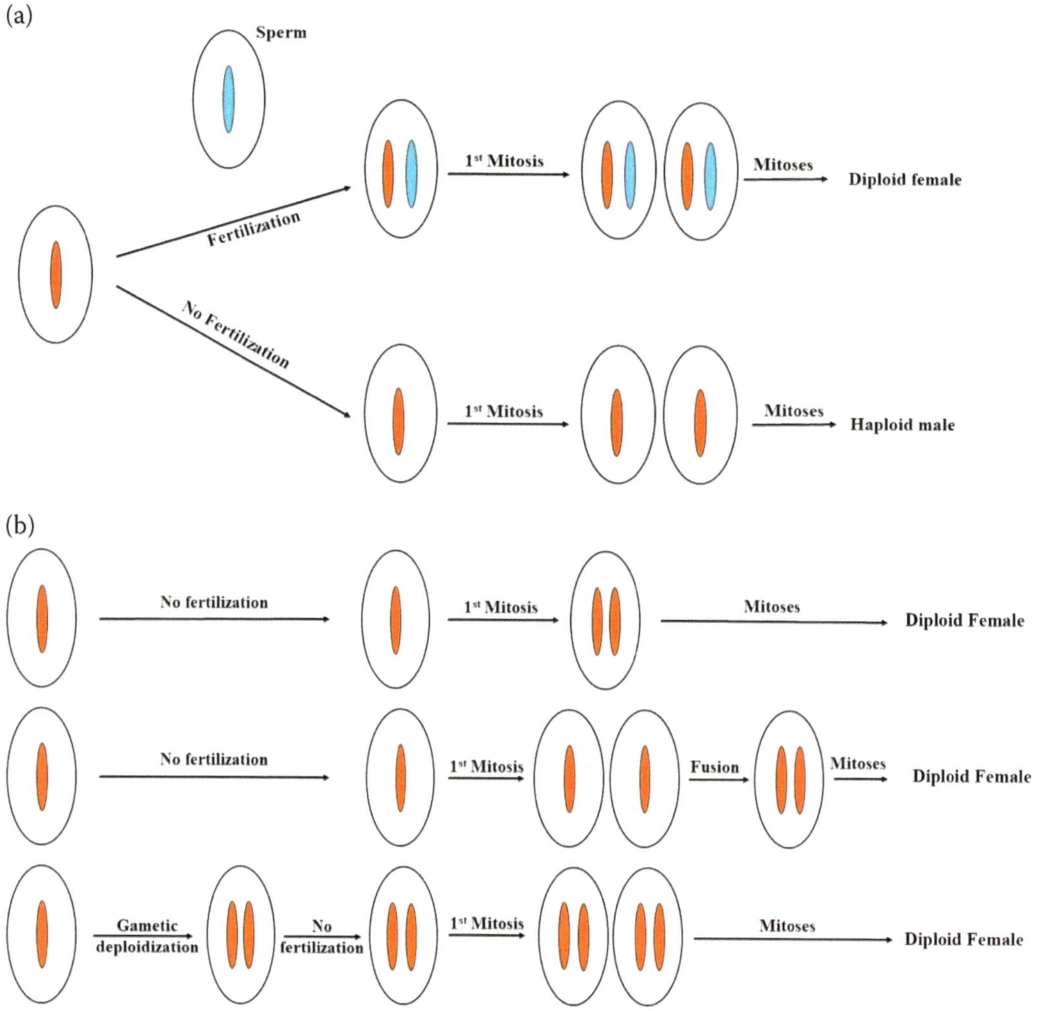

FIGURE 1.4 Schematic illustration of mechanisms of male and female development in haplodiploid species lacking or carrying parthenogenesis-inducing endosymbionts. (a) Reproduction in absence of endosymbiont. (b) Reproduction in presence of parthenogenesis-inducing endosymbionts. Maternal (red) and paternal (blue) sets of chromosomes.

1.7 Physiological Regulation of Insect Reproduction

The initiation and termination of some reproductive events often depend on environmental factors, such as temperature, humidity, photoperiod, or availability of food or a suitable egg-laying site. Additionally, these external influences may be modified by internal factors, such as nutritional condition and state of maturation of oocytes. Copulation also may trigger oocyte development, oviposition, and inhibition of sexual receptivity in female via enzymes or peptides transferred to female reproductive tract in male accessory gland secretions. Fertilization following mating normally triggers embryogenesis via egg activation. Regulation of reproduction is complex and involves sensory receptors, neuronal transmission, and integration of messages in brain, as well as chemical messengers (hormones) transported in haemolymph or via nerve axons to target tissues or to other endocrine glands. Certain parts of nervous system, particularly neuro-secretory cells in brain, produce neurohormones or neuropeptides (proteinaceous messengers) and also control synthesis of two groups of insect hormones – the ecdysteroids and juvenile hormones (JH).

TABLE 1.2

Role of neuropeptides in insect reproduction

Sl. No.	Neuropeptides	Function
1	Antigonadotropin (e.g. oostatic hormone, OH)	Suppresses oocyte development
2	Ovarian ecdysteroidogenic hormone (OEH = EDNH)	Stimulates ovarian ecdysteroid production
3	Ovary maturing peptide (OMP)	Stimulates egg development
4	Oviposition peptides	Stimulate egg deposition
5	Prothoracicotropic hormone (PTTH)	Affects egg development
6	Pheromone biosynthesis-activating neuropeptide	Regulates pheromone production (PBAN)

JH and/or ecdysteroids are essential to reproduction, with JH mostly triggering functioning of organs, such as ovary, accessory glands, and fat body, whereas ecdysteroids influence morphogenesis as well as gonad functions. Neuropeptides play various roles at different stages of reproduction, as they regulate endocrine function (via corpora allata and prothoracic glands) and also directly influence reproductive events, especially ovulation and oviposition or larviposition. The role of neuropeptides (Table 1.2) in control of reproduction is an expanding area of research, made possible by new technologies, especially in biochemistry and molecular biology. To date, most studies have concentrated on Diptera (especially *Drosophila,* mosquitoes, and houseflies), Lepidoptera (especially tobacco hornworm, *Manduca sexta*), locusts, and cockroaches.

1.8 Conclusions

Understanding evolutionary perspectives of insect's modes of reproduction may explain phylogenetic relationship between different groups of insects. Detailed investigations on impact of ecological factors on adaptation of insects and their influence on modes of reproduction is an interesting subject matter for investigation in context of climate change scenario or global warming. Understanding molecular basis of sex manipulation by endosymbionts and especially parthenogenetic induction by reproductive parasites offers a great scope for applied pest control in a variety of ways.

REFERENCES

Anderson, D.T. (1972). The development of holometabolous insects. In J. Counce and C.H. Waddington (Eds.), *Developmental systems: Insects* (*Vol. 1*, pp. 165–242). New York: Academic Press.

Baehrecke, E.H., & Strand, M.R. (1990). Embryonic morphology and growth of polyembryonic parasitoid *Copidosoma floridanum* (Ashmead) (Hymenoptera: Encyrtidae). *International Journal of Insect Morphology and Embryology*, *19*(3–4), 165–175.

Baehrecke, E.H., Aiken, J.M., Dover, B.A., & Strand, M.R. (1993). Ecdysteroid induction of embryonic morphogenesis in a parasitic wasp. *Developmental Biology*, *158*(2), 275–287.

Blackburn, D.G., (1999). Viviparity and oviparity: Evolution and reproductive strategies. In E. Knobill and J.D. Neil (Eds), *Encylopedia of reproduction* (pp. 994–1003). London: Academic press.

Davidson, E.H. (1990). How embryos work: A comparative view of diverse mode of cell fate specification. *Development*, *108*, 365–389.

Dowton, M., & Austin, A.D. (1994). Molecular phylogeny of insect order Hymenoptera: Apocritan relationships. *Proceedings of National Academy of Sciencesof USA*, *91*(21), 9911–9915.

French, V. (1996). Segmentation (and eve) in very odd insect embryos. *BioEssays*, *18*, 435–438.

Grbic, M., Nagy, L.M., Carroll, S.B., & Strand, M. (1996). Polyembryonic development: Insect pattern formation in a cellularized environment. *Development*, *122*(3), 795–804.

Gurdon, J.B. (1992). The generation of diversity in pattern in animal development. *Cell*, *68*, 185–199.

Hagan, H.R. (1951). *Embryology of viviparous insects*. New York: Ronald Press Co. p. 472.

Hall, S.R., Becker, C., & Cáceres, C.E. (2007). Parasitic castration: A perspective from a model of dynamic energy budgets. *Integrative and Comparative Biology*, *47*(2), 295–309.

Ivanova-Kasas, O.M. (1972). Polyembryony in insects. In S.J. Counce and C.H. Waddington (Eds.), *Developmental systems, insects* (*Vol. I*, pp. 243–271). New York: Academic Press.

Kerr, W.E. (1962). Genetics of sex determination. *Annual Review of Entomology*, *7*, 157–176.

Kuris, A.M. (1974). Trophic interactions: Similarity of parasitic castrators to parasitoids. *Quarterly Review of Biology*, *49*, 129–148.

Lafferty, K.D., & Kuris, A.M. (2002). Trophic strategies, animal diversity and body size. *Trends in Ecology & Evolution*, *17*, 507–513.

Lafferty, K.D., & Kuris, A.M. (2009). Parasitic castration: Evolution and ecology of body snatchers. *Trends in Parasitology*, *25*(12), 564–572.

Marchal, P. (1898). Dissociation de l'oeuf en un cycle evolutif chez l'Encyrtus fuscicollis (Hymenoptera). *Comptes rendus de l'Académie des Sciences Paris*, *126*, 662–664.

Noskiewicz, J., & Poluszynski, I. (1935). Embryologische Untersuchungen an Strepsipteren. *Zool Pol*, *1*, 53–92.

Nakano, M., Morgan-Richards, M., Godfrey, A. J. R. (2019). Parthenogenetic females of the stick insect Clitarchus hookeri maintain sexual traits. *Insects*, *10*(7), 202–218.

Nur, U. (1971). Parthenogenesis in coccids (Homoptera). *American Zoologist*, *11*, 301–308.

Peacock, A.D., & Sanderson, A.R. (1939). The cytology of thelytokous parthenogenetic sawfly *Thrinax macula. Transactions Royal Society of Edinburgh*, *59*:647–660.

Pearcy, M., Hardy, O., & Aron, S. (2006). Thelytokus parthenogenesis and its consequences on breeding in an ant. *Heredity*, *96*, 377–382.

Sander, K. (1976). Specification of basic body pattern in insect embryogenesis. In *Advances in insect physiology* (*Vol. 12*, pp. 125–238). Cambridge, Massachusetts, USA: Academic Press.

Simon, J., Rispe, C., & Sunnucks, P. (2002). Ecology and evolution of sex in aphids. *Trends in Ecology & Evolution*, *17*, 34–39.

Strand, M.R. (1986). The physiological interactions of parasitoids with their hosts and thier influence on reproductive strategies. In Insect parasitoids: 13th symposium of the Royal Entomological Society of London, 18–19 September 1985 at the Department of Physics Lecture Theatre, Imperial College, London/edited by Jeff Waage, David Greathead. London: Academic.

Strand, M.R. and Grbic, M. (1996). Development and life history of polyembryonic wasps. In N. E. Beckage (Eds.), *Parasites and Pathogens: Effects on Host Hormones and Behavior*. pp. 37–56. New York: Chapman and Hall.

Strand, M.R., & Obrycki, J.J. (1996). Host specificity of insect parasitoids and predators. *BioScience*, *46*(6), 422–429.

Suomalainen, E. (1962). Significance of parthenogenesis in evolution of insects. *Annual Review of Entomology*, *7*, 349–366.

Suomalainen, E., Saura, A., & Lokki, J. (1976). Evolution of parthenogenetic insects. In M.K. Hecht *et al.* (Eds.), *Evolutionary biology*. pp. 209–257.

White, M.J.D. (1964). Cytogenetic mechanisms in insect reproduction. *Symposia of Royal Entomological Society London*, *2*, 1–12.

Whitfield, J.B. (1992). Phylogeny of non-aculeate Apocrita and evolution of parasitism in Hymenoptera. *Journal of Hymenoptera Research*, *1*(1), 3–14.

Whiting, P.W. (1945). The evolution of male haploidy. *The Quarterly Review of Biology*, *20*(3), 231–260.

2

Sexual Reproduction

Bhupendra Kumar[1] **and Omkar**[2]
[1]*Department of Zoology, Banaras Hindu University, Varanasi, India*
[2]*Ladybird Research Laboratory, Department of Zoology, University of Lucknow, Lucknow India*

CONTENTS

2.1 Introduction

The wide range of reproductive strategies in insects alongwith their physiological systems has resulted in their enormous success on earth. Most insects have very few reproductive opportunities due to a short lifespan (Arnqvist *et al.*, 2000), for which they compensate by displaying numerous reproductive strategies (parthenogenesis, pedogenesis, polyembryony, functional hermaphroditism, viviparity, or sexual reproduction) facilitating production of phenomenal numbers of offspring under varied ecological circumstances (Klowden, 2013). The basic organization of reproductive systems of both male and female insects is quite similar (Figure 2.1).

The germ cells of reproductive organs arise from pole cells, which are amongst foremost to be differentiated during embryogenesis. Along with mesodermal tissues, germ cells develop into adults' reproductive organs. A pair of gonads is connected by individual ducts to a common duct. This common duct is both mesodermal as well as ectodermal in origin. The activities of both male and female

DOI: 10.1201/9781003043195-2

reproductive systems are coordinated by hormones and transcription factors, and are eventually regulated by physiological and environmental factors (Klowden, 2013).

The male reproductive organs consist of a pair of testes, with a series of testicular tubes/follicles where spermatozoa are produced. The testes open separately into sperm duct or vas deferens of mesodermal origin, which extends posteriorly and forms seminal vesicle or sperm-storage organ (Snodgrass, 1957). Most insects have a number of accessory glands that open into the ejaculatory duct. These glands are either formed as diverticula of vasa deferentia, or vasa deferentia themselves become glandular and carry out the functions of accessory glands. The paired vasa deferentia unite and lead into ectodermally derived ejaculatory duct, the tube that transports semen or sperm package to female gonopore/genital pouch (Snodgrass, 1957; Gullan & Cranston, 2014).

The female reproductive system consists of a pair of ovaries, which are connected with a pair of postero-lateral oviducts that join to form a median/common oviduct. This median oviduct opens posteriorly into a genital chamber that sometimes forms a tube, vagina, and is often developed to form a bursa copulatrix for reception of penis/aedeagus. The genital chamber (vagina) opens into a spermatheca that stores sperms, and, frequently, a pair of accessory glands (Klowden, 2013) (Table 2.1).

In this chapter, attempts have been made to provide a detailed description of male and female reproductive systems of insects.

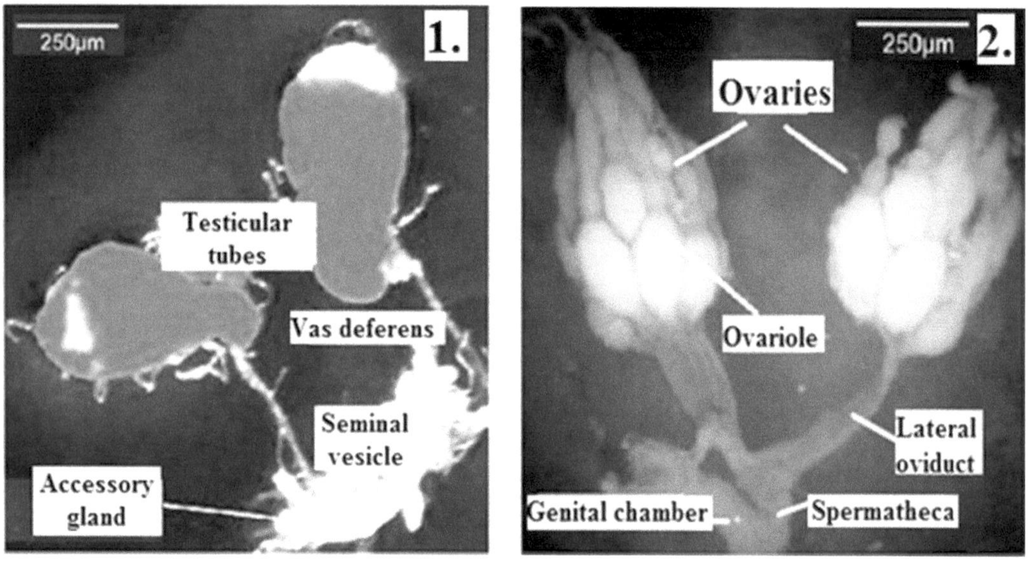

FIGURE 2.1 (1) Male and (2) female reproductive systems of an insect.

TABLE 2.1

Comparison between male and female reproductive systems (Richards & Davies, 1979; Gullan & Cranston, 2014)

Female Reproductive System	Male Reproductive System
Paired ovaries composed of ovarioles	Paired testes composed of follicles
Paired oviducts	Paired vasa deferentia
Egg calyces (reception of eggs)	Seminal vesicles (sperm storage)
Common/median oviduct and vagina	Median ejaculatory duct
Accessory glands (ectodermal origin) colleterial or cement glands	Accessory glands (ectodermal or mesodermal in origin)
Bursa copulatrix (copulatory pouch) and spermatheca (sperm storage)	No equivalent
Ovipositor	Male genitalia/aedeagus

2.2 Evolution of Sex

Previous studies (Maynard Smith, 1978, 1982) have predicted that ancestral gametes were small and isogamous. The evolution of anisogamy was an important transition in evolution, as it caused separation of sexes (Maynard Smith & Szathmary, 1997). Physiologically, males produce sperms and females produce ovum. Anisogamy or sexual asymmetry in gamete size plays central role in regulating all pre- and postcopulatory processes of Darwinian sexual selection. Parker *et al.* (1972) proposed a theory, often known as PBS (Parker, Baker, and Smith) model (Bell, 1978; Lessells *et al.*, 2009) for evolution of anisogamy and two sexes by gamete competition. Based on PBS model, Charlesworth (1978) and Bell (1978) suggested that anisogamy may evolve only if there is sufficient size variation in ancestral population to generate sexual asymmetry in gamete size or when gamete size mutations have sufficiently large effects (Maire *et al.*, 2002).

Parker *et al.* (1972) proposed two selection pressures that act on gamete production, i.e. number of gametes and fitness of resulting zygotes. They further argued that whatever gamete size ancestral isogamous population showed, selection generally pushes gamete size towards a zone of zygote fitness versus size ratio relationship. Anisogamy will occur depending on zygote fitness versus size ratio relation for ancestral state. The analysis of Bulmer and Parker (2002) suggested that ancestral stage would have been isogamous unicells (microgametes). As adult size increased through evolution of muticellularity, zygote fitness versus size ratio increased away from its early position near gamete survival. The isogamous microgamete size initially increased and at some threshold, isogamy became unstable, and was replaced by anisogamy through invasion of macrogametes (Parker, 2011).

The PBS model (Parker, 1979) also suggests that once anisogamy evolved, disassortative fusions of ova with sperm further led to loss of motility in ova. Due to vast numerical predominance of sperm arising through size-number trade-off at production, selection quickly favoured avoidance of sperm-sperm fusions, while ova fused much quickly with sperm. Although loss of motility in ova interfered with ability of ova to fuse with other ova, it allowed energy spent on motility to be concentrated on to increased productivity (Parker, 1978). The PBS model further argued that gamete competition between sperm of different males over fertilization of a given set of ova was responsible for evolutionary origin and maintenance of anisogamy (Parker, 1970, 1982). As multicellularity increased, increased size of zygote further led to an evolutionary increase in size of ovum. Simultaneously, sperm competition, maintained sperm at a minimal size with increased fertilization success. Consequently, anisogamy ratio (ovum size/sperm size) became very large and became the fundamental reason why sperm are small and adapted solely to function in succesfull fertilization without making any contribution to zygote provisioning (Parker, 2011).

The production of a sperm attraction pheromone is another adaptation that has effectively favoured increased size of egg. The large size and extra volume of egg is not only solid matter, but also a zone of attraction around egg, which increases its probability of fertilization. Dusenbery (2000) assumed that rate of pheromone production is proportional to egg volume. Pheromone-releasing eggs show greater than 50-fold increase in fertilization success (Jantzen *et al.*, 2001). Gamete motility is also a possible factor in determining anisogamy, with faster sperm having fertilization advantage under sperm competition (Levitan, 2000). Dusenbery (2000) predicts that anisogamy offers advantage of increased gamete encounter rates, and evolution towards anisogamy occurs by creating one gamete smaller and motile and the other larger and non-motile.

Theory of disruptive selection by gamete competition is the most powerful explanation of origin of anisogamy that has resulted in sexual selection and sexual conflict (Lessells *et al.*, 2009).

2.3 Evolution of Sexual Reproduction

The term 'mating system' includes nature of gametes, mate-choice strategies, patterns of male–female pairings, and behavioural strategies displayed by individuals to attain reproductive success and mate, even if they are not chosen (Breed & Moore, 2016); all dimensions of sexual reproduction. While an asexually reproducing animal passes all of its genes to each offspring, however, in sexual reproduction only half are passed on to each offspring. Despite the cost associated with producing male and female offspring during sexual reproduction (Otto, 2008), yet selection favours sexual reproduction (Hartfield & Keightley, 2012). This mismatch between theoretical prediction and empirical reality is generally called as 'the paradox of sex' (Maynard Smith, 1978). Selective advantages from genetic recombination and genetic diversity, i.e. non-accumulation of deleterious genes among offspring outweigh costly nature of sexual reproduction (Lehtonen *et al.*, 2012).

In a strict sense, sexual reproduction refers to mixing of genetic material, either within an organism (e.g. recombination) or between organisms (Alcock, 2001). While meiosis eliminates harmful changes and mutations, and favours repair of damage to DNA; it also promotes reorganization of accumulated epigenetic signals that regulate gene expression (Dugatkin, 2014). The main function of sexual reproduction could possibly be genetic editing. In addition, sexual reproduction produces offspring that are better suited for changing environment.

When Valen introduced 'Red Queen hypothesis' to evolutionary biology, it initially emphasized on the concept of co-evolution, stating that when a prey species evolves a defense, predator evolves to defeat that defense and therefore evolves another defense (Alcock, 2001). However, the hypothesis can be equally applied to evolution of sexual reproduction.

Sexual reproduction evolves more quickly when a species' environment changes rapidly, such as a shift in community of interacting species, especially host and parasite species (Dugatkin, 2014). The genetic associations built up by past selection, along with sex and recombination improve fitness of offspring, thereby turning recombination load into an advantage. Genetic recombination gained from sexual reproduction produces new defenses against diseases and parasites (Alcock, 2001). When disease and parasite pressures are low, cost of meiosis and benefit of proven genetic combinations favour asexual reproduction, however, in case of a surge sexual reproduction is favoured (Dugatkin, 2014).

In animals, meiosis and gametogenesis occur sequentially in germline cells and sex differentiation occurs in somatic gonads into which the germline cells migrate, which is in contrast to plants, where meiosis and gametogenesis occur separately from somatic cells of sporophyte and gametophyte, while sex differentiation occurs in them (Bai, 2015). Meiosis, gametogenesis, and sex differentiation seem to have evolved independently, integrated later by chance, and genetically fixed as a program in protists. Each cell can thus behave independently for emergence of meiosis and gametogenesis, and can also live together closely enough to make cell fusion and cell–cell recognition possible. Integration of these three events have brought selective advantages together and the integration is selected for and referred to as 'sexual reproduction' (Bai, 2015).

During sexual reproduction, individual animals combine their genes with genes of another animal, via gametes of different sizes, to give their offspring greatest possible genetic advantage. Sexual reproduction creates both a social atmosphere and competition among individuals as each strives to maximize its genetic contribution to subsequent generations (Alcock, 2001). Males usually make multiple small, relatively inexpensive gametes and try to fertilize as many eggs as possible, while providing little or no care for their offspring. On other hand, production of an egg by a female is more costly than that of sperm, leading to production of few individually expensive gametes (Breed & Moore, 2016). The value of these gametes suggests that perhaps most eggs should be fertilized, while sperm may go waste (Dugatkin, 2014), thereby making females more protective of eggs. Because females make fewer, larger gametes and often also provide parental care, they usually have a lower potential rate of reproduction than males (Alcock, 2001). As a result, receptive females are scarce, and males typically compete for access to them, while females choose among many potential mates. Mate choice and parental investment follow from these differences in investment in gametes (Dugatkin, 2014).

2.4 Male Reproductive System

In males, paired testes produce spermatozoa. Each testis comprises a series of tubular follicles that may vary from one in some apterygotes and Diptera to over 100 in Orthoptera and 300 in Hymenoptera.

The follicles are enclosed by a peritoneal sheath. Within each follicle, developing sperm in successive stages of maturation are present (Phillips, 1970). Each follicle is divided into a series of zones: (i) *Germarium:* contains primordial germ cells (spermatogonia) that multiply; (ii) *Zone of growth:* where spermatogonia increase in size and develop into spermatocytes after repeated mitosis; (iii) *Zone of division and reduction:* where spermatocytes undergo meiosis, giving rise to spermatids; (iv) *Zone of transformation:* where spermatids get transformed into spermatozoa (Richards & Davies, 1979). The follicles are connected to a main duct, *vas deferens*, through individual *vas deferens* tubes. A portion of *vas deferens* is enlarged as *seminal vesicle*, and serves as a storage reservoir for sperm before being transferred to female reproductive system. The two *vasa deferentia* are enclosed by circular muscles and are connected to *ejaculatory duct*, which is ectodermal in origin (Klowden, 2013; Gillot, 2018).

The terminal portion of ejaculatory duct forms intromittent organ known as *aedeagus*. Aedeagus is absent in Apertygota males, and in Odonata a secondary copulatory structure at anterior of abdomen performs its function. Some ancestral pterygotes bear a pair of intromittent organs. Other accessory structures, known as claspers/parameres, maybe present on neighbouring segments and are used to grasp female during copulation. A pair of male accessory glands may also be present that can open into either vas deferens or ejaculatory duct (Richards & Davies, 1979; Chapman, 2009). Male accessory glands perform a variety of functions, viz. (i) production of seminal fluid that transports and/or activates sperm, (ii) as vaginal mating plug that temporarily blocks entry of another males' sperm, and (iii) formation of spermatophores that enclose sperm (Leopold, 1976). Apterygote males produce and deposit spermatophores on moist ground, which are taken up by females (Mann, 1984). However, in more advanced insect orders, sperm are transferred directly in seminal fluid by internal fertilization, without production of spermatophores (Hinton, 1974) (Figure 2.2).

Male accessory gland proteins (AGPs) can supplement nutritional reserves of females and stimulate egg production/ripening, post mating (Happ King, 1984). AGPs are known to remove physiological block preventing ovulation and egg deposition until mating takes place, in many insect species (Gillott, 2003; Klowden, 2013). AGPs reduce oxidative stress in females and sustain sperm quantity necessary for fertilization. Antibacterial proteins present in AGPs even protect sperm, and male and female reproductive tracts from infection (Lay *et al.*, 2004). Although male accessory glands cannot synthesize juvenile hormone (JH), they are capable of methylating already-synthesized JH acid and subsequently transferring active JH to females (Engelmann, 1970; Koeppe *et al.*, 1985). This transferred JH may facilitate patency of developing oocytes and thereby increase female fecundity (Lay *et al.*, 2004; Chapman, 2009; Klowden, 2013).

2.4.1 Spermatogenesis

In male insects, a small number of stem cells give rise to cells having two different fates: a daughter spermatogonial cell that subsequently differentiates into spermatozoa and pluripotent stem cell that is

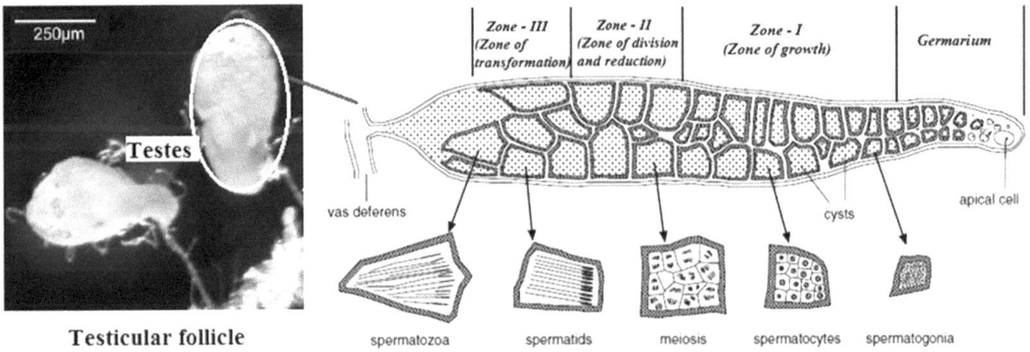

FIGURE 2.2 Testicular follicle of an insect showing different stages of development of sperm (modified from Chapman, 2009).

capable of regenerating. The latter allows male insects to continue producing mature sperm throughout its reproductive life. The spermatogonial cell undergoes several mitotic divisions to produce spermatocytes and subsequently divide meiotically to produce spermatids that further differentiate into spermatozoa. The production of spermatozoa occurs within testes follicles (Klowden, 2013). Two primary cell types, viz. primordial germ cells and somatic precursor cells, occupy microenvironment of a stem cell niche at its apical end (Chen & McKearin, 2005). Constituting niche are 10–15 specialized non-dividing somatic cells that comprise hub. The somatic cells are surrounded by 6–9 apical germline stem cells (GSCs), which are further surrounded by somatic cyst stem cells. Both germline and somatic cyst stem cells directly contact hub. JAK-STAT signaling pathway from niche regulates renewal, maintenance, and survival of stem cells and provides cues that prevent differentiation and self-renewal of stem cells (Klowden, 2013).

Insulin-like peptides are also involved in division of both germline and somatic cyst stem cells. After receiving niche signal, somatic cyst stem cells divide mitotically to produce one gonialblast and another stem cell. The daughter cells that remain within niche post stem cell division continue to receive signals and self-renew. However, daughter cells that move out of niche do not receive signals, and began to differentiate as gonialblasts. The gonialblasts initiate four mitotic divisions with incomplete cytokinesis. These divisions result in a syncytium of 16 spermatogonia connected by ring canals (lined with anillin, an actin-binding protein that attaches it to membrane). The spermatogonia divide mitotically, grow in volume, and become primary spermatocytes. While in *Rhodnius* spermatocyte stage is reached at eight divisions (128 cells per cyst), in *Drosophila* four divisions result in cysts with 64 cells (Chapman, 2009; Klowden, 2013).

The subsequent meiotic division of spermatocytes yields haploid spermatids. Gene transcription ceases and morphogenesis of spermatids continues within syncytium, where all spermatid nuclei that are descended from a primary spermatocyte remain connected by cytoplasmic bridges. The mitochondria of these postmeiotic spermatids aggregate alongside each haploid nucleus and fuse into two large mitochondrial derivatives, forming spherical *nebenkern*. These mitochondrial lobes begin to unfold and elongate parallelly to develop axoneme and become the source of sperm motility (Klowden, 2013).

The sperm maturation further involves individualization of syncytium into individual spermatozoa. One meiotic and four mitotic divisions lead to four haploid spermatids for each diploid spermatocyte within cyst. The spermatids further differentiate into mature spermatozoa through developmental events called spermiogenesis. Events of differentiation through individualization process occur as a cytoskeletal membrane complex assembles at end of cyst containing nuclei and moves down its length. The individualization complex moves down cyst, becomes visible as a cystic dilation, encases each spermatid in its own cell membrane and extrudes ring canals and syncytial cytoplasm (Klowden, 2013). These components are eliminated from cyst tail as complex detaches at the end of individualization process. The acrosomal complex develops at sperm head and is associated with enzymes acrosin and hyaluronidase. The individualized bundles of sperm become coiled before their release. As differentiation proceeds, cysts elongate and eventually rupture, releasing spermatozoa into vas efferens. The spermatozoa move to seminal vesicles where they remain until mating takes place (Chapman, 2009; Klowden, 2013).

2.4.1.1 Spermatozoa

Insect sperm comprise a head region and a long flagellum that is used for locomotion. Within head are a haploid nucleus and acrosomal complex at tip that arises from Golgi apparatus during differentiation. The acrosome contains enzyme acrosin that dissolves egg membranes for fertilization. Sperm are only insect cells that bear flagella, which may be absent in proturans (disc-like sperm). The motor portion of flagellum is known as axoneme. The axoneme consists of microtubules originating from centrosome that forms flagellum at the base of sperm nucleus (Phillips, 1970; Baccetti, 1972). The centrosome is the organizer of microtubular network and influences formation of mitotic spindle and consists of two centrioles surrounded by a mass of protein, with a centriolar adjunct forming a collar around it. Although centrosome is inactive in female gamete, it persists in sperm to organize first mitotic spindle in fertilized zygote.

The arrangement of microtubules in collembolans consists of a cylindrical array of doublets connected by cross-bridges of motor protein dynein and a pair of single microtubules running up the center, known as 9 + 2 arrangement. However, other insect sperm have a modified 9 + 2 arrangement with an additional ring of nine accessory microtubules, a 9 + 9 + 2 arrangement. A 9 + 9 + 3 pattern, with three central tubules, is found in some dipterans and neuropterans (Geyer, 1951; Baccetti *et al.*, 1973).

The microtubules are made of dimeric protein tubulin, and dynein arms that form cross-bridges between doublet fibers allowing them to slide past each other in a sliding filament model to cause bending of flagellum. Accessory tubules further strengthen axoneme and amplify its motion. Comparative studies indicate that 9 + 9 + 2 insect sperm show helicoidal pattern compared to planar beat of species with 9 + 2 sperm. The tail length of flagella may vary within a species (Baccetti *et al.*, 1973). In *D. melanogaster,* sperm ranges from 1.6 to 2 mm, but males of *D. bifurca* produce sperm over 58-mm long. *Drosophila* species that produce longer sperm also take longer to reproductively mature (Parker, 1993). This variation in individual size of sperm may result from cross-linking and sliding of microtubules against mitochondria during individualization process (Klowden, 2013; Gillot, 2018).

Lepidopteran males produce two different types of sperm, apyrene and eupyrene, with former having no nuclei and being genetically non-functional. Both are transferred to females, but only eupyrene sperm fertilize oocytes, while apyrene sperms assist former to move through female reproductive tract, and/or provide essential nutrients (Friedländer, 1997). About 70–90% of transferred sperm may be apyrene, and may displace eupyrene sperm from previous matings. In *Bombyx mori,* absence of apyrene sperm results in lack of fertilization. Thus, both types of sperm are necessary for fertilization to occur (Friedländer, 1997; Chapman & Davies, 2004; Klowden, 2013).

2.4.2 Endocrine Control of Male Reproductive System and Spermatogenesis

Testes of several lepidopteran species produce ecdysteroids when stimulated by an ecdysiotropin, produced by medial neurosecretory cells of brain. In addition, insulin signaling stimulates production of gonial blasts from GSCs and stimulates growth of spermatocytes (Engelmann, 1970). In some insects, rate of mitotic divisions of spermatogonia to form spermatocytes increases with 20-hydroxyecdysone; however, the increase is restricted by high titers of JH (Gullan & Cranston, 2014; Hardie, 2018). Spermatocytes then begin meiotic divisions that are arrested at prophase until end of larval development. The post-wandering peak of 20-hydroxyecdysone unblocks meiosis and subsequently allows cells to proceed to metaphase. In many insect species, JH stimulates protein synthesis in male accessory gland and enhances the process of spermatogenesis (Wigglesworth, 1964; Gillott & Gaines, 1992). However, in some lepidopterans stimulation of protein synthesis is done by 20-hydroxyecdysone. During mating, JH is also transferred to females and augments vitellogenesis (Chapman, 2009; Klowden, 2013).

2.5 Female Reproductive System

Female reproductive system consists of a pair of ovaries, which are connected to a median oviduct by a pair of posterolateral oviducts. Ovaries develop from splanchnic mesoderm during embryogenesis (Hoffmann, 2018). The portions that bear germ cells develop into ovary, while other anterior cells give rise to a suspensatory ligament, which unites all ovarioles and anchors ovary to body wall. At posterior ends, ovaries are connected to posterolateral oviducts that combine to form ectodermally derived common oviduct. Ovarioles are functional units of ovary-containing developing oocytes. These oocytes grow sequentially within ovarioles in an assembly line fashion (Cruickshank, 1973). The number of ovarioles per ovary varies enormously and depends on size and reproductive strategies of insect (Chapman, 2009). While apterygotes have single ovariole per ovary; some social insects may have over 2,000 ovarioles per ovary. However, the most common range of ovariole number in majority of insect species is between 4 and 10 per ovary (Martoja, 1977; Klowden, 2013).

2.5.1 Germarium

The separation of somatic and germline cells occurs during early embryogenesis (Williamson & Lehmann, 1996). At the tip of mature ovariole, 2–3 GSCs reside within a special microenvironment, stem cell niche (Chen & McKearin, 2005). Signaling within niche acts on GSCs and promote either their self-renewal or differentiation.

Three types of somatic cells that surround GSCs, i.e. terminal filament, cap, and escort stem cells provide signaling to GSCs through physical contact and molecular signaling. The GSCs contain a cytoplasmic organelle known as spectrosome (Telfer, 1975; Williamson & Lehmann, 1996). The spectrosome is composed of cytoskeletal proteins α- and β-spectrin and ankyrin, and is attached to cap cells by E-cadherin adherens junctions.

The branching of spectrosome in later stages of development determines which differentiating cell will become oocyte (Chapman, 2009; Klowden, 2013). During asymmetrical divisions within GSC, daughter cell closest to terminal filament and cap cells remains attached to cap cells as a stem cell. However, other cell differentiates into a cystoblast that begins oogenesis, ultimately forming a 16-cell cluster of interconnected cystocytes surrounded by escort cells (Chapman, 2009; Hoffmann, 2018) (Figure 2.3).

Although JAK-STAT signaling pathway is required to maintain self-renewal, other signals from surrounding cells also play crucial roles. Ecdysone modulates intensity of signaling, while insulin-like peptides regulate the rate at which GSCs divide. With continuation of mitotic division, spectrosome becomes branched and gives rise to fusome of daughter cells that anchor mitotic spindle during cell division. The escort stem cells produce follicle cells. The latter surround newly formed cystoblast, as it undergoes cell division with incomplete cytokinesis (Gullan & Cranston, 2014; Klowden, 2013).

The cystoblast ultimately gives rise to a cyst consisting of 16 cystocytes that are connected to each other by cytoplasmic bridges or ring canals. The fusome continues through ring canals as cyst enlarges, and is distributed asymmetrically to daughter cells. As cystocyte complex moves down ovariole and is completely surrounded by follicle cells, it enters vitellarium region (Telfer, 1975; Klowden, 2013). Follicle cells surround cystocyte complex and provide cell signaling interactions that determine polarity of developing oocyte. In certain insect orders, follicle cells also express vitellogenin gene. Ring canals are means for a number of proteins and mRNAs that provide a pattern for development of oocyte. One of the two cystocytes that has inherited more fusome and contains four ring canals subsequently becomes oocyte (Spradling *et al.*, 1997). The developing oocyte enters meiosis-I, but is arrested at prophase-I until fertilization occurs. The remaining cystocytes develop as nurse cells at anterior end of oocyte (Chapman, 2009; Klowden, 2013).

2.5.2 Additional Structures of Female Reproductive System

The calyx joins the posterior end of ovariole to lateral oviduct. The common oviduct is lined with cuticle and is ectodermal in origin. However, lateral oviducts are mesodermal in origin, similar to follicle cells and ovariole wall. A thin acellular membrane, commonly known as tunica propria, covers ovariole from terminal filament to calyx. The female accessory glands or collaterial glands produce a cement that allows deposited eggs to be attached to substrate or glued together (Mercer & Brunet, 1959). In certain insect species that retain their eggs after hatching, such as tsetse fly, *Glossina*, and cockroach, *Diploptera*, accessory glands produce a nutritive secretion that nourishes larvae during their development. The poison glands of Hymenoptera are also modified accessory glands (Richards & Davies, 1979). However, female accessory glands of cockroaches and mantids also produce hard egg case, or ootheca. In insects, females also contain capsule-like spermathecae that temporarily store sperm for subsequent fertilization.

Spermathecae are ectodermal in origin and open into common oviduct of female to release stored sperm. Spermathecae are often present as multiple storage organs and allow females to better segregate and manipulate sperm. Depletion of stored sperm may limit fertility of female. Under such circumstances, multiple matings in females are necessary to replenish sperm numbers. Presence of spermathecal duct, containing glycogen deposits, serves as an energy source for sperm when they pass

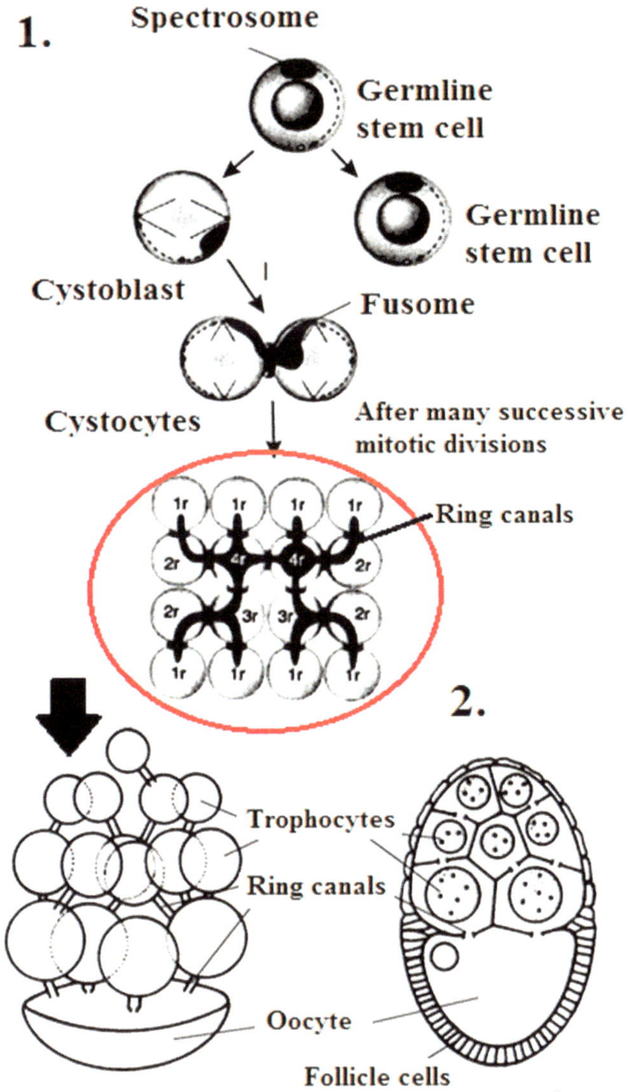

FIGURE 2.3 (1) Division of germline stem cell and formation of cystocytes. (2) Differentiation of cystocytes into trophocytes and an oocyte (modified from Klowden, 2013).

through to egg. The secretions of spermathecal gland, which are rich in proteins and carbohydrates, are responsible for maintaining sperm viability within spermathecae (Gillott, 1988). A honey bee queen can store about 6 million sperm for up to 7 years, and presence of high levels of antioxidative enzymes within spermathecae protects sperm from oxidative damages. While opening to outside, common oviduct becomes modified to a genital chamber, which is capable of incubating eggs internally. The bursa copulatrix is an additional pouch present within genital chamber. It is the place where sperm are first deposited after mating. Thereafter, sperm leave bursa copulatrix, move to spermatheca, and are stored permanently (Klowden, 2013).

In primitive insects that produce spermatophore to enclose sperm, bursa copulatrix bears a series of tooth-like structures that disrupt spermatophore and facilitate release of sperm. Bursa copulatrix secretes certain chemicals into haemolymph when it is filled with sperm so as to signal that mating has occurred. In certain tephritid flies, an additional sperm storage organ, fertilization chamber, is present as a cuticular extension of ventral wall of bursa copulatrix. The chamber is initially filled with sperm, but sperm are then transferred to spermathecae for their long-term storage. Only a smaller number of sperm remain in fertilization chamber and are used for fertilization of eggs. The fertilization chamber also behaves as a staging area for those sperm that are used for immediate fertilization, rather than releasing and potentially wasting a large number of sperm from spermathecae when an egg is ovulated (Chapman, 2009; Klowden, 2013).

2.5.3 Types of Ovarioles

The ovarioles in insects are classified into two major types that differ in ways by which nutrients are supplied to oocyte. In panoistic ('all eggs') ovariole, germarium contains only oogonia, primary oocytes, and mesodermal prefollicular tissue. The developing oocyte receives all nourishment from follicular epithelium (Cruickshank, 1973; Büning, 1998). The ring canals are absent. Panoistic ovarioles are ancestral, and found in orders Archaeognatha, Thysanura, Odonata, Orthoptera, Plecoptera, and Isoptera. In more advanced meroistic ('part eggs') ovariole, germarium contains stem cells, primary oocytes, prefollicular tissue, and nurse cells or trophocytes. Meroistic ovarioles are divided into telotrophic meroistic and polytrophic meroistic. In both of these, nurse cells or trophocytes provide developing oocyte with RNA, proteins, and ribosomes, which in panoistic ones are provided by oocyte (Mahowald, 1972; Wightman, 1973) (Figure 2.4).

In telotrophic meroistic ovariole, nurse cells are located at apex of ovariole and feed oocyte through a nutritive cord. After maturation of oocyte, cytoplasmic (nutritive) cord is broken (Huebner, 1984). Such ovarioles are found in insect orders, Ephemeroptera, Hemiptera, and some species of Coleoptera. In contrast, most holometabolous insect orders have polytrophic meroistic ovariole (Büning, 1998). Here oocyte is one amongst interconnected cells, or cystocytes, that all descend together within the follicle surrounded by follicular epithelium. The nurse cells are located at anterior end of follicle while oocyte is present at posterior end. Since nurse cells are sister cells of oocyte, they remain connected to oocyte and with each other by cytoplasmic bridges called ring canals, forming a cystocyte complex (Büning, 1993, 1994).

When nurse cells have passed all their contents into oocyte, they degenerate, leaving quiescent oocyte in follicle prior to vitellogenesis. The polytrophic meroistic ovariole is primitive than telotrophic type. Polytrophic meroistic types have evolved from panoistic form when some cystocytes differentiated into

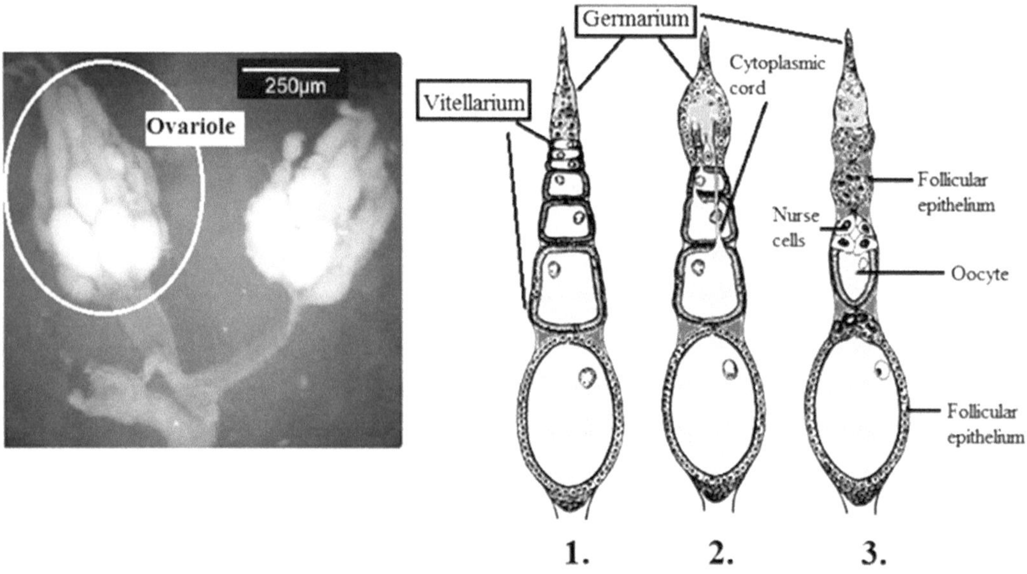

FIGURE 2.4 Types of ovarioles- 1. Panoistic ovariole; 2. Telotrophic meroistic ovariole; 3. Polytrophic meroistic ovariole (modified from Klowden, 2013).

nurse cells that amplified their genome, and cystocytes underwent incomplete cytokinesis. However, telotrophic ovarioles have evolved from polytrophic type when nurse cells were retained within germarium. Another type of ovariole, neopanoistic type, has secondarily lost nurse cells and all cystocytes have developed into oocytes or follicle cells. Neopanoistic ovarioles are found in Protura, Thysanoptera, Mecoptera, and Siphonaptera (Klowden, 2013; Hoffmann, 2018).

2.5.4 Vitellogenesis

The major source of nutrients for developing oocyte is vitellogenin. It is usually produced in fat body, secreted into haemolymph and taken up specifically by developing oocytes and deposited in their cytoplasm as vitellin. The vitellophages, which get differentiated from energids early during development, break down yolk and make it available for embryogenesis. Besides provisioning embryo, vitellogenins are also used in other ways. While feeding brood, worker bees produce royal jelly from vitellogenin they synthesize that binds to hypopharyngeal glands, thereby supplementing metabolism of workers and synthesis of royal jelly when pollen sources are scarce (Bownes *et al.*, 1993; Klowden, 2013). Burrower bug, *Canthophorus*, produces trophic eggs filled with vitellogenins for consumption of immature stages. Within oocytes, vitellogenins are stored in a crystalline form as vitellins. Besides vitellogenins, lipids are also deposited in oocyte and make up 40% of the dry weight of an egg. Majority of these lipids are triacylglycerol and are main sources of energy for developing embryo (Chapman, 2009).

Small amounts of cholesterol, phospholipids, and free fatty acids are also present within oocytes. These are all transported by lipophorins from one tissue to another through haemolymph and are deposited within oocyte. Carbohydrate may be present in small quantity within the egg as glycogen stores. Yolk is deposited in oocyte when it moves down through vitellarium and increases volume of oocyte cell. In autosynthetic vitellogenesis, found in primitive apterygotes, egg synthesizes its own yolk from haemolymph proteins (Bownes *et al.*, 1993). However, in heterosynthetic vitellogenesis found in advanced insects, yolk is synthesized elsewhere and transported to oocyte through haemolymph. The fat body of an insect is stimulated by hormones to secrete vitellogenins into haemolymph, and follicle cells in addition synthesize and secrete yolk proteins into oocyte. The surrounding follicle cells regulate passage of yolk proteins into oocyte (Davey, 1965; Klowden, 2013).

At end of vitellogenesis, follicle cells produce an acellular membrane under the influence of 20-hydroxyecdysone, vitelline envelope, that surrounds oocyte. The last synthetic act of follicle cells is choriogenesis (secretion of chorion), which is a complex assemblage of proteins laid down sequentially to form an impermeable envelope around oocyte that restricts movement of substances in or out, except sperm, which enter through specialized micropyle. After synthesis of chorion, follicle cells degenerate and leave chorion on outer surface of egg (Davey, 1965; Chapman, 2009).

2.5.5 Endocrinology of Female Reproduction

In some holometabolous and all hemimetabolous insects, production of vitellogenins is dependent on JH. Insects having this strategy include earwigs, grasshoppers, cockroaches, and a few butterflies and moths. A few apterygotes also show JH-influenced vitellogenesis (Engelmann, 1970). JH acts directly on fat body cells, causing them to secrete vitellogenin and its uptake by oocytes. However, in tenebrionid beetles, JH has no role in ovarian growth and oocyte maturation, and 20-hydroxyecdysone is required (Wigglesworth, 1964; Hardie, 2018). When oocytes mature, signals are released from ovary to fat body that triggers JH to initiate expression of vitellogenin gene and synthesis of vitellogenin. In such a case, JH is involved in oocyte maturation while 20-hydroxyecdysone is engaged in vitellogenin synthesis. However, in many insects, including most dipterans, both JH and 20-hydroxyecdysone are involved in maturation and synthesis of vitellogenin (Klowden, 2013; Gullan & Cranston, 2014).

The mechanism of vitellogenesis also depends on insulin-like peptide signaling. In *Bombyx,* 38 genes for insulin-like peptides, and insulin receptors have been identified from both nurse cells and follicle cells. Insulin receptors have also been reported from nurse cells and follicle cells of mosquitoes.

Drosophila species that are mutant for insulin-like receptors are non-vitellogenic and fail to reproduce. In addition, production of oostatic hormone during later stages of egg development prevents maturation of additional secondary follicles until maturing eggs have been laid, and avoids development of large number of eggs which would restrict flight (Engelmann, 1970; Chapman, 2009).

2.6 Ovulation and Fertilization

When egg development is completed, mature eggs are moved into oviduct via ovulation. Ovulation takes place when follicle cells surrounding oocytes degenerate and leave egg free to move out of the ovariole. Once central nervous system receives signals that eggs are fully mature post mating, neurosecretory cells secrete myotropins that coordinate muscular contractions of ovariole and oviduct to propel egg through reproductive tract (Gullan & Cranston, 2014). As ovulation proceeds, spermathecal muscles contract and release sperm that are stored within spermatheca. As anterior portion of spermathecae contracts, sperm move down spermathecal duct, but are held there until sensory cells of spermathecal duct indicate that egg has achieved proper position in genital chamber (Klowden, 2013). Thereafter, posterior duct of spermathecae contracts and moves sperm onto micropyle of egg. The degree of complexity with which passage of eggs and sperm are coordinated is necessary for their successful union during fertilization (Ridley, 1988, 1990).

After ovulation and fertilization, eggs are deposited outside female's body via oviposition/egg deposition; eggs move down common oviduct by muscular contractions and out of body through ovipositor (Chapman, 2009). Once eggs are ovulated, a behavioural sequence is initiated via nervous system and sensory receptors of oviduct and external tip of ovipositor, that allows passage of egg and suitability of substrate to be monitored (Ridley, 1988, 1990; Chapman, 2009; Klowden, 2013).

The ovipositor allows the egg to be placed in a secure and appropriate environment. The brain triggers the release of an oviposition-stimulating hormone that activates ovipositor, ovariole and oviduct muscles, and terminal abdominal ganglion. As soon as the female insect evaluates substrate with its ovipositor, tactile sensations from sensillae are sent to terminal abdominal ganglion that subsequently controls the movement of egg and secretions by accessory glands (Klowden, 2013).

2.7 Mating and Oviposition of Eggs

2.7.1 Courtship Behaviour and Sexual Selection

Insects generally use species-specific signals to attract members of opposite sex of same species for mating. The flashing lights of fireflies, singing of crickets, and cacophony of cicadas are some common examples. Swarming brings opposite sex together in ancient insect lineages (Gullan & Cranston, 2014). Swarming sites are identified by visual markers and are usually species-specific. However, mixed-species swarms have also been reported, especially in insect species of tropics or subtropics. Swarms are mostly of 'male sex' only, but 'female-only' swarms also occur (Chapman, 2009). The formation of a swarm allows insects of differing genotypes to meet and outbreed, and avoid inbreeding, which will occur if adults do not disperse. Besides aerial aggregations, many male insects form substrate-based aggregations where they defend a territory against conspecific males and/or court arriving females (Klowden, 2013; Gullan & Cranston, 2014). Apart from swarming, both sexes may employ odoriferous communication via sex pheromones to facilitate meetings. These are usually produced by female to announce her presence and sexual availability to conspecific males, which on receipt attempt to locate female (Gullan & Cranston, 2014).

When sexes are in close proximity, immediate mating sans courtship or elaborate mating with courtship may take place (Gullan & Cranston, 2014). Courtship behaviour may include visual displays, chiefly by males, that include movements of adorned body parts, such as antennae, eyestalks, and 'picture' wings, and ritualized movements ('dancing'). Tactile stimulation also takes place during courtship, immediately prior to copulation, and may continue during mating. Antennae, palps, head horns, external genitalia, and legs are used during tactile stimulation (Chapman, 2009). However, in some predatory insects (like empidid flies and mecopterans), males may offer a prey item as a nuptial gift. If courtship behaviour proceeds correctly, it results in mating; however, it may even be unsuccessful (Klowden, 2013; Gullan & Cranston, 2014).

In many insects, courtship maybe viewed as intraspecific competition for mates with prize being increased mating success (Gullan & Cranston, 2014). The result of this competitive nature of courtship is decided via mate choice; an evolutionary process leading to members of one sex to mate non-randomly with respect to one or more varying traits in other sex (Andersson, 1994). The classical view of mate choice mechanism is that males are indiscriminate, while females are choosy (Bateman, 1948). Mate choice, either by males or females, has several costs, such as searching time (Johnstone *et al.*, 1996), risks of rejecting good mates (Real, 1990), energy for searching and assessing mates (Watson *et al.*, 1998), risks of predation (Crowley *et al.*, 1991; Rowe, 1994), and risks of injury (Bonduriansky & Brooks, 1998). Male mate choice is rarer than female mate choice, but occurs when: (i) male has substantial involvement in parental care, i.e. male transfer nutrients during copulation, and (ii) female mates vary greatly in quality (Jennions & Petrie, 1997; Bonduriansky, 2001). Male mate choice, may be precopulatory or via sperm competition, while female choice may be precopulatory or via cryptic female choice; details of all are discussed in further chapters (Bonduriansky, 2001).

2.7.2 Copulation

Except for Collembola, Diplura, and apterygote insects that deposit a spermatophore for female to pick up from substrate, all insects are free from such an indirect method of sperm transfer. Evolution of male external genitalia has made it possible for male insects to transfer their sperm directly to female body during copulation (Klowden, 2013; Gullan & Cranston, 2014). In pterygote insects, copulation involves physical apposition of male and female genitalia, usually followed by insemination. Male terminalia of many insect species even have specially modified claspers, which lock with specific parts of female terminalia to maintain connection of their genitalia during sperm transfer. Males may stroke, tap, or bite body or legs of female, wave antennae, produce sounds, or thrust or vibrate parts of his genitalia (Chapman, 2009; Gullan & Cranston, 2014). Ultimately sperm are received by female in a copulatory pouch (genital chamber, vagina, or bursa copulatrix) or directly into a spermatheca or its duct. Although spermatophore is means of sperm transfer in most orders, but some species belonging to Heteroptera, Coleoptera, Diptera, and Hymenoptera deposit unpackaged sperm. Sperm transfer or packaging of sperms within spermatophore requires lubrication, and secretions of male accessory glands not only serve these functions but they even facilitate final maturation of sperm, supply energy for sperm maintenance, regulate female physiology and, in a few species, provide nourishment to females (Gullan & Cranston, 2014).

2.7.3 Oviposition

After mating, egg-laying occurs. Except larviparous insects or insects providing parental care to their progeny, deposition of an egg on a specific substrate represents last decision of a female in favour of her progeny. The need to tightly regulate when and where to lay an egg demands neuronal control and coordination across multiple levels and timescales (Cury *et al.*, 2019). A gravid female must find a suitable site for egg deposition, and is guided in her search by cues, which is discussed in much detail in Chapter 15. The timing of oviposition is largely conditioned by numerous changes in a female's internal state.

Mated females develop a strong appetite for certain nutrients necessary for egg production, like polyamines (Hussain *et al.*, 2016). Male ejaculate contains spermatozoids and hundreds of proteins, peptides and transcripts (Kelleher *et al.*, 2009; Bono *et al.*, 2011), and pheromones like *cis*-vaccenyl acetate (Brieger & Butterworth, 1970). These proteins perform a variety of molecular functions that facilitate sperm transfer and influence post-copulatory behaviour of female (Chen *et al.*, 1988; Ameku & Niwa, 2016).

The external genitalia of female insects are covered with mechanosensory bristles and sensillae (Snodgrass, 1935; Taylor, 1989), which are stimulated during copulation, but play their role during process of oviposition. Females look for substrates that may protect their eggs and larvae from predators and pathogenic microorganisms, and sustain larval development. The assessment of oviposition sites involves a complex interplay of a variety of long- and short-range cues, which are discussed in Chapter 15. These may range from shape, colour, and texture of substrates, substrate-released chemicals, host plant-released chemicals, and conspecific and heterospecific cues on substrate (Del Solar *et al.*, 1974; Kelber, 1999; Zhu *et al.*, 2014; Seada *et al.*, 2016; Karageorgi *et al.*, 2017; Yadav & Borges, 2017)

Some insects lay a large clutch of eggs at a single site while others deposit just a single egg. Insect oviposition is a powerful example to address fundamental questions about decision-making, multimodal integration, and evolution of behaviour (Cury *et al.*, 2019).

2.8 Significance of Sexual Reproduction

Sexual reproduction is dominant and the most evolved mode of reproduction in nature. Since formulation of theory of natural selection, naturalists have repeatedly examined adaptive significance of both sexual and asexual reproduction. Sexual reproduction seems to require far more energy than asexual reproduction and diverts organisms from their routine activities and also makes them susceptible to predation and infection (Daly, 1978; Lloyd, 1980). There may be costs arising from conflicting male and female strategies, including damage imposed by males and costs of escaping from unwanted sexual attentions. Such factors may reduce current or future fecundity of individuals or shorten their lives (Burke & Bonduriansky, 2017). Further, sexual reproduction requires gametes of potential mates to come into contact, either through movements of mates themselves or of their mobile gametes. Mating success is also not assured in sexual reproduction. Maynard Smith (1978) and Williams (1971) suggest that advantage of asexual reproduction is advantage of not producing males. An asexual female transmits all her genes to her offspring, whereas only half of genes are passed on from a sexual female to her offspring.

Weismann (1889) on the other hand has suggested that while sexual reproduction lowers rate of reproduction and requires special morphological structures, it also facilitates evolution by providing genetic variability that natural selection depends upon. Reduction in accumulation of deleterious mutations (Agrawal, 2001; Siller, 2001), increasing the rate of adaptation to changing environments (Ladle, 1992; Agrawal, 2006), dealing with competition, DNA repair, and masking deleterious mutations (Burt, 2000; Roze, 2012) are major advantages of sexual reproduction. It also increases fitness of parents by generating genetically diverse offspring and by reducing competition between opposite-sex siblings. Cosexual individuals are also benefitted greatly from assistance from their mates, which may further lead to an increased parental care to their offspring (Lloyd, 1980).

2.9 Conclusions

Insects generally reproduce sexually. The reproductive organs in insects exhibit a wide range of variations. The sexually mature males and females must be present at same time and place for reproduction to be achieved. The male reproductive organ consists of a pair of testes that open separately into vas deferens. The latter generally expands posteriorly and forms seminal vesicle or sperm storage organ. The male insects also have numerous accessory glands that open into ejaculatory duct and transport semen to female gonopore/genital pouch. However, female reproductive organ consists of a pair of ovaries, connected with a pair of postero-lateral oviducts that join to form a median oviduct. The median oviduct opens posteriorly into a genital chamber, which is often developed to form a bursa copulatrix for reception of aedeagus. Genital chamber consequently opens into a spermatheca that temporary stores sperm utilized to fertilize eggs. As insects have short lifespan, their life history, behaviour, and reproductive condition need to be synchronized. This further requires finely tuned and complex physiological responses to external environment. In addition, reproduction also depends on coordination between internal physiological stimuli and neuroendocrine system. Although mating and egg production in insects are controlled by a sequence of hormonal and behavioural changes, yet there is much more to be explored about control and regulation of insect reproduction, especially when compared with vertebrate reproduction.

REFERENCES

Agrawal, A.F. (2001). Sexual selection and maintenance of sexual reproduction. *Nature*, *411*, 692–695.

Agrawal, A.F. (2006). Evolution of sex: why do organisms shuffle their genotypes?. *Curr. Biol.*, *16*, R696–R704.

Alcock, J. (2001). *Animal Behavior: An Evolutionary Approach*. USA: Sinauer Publishers.

Ameku, T., & Niwa, R. (2016). Mating-induced increase in germline stem cells via neuroendocrine system in female *Drosophila*. *PLoS Genet.*, *12*, e1006123.

Andersson, M. (1994). *Sexual Selection*. Princeton: Princeton University Press.

Arnqvist, G., Edvardsson, M., Friberg, U., & Nilsson, T. (2000). Sexual conflict promotes speciation in insects. *Proc. Natl. Acad. Sci. USA*, *97*, 10460–10464.

Baccetti, B. (1972). Insect sperm cells. *Adv. Insect Physiol.*, *9*, 315–397.

Baccetti, B., Dallal, R., & Fratello, B. (1973). The spermatozoon of Arthropoda. XXII. The '12 + 0, '14 + 0' or aflagellate sperm of Protura. *J. Cell Sci.*, *13*, 321–335·

Bai, S.N. (2015). The concept of sexual reproduction cycle and its evolutionary significance. *Front. Plant Sci.*, *6*(11), 1–6.

Bateman, A.J. (1948). Intra-sexual selection in Drosophila. *Heredity*, *2*, 349–368.

Bell, G. (1978). The evolution of anisogamy. *J. Theor. Biol.*, *73*, 247–270.

Bonduriansky, R. (2001). The evolution of male mate choice in insects: a synthesis of ideas and evidence. *Biol. Rev.*, *76*, 305–339.

Bonduriansky, R., & Brooks, R.J. (1998). Male antler flies (*Protopiophila litigata*; Diptera: Piophilidae) are more selective than females in mate choice. *Can. J. Zool.*, *76*, 1277–1285.

Bono, J.M., Matzkin, L.M., Kelleher, E.S., & Markow, T.A. (2011). Postmating transcriptional changes in reproductive tracts of con- and heterospecifically mated *Drosophila mojavensis* females. *Proc. Natl. Acad. Sci. USA*, *108*, 7878–7883.

Bownes, M., Ronaldson, E., Mauchline, D., & Martinez, A. (1993). Regulation of vitellogenesis in *Drosophila*. *Int. J. Insect Morphol. Embryol.*, *22*, 349–367.

Breed, M.D., & Moore, J. (2016). *Animal Behavior*. USA: Academic press.

Brieger, G., & Butterworth, F.M. (1970). Drosophila melanogaster: identity of male lipid in reproductive system. *Science*, *167*, 1262.

Bulmer, M.G. & Parker, G.A. (2002). The evolution of anisogamy: a gametheoretic approach. *Proc. R. Soc. London B*, *269*, 2381–2388.

Büning, J. (1993). Germ cell cluster formation in insect ovaries. *Int. J. Insect Morphol. Embryol.*, *22*, 237–253.

Büning, J. (1994). The insect ovary. *Ultrastructure, Previtellogenic Growth and Evolution*. London: Chapman & Hall, 400 pp.

Büning, J. (1998). The ovariole: structure, type and phylogeny. In: Harrison, F.W., & Locke, M. (eds.). *Microscopic Anatomy of Invertebrates*. New York: Wiley-Liss, pp. 897–932.

Burke, N.W., & Bonduriansky, R. (2017). Male coercion, female resistance and evolutionary trap of sexual reproduction. *BioRxiv*, 1, 146076.

Burt, A. (2000). Perspective: sex, recombination, and efficacy of selection--was Weismann right? *Evolution*, *54*, 337–351.

Chapman, R.F. (2009). *The Insects: Structure and Function*. Cambridge, UK: Cambridge University Press, pp. 770.

Chapman, T., & Davies, S.J. (2004). Functions and analysis of seminal fluid proteins of male *Drosophila melanogaster* fruit flies. *Peptides*, *25*, 1477–1490.

Charlesworth, B. (1978). The population genetics of anisogamy. *J. Theor. Biol.*, *73*, 347–357.

Chen, D., & McKearin, D. (2005). Gene circuitry controlling a stem cell niche. *Curr. Biol.*, *15*, 179–184.

Chen, P.S., Stumm-Zollinger, E., Aigaki, T., Balmer, J., Bienz, M., & Bohlen, P. (1988). A male accessory gland peptide that regulates reproductive behavior of female *D. melanogaster*. *Cell*, *54*, 291–298.

Crowley, P.H., Travers, S.E., Linton, M.C., Cohn, S.L., Sih, A., & Sargent, R.C. (1991). Mate density, predation risk, and seasonal sequence of mate choices: a dynamic game. *Am. Nat.*, *137*, 567–596.

Cruickshank, W.J. (1973). The ultrastructure and functions of ovariole sheath and tunica propria in flour moth. *J. Insect Physiol.*, *19*, 577–592.

Cury, K., Prud'homme, B., & Gompel N. (2019). A short guide to insect oviposition: when, where and how to lay an egg. *J. Neurogenet.*, *33*, 75–89.

Daly, M. (1978). The cost of mating. *Am. Nat.*, *112*, 771–774.

Davey, K.G. (1965). *Reproduction in Insects*. Edinburgh, London: Oliver & Boyd, pp. 96.

Del Solar, E., Guijon, A.M., & Walker, L. (1974). Choice of colored substrates for oviposition in *Drosophila melanogaster*. *Boll. Zool.*, *41*, 253–260.

Dugatkin, L.A. (2014). *Principles of Animal Behavior*. USA: Norton and Company.

Dusenbery, D.B. (2000). Selection for high gamete encounter rates explains success of male and female mating types. *J. Theor. Biol.*, *202*, 1–10.

Engelmann, F. (1970). *The Physiology of Insect Reproduction*. Oxford: Pergamon Press, pp. 307.

Friedländer, M. (1997). Control of eupyrene-apyrene sperm dimorphism in Lepidoptera. *J. Insect Physiol.*, *43*, 1085–1092.

Geyer, J.W.C. (1951). The reproductive organs of certain termites, with notes on hermaphrodites of *Neotermes*. *Ent. Mem. Dept. Agric.* S. *Africa*, *2*, 233–325.

Gillot, C. (2018). Insect male mating systems. In: Leather, S.R., & Hardie, J. (eds.). *Insect Reproduction*. USA: CRC Press, pp. 266.

Gillott, C. (1988). Arthropoda – insecta. In: Adiyodi, K.G., & Adiyodi, R.G. (eds.), *Reproductive Biology of Invertebrates. III. Accessory Sex Glands*. Chichester: Wiley & Sons, pp. 319–471.

Gillott, C. (2003). Male accessory gland secretions: modulators of female reproductive physiology and behavior. *Annu. Rev. Entomol.*, *48*, 163–184.

Gillott, C., Gaines, S.B. (1992). Endocrine regulation of male accessory gland development and activity. *Can. Entomol.*, *124*, 871–886.

Gullan, P.J., & Cranston, P.S. (2014). *The Insects: An Outline of Entomology*. USA: John Wiley & Sons, pp. 624.

Happ,G.M. (1984). Structure and development of male accessory glands in insects. In: King, R.C., & Akai, H. (eds.), *Insect Ultrastructure*. New York: Plenum Press, pp. 365–396.

Hardie, J. (2018). Hormones and reproduction. In: Leather, S.R., & Hardie, J. (eds.), *Insect Reproduction*. USA: CRC Press, pp. 266.

Hartfield, M., & Keightley, P.D. (2012). Current hypotheses for evolution of sex and recombination. *Integr. Zool.*, *7*, 192–209.

Hinton, H.E. (1974). Accessory functions of seminal fluid. *J. Med. Entmol.*, *2*, 19–25.

Hoffmann, K.H. (2018). Oogenesis and female reproductive system. In: Leather, S.R., & Hardie, J. (eds.), *Insect Reproduction*. USA: CRC Press, pp. 266.

Huebner, E. (1984). The ultrastructure and development of telotrophic ovary. In: King, R.C., & Akai, H. (eds.), *Insect Ultrastructure*. New York: Plenum Press, pp. 3–48.

Hussain, A., Ucpunar, H.K., Zhang, M., Loschek, L.F., & Grunwald Kadow, I.C. (2016). Neuropeptides modulate female chemosensory processing upon mating in Drosophila. *PLoS Biol.*, *14*, e1002455.

Jantzen, T.M., De Nys, R., & Havenhand, J.N. (2001) Fertilization success and effects of sperm chemoattractants on effective egg size in marine invertebrates. *Marine Biol.*, *138*, 1153–1161

Jennions, M.D., & Petrie, M. (1997). Variation in mate choice and mating preferences: a review of causes and consequences. *Biol. Rev. Camb. Philos. Soc.*, *72*, 283–327.

Johnstone, R.A., Reynolds, J.D., & Deutsch, J.C. (1996). Mutual mate choice and sex differences in choosiness. *Evolution*, *50*, 1382–1391.

Karageorgi, M., Bracker, L.B., Lebreton, S., Minervino, C., Cavey, M., Siju, K.P., Grunwald Kadow, I.C., Gompel, N., & Prud'homme, B. (2017). Evolution of multiple sensory systems drives novel egg-laying behavior in fruit pest *Drosophila suzukii*. *Curr. Biol.*, *27*, 847–853.

Kelber, A. (1999). Ovipositing butterflies use a red receptor to see green. *J. Exp. Biol.*, *202*, 2619–2630.

Kelleher, E.S., Watts, T.D., LaFlamme, B.A., Haynes, P.A., & Markow, T.A. (2009). Proteomic analysis of *Drosophila mojavensis* male accessory glands suggests novel classes of seminal fluid proteins. *Insect Biochem. Mol. Biol. 39*, 366–371.

Klowden, M.J. (2013). *Physiological Systems in Insects*. USA: Academic Press, pp 682.

Koeppe, J.K., Fuchs, M., Chen. T.T., Hunt, L.M., Kovalick, G.E., & Briers, T. (1985). The role of juvenile hormone in reproduction. In: Kerkut, G.A., & Gilbert, L.I. (eds.), *Comprehensive Insect Physiology, Biochemistry and Pharmacology*. Oxford: Pergamon Press, pp. 165–203.

Ladle, R.J. (1992). Parasites and sex: catching Red Queen. *Trends Ecol. Evol.*, *7*, 405–408.

Lay, M., Loher, W., & Hartmann, R. (2004). Pathways and destination of some male gland secretions in female *Locusta migratoria migratoriodes* (R&F) after insemination. *Arch. Insect Biochem. Physiol.*, *55*, 1–25.

Lehtonen, J., Jennions, M.D., & Kokko, H. (2012). The many costs of sex. *Trends Ecol. Evol.*, *27*, 172–178.

Leopold, R.A. (1976). The role of male accessory glands in insects. *A. Rev. Ent.*, *21*, 199–221.

Lessells, C.M., Snook, R.R., & Hosken, D.J. (2009). The evolutionary origin and maintenance of sperm: selection for a small, motile gamete mating type. In: Birkhead, T.R.D., Hosken, & Pitnick, S. (eds.), *Sperm Biology: An Evolutionary Perspective*. London: Academic Press, pp. 43–67.

Levitan, D.R. (2000). Sperm velocity and longevity trade off each other and influence fertilization in sea urchin *Lytechinus variegatus*. *Proc. R. Soc. London B Biol. Sci.*, *267*, 531–534

Lloyd, D.G. (1980). Benefits and handicaps of sexual reproduction. In: Hecht, M.K., Steere, W., & Wallace, B. (eds.), *Evolutionary Biology*. New York: Plenum Pess, pp. 69–111.

Mahowald, A.P. (1972). Oogenesis. In: Counce, S.J., & Waddington, C.H. (eds.), *Developmental Systems: Insects*, *I*, New York: Academic Press, pp. 1–47.

Maire, N., Ackermann, M., & Doebeli, M. (2002). Evolutionary branching and evolution of anisogamy. *Selection*, *2*, 119–131.

Mann, T. (1984). *Spermatophores. Development, Structure, Biochemical Attributes and Role in Transfer of Spermatozoa*. Berlin: Springer-Verlag.

Martoja, R. (1977). Les organes génitaux femelles. In: Grassé, P.P. (ed.), *Traité de Zoologie*. Paris: Masson et Cie, pp. 1–123.

Maynard Smith, J. (1978). *The Evolution of Sex*. Cambridge: Cambridge University Press.

Maynard Smith, J. (1982). *Evolution and Theory of Games*. Cambridge, UK: Cambridge University Press.

Maynard Smith, J., & Szathmary, E. (1997). *The Major Transitions in Evolution*. Oxford: Oxford University Press.

Mercer, E.H., & Brunet, P.C.J. (1959). The electron microscopy of left colleterial gland of cockroach. *J. Biophys. Biochem. Cytol.*, *5*, 257–261.

Otto, S. (2008). Sexual reproduction and evolution of sex. *Nature Edu.*, *1*, 182.

Parker, G.A. (1970). Sperm competition and its evolutionary consequences in insects. *Biol. Rev.*, *45*, 525–567.

Parker, G.A. (1978). Selection on non-random fusion of gametes during evolution of anisogamy. *J. Theor. Biol.*, *73*, 1–28.

Parker, G.A. (1979). Sexual selection and sexual conflict. In: Blum, M.S., & Blum, N.A. (eds.), *Sexual Selection and Reproductive Competition in Insects*. New York: Academic Press, pp. 123–166.

Parker, G.A. (1982). Why are there so many tiny sperm? Sperm competition and maintenance of two sexes. *J. Theor. Biol.*, *96*, 281–294.

Parker, G.A. (1993). Sperm competition games: sperm size and sperm number under adult control. *Proc. R. Soc. Lond. B*, *253*, 245–254.

Parker, G.A. (2011). The origin and maintenance of two sexes (anisogamy), and their gamete sizes by gamete competition. In: Togashi, T., & Cox, P.A. (eds.), *The Evolution of Anisogamy: A Fundamental Phenomenon Underlying Sexual Selection. Cambridge: Cambridge University Press*, pp. 17–74.

Parker, G.A., Baker, R.R., & Smith, V.G.F. (1972). The origin and evolution of gamete dimorphism and male-female phenomenon. *J. Theor. Biol.*, *36*, 529–553.

Phillips, D.M. (1970). Insect sperm: their structure and morphogenesis. *J. Cell Biol.*, *44*, 243–277

Real, L. (1990). Search theory and mate choice. I. Models of single sex discrimination. *Am. Nat.*, *136*, 376–404.

Richards, O.W., & Davies, R.G. (1979). *Imm's General Textbook of Entomology*. London: Chapman & Hall, pp. 418.

Ridley, M. (1988). Mating frequency and fecundity in insects. *Biol. Rev.*, *63*, 509–550.

Ridley, M. (1990). The control and frequency of mating in insects. *Funct. Ecol.*, *4*, 75–84.

Rowe, L. (1994). The costs of mating and mate choice in water striders. *Anim. Behav.*, *48*, 1049–1056.

Roze, D. (2012). Disentangling benefits of sex. *PLoS Biol.*, *10*, e1001321.

Seada, M.A., Ignell, R., & Anderson, P. (2016). Morphology and distribution of ovipositor sensilla of female cotton leaf worm *S podoptera littoralis* (Lepidoptera: Noctuidae), and evidence for gustatory function: *Spodoptera* ovipositor sensilla. *Entomol. Sci.*, *19*, 9–19.

Siller, S. (2001). Sexual selection and maintenance of sex. *Nature*, *411*, 689–692.

Snodgrass, R.E. (1935). *Principles of Insect Morphology*. New York: McGraw-Hill, pp. 667.

Snodgrass, R.E. (1957). A revised interpretation of external reproductive organs of male insects. *Smithson. Misc. Collect.*, *135*, 1–60.

Spradling, A.C., de Cuevas, M., Drummond-Barbosa, D., Keyes, L., Lilly, M., Pepling, M., & Xie, T. (1997). The *Drosophila germarium*: stem cells, germ line cysts, and oocytes. *Cold Spr. Harb. Symp. Quant. Biol.*, *62*, 25–34.

Taylor, B.J. (1989). Sexually dimorphic neurons of terminalia of *Drosophila melanogaster*: II. Sex-specific axonal arborizations in central nervous system. *J. Neurogenet.*, *5*, 193–213.

Telfer, W.H. (1975). Development and physiology of oocyte nurse cell syncytium. *Adv. Insect Physiol.*, *11*, 223–319.

Watson, P.J., Arnqvist, G., & Stallman, R.R. (1998). Sexual conflict and energetic costs of mating and mate choice in water striders. *Am. Nat.*, *151*, 46–58.

Weismann, A. (1889). The significance of sexual reproduction in theory of natural selection. In: Poulton, S. Schonland, & Shipley, A.E. (eds.), *Essays upon Heredity and Kindred Biological Subjects*. Oxford: Clarendon Press, pp. 254–338.

Wigglesworth, V.B. (1964). The hormonal regulation of growth and reproduction in insects. *Adv. Insect Physiol.*, *2*, 247–336.

Wightman, J.A. (1973). Ovariole microstructure and vitellogenesis in *Lygocoris pabulinus* (L.) and other Mirids (Hemiptera: Miridae). *J. Entmol.* (A), *48*, 103–115.

Williams, G.C. (1971). Introduction. In: Williams, G.C. (ed.), *Group Selection*. Chicago: Aldine Atherton, pp. 1–15.

Williamson, A., & Lehmann, R. (1996). Germ cell development in *Drosophila*. *Annu. Rev. Cell Dev. Biol.*, *12*, 365–391.

Yadav, P., & Borges, R.M. (2017). The insect ovipositor as a volatile sensor within a closed microcosm. *J. Exp. Biol.*, *220*, 1554–1557.

Zhu, E.Y., Guntur, A.R., He, R., Stern, U., & Yang, C.H. (2014). Egg-laying demand induces aversion of UV light in Drosophila females. *Curr. Biol.*, *24*, 2797–2804.

3

Parthenogenesis

Marcela S. Rodriguero[1,2]
[1]*Departamento de Ecología, Genética y Evolución, Facultad de Ciencias Exactas y Naturales, Universidad de Buenos Aires, Av Intendente Güiraldes 2160, Pabellón II, Ciudad Universitaria, 1428, Ciudad Autónoma de Buenos Aires, Argentina*
[2]*Instituto de Ecología, Genética y Evolución de Buenos Aires (IEGEBA, CONICET-UBA), Buenos Aires, Argentina*

CONTENTS

DOI: 10.1201/9781003043195-3

3.1 Introduction

> *"We do not even in least know final cause of sexuality; why new beings should be produced by union of two sexual elements, instead of by a process of parthenogenesis." Charles* Darwin (1862)

Among eukaryotes, sexual reproduction is the predominant reproductive system. However, sexually reproducing females suffer from a twofold cost (meiosis; Williams, 1975 and males; Maynard Smith, 1971) or higher (Brandeis, 2018) when competing with related females that reproduce unisexually. As Crow (1994) funnily asserted: '*Reproduction in [hetero]sexual species requires two to do work of one*'. With all these disadvantages, sexual reproduction thus must have a large advantage to become established and to resist invasion by a parthenogenetic lineage (Crow, 1994).

'*Parthenogenesis*' (*partheno* – virgin, *genesis* – origin, from *gen* – to be produced) is the process by which progeny can develop from an unfertilized egg. The study of parthenogenetic lineages in tandem with closely related sexual lineages is a promising approach for determining benefits and/or costs of sex.

Insecta and other members of Hexapoda like Collembola, Protura, and Diplura show 'immaculate conception' in about 1,100 species, that is less than 0.1% (Normark, 2014).

3.2 Discovery of Parthenogenesis: An Entomological Journey

The earliest observations on parthenogenesis were by Aristotles in the honey bee (Phillips, 1903). Much later, Dutch naturalist, entomologist, and painter Johannes Goedart bred larvae from unfertilized eggs of moth *Orgyia gonostigma* (Plateau, 1868). Later, Leeuwenhoek noted developing embryos, rather than oocytes, within aphid ovaries (van Leeuwenhoek, 1695) and suggested that this reproduction was without copulation (van Leeuwenhoek, 1700). Italian naturalist Giacinto Cestoni suggested that these may be hermaphrodites owing to absence of mating[1] (Berçot & Prestes, 2017), absence of which was also observed by French entomologist René Antoine de Réaumur, who assumed them to be either hermaphrodites or completely females (3rd volume of *Mémoires pour servir à l'histoire des insectes* Amsterdam, 1734–1742).

Charles Bonnet, inspired by Réaumur's suggestion, selected an aphid and isolated it immediately after birth and raised it with food and water, which resulted in 95 offspring within a month; he continued his observations for over 9 generations of isolated virgin females. He continued to demonstrate parthenogenesis in plant lice (Phillips, 1903). A century later, Dzierzon (1845), a Polish apiarist, theorized parthenogenetic development of drone eggs of *Apis mellifica,* considered the starting point of theory of parthenogenesis (Phillips, 1903).

Earlier known as '*lucina sine concubitu nulla*', this phenomenon was termed '*parthenogenesis*' by Richard Owen[6] in 1849, earlier in relation to alternation of generations or '*metagenesis*' (Owen, 1849;

Phillips, 1903), defining it as '*procreation without immediate influence of a male*'. The usage of word 'parthenogenesis' in sense of development of eggs without fertilization is however due to von Siebold (1856), who called it 'true parthenogenesis'.

3.3 Parthenogenesis: A Plethora of Definitions

Despite after almost three centuries of studies on parthenogenesis, there are still a lot of heterogeneity and confused terminologies across scientific literature with regard to this process (Martin, 2015)[2]. In the subsequent sections, we try and define parthenogenesis (the stage), reproducing individuals (the actors), and type of parthenogenesis (the plays).

3.3.1 The Stage

Sex is a two-act play: ploidy reduction through meiosis, and ploidy restitution through syngamy. Bell (1982) in chapter '*Reproduction without sex*' in '*The Masterpiece of Nature: The Evolution and Genetics of Sexuality*', distinguished between '*mixis*' (fusion of haploid nuclei for zygote formation) and '*amixis*' (reproductive modes having syngamy without meiosis; gametes being from same or different individuals). Fusion of two meiotcally generated haploid nuclei from same diploid genome is '*automixis*' see '*Thelytokous parthenogenesis: automixis*'). *Amphimixis* then would be sygamy of gametes from different sexes (sexuality per convention).

Parthenogenesis is different from '*asexuality*' as in latter new individuals are formed from parental somatic cells (*amixis*, Bell, 1982). Thus, parthenogenesis is more like '*unisexuality*', with identifiable sex/gender and meiosis or '*apomeiosis*' (deregulation of meiosis resulting in mitotic-like division (d'Erfurth *et al.*, 2009). A successful transition from reproduction by fertilization to parthenogenesis requires: (i) females to be able to lay eggs without mating and (ii) a high proportion of unfertilized eggs have to be able to hatch (Degrange, 1960).

Parthenogenesis is better regarded as an incomplete form of sexual reproduction (Mittwoch, 1978); however, when meiosis is deregulated and mother and offspring have same genetic constitution, parthenogenesis may be equated to '*clonality*'.

Suomalainen (1950) defines parthenogenesis as the '*development of egg cell into a new individual without fertilization*'; which is adequate for many organisms; but unfertilized embryos undergo heavy mortality in some others. Beatty (1967) remodified his definition of parthenogenesis as '*production of an embryo from a female gamete without any genetic contribution from a male gamete, and with or without eventual development into an adult*';'*without any genetic contribution from a male gamete*' substituting for initial '*concurrence of a male gamete*' to include the special case of gynogenesis (see '*Sperm dependent thelytoky: pseudogamy or gynogenesis*').

3.3.2 The Actors

An individual developing from an unfertilized egg is referred to as '*parthenogenone*', '*parthenogen*', or '*parthenote*' (Mittwoch, 1978); *parthenogen* is most used in cases of spontaneous parthenogenesis.

3.3.3 The Plays

Parthenogenesis can be classified according to cytological mechanisms, offspring sex, dependence on sperm presence, and so on.

The classification designed by Normark (2003) takes into account all the variations in parthenogenesis;

- I. Obligate amphimixis (sex, sexuality)
  - A. ***Diplo-diploidy*** (diploid-male systems)
  - B. ***Haplo-diploidy*** (haploid-male systems, male haploidy, uniparental-male systems)
    1. *Arrhenotokous haplo-diploidy (Arrhenotoky)*
    2. *Paternal genome elimination [(PGE), paternal genome loss]*
    3. *Parallel clonality*
- II. ***Thelytoky*** (all-female systems, uniparental reproduction, parthenogenesis)
  - A. ***Thelytokous parthenogenesis*** (ameiotic parthenogenesis, strict clonality)
    1. *Apomixis* (ameiotic parthenogenesis, strict clonality)
    2. *Apomixis* (meiotic parthenogenesis)
  - B. ***Sperm-dependent thelytoky*** (obligate mating)
    1. *Pseudogamy* (gynogenesis)
       - a. Apomictic pseudogamy
       - b. Automictic pseudogamy
    2. *Hybridogenesis* (hemiclonal inheritance)
- III. **Mixed systems**
  - A. ***Thelytoky (or polyembryony) alternating with haplo-diploidy***
    1. *Cyclic alternation*
       - a. Polyembryony
       - b. Cyclic haplo-diploidy-thelytoky (cyclic parthenogenesis)
    2. Facultative haplo-diploidy-thelytoky (facultative parthenogenesis)
  - B. ***Thelytoky alternating with diplo-diploidy***
    1. *Cyclic diplo-diploidy-thelytoky* (cyclic parthenogenesis)
    2. *Facultative diplo-diploidy-thelytoky* (facultative parthenogenesis)

We further define some of the terms according to the scenario involved.

3.4 Types of Parthenogenesis

3.4.1 Obligate Amphimixis Scenario

3.4.1.1 Diplo-Diploidy

Male inherits one haploid genome from mother and one from his father, and both haploid genomes have equal probability of transmission through his sperm. This is found in majority of all insect species and in ancestral system is all orders except Hymenoptera and Thysanoptera.

3.4.1.2 Haplo-Diploidy

Male transmits only mother's genome via *'arrhenotoky' or 'paternal genome elimination' (PGE).*

3.4.1.2.1 Arrhenotokous Haplo-Diploidy (Arrhenotoky)

Males develop from unfertilized eggs inheriting haploid maternal genome.

Although males arise parthenogenetically, a haplo-diploid species reproduces sexually, and male genotype is a unique recombinant product of meiosis. This is the only known sexual system in Hymenoptera and Thysanoptera, and in some scale insects (Hemiptera: Margarodidae), whiteflies (Hemiptera: Aleurodidae), beetles (Coleoptera: Micromalthidae (but see Perotti *et al.*, 2016 and section '*Mixed systems*'), and bark beetles (Coleopter: Curculionidae) (Vershinina & Kuznetsova, 2016; Gokhman & Kuznetsova, 2018).

The term '*deuterotoky*' or '*amphitoky*' (parthenogenetic production of males and females) is not much in use since its production of offspring thelytokously is often temperature-dependent (Luck *et al.*, 1992), or on coevolution with a parthenogenesis inductor agent (Reumer *et al.*, 2012) (see '*Modes of origin of parthenogenetic lineages: infectious origin*'). Bergman and Hilsenhoff (1978) and Funk *et al.* (2010) reported this reproductive mode in different species of mayflies, but it may be sexual coupled to facultative parthenogenesis and offspring of gynandromorphs, respectively. Thus, deuterotoky is sometimes applied to cyclic and facultative parthenogenesis; for rest of this text, parthenogenesis will refer to thelytoky.

3.4.1.2.2 Paternal Genome Elimination

In this system, males are not fatherless but their paternally inherited chromosomes are actively *eliminated* during i) embryonic development, ('*embryonic elimination*'), or ii) spermatogenesis, i.e. males are somatically diploid but exhibit non-Mendelian patterns of meiosis (like B-chromosomes, van Vugt *et al.*, 2009) and of spermiogenesis (inactivation by chromatin condensation) ('*germ-line elimination*') (Herrick & Seger, 1999).

This is found in 14 families of scale insects (Ross *et al.*, 2010), *coffee berry borer* (Borsa & Kjellberg, 1996), *a few arrhenotokous parasitic wasps* (Beukeboom & Werren, 2000; van Vugt *et al.*, 2003), *human body lice* (McMeniman & Barker, 2006), *booklice* (Hodson *et al.*, 2017), *dipterans of Cecidomyiidae* (Painter, 1966), *Sciaridae* (Goday & Esteban, 2001), and *Chironomidae* (Staiber & Schiffkowski, 2000). The embryonic and germ-line elimination systems are often characterized as '*parahaplo-diploid*' or '*pseudoarrhenotokous*' (Herrick & Seger, 1999). However, *PGE* is identical to arrhenotokous haplo-diploidy in terms of transmission genetics (Normark, 2003, 2014).

3.4.1.2.3 Parallel Clonality

This reproductive mode (name by Normark, 2014) is only known in fire ant *Wasmannia auropunctata* (Fournier *et al.*, 2005) and Japanese ant *Vollenhovia emeryi* (Ohkawara *et al.*, 2006). Sexually produced individuals are sterile workers, while males and females are clonally propagated. Although queens are clones of their mother, in a proportion of fertilized eggs father's genome replaces mother's completely ('*androgenesis*'), resulting in haploid male clones of father.

3.4.2 Thelytoky

3.4.2.1 Thelytokous Parthenogenesis

Thelytoky results in female-only offspring, via meiosis suppression or fusion of haploid meiotic products. It allows females to: i) pass along their successful genotypes to all offspring; ii) produce only daughters; and iii) eliminate expenditure in finding a mate.

Female parthenogenesis or Thelytoky is found in most orders of Insecta; Thysanoptera, Psocoptera, Hemiptera (suborder Sternorrhyncha), and Phasmatodea have highest frequency. Lepidoptera, Diptera, Coleoptera, and Hymenoptera have low incidence of thelytoky but high rates in families (Coleoptera: Curculionidae; Lepidoptera: Psychidae; Diptera: Chironomidae). Thelytoky has never been reported in Diplura, Megaloptera, Raphidioptera, Mecoptera, Siphonaptera, Coleorrhyncha, Notoptera, and unsubstantiated reports exist for Dermaptera, Neuroptera, and Strepsiptera (Suomalainen *et al.*, 1976; Vershinina & Kuznetsova, 2016; Rafael *et al.*, 2017; Gokhman & Kuznetsova, 2018).

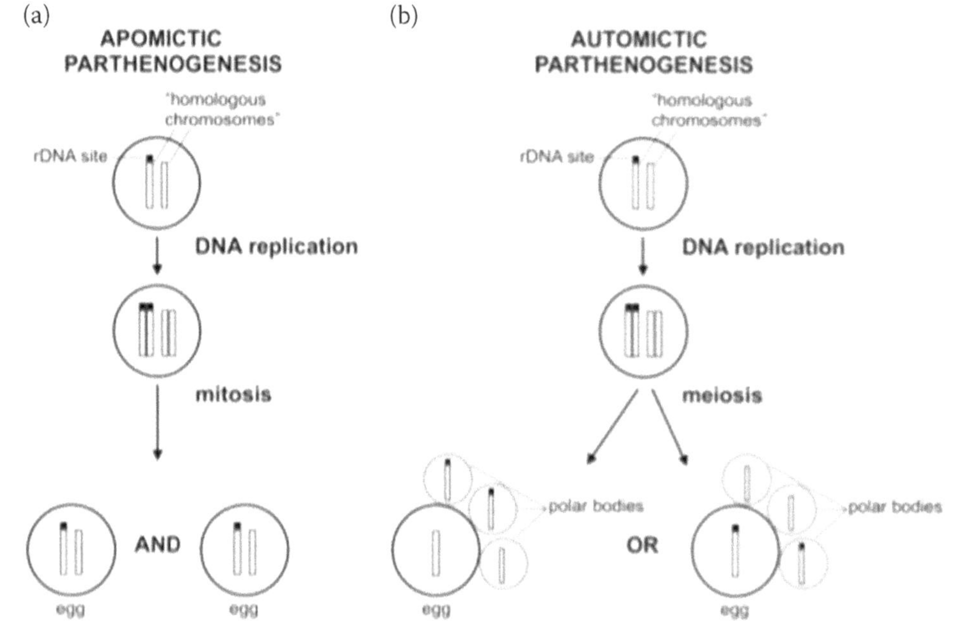

FIGURE 3.1 Eggs produced by (a) apomictic and (b) automicitic females (taken from Schneider & Cella, 2010).

It is particularly common in solitary Hymenoptera and many eusocial species also regularly reproduce thelytokously (Rabeling & Kronauer, 2013). In eusocial species thelytoky probably has a nuclear genetic basis with underlying cytological mechanism retaining high levels of heterozygosity, while in many solitary wasps, it is often induced by cytoplasmic bacteria and with immediate loss of heterozygosity. An unusual case of thelytoky, without PGE but with diploid and haploid embryos inside same female was described for fat grass scale *Aclerda takahashii* (Gavrilov-Zimin, 2012).

Thelytokous parthenogenesis has two cytological mechanisms (i) a meiotic phase or automixis, and (ii) deregulated meiosis (or apomeiosis), or apomixis.

3.4.2.1.1 Apomixis

The simplest way to achieve diploidy in parthenogenesis is by replacing meiosis by mitosis (implies clonal offspring except for new mutations) (Figure 3.1). Genetically speaking, this is simplest form of parthenogenesis (Normark & Kirkendall, 2009), is most widespread in insects in at least six different orders (Suomalainen *et al.*, 1987). For a thorough enumeration of cases, see Suomalainen *et al.* (1976), Suomalainen *et al.* (1987), Vershinina and Kuznetsova (2016) and Gokhman and Kuznetsova (2018).

3.4.2.1.2 Automixis

Restoration of diploidy post-meiosis is slightly more complicated requiring fusion of haploid nuclei from same gametic sex or ploidy restoration (Lamb & Willey, 1987). There are different mechanisms for diploidy restoration that involve fusion of feminine nuclei, which may be premeiotic or postmeiotic.

3.4.2.1.2.1 Premeiotic Restoration. 3.4.2.1.2.1.1 Premeiotic Doubling Chromosome number double in every oogonial cell prior to meiosis ('*endomitosis*'). These daughter chromosomes pair with their genetically identical counterpart in first meiotic prophase and produce a diploid cell. White *et al.* (1963) inferred this mechanism for grasshoppers, though no direct demonstration has been made yet.

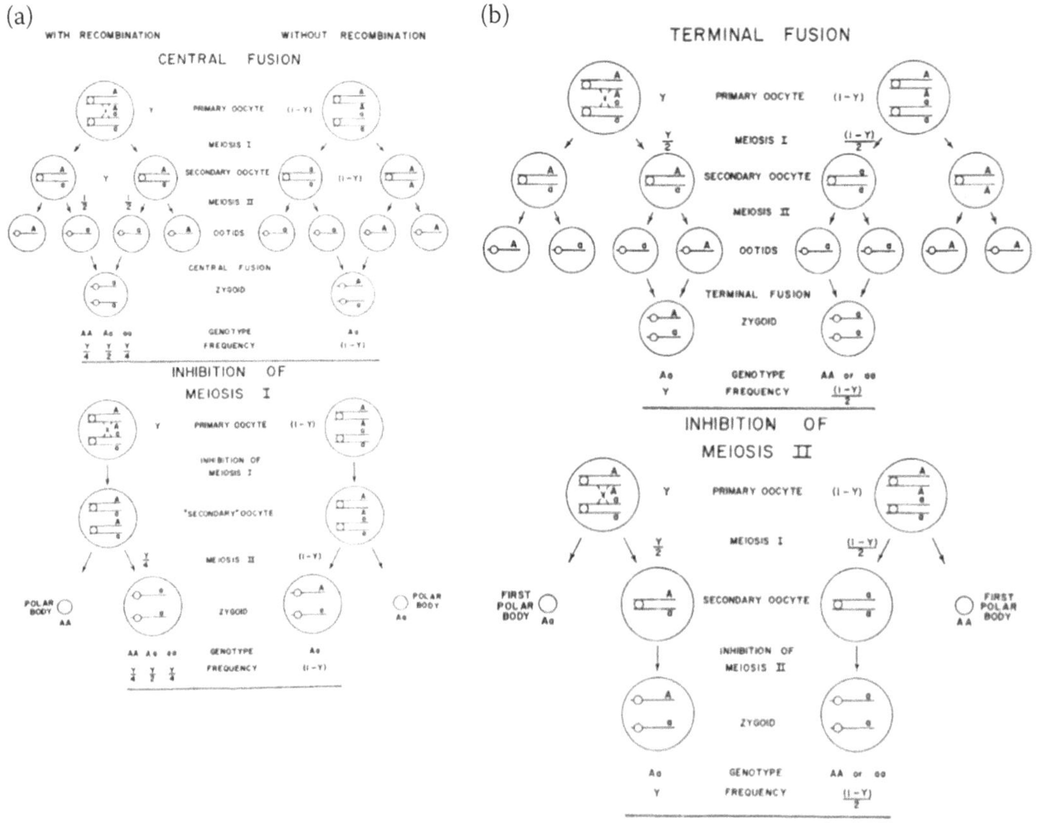

FIGURE 3.2 Diagrammatic representation of (a) Central fusion and (b) terminal fusion (taken from Asher 1970, with permission).

3.4.2.1.2.1.2 Pachyreduplication Doubling of DNA content occurs just after pachytene, and is reported in stick insects (Pijnacker & Ferwerda, 1978).

3.4.2.1.2.2 Perimeiotic Restoration. 3.4.2.1.2.2.1 Terminal Fusion Here egg nucleus fuses with second polar nucleus (Figure 3.2) and is also known as *terminal fusion.* Some examples are: *Drosophila parthenogenetica* (Stalker, 1954), termites (Matsuura *et al.*, 2004), acridids (Suomalainen *et al.*, 1987), and many more.

3.4.2.1.2.2.2 Central Fusion In '*central fusion*' polar nuclei from different secondary oocytes combine to restore ploidy (Figure 3.2) and is found in moth *Solenobia triquetrella* (Seiler, 1960) and *Drosophila mangabeirai* (Murdy & Carson, 1959), *Apis mellifera capensis* (Suomalainen *et al.*, 1987) and many others.

3.4.2.1.2.2.3 Instant Triploidy Any three products of second meiotic division can combine in certain organisms to form an '*instant triploid*' or gonoid thelytoky (Stenberg & Saura, 2009): parthenogenetic egg undergoes two meiotic divisions and chromosomes pair at second meiotic metaphase. The resulting organism, although triploid, is created each generation from a diploid parthenogenetic mother (Stalker, 1954; Nur, 1979).

3.4.2.1.2.3 Postmeiotic. 3.4.2.1.2.3.1 Pronucleus Duplication or Gamete Duplication Ploidy is restored by fusion of two nuclei resulting from first cleavage division. It is found in coccid, phasmids, aleyrodids, lepidopterans, in some parthenogenetic *Drosophila* and cynipids (Nur, 1971; Suomalainen *et al.*, 1987; Norton *et al.*, 1993). '*Gamete duplication*' is frequently induced by cytoplasmically inherited bacteria like *Wolbachia* in hymenopterans (Gottlieb *et al.*, 2002; Pannebakker *et al.*, 2004). *Wolbachia* is responsible for transition from arrhenotoky (and haplo-diploidy) to thelytoky by duplicating genome of unfertilized egg converting a potential haploid male into a diploid unfertilized female, either by segregation failure in first mitotic anaphase or through fusion of mitotic nuclei in second prophase (Braig *et al.*, 2002; Pannebakker *et al.*, 2004).

3.4.2.1.2.3.2 Multiple Conversions This system is similar to that above except many, if not all nuclei that reach syncytial blastoderm or segmentation by haploid mitosis, become diploid before embryogenesis continues (Lamb & Willey, 1987) While genetic consequence is same as in gamete duplication, this should be placed in a separate category because ploidy restoration occurs at a much later stage. This process is known only in two species of stick insects (Pijnacker, 1969) and one species of gryllacridid (Lamb & Willey, 1989).

Final Considerations of Automixis Insects probably show widest range of parthenogenetic automictic processes of any taxonomic group (Mogie, 1986). Because automixis includes syngamy, some support automixis as a process of fertilization and thus should be accounted within amphimixis. They question this kind of reproduction as a sub-division of parthenogenesis (Bell, 1982).

3.4.2.2 Sperm-dependent Thelytoky

3.4.2.2.1 Pseudogamy or Gynogenesis

Some insects despite being apo/automictic, require sperm to trigger onset of embryogenesis (Normark, 2014). Egg division is activated by contact with or penetration by sperm with no genetic contribution due to parental chromosome degeneration also known as pseudofertilization (Beukeboom & Vrijenhoek, 1998). This is genetically parthenogenesis (Normark & Kirkendall, 2009) and called '*gynogenesis*' or '*pseudogamy*' requiring coexistence males from conspecific sexual lineages, conspecific hermaphrodites or males from a closely related sexual species (Beukeboom & Vrijenhoek, 1998).

This has been reported for a few species of Collembola (springtails), Coleoptera (beetles), Lepidoptera (moths), Hemiptera (planthoppers) (see Beukeboom & Vrijenhoek, 1998; Normark, 2003), *Culex pipiens* (Laven, 1956) and *Drosophila melanogaster* (Fuyama, 1986). It may be more widespread as it is usually discovered by noting that mated females regularly produce all-female broods (Normark & Kirkendall, 2009).

3.4.2.2.2 Hybridogenesis

In '*hybridogenesis*', egg nucleus fuses in each new generation with spermatozoon nucleus from a male of a sexually reproducing species, but paternal genome is excluded during oogenesis. Although both maternal and paternal genomes are expressed in somatic tissues, germ-line is purely maternal and male genetic material is not heritable. This is a '*hemiclonal*' mode of reproduction because half of genome is transmitted sexually while other half is transmitted clonally. Hybridogenesis is only known in *Bacillus* (order Phasmatodea) (Tinti & Scali, 1995, 1996)

Although classed as '*sperm-dependent parthenogenesis*', it is actually a hybrid form of reproduction because it combines sexual reproduction (spermatogenesis), with non-recombinant, clonal maternal genomes (Normark & Kirkendall, 2009). *Bacillus rossius grandii* also presents '*androgeny*'; replacement of female pronucleus by male one during fertilization and subsequent fusion of two (now) male gametes, or alternatively, fusion of a replicated spermatozoon pronucleus, what is called '*paternal apomixis*' (Pigneur *et al.*, 2012).

In desert ant *Cataglyphis hispanica* workers are produced exclusively from interbreeding between two distinct genetic lineages, whereas sexual males and females are produced through parthenogenesis. This is called 'social hybridogenesis" (Leniaud *et al.,* 2012). All workers are thus pure hybridogens, and only maternal genes are perpetuated from one generation to next. Queens of *C. hispanica* use sexual reproduction for colony growth, but use parthenogenesis for germ-line production.

Final Considerations of Sperm-dependent Parthenogenesis The requirement for sperm forces all-female gynogens and hybridogens into a parasitic relationship with a sexually reproducing host (Beukeboom & Vrijenhoek, 1998), with sperm becoming limiting (Vrijenhoek, 1989) and mitigating twofold reproductive advantage of all-female lineages (Kirkendall, 1990). Conspecific mating preferences by sexual males may also limit all-female reproduction (Moore, 1975, 1976). Males may learn to discriminate against common gynogenetic and hybridogenetic clones giving phenotypically rare clones a temporary advantage facilitating maintenance of clonal diversity (Keegan-Rogers, 1984). Niche separation between sexual and clonal forms may be responsible for stable coexistence in mixed reproductive complexes (Kirkendall & Stenseth, 1990). Sperm-dependent parthenogens seem to have recently evolved from sexually reproducing progenitors (Sanderson & Jacob, 1957).

3.4.3 Mixed Systems

This category covers all those complex systems in which there is an '*alternation*' between thelytoky and diplo-diploidy or haplo-diploidy (Normark, 2003). An obligate parthenogentic lineage is permanent but many insects have life cycles that encompass sexual and parthenogenetic reproduction, either '*cyclically*' or '*facultatively*'.

'*Cyclical parthenogenesis*' involves an alternation of one generation of sexual reproduction with one or more generations of parthenogenetic reproduction. It is seen in aphids (Dixon, 1998), oak and maple gall wasps (Hebert, 1987) and two divergent groups of gall midges (Baxendale & Teetes, 1981).

Micromalthus debilis, only extant species of Micromalthidae, usually reproduces by thelytokous parthenogenesis, with rare arrhenotokous phase that apparently represents an evolutionary dead end (Pollock & Normark, 2002). However, Perotti *et al.* (2016), dismissed haplo-diploid cyclical parthenogenetic life history in favour of '*parthenogenetic paedogenesis*'transition with thelytoky as only functional mode of reproduction in this species.

Some insects abbreviate their life cycles by loss of adult stage, or even both adult and pupal stages; thus only way of reproduction in a precocious stage is by parthenogenesis, and is known as '*paedogenesis*' or '*larval thelytoky*' (Perotti *et al.*, 2016). It has been reported for a wide variety of insects (Johannsen, 1910; Bezzi, 1922; Scott, 1938; Hauschteck, 1962; Went, 1971; Wood & Starks, 1975; Ibrahim & Gad, 1975; Achterkamp, 1999). The reproductive stage, larval or pupal, varies between species. In '*pupal paedogenesis*' ('*pupal thelytoky*') (gall midges; Wyatt, 1961, 1963, 1967), embryos are formed in hemocoel of a paedogenetic mother-pupa, termed a '*hemipupa*', which is destroyed by offspring from within, either by larval perforation or consumption of mother. This reproductive mode appears to have evolved to allow maximum use of locally abundant but ephemeral larval habitats; with adults developing only in response to adverse larval conditions such as food depletion and over-crowding.

Continuing further with alternation of reproductive modes, '*facultative parthenogenesis*' or '*tychoparthenogenesis*' implies that a given individual can reproduce either sexually or unisexually. In some insects, sexual reproduction is norm, but if a clutch of eggs is left unfertilized for sufficiently long, a small proportion of them will begin development. It has been reported in Ephemeroptera (Ball 2001, 2002), Orthoptera, Blattodea, Phasmatodea, and Mantodea (Vershinina & Kuznetsova, 2016) and Psocoptera, Lepidoptera, and Diptera (Gokhman & Kuznetsova, 2018). The cytology of tychoparthenogenesis is always automictic (Normark, 2003).

Another unusual system is '*polyembryony*', which is not strictly speaking a form of parthenogenesis, but rather a process of "vegetative" fission of multicellular embryos (Grbic, 2003; Normark, 2003).

Polyembryony occurs regularly in parasitoids of a few families of Hymenoptera and in Strepsiptera, and has been reported to occur sporadically in grasshoppers (Noskiewicz & Poluszynski, 1935; Slifer & Shulow, 1947; Prévost & McFarlane, 1979; Hardy *et al.,* 1993; Grbic, 2003).

Thus, a species may have a permanent genetic system or alternate between two. For latter, parthenogenesis may be either facultative or obligate and thus cyclic for only one generation or maybe more. Now, let's put some order of which details are available in Normark (2003).

A. **Thelytoky (or polyembryony) alternating with haplo-diploidy**
   1. Cyclic alternation
      a. Cyclic polyembryony: parasitic wasps.
      b. Cyclic parthenogenesis
         i. Fixed (single-generation) cycling. A single generation of amphimixis alternates with a single generation of thelytoky: gall wasps.
         ii. Flexible (several- to many-generation) cycling. Reproduction is typically thelytokous, but an environmental cue can trigger a switch to amphimixis: beetles and *gall midges.*
   2. Facultative parthenogenesis. Any of various systems in which reproduction may be either amphimictic or thelytokous. In most of these systems, reproduction is typically amphimictic, but unmated females may produce some viable offspring by thelytoky. Widespread in parasitic wasps, bees and ants and in a few giant scales.

B. **Thelytoky alternating with diplo-diploidy**
   1. Cyclic parthenogenesis: aphids
   2. Facultative parthenogenesis: mayflies, termites, stick insects, mantids, cockroaches, moths (like silkworm and green oak tortrix), grasshoppers and a few barklice and flies.

3.5 Modes of Origin of Parthenogenetic Lineages

3.5.1 Spontaneous Origin

'*Spontaneous parthenogenesis*' may occur through mutations in genes producing sexual forms and meiosis resulting in either unreduced gamete formation or facilitating spontaneous development of unfertilized gametes. Linkage among loci affecting these two traits may facilitate transition to obligate unisexuality (Neiman *et al.,* 2014).

Unisexuality is less likely to evolve if these two traits are controlled by separate loci. This predicted connection between genetic architecture and likelihood of transition to unisexuality is possible because spontaneous development of meiotically reduced gametes would result in nonviable (haploid) embryos for most species (Beukeboom *et al.,* 2007; Soriano *et al.,* 2013), constraing abandonment of sex (Neiman *et al.,* 2014). Similarly, production of unreduced gametes that cannot develop spontaneously is also expected to be costly as it would result in increased ploidy upon fertilization (Neiman *et al.,* 2014).

Multiple evidences suggest that unreduced gamete formation and spontaneous development are often regulated by distinct genetic factors, and evolution of pseudogamy is support for this hypothesis (Beukeboom & Vrijenhoek, 1998; Schlupp, 2005). The ability to produce unreduced gametes should be independent from that of producing spontaneously developing embryos; plants provide most evidential support (Catanach *et al.,* 2006; Noyes *et al.,* 2007; Matzk *et al.,* 2005). A few studies however talk about physical linkage and co-segregation in two loci (Bicknell & Koltunow, 2004). Co-expression or co-segregation of separate adaptations to parthenogenesis would favour evolution of novel unisexual lineages in a sexual population; control of dependent traits by a single master regulatory factor or control by multiple factors that are physically linked in a genomic region that undergoes little or no recombination may also play a role (Grossniklaus *et al.,* 2001).

Meyer *et al.* (2010) reported a mutant allele of *yem-alpha* gene in *D. melanogaster* which to some extent affects diploidy reduction and its restoration. This point mutation affects chromosome segregation and results in development of parthenogenetic progeny, which did not display paternal chromosome markers, suggesting that their diploid gametes underwent gynogenesis.

In cereal aphid *Sitobion avenae* two autosomal loci seem to control obligate parthenogenesis (Dedryver *et al.,* 2013). In pea aphid *Acyrthosiphon pisum* a 9 cM-wide recessive X-linked region was found to control reproductive mode (Jaquiéry *et al.,* 2014). A single-locus recessive trait was also identified in parasitic wasp *Lysiphlebus fabarum,* a case of contagious parthenogenesis (Sandrock & Vorbuger 2011).

In *Apis mellifera capensis*, the ability of sterile workers to produce female progeny via thelytokous parthenogenes is owing to an incomplete meiotic division, probably controlled by a recessive allele at *thelytoky* locus (*th*) on chromosome 13. This locus spans 11.4 cM (180 kb), comprises 15 genes, and is genetically linked to rapid ovary activation and production of high levels of queen substance pheromone 9-oxo-2-(E)-decenoic acid (9-ODA) (Ruttner, 1988; Lattorff *et al.*, 2005, 2007). A candidate gene approach revealed that alternative splicing of a gene homologous to *gemini* transcription factor of *Drosophila* controls worker sterility: a rapid worker ovary activation may be induced by removal of a short intronic splice enhancer motif nine nucleotides long attached to an alternative splice site, turning an altruistic worker into a social parasite (Jarosch *et al.*, 2011). However, subsequent studies (Chapman *et al.,* 2015; Aumer *et al.*, 2017) showed that it was not the ultimate gene switch controlling mode of parthenogenesis. Aumer *et al.* (2019) showed that a single nucleotide polymorphism (SNP) at thelytoky locus (*Th*) is associated with thelytokous phenotype. *Th* forms a linkage group with ecdysis-triggering hormone receptor (*Ethr*) within a non-recombining region under strong selection in genome.

Yagound *et al.* (2020) rejected all previous *th* gene candidates, except for gene GB45239 on chromosome 11, which has a putative role in chromosome segregation. It is expressed in ovaries and is downregulated in thelytokous bees, likely because of polymorphisms in promoter region. Alleles associated with thelytoky are present in all *A. melliferamellifera capensis* genomes and absent from all other honeybees species worldwide (Cole-Clark *et al.,* 2017).

In bird cherry-oat aphid *Rhopalosiphum padi*, obligate parthenogenesis appears to be determined by a single locus, and is recessive to cyclical parthenogenesis (Frantz *et al.,* 2005). In contrast, predisposition for parthenogenesis in *Drosophila mercatorum* was induced by genes at a number of independent loci (Carson, 1967). Hence, it is likely that it will be proven that genetic architecture of parthenogenesis is as varied as are mechanisms of this reproductive mode. Non-coding regions may also facilitate loss of sex as seen by differential expression of miRNAs and piRNAs in alternate reproductive morphs in pea aphid *A. pisum* (Lu *et al.*, 2011).

3.5.2 Hybrid Origin

Interspecific hybridization can disrupt meiosis and development and create opportunities for selection of cytological and developmental processes that rescue egg production and division (Vrijenhoek, 1998). Hybridization is a major route to parthenogenesis in animals, and most unisexual vertebrates have a hybrid origin (Neaves & Baumann, 2011). Hybridization as factor leading to parthenogenetic lineages has been demonstrated in a number of insects (Lokki & Saura, 1980; Delmotte *et al.,* 2003; Milani *et al.,* 2010; Montelongo & Gomez-Zurita, 2014; Kotásková *et al.,* 2018; Rodriguero *et al.*, 2018).

Parthenogenetic hybrids resulting from crosses between two bisexual species possess high heterozygosity and alleles typical of two parental species initially resulting in diploid parthenogenetic lineages (Stenberg *et al.,* 2000). However, higher ploidy levels may result via intra or interspecific cross or backcrossing with polyploid parthenogen (Suomalainen, 1962). This is thought to be a major route to polyploidy in unisexuals (Stenberg *et al.,* 2003).

3.5.3 Infectious Origin

Parthenogenesis may originate through infection with vertically inherited endosymbionts; '*infectious parthenogenesis*' (Huigens *et al.*, 2000). Because males are an evolutionary dead end form microorganisms that are maternally transmitted via egg cytoplasm inducting thelytokous parthenogenesis in their host will increase their rate of transmission (Werren *et al.*, 2008). Werren *et al.* (1995) coined the term '*reproductive parasites*' to indicate their selfish nature. All currently known cases of '*endosymbiont-induced parthenogenesis*' (EIP) are caused by bacteria, primarily *Wolbachia*, and at a much lesser extent by *Rickettsia* and *Cardinium* (Ma & Schwander, 2017). However, percentage of EIP due to *Wolbachia* may be overestimated, as testing is largely biased in its favour without due consideration of other endosymbionts. Metagenomics or massive parallel sequencing of 16S rRNA sequences may help in identification of parthenogenesis-inducing endosymbiont by distinguishing those involved in reproductive manipulation from other processes (e.g. Hosokawa *et al.*, 2010). Detection of *Wolbachia* does not necessarily imply its role in parthenogenesis induction (see Zug & Hammerstein, 2015, 2018).

Confirmed cases of EIP have only been reported in insects with haplo-diploidy as sex determination (SD) system (Ma & Schwander, 2017). Confirming EIP is much more difficult in insects with diplo-diploid SD system (see references in Ma & Schwander, 2017; Magro *et al.*, 2020).

Some experiments in diplo-diploid species like *Wolbachia*-infected weevils and springtails (Rodriguero, 2009; Chen *et al.*, 2012; Rodriguero *et al.*, 2021) while revealing dependence on endosymbionts for production of viable oocytes, cannot distinguish whether lack of egg viability is because of EIP or because oogenesis in this species is dependent on *Wolbachia* infection. In springtail *Folsomia candida, Wolbachia* may be replacing paternally provided centrioles needed for centrosome formation in sexual species (Riparbelli *et al.*, 2006). This finding constitutes a strong support for hypothesis of EIP in a diplo-diploid species; however, further additional and subsidiary evidence for supporting EIP is required.

Species with EIP generally display reproductive polymorphism: at least 50% of species comprise both parthenogenetic (infected) and sexual (uninfected) females. Although all species with confirmed EIP are haplo-diploid, reproductive polymorphism also appears widespread among diplo-diploid species in which EIP is suspected but speculative (see Elias-Costa *et al.*, 2019; Magro *et al.*, 2020). The reasons why polymorphisms are so common may be: i) recent infection, so that endosymbiont has not invaded all populations yet (Reumer *et al.*, 2012); ii) evolution of resistance mechanisms (or 'suppressor gene') in host population during spread of initial infection (Stouthamer *et al.*, 2010); iii) reversal from parthenogenetic to sexual reproduction upon loss of infection or incomplete endosymbiont transmission (Huigens *et al.*, 2000).

Overall percentages suggest that EIP may be a relatively frequent event; in hymenopterans, frequency of EIP is 15–31%. EIP should not be confused with endosymbiont-induced feminization (EIF). For latter, endosymbionts feminize genetic males which then develop as females, thereby increasing their transmission to future generations (Werren *et al.*, 2008).

3.5.4 Contagious Origin

When reproductive isolation between sexual and parthenogenetic lineages that retain ability to produce fertile males is incomplete, gene flow may allow spread of unisexuality genes that could convert sexual into parthenogenetic lineages. Males would inherit genes responsible for thelytoky from their mothers, converting sexual lines into thelytokous ones. Thus, novel parthenogenetic lineages may arise secondarily from pre-existing parthenogenetic lineages. Simon *et al.* (2003) called this '*contagious parthenogenesis*'. New parthenogenetic genotypes are generated upon recombination, slowing accumulation of deleterious mutations and increasing genetic diversity in parthenogenetic populations, thus increasing chances to face environmental changes. Therefore, contagious parthenogenesis has important evolutionary consequences. It is important to notice that more than a mode of origin of parthenogenetic lineages, rather, this is a mode of increasing genotypic diversity in already existing unisexual lineages.

Contagious parthenogenesis has been reported for wasps (Sandrock *et al.*, 2011), jumping plant lice (Nokkala *et al.*, 2015), and aphids (Blackman, 1972; Jaquiéry *et al.*, 2014). There is little evidence of this phenomenon in natural populations of species other than cyclical parthenogens or hymenopteran-producing haploid males via failed fusion of meiotic products during automixis (1 male:3000 females). Incidence in wild could be limited by: (i) functionality of parthenogenetically produced males; (ii) parthenogenetically produced males must successfully mate with conspecific sexual females; and (iii) parthenogenetically produced males must transmit their parthenogenetic genes to their offspring, creating new parthenogenetic lineages.

3.5.5 Multiple Modes of Origin

Parthenogenetic lineages can arise through several mechanisms (Tomiuk *et al.*, 1994; Normark & Kirkendall, 2009). Parthenogenesis may arise through hybridization (Saura *et al.*, 1993), spontaneously (Suomalainen, 1962), EIP (Elias-Costa *et al.*, 2019), or a mixture of various methods (Delmotte *et al.*, 2001; Delmotte *et al.*, 2003).

Sexual reproduction is thus a labile genetic system that could lead to many situations where sexual populations coexist with a pool of polyphyletic and diverse parthenogenetic lineages (Simon *et al.*, 2003).

3.6 Constraints on Evolution of Unisexuality

As already discussed, parthenogenesis is a rarity, which may be due to propensity of parthenogenetic lineages to go extinct. However, how easy it is to abandon sexual reproduction when a complicated process like meiosis has to be abolished or altered and still be capable of giving rise offspring? As Normark (2014) accurately pointed out, '*if meiosis is key to [bi]sexuality, then it is ruin of parthenogenesis*'. This being one of many hurdles to overcome. So, rarity of parthenogenesis may indeed be due to severe constraints upon its rate of origin (Engelstädter, 2008; Normark, 2014).

Engelstädter (2008) argued that investigation of genetic and developmental mechanistic barriers for parthenogenetic reproduction might help identify constraints in evolution of parthenogenesis.

3.6.1 Maintenance of Ploidy Levels

In absence of a male, alteration in meiosis is required to retain maternal ploidy level in eggs of parthenogenetic females. While apomictic females completely suppress meiosis, in automictic females diploidy must be restored during or after meiosis, a process much discussed above. The large extent of these phenomena in insects shows that this constraint is relatively easy to overcome, sometimes with help *Wolbachia*.

3.6.2 Egg Activation

The activation of meiotic arrest of mature oocytes is required to re-enter cell cycle; which may require sperm. In species with sperm-dependence egg activation, mutations for apomictic or automictic egg production will lead to parthenogenesis coupled with sperm stimulation of embryonic development, or pseudogamy or gynogenesis. While evolutionarily and ecologically unstable (because of male requirement in a female dominated population)sperm may require a second mutation, it is plausible (Kirkendall & Stenseth, 1990; Schlupp, 2005).

Eggs not requiring activation by sperm may require a second mutation to suppress sperm-dependent egg activation, though it is highely unlikely (Went, 1982). Mechanical stress imparted upon egg as it is squeezed through oviductor or ovipositor during ovulation or egg laying is a common activation mechanism (King & Rafai, 1970; Went & Krause, 1974; Horner & Wolfner, 2008). *Drosophila melanogaster* and *Drosophila mercatorum* may require a seminal fluid molecule or mechanical stimulation of copulation to oviposit, whereas in others like obligate parthenogenetic *Drosophila mangabeirai* they do not (Srdic & Jacobs-Lorena, 1978; Srdic *et al.*, 1979). In *Carausius morosus* activation factor is oxygen from air, which penetrates to egg through micropyle immediately post-oviposition (Went, 1982). It is likely that parthenogenesis would arise more frequently in groups in which egg activation is not dependent upon sperm.

3.6.3 Centrosomes

A major obstacle that parthenogens must surpass is requirement of paternal centrioles (Manandhar *et al.*, 2005). In oocytes, centrosomes disintegrate and centrioles are lost during oogenesis, with centriole being contributed by sperm, which further recruits maternal proteins for second centriole formation. Thus, for parthenogenesis to evolve, centrioles should either not be degraded or synthesized *de novo*. The former scenario has not been reported (Engelstädter, 2008), *de novo* synthesis from maternal proteins has been extensively observed (Riparbelli *et al.*, 2010, 2017; Ross & Normark, 2015).

3.6.4 Inbreeding Depression

Inbreeding depression via loss of heterozygosity is likely to affect transition from sexual to unisexual reproduction, which may range from completely preserved (apomixis) to completely disappeared (automixis through gamete duplication) and all possibilities in between (central fusion, terminal fusion, etc.) (see below "Consequences of unisexuality for genomes: Heterozygosity).

Only for apomixis, inbreeding depression will not affect transition to unisexuality. Fitness costs may explain why most obligate or cyclic parthenogenetic species reproduce by apomixis (Delmotte *et al.*, 2001; Tsutsui *et al.*, 2014) while automixis is mainly observed among facultative parthenogens or in certain populations only (Matsuura *et al.*, 2004; Kellner & Heinze, 2011; Alavi *et al.*, 2018).

3.6.5 Sex Determination Systems

Some SD systems pose a severe restriction for evolution of unisexuality. Female heterogamety may produce offspring that develop into males rather than females. In butterflies, females are ZW; parthenogenesis through some forms of automixis like gamete duplication and terminal fusion will result in all-male offspring (50% will be ZZ males and 50% will be WW, and thus not viable). Another system is complementary sex determination (CSD); sex is determined by heterozygosity at one or several loci, heterozygous individuals develop into females and hemizygous (haploid) into males. Thus, gamete duplication and terminal fusion will render an all-male offspring. It is worth to say that parthenogenesis induced by *Wolbachia* through gamete duplication in Hymenoptera is restricted to taxa with other SD systems than CSD (Heimpel & De Boer, 2008).

Thus, while some SD systems may impede evolution of parthenogenesis they will never impede apomictic parthenogenesis (Engelstädter, 2008).

Based on above discussions, it is expected that parthenogenesis will be most prevalent in insects groups with XY SD system, little inbreeding depression, no paternal centriole contribution, and sperm independent egg activation.

3.7 Consequences of Unisexuality for Genomes

Unisexual lineages might be at risk of extinction because of impossibility of purging deleterious mutations ('*Muller's ratchet*' and '*Kondrashov's hatchet*') (Muller, 1964; Kondrashov, 1993) and generating novel, rare, transient genotypes that might confer resistance to parasite attacks (Bell, 1982). Accordingly, unisexuality may print distinctive genomic patterns that are likely to influence evolutionary trajectory of a lineage.

3.7.1 Heterozygosity

Greeff and Villet (1993) and Pearcy *et al.* (2006) raised a series of coefficients to ascertain mode of parthenogenesis through comparison of rate of transition to homozygosity of a heterozygous locus according to distance to centromere (also Templeton, 1982; Pearcy *et al.*, 2011).

In apomictic parthenogenesis maternal genotype is passed on to offspring without change, thereby retaining heterozygosity (Soumalainen *et al.*, 1987; Vavre *et al.*, 2004), which continues to steadily increase (Gustafsson, 1947; Suomalainen, 1950). Such organisms become gradually heterozygous in regard to more and more gene pairs, making elimination of recessive mutations impossible (Suomalainen, 1962).

In insects with automictic parthenogenesis, crossing-over and segregation may occur. Heterozygosity is obligatorily replaced by homozygosity (regardless of crossing-over occurrence) in automictic parthenogens in which diploid chromosome number is restored by gametic duplication, with heterozygosity only occurring through mutations (Suomalainen, 1962).

When ploidy is restored through terminal fusion, offspring heterozygosity depends on occurrence of crossing-over (Figure 3.2). If mother is heterozygous for two alleles but there is no crossing-over, terminal fusion enforces homozygosity; however, if there is crossing-over between locus and centromere, heterozygosity can be maintained (Asher, 1970; Suomalainen *et al.*, 1987). Each heterozygous locus has a probability ranging from 1/3 (far from centromere) to 1 (close to centromere) of becoming homozygous (Pearcy *et al.*, 2006).

For ploidy restoration through central fusion, two fusing nuclei are non-sister ootids (Figure 3.2). Then, if mother is heterozygous for a given locus, given that there is no crossing-over, all her offspring will be heterozygous (Figure 3.2). In case of recombination terminal fusion probabilities will hold (Pearcy *et al.*, 2006).

The other mechanisms of ploidy restoration given before can be grouped under term '*random fusion*'. In this case, it may be considered that all four chromatids segregate independently, and each heterozygous locus has a probability of 1/3 of becoming homozygous, independently of its position on chromosome (Pearcy *et al.*, 2006).

The rate of transition from heterozygosity to homozygosity associated with recombination events has been observed and compared with theoretical expectation in fire ant *Wasmannia auropunctata* (Rey *et al.*, 2011) and ant *Cataglyphis cursor* (Pearcy *et al.*, 2006), parasitoid wasp *Meteorus pulchricornis* (Tsutsui *et al.*, 2014), and termite *Reticulitermes speratus* (Matsuura *et al.*, 2004). In *A. mellifera capensis*, to avoid extinction risk, a high proportion of unmated workers can lay eggs that develop into females via thelytokous parthenogenesis involving central fusion of meiotic products. This ability allows pseudo-clonal lineages of workers to establish, which are presently widespread as reproductive parasites within honeybee populations of South Africa (Oldroyd *et al.*, 2008). In successful lineages of parasitic workers, recombination events are reduced by an order of magnitude relative to meiosis in queens of other honeybee subspecies. These authors showed that in unmated *A. mellifera capensis* queens treated to induce oviposition, no such reduction in recombination occurs, indicating that thelytoky and reduced recombination are not controlled by same gene. Virgin queens were able to lay both arrhenotokous male-producing haploid eggs and thelytokous female-producing diploid eggs at same time, with evidence that they have some voluntary control over which kind of egg was laid. If so, they are able to influence kind of second-division meiosis that occurs in their eggs post-partum.

Summarizing, automixis typically leads to reduction in heterozygosity in offspring compared to mother. However, models that account for rate at which heterozygosity is lost at one and two loci and number of crossovers between loci and centromere suggest that although automictic populations are expected to suffer from lack of gene shuffling with other individuals, they are nevertheless, superior to both clonal and outbreeding sexual populations in way they respond to beneficial and deleterious mutations (Engelstädter, 2017).

3.7.2 Accumulation of Deleterious Mutations

Linkage between genetic loci reduces efficiency of natural selection, which is avoided via decoupling of alleles through recombination, formalized by the '*Model of Fisher-Muller*' (Fisher, 1930; Muller, 1932) or '*Vicar of Bray*' hypothesis which accounts for advantage of sexual reproduction over unisexuality despite expenditures. Hill & Robertson (1966) showed that selection of an allele at one locus increases genetic drift at a second linked locus, decreasing effectiveness of selection at both loci (*Hill-Robertson effect*). While mathematically identical, verbal arguments of both models were different (Felsenstein, 1974) with latter predicting '*ratchet mechanism*' or '*Muller's ratchet*', by which deleterious mutants would more readily increase in a population without recombination (Muller, 1964; Felsenstein, 1974).

If too many harmful mutations accumulate in a unisexual genome, then entire population could undergo '*mutational meltdown*', which will lead to its extinction (Lynch *et al.,* 1993). However, only small amounts of recombination are needed to stop otherwise unassailable ratchet, especially in large populations (Bell, 1982; Charlesworth *et al.*, 1993). In low-recombining areas of genome and low-recombining organisms, selection at a focal locus can be affected by selection acting on linked sites (Bachtrog *et al.,* 2008).

Unisexual lineages may be exceptionally prone to deleterious-mutation accumulation in both nuclear and organelle genes. The predicted signature is an increased evolutionary rate at amino acid level (i.e. non-synonymous mutations) for genes subject to purifying selection. Assuming that most mutations are deleterious, ratio of non-synonymous to synonymous substitutions (dN/dS) is predicted to be higher in unisexuals (Glémin *et al.,* 2019).

Undeniably, studies of associations between reproductive mode and mutation accumulation have had mixed results (Tvedte *et al.,* 2019). Small datasets of nuclear and mitochondrial loci showed higher mutation loads in parthenogenetic *Timema* stick insects compared to sexual relatives (Henry *et al.*, 2011) and elevated rates of synonymous and non-synonymous substitutions were demonstrated in unisexual *Timema* using whole transcriptome datasets (Bast *et al.*, 2018). Conversely, reproductive mode is not apparently associated with mutation accumulation neither in nuclear or mitochondrial genes, with exception of one unisexual lineages assayed, which showed accelerated mutation accumulation in nuclear sequence (Normark & Moran, 2000). Further comparisons among complete *A. pisum* genome and large expressed sequence tag (EST) datasets for other aphids showed that overall mutation rates and non-synonymous substitution rates did not differ between sexual and unisexual aphid lineages, and differences in dN/dS values appeared to be the consequence of changes in a small number of genes (Ollivier *et al.,* 2012) and also in eight groups of hexapods (Brandt *et al.*, 2019). Thus, at least for insects, there are conflicting evidences in favour or against Muller's ratchet.

Overall, most evidences indicate that deleterious mutation accumulation can differentially affect sexual and unisexual lineages and support the idea that deleterious mutation accumulation plays an outstanding role in limiting long-term persistence of all-female lineages both in plants and animals (Neiman & Koskella, 2009; Hollister *et al.*, 2015), with deleterious amino acid substitutions accumulating 2.4 to 13.4-fold higher in unisexual lineages (Henry *et al.,* 2011; Tucker *et al.,* 2013; Bast *et al.,* 2018). However, Brandt *et al.* (2017) found evidence for strong purifying selection being more effective in unisexual oribatid mite lineages as compared to sexual ones. They speculated about the role of population size as the agent sustaining effective purifying selection and facilitating the escape of mutational meltdown in the absence of sex, concluding that sex per se is not a prerequisite for the long-term persistence of animal lineages (see "Ecological consequences of parthenogenesis: population size").

3.7.3 Fate of Reproductive Traits

Abandonment of sexual reproduction may be the kick start for neutral evolutionary processes to degrade general sexual structures and mechanisms, if they become non-functional. However, if they become an unnecessary metabolic burden, they may undergo strong selection for loss (Kraaijeveld *et al.,* 2016; Tvedte *et al.,* 2019). Since male-specific traits are not expressed in thelytokous strains, and thus not exposed to selection, their decay would occur as a slow consequence of genetic drift. Evidence for males of parthenogenetic lineage retaining their normal reproductive behaviours (Schwander *et al.*, 2013) or only slight deterioration is present, or that show some slight deterioration of male sexual function (Russell & Stouthamer, 2011). In cases where males derived from a thelytokous strain were incapable to produce mature sperm (Gottlieb & Zchori-Fein, 2001), an ancient unisexuality was inferred.

However, reverse would be true for sexual traits expressed in females, which are expected to regress rapidly with evidentiary support available (Schwander *et al.*, 2013; Parker *et al.*, 2018; Tvedte *et al.*, 2019). However, in some cases female sexual traits may remain functional, either due to recent evolution of parthenogenesis, low cost of maintenance, or because there might be an advantage in maintaining them (Lehmann *et al.*, 2007, 2010).

It remains largely unknown if decay of sexual traits would stem from multiple mutations with small effects (Ma *et al.*, 2014) or rather involve few loci with major phenotypic effects (Russell & Stouthamer, 2011). More data are needed to reach a consensus on unisexual lineages undergoing '*mutational meltdown*'.

3.7.4 Codon Usage

The '*codon usage*' bias refers to differences in frequency of occurrence of synonym codons in coding DNA. Theory predicts sub-optimization in unisexuals because of difference in base compositions between recombining and non-recombining genomes (Glémin *et al.*, 2019). Since recombination-associated gene conversion tends to increase GC content, it may be non-random with respect to allele base composition, and thus bases A and T may be converted into G and C more frequently than expected by chance (Mugal *et al.*, 2015). This mechanistic effect (independent of selection or drift) can rapidly increase GC content, especially around recombination hotspots (Yang & Nielsen, 2008). In non-recombining regions of sexual genomes this '*GC-biased gene conversion*' (gBGC) should occur at greatly reduced rates, and base composition of neutral sequences should shift towards mutational equilibrium, somewhat "arresting" gBGC (Galtier *et al.*, 2018). Empirical evidence coming from non-recombining portions of genome of sexually reproducing species supports this hypothesis (Haddrill *et al.*, 2007; Choudhury *et al.*, 2017, but see Sun & Heitman, 2012) and it remains largely unknown if this phenomenon extends to a whole genome scale in unisexual organisms. Recently, Bast *et al.* (2018) tested for evidence of arrested gBGC in unisexual stick insects and found a reduced impact of GC accumulation in their genomes. This finding represents the first documented case of arrested gBGC in unisexual eukaryotes.

3.7.5 Endosymbiont Infection

Maternally inherited endosymbionts are known to influence population genetics and molecular evolution of host's mitochondrial DNA (mtDNA) and both genomes will be more strongly correlated with each other than with nuclear alleles of host species (Shoemaker *et al.*, 2004). This genetic hitchhiking will confound interpretation of history of a host species (Hurst & Jiggins, 2005; Galtier *et al.*, 2009) and DNA barcoding (Gerth *et al.*, 2011, but see Smith *et al.*, 2012 and Stahlhut *et al.*, 2012) when inferred through mitochondrial molecular markers.

Spread of these maternally transmitted microbes will: i) be regular driver of selective sweeps on arthropod mitochondrial genomes via indirect positive selection, leading to low mtDNA diversity within population; ii) produce balancing selection on different mtDNA variants when different strains occurs in a host species, leading to high mtDNA diversity within population and relatively deep nodes in mtDNA divergence within a species. These patterns also may throw interpretations when several infected populations (either with same or with different endosymbiont strains) are analyzed simultaneously (see Hurst & Jiggins, 2005).

Spread of maternally transmitted endosymbionts inducing thelytokous parthenogenesis results in hitchhiking not only of all maternally inherited organelles, but also of nuclear genome, which will be associated with initially infected female. Rodriguero *et al.* (2010) showed phylogenetic congruence between nuclear and mitochondrial variants, and then coevolution of bacterial, mitochondrial, and nuclear genomes in *Wolbachia*-infected weevil *Naupactus cervinus*. High diversity on both mitochondrial and nuclear markers and fact that many phylogenetically related mitotypes share same nuclear genotype suggest an ancient linkage disequilibrium and long history of mito-nuclear genomic association as a consequence of an ancient infection. The fact that this pattern was found in several highly divergent lineages of this nominal species (which in fact is a species complex), indicates that more than one infection with same strain might have occurred (Rodriguero *et al.*, 2013). Kraaijeveld *et al.* (2011) also showed that both nuclear and mitochondrial markers of parasitoid *Leptopilina clavipes* displayed significant and congruent variation among thelytokous wasp lineages, showing that multiple lineages have become infected with *Wolbachia.* Both contributions show that *Wolbachia* has been transferred horizontally to infect multiple female lineages during early spread in these hosts.

Another consequence of initial infection in EIP cases is that effective population size of mtDNA will suffer a drastic decrease because it is forced through a bottleneck; all mtDNA haplotypes in population will descend from that molecule of initially infected host female (Turelli, 1994). Then, mtDNA may experience accelerated rates of molecular evolution relative to those of uninfected populations. This effect was proved for mtDNA of some *Drosophila* species with cytoplasmic incompatibility induced by *Wolbachia* (Ballard, 2000; Shoemaker *et al.,* 2004). It would be interesting to see if maternally transmitted endosymbionts can accelerate rate of divergence of both mtDNA and nDNA of EIP lineages vs. sexually reproducing lineages in different species.

Indirect selection on maternally inherited genomes arising from disequilibrium with maternally inherited microorganisms is very common in insects (Hurst & Jiggins, 2005). But once again, results may be *Wolbachia*-biased (Hurst & Jiggins, 2005). In a recent meta-analysis, Cariou *et al.* (2017) concluded from a big dataset (10,929 arthropod specimens, 1,110 like-species and 26 orders) that on a large phylogenetic scale other sources of variation, such as mutation rates, tend to hide effects of *Wolbachia.* However, comparisons between closely related infected and uninfected taxa reveal that *Wolbachia* is associated with a twofold reduction in silent mitochondrial polymorphism and a 13% increase in non-synonymous substitution rates (because of Muller's ratchet). These findings confirm that widespread distribution of *Wolbachia* infection throughout insects impacts effective population size of mitochondria (and for species with EIP, also nucleus).

Smith *et al.* (2012) examined the extent to which these endosymbionts are detected in routine DNA barcoding, assessed their impact upon insect sequence divergence and identification accuracy, and considered variation present in *Wolbachia* COI sequence (named coxA). Fortunately, these authors concluded that *Wolbachia* is unlikely to compromise accuracy of DNA barcode library in insects because there is limited overlap between *Wolbachia* coxA and eukaryotic DNA barcode region (Stahlhut *et al.,* 2012). However, they suggest that insect barcoding projects would do well to incorporate additional steps that standardize collection of ancillary data present in whole genome extracts, including *Wolbachia* MLST analyses, and increasing number of extractions based on abdomens rather than somatic tissue (Smith *et al.,* 2012).

Summarizing, *Wolbachia*-insect interactions occur commonly enough to hazard inference of patterns about population history based on mtDNA (and possibly nDNA) sequence data. However, there is a second red flag: insects may harbor a myriad of still unidentified symbionts (Duron *et al.,* 2008), with known microbial diversity growing. And this means one cannot 'test against' presence of microbial agents with any certainty. Hence, absence of evidence (which is not same as evidence of absence) of known maternally inherited bacteria in a particular arthropod species does not mean that other (unknown) ones are not present. Furthermore, rate of gain and loss of these elements on an evolutionary timescale is such that current infection state is not a good indicator of past state (Bailly-Bechet *et al.,* 2017). Thus, one can never know if mtDNA evolution has not been influenced by past or present endosymbionts (Galtier *et al.*, 2009).

3.7.6 Accumulation of Transposable Elements

Mode of reproduction has been predicted to affect proliferation of transposable elements (TEs): TEs should be less frequent in parthenogens because they cannot spread from one genomic background to another through syngamy. However, parthenogenesis may lead to proliferation of TEs due to inability of unisexual lineages to purge transposons. Reproductive parasites may also affect TEs dynamics as a side effect (Kraaijeveld & Bast, 2012; but see Bast *et al.*, 2015).

TEs distribution in parthenogentic lineages will depend on the balance between additional transposition within clonal lineages on the one hand and selection or excision on the other. Large unisexual populations are expected to purge their load of TEs, which can take a long time, and in absence of excision mechanism TEs should slowly accumulate, not decline (Igic & Kohn, 2006; Orr & Unckless, 2008).

Genome-scale analyses of sexual versus unisexual lineages indicated similar levels of TEs in both reproductive modes (Bast *et al.*, 2015; Lindsey *et al.*, 2018). This contrasts with increasing evidence for accumulation of other types of deleterious mutations in unisexuals (Henry *et al.*, 2011; Hollister *et al.*, 2014; but see Tucker *et al.*, 2013) and enrichment of TEs in non-recombining genome portions of sexual species (Erlandsson *et al.*, 2000; Bachtrog, 2013).

Current evidence conveys that TEs are unlikely to drive decay or extinction of extant unisexual lineages and suggest that genome-wide lack of recombination generates different TE dynamics than lack of recombination in isolated genome portions. Overall, these results and some more coming from other taxa than insects suggest that early extinction of unisexual populations is unlikely to be caused by an overall expansion of TE copies in their genomes, but potentially by certain TE types if not contained. Furthermore, TEs might be cleared from unisexual species over time in at least some cases (Flot *et al.*, 2013).

3.7.7 Hybridization and Polyploidy

Hybrid-parthenogenetic lineages are expected to display a large amount of genomic heterozygosity, alleles typical of two parental species and reticulate evolution revealed by incongruence between mtDNA and nDNA. This can give a hint about hybrid origin of parthenogenetic lineages, and can also allow identification of parental species as well as direction of crosses (Simon *et al.*, 2003).

In *Timema*, next-generation sequencing (NGS) data demonstrated high levels of heterozygosity in a unisexual hybrid, and identified paternal variants nested as a subset within parthenogentic lineage (Morgan-Richards *et al.*, 2016). Rodriguero *et al.* (2018) found unusual intraindividual variation in ribosomal sequences in some parthenogenetic populations of Fuller's rose weevil (the already commented *Naupactus cervinus*), which revealed out-crossing between two highly divergent lineages of this insect as was also seen in stick insect, *Acanthoxyla* (Morgan-Richards & Trewick, 2005). Heterozygosity at most nuclear loci suggested hybrid origin of parthenogenesis in unisexual lineages of aphid *Rhopalosiphum padi* (Delmotte *et al.*, 2003).

In stick insect *Clonopsis* (Milani *et al.*, 2010), leaf beetles *Calligrapha* (Gómez-Zurita *et al.*, 2006), and weevil Aramigus (Normark and Lanteri, 1998), incongruence between morphological and mitochondrial markers (biparentally inherited vs. monoparentally inherited markers, respectively) was revealed. Old World weevil *Otiorhynchus scaber* showed great allelic diversity in triploid and tetraploid parthenogenetic forms, which may be a by-product of hybridization (Stenberg *et al.*, 2000).

Polyploidy in parthenogens may be a consequence of: i) minority cytotype disadvantage (one cytotype being dominant over other, with less common not finding mates; Levin 1975); ii) unfertile crosses between polyploids and diploids; iii) inability of even-numbered polyploids (Stenberg *et al.*, 2000). Thus, polyploid mutants might be more likely to establish as new lineages in unisexuals (Glémin *et al.*, 2019).

Polyploidy can lead to *genomic shock* caused by genomic instability, mitotic and meiotic abnormalities, and gene expression and epigenetic changes following whole-genome duplication, which can lead to increased sterility and decreased fitness (van de Peer *et al.*, 2017). Merging and doubling of two genomes triggers extensive modifications of novel genome and/or transcriptome, and considerably affects duplicate gene expression (including silencing and up- or down-regulation of one of duplicated genes). Genome doubling can also create cascades of novel expression patterns, regulatory interactions, and new phenotypic variation for evaluation by natural selection, inter-chromosomal exchanges, cytosine methylation, gene repression, novel activation, genetic dominance, subfunctionalization, and transposon activation (Adams & Wendel, 2005; Chen & Ni, 2006). However, some advantages may arise from whole-genome duplication, which may lead to diversification of gene function by altering redundant copies of important or essential genes and subsequent acquisition of different functions or production of novel results in a particular tissue or cell type (neofunctionalization processes) (Comai, 2005; Wertheim *et al.,* 2013). Thus, polyploidization has potential to be an important source of evolutionary novelty (Levin, 1983). Earlier considered uncommon in animals (Lushai *et al.*, 2003), polyploidy in this group is not as rare as assumed (Gregory & Mable, 2005). All well-documented polyploid insects reproduce by parthenogenesis (Vershinina & Kuznetsova, 2016; Gokhman & Kuznetsova, 2018).

Hybridization might cause immediate transition to obligate parthenogenesis if two progenitor genomes are so divergent that meiosis is impaired. Hybridization would therefore appear as a potential cause, and polyploidy a potential consequence of parthenogenesis, although many genome-wide studies are needed to draw conclusions on these complex relationships (Glémin *et al.*, 2019).

3.8 Ecological Consequences of Parthenogenesis

3.8.1 Geographic Parthenogenesis

Parthenogens often have different geographic distributions from closely related sexual lineages (though with potential overlap), also known as '*geographic parthenogenesis*' (Vandel, 1928, 1931) in place of '*geographic spanandry*', a term previously used to describe that males of some species became rarer with higher latitudes (Tilquin & Kokko, 2016).

In general, parthenogens occupy areas of lower biotic diversity and subject to greater seasonality of abiotic factors, mainly temperature (Jensen *et al.,* 2002), which allows them a much wider distribution and greater ecological tolerance that their ancestral sexual forms (Schön *et al.*, 2000; Stenberg *et al.*, 2003).

Geographic parthenogenesis has been proposed for weevils, aphids, moths, damselflies, termites, stick-insects, cockroaches, leafhoppers, mayflies, grasshoppers, dancing midges, and psocids (Vershinina & Kuznetsova, 2016; Gokhman & Kuznetsova, 2018; Tvedte *et al.,* 2019). Insect parthenogenesis seems to be associated with low dispersal capabilities (Lanteri & Normark, 1995), disturbed or ephemeral habitats (Normark, 2014), and soil-dwelling (Chernova *et al.,* 2010). A satisfactory explanation would require a combination of geographical, historical, ecological, and geological factors (Kawecki 2008).

Several non-exclusive hypotheses have been proposed, but there is no general line of conclusions (Parker & Niklasson, 2000). One group of hypotheses relies on intrinsic features of unisexuality such as uniparental reproduction ('*demographic hypotheses*') (Mogie *et al.,* 2007); another theoretical approach regards interactions between parthenogenetic and sexuals forms in mixed populations under certain ecological conditions like higher latitudes and altitudes, on islands, and in disturbed habitats as main cause for geographical parthenogenesis ('*ecological competence hypotheses*') (Mogie *et al.,* 2007).

3.8.2 Asexual Speciation

A unisexual population is an independent arena for evolutionary processes of mutation, selection, and random drift because it is separate from other groups and does not share evolutionary processes with them (Hey, 2001). The process by which a unisexual population splits from an ancestral sexual population is termed '*asexual speciation*' (Brucker & Bordenstein, 2012). EIP can result in reduced or no gene flow between infected and uninfected populations and cause evolution of reproductive isolation (Brucker & Bordenstein 2012; Shropshire & Bordenstein, 2016). Speciation between sexual and unisexual populations may proceed because of: i) degeneration of sexual traits in infected population; ii) relaxed sexual selection, which involves evolution of novel sexual traits in uninfected sexual population that prevent mating infected parthenogenetic population (Shropshire & Bordenstein, 2016).

Adachi-Hagimori *et al.* (2011) demonstrated that both *Rickettsia*-infected thelytokous and uninfected arrhenotokous strains of parasitic wasp *Neochrysocharis formosa* speciated sympatrically. Additionally, both groups of wasps were genetically isolated, as divergence accumulated at both nuclear and mitochondrial loci suggested in spite of their sister-group relationship. Deterioration of genes involved in sexual reproduction in parthenogenetic lineages has also been seen in other haplodiploid systems (Zchori-Fein *et al.*, 1995; Gottlieb & Zchori-Fein, 2001). In diplo-diploid species (and then, speculative cases of EIP), high divergence between parthenogenetic / Wolbachia-infected populations and sexual/uninfected populations, and possibly an incipient process of speciation mediated by EIP, have been reported for the springtail Folsomia candida (Frati *et al.*, 2004) and the weevil Pantomorus postfasciatus (Elias-Costa *et al.*, 2019). It would be interesting to test decay of the sexual function in these species.

It is important to note that asexual speciation is not 'final destination' whenever EIP has occurred. Mixed populations of sexual and unisexual forms (and gene flow between two) are common in *Trichogramma* genus. Genetic exchange between both arrenothokous and thelytokous populations would prevent decay of sexual reproduction in latter and accumulation of genetic divergence between two types.

3.8.3 Diversification within Parthenogenetic Populations

Speaking about species and speciation, one central question concerning nature of unisexual organisms for a long time has been whether these entities diversify into species (Birky & Barraclough, 2009), due to lack of interbreeding, each individual being monophyletic and reproductively isolated. Population and evolutionary genetic theories show that they can undergo speciation to form inclusive populations that evolve independently of each other; in other words, evolutionary species (Birky & Barraclough, 2009). If other factors were as much or more important than gene flow for maintaining cohesion, then unisexual taxa might diversify in a similar manner to sexually reproducing lineages. A good starting point for this process to occur is when members of a parthenogenetic species become physically isolated from each other for a long time or undergo divergent selection for adaptation to different niches (Fontaneto *et al.*, 2007).

Cohesion Species Concept (CSC) (Templeton, 1989) defines evolutionary groups in terms of mechanisms that yield cohesion (not only genetic, but also demographic exchangeability) rather than manifestation of cohesion over evolutionary times leading to diversification of unisexual taxa.

A few examples of diversification of unisexual taxa can be found across entomological literature: Fuller's rose weevil *Naupactus cervinus* has occurrence of several independently evolutionary units (a 'species complex') that have nothing to do with *Wolbachia* implying a process of incipient speciation (Rodriguero *et al.*, 2013). Similarly, black scale insect, *Parasaissetia nigra* comprises at least two lineages ecologically differentiated, particularly, in relation to temperature and moisture (Lin *et al.*, 2017).

3.8.4 Population Size

One largely neglected ecological factor linked to parthenogenesis is '*population size*', although its potential importance is coming to light (Normark 2014). Poon & Chao (2004) demonstrated that advantage of sex is attenuated with increasing effective population size, i.e. advantage of sex during adaptation increases with intensity of drift, being sex less advantageous at larger population sizes. Thus, we may expect that very large populations were prone to acquire parthenogenesis. Hoffmann *et al.* (2008) showed that pest insects tend to be parthenogenetic and Normark and Johnson (2011), using pest status as an index of high abundance, and abundance as a proxy for population size, concluded that abundant species have more chances of transitions to unisexuality and that parthenogenetic pests have a longer expected mean time to extinction, being more probable to be observed before they go extinct.

By this logic, they reversed interpretation of causal connection between abundance and parthenogenesis (parthenogenesis causing an increase in abundance).

Normark and Johnson (2011) proposed that several groups of phytophagous insects have experienced '*niche explosion*'. This phenomenon was defined as a positive feedback loop that connects large population size to broad host range and allows a species, as a consequence of this explosive expansion, to be highly polyphagous, abundant, and widespread, among other features. Because of positive loop, there will be positive effects of host range on population size and vice versa. In their contribution, and following Hamilton (1967) and his '*outline of an ideal extreme biofacies*' for inbreeding and haplodiploidy, they sketched an outline of an '*ideal syndrome of extreme polyphagy*', which includes, namely: flightless adult females; dispersal of first-instar larvae by wind or through soil; most hosts being woody plants; species as a whole feeds on a very great taxonomic diversity of host plants; very high abundance (species considered as economic pest); cosmopolitan or invasive; **obligate parthenogenesis** in some populations. This syndrome is found convergently in several species of bagworm moths, tussock moths, root weevils, and five families of scale insects. Thus, Normark and Johnson (2011) hypothesized that parthenogenesis would be a consequence of niche explosion, rather than cause of an increased population size. The niche explosion has a demographic component due to increased probability of colonizing new hosts and new areas, as well as a population-genetic component, mainly due to increased effectiveness of natural selection in larger populations. Hence, frequent origin of parthenogenetic lineages in extreme polyphagous species would be a consequence of this increased effectiveness of natural selection and consequent reduced importance of sexuality.

Some associations between parthenogenesis and some features of syndrome have been pointed out before like female flightlessness (Roff, 1990; Lanteri & Normark, 1995), broad niche breadth (Lynch, 1984), invasiveness (Bell, 1982), and economic pest status (Hoffmann *et al.*, 2008). These authors argued about parthenogenesis as causal factor in these associations. Yet, many extremely polyphagous invasive pest species are sexual, and those that have parthenogenetic populations often also have sexual populations that are equally polyphagous and invasive. By this argument Normark and Johnson (2011) inferred that parthenogenesis is not a major cause of syndrome, but just a mere consequence. Although data on abundance in natural habitats is unavailable for most parthenogenetic pests, for many of them abundance appears to be quite high, including moths (Nothnagle & Schultz, 1987; Barbosa *et al.*, 1989), adelgids (Blackman 1972), and weevils (Normark, 1994; Stenberg *et al.*, 2003; Rodriguero, 2009).

Moreover, the fact that set of traits encompassed under syndrome detailed above occurs with or without parthenogenesis implies that parthenogenesis is not a critical factor in origin of syndrome, but rather that these traits present in ancestral sexual population have facilitated success of unisexual descendants. This view is supported by meta-analysis accomplished by Ross *et al.* (2013), who detected that sexual species in genera that also contain unisexuals have larger geographic ranges than sexual species in genera where parthenogenesis is absent. These authors added geographic range to loop because it has been argued that large population size is causally connected to broad host range in a positive feedback loop (Normark & Johnson, 2011). A similar argument could be made for synergy between large population size and broad geographic range. And there is a plausible mechanism connecting large population size to high rate of generation of parthenogenetic lineages: if origin of parthenogenesis is rare, it is more likely to occur in large populations than in small ones (Normark & Johnson, 2011). Because of strong correlation between host range and geographic range, it is not possible to determine which of these factors is most important.

Empirical support for hypothesis of Normark and Johnson (2011) is still lacking despite strong theoretical grounds (Otto, 2009; Rice & Friberg, 2009). Recently, Ross *et al.* (2013), using a comparative approach with scale insects, concluded that population size also matters both at broadest scales (small vs. large organisms) and within particular groups, and acknowledged that association of parthenogens with 'marginal' habitats (e.g. high altitude, high latitude, xeric, island) (Bell, 1982; Bierzychudek, 1985) may reflect large geographic ranges, which are associated with large population sizes.

3.8.5 Parasites

In Lewis Carroll's novel '*Through looking glass*', Red Queen, a living chess piece, responds to Alice's question why, despite running so fast as she could in Wonderland, she is not moving forward and never gets anywhere: '*Now, here, you see, it takes all running you can do, to keep in same place*'. Van Valen (1973) took this metaphor to describe why species within a taxonomic group tend to go extinct at a constant rate. He argued that this pattern could be due to 'stochastically constant' deterioration of environmental conditions (caused by either biotic or abiotic forces, though he emphasized antagonistic biotic interactions) from perspective of species adapted to earlier conditions. Bell (1982) recognized parallels between Van Valen's view of extinction rates and constant adaptation that should characterize host-parasite coevolution. He applied Red Queen moniker to a new hypothesis that invoked pressure from coevolving parasites to generate an advantage for sex: continual adaptation is necessary in order to stay in same place and not go extinct. In other words, parasites coevolve with their targets, causing adaptive landscape to constantly change, and that is why host populations can never remain on an adaptive peak (Kover & Caicedo, 2001).

The RQH states that, in defending against highly variable parasites, sexual host populations would be favoured over related unisexual lineages because sexuality and recombination enables them to generate variable offspring, including rare phenotypes. Because parasites coevolve mainly with most common phenotype, rare variants will have a higher fitness due to a lower pathogen load and can therefore evolve resistances, leading to a negative frequency-dependent selection. However, after becoming common, former rare phenotype comes into focus of parasite and another rare variant has chance of escape attack (Martin, 2016). This coevolution leads to a fluctuating equilibrium between host and parasite with time-tagged oscillations abundances, what is known as '*Red Queen Dynamics*' (RQD). In contrast to sexually reproducing individuals, unisexuals are unable to produce rare phenotypes and, thus, once becoming common, pathogen will evolve the means to attack parthenogens, which will become extinct (Neiman & Koskella, 2009; Martin, 2016). Two closely linked predictions derived from RQH: i) sexual reproduction should be favoured when infection risk is high; ii) given lower genetic diversity associated with parthenogenesis, higher parasite loads are expected in unisexual lineages in sympatry with sexual lineages. Testing predictions of RQH is difficult because ideal situation for such research involves the comparison of parasitism in related sexual and unisexual forms that compete directly. Many studies have attempted to test RQH indirectly (Salathé *et al.*, 2008). Ideally, indirect tests should compare both parasite loads and levels of genetic diversity in coexisting sexual and unisexual populations to see if any of these factors might increase actual risk of parasite infection. However, both factors are rarely assessed simultaneously.

Kumpulainen *et al.* (2004) indicated an association between reproductive mode and parasitoid prevalence in closely related species of bagworm moths, *Siederia rupicolella* (sexual), and *Dhalica fennicella* (unisexual), so that in areas with extremely low parasitoid prevalence unisexuality was widespread, whereas in areas with a high parasitoid load sexuality was predominant. Although this work showed that parasitoids prevented parthenogenetic species from out-competing sexual one allowing sympatry, it did not prove that former had a higher risk of infection in sympatric sites. Subsequently, Elzinga *et al.* (2012) tested whether parasitoids preferentially attack parthenogenetic species of bagworm moths in sites with both reproductive modes. They found that parthenogenetic forms were less parasitized than sexuals in sympatric sites, what might be explained by differences in phenology between both reproductive modes. So, parasitoids do not seem to drive parthenogens to extinction. Rather, parasitoids preferentially attack sexual organisms, most likely due to their different phenology.

Lorenzo-Carballa *et al.* (2011) attempted to prove RQH in the damselfly *Ischnura hastata*; the low genetic diversity of parthenogenetic populations (Lorenzo-Carballa *et al.*, 2012) made this species ideal for testing predictions of RQH. Extremely low prevalence of several mite species was found, and although sexual congener *Ischnura pumilio* can be found in *Azores archipelago*, they only coexisted in a small fraction of ponds, preventing any comparison of parasitism rates. Lorenzo-Carballa *et al.* (2011)

argued that low prevalence of mite parasitism found in *Azores* (that might be a consequence of overall lower abundances of parasites in archipelago, rather than a heightened resistance to mite parasitism in these unisexual populations), coupled with the fact that most of the populations in these islands are almost free from competitors and predators, could explain persistence of these *I. hastata* parthenogenetic populations despite their low levels of genetic variation.

There is only a single direct test of RQH, which demonstrated that exposure of flour beetles to coevolving microsporidian parasites led to an increase in recombination rate (Fischer & Schmid-Hempel, 2005). These results suggested that short-term benefits of recombination (as a proxy of evolution of sex) may be related to parasitism. Largely studies have shown that unisexuality is disadvantageous under parasitic selection forces, though there are studies to contrary.

3.8.6 Polyploidy

There is growing evidence about correlation of polyploidy with environmental change or stress (van de Peer *et al.*, 2017). In *O. scaber,* diploid sexuals and diploid, triploid, and tetraploid parthenogens coexist in a small central area, while only polyploid parthenogens have extended to marginal areas (Stenberg *et al.,* 2003). Thus, polyploid clones seem to be superior colonizers when compared with diploid clones and sexuals. The evident benefit of polyploidy cannot be attributed to usual ecological or demographic positive effects traditionally attributed to unisexuality (Maynard Smith, 1978), since diploid clones share such advantages. Thus, discovery of diploid clones is a proof that this pattern, which Stenberg *et al.* (2000) had recognized as geographic parthenogenesis, was mainly a consequence of increase in ploidy level and not of parthenogenesis itself.

Stenberg *et al.* (2003) termed this phenomenon '*geographical polyploidy*', which has been observed in insects (Suomalainen, 1962; Parker & Niklasson, 1995; Gade & Parker, 1997; Scali *et al.,* 2003). In all cases, higher ploidy level is correlated to wider distribution (Suomalainen *et al.*, 1987).

It is predicted that polyploidy should be more common in temperate than in tropical breeders because there is a higher probability of their initial formation (Mable, 2004). Later, Takenouchi (1980, 1983, 1984, 1985, 1986a, 1986b) and Takenouchi *et al.* (1981) using weevils provided experimental support for this hypothesis. This author (and collaborators) managed to increase ploidy levels in eggs from cold-shocked sexual and polyploid parthenogenetic females. While sexual species rendered tetraploid, triploid, diploid, as well as haploid eggs, tetraploid parthenogenetic species yielded triploid, diploid, and octaploid eggs, and pentaploid parthenogens yielded pentaploid but also numerous decaploid, tetraploid, triploid, diploid, and haploid eggs. Tada *et al.* (2014) emphasized that fate of eggs laid by parthenogens should be traced through careful rearing experiments to fully understand nature and biological role of plasticity in ploidy. Still, high frequency of polyploids in temperate and arctic regions (and at higher altitudes) may be due to higher chance that individuals will experience cold shocks during their breeding season (Gregory & Mable, 2005); production of unreduced gametes could increase during periods of environmental instability typical of aforementioned regions, or even in temporary periods in which severe global climate changes occur, like glaciations.

3.9 Conclusions

Although parthenogenesis is a rarity in insects as much as in other taxa, it has impacted in life history of many species, printing notorious signatures in their genome and ecology. Its study has stimulated imagination of many scholars leading to new interpretations of old patterns (e.g. Normark & Johnson idea of niche explosion), and as data and models accumulate, new explanations arise for fascinating paradox of sex. However, no single theory is able to explain why sex is prevalent reproductive mode in eukaryotes. Insects provide a set of examples to compare sexual and unisexual lineages and a diversity of modes of transitions to a life without sex and several constraints to this resignation. It is our task as entomologists to propose right questions to help solve queen of problems.

Acknowledgements

This research is funded by Agencia Nacional de Promoción Científica y Tecnológica and Universidad of Buenos Aires. Gratitude to Research Career of CONICET, Argentina.

REFERENCES

Achterkamp, B. (1999). Paedogenesis in eristales arbustorum. Master's Thesis. Leiden University, Leiden. The Netherlands, Biology.

Adachi-Hagimori, T., Miura, K. & Abe, Y. (2011). Gene flow between sexual and asexual strains of parasitic wasps: a possible case of sympatric speciation caused by a parthenogenesis-inducing bacterium. *Journal of Evolutionary Biology*, *24*(6): 1254–1262.

Adams, K.L. & Wendel, J.F. (2005). Polyploidy and genome evolution in plants. *Current Opinion in Plant Biology*, *8*(2): 135–141.

Alavi, Y., van Rooyen, A., Elgar, M.A., Jones, T.M. & Weeks, A.R. (2018). Novel microsatellite markers suggest the mechanism of parthenogenesis in *Extatosoma tiaratum* is automixis with terminal fusion. *Insect Science*, *25*(1): 24–32.

Asher Jr., J.H. (1970). Parthenogenesis and genetic variability. II. One-locus models for various diploid populations. *Genetics*, *66*(2): 369–391.

Aumer, D., Allsopp, M.H., Lattorff, H.M.G., Moritz, R.F. & Jarosch-Perlow, A. (2017). Thelytoky in Cape honeybees (*Apis mellifera capensis*) is controlled by a single recessive locus. *Apidologie*, *48*(3): 401–410.

Aumer, D., Stolle, E., Allsopp, M., Mumoki, F., Pirk, C.W. & Moritz, R.F. (2019). A single SNP turns a social honey bee (*Apis mellifera*) worker into a selfish parasite. *Molecular Biology and Evolution*, *36*(3): 516–526.

Bachtrog, D. (2013). Y-chromosome evolution: emerging insights into processes of Y-chromosome degeneration. *Nature Reviews Genetics*, *14*(2): 113–124.

Bachtrog, D., Hom, E., Wong, K.M., Maside, X. & de Jong, P. (2008). Genomic degradation of a young Y chromosome in *Drosophila miranda*. *Genome Biology*, *9*(2): R30.

Bailly-Bechet, M., Martins-Simões, P., Szöllősi, G.J., Mialdea, G., Sagot, M.F. & Charlat, S. (2017). How long does *Wolbachia* remain on board? *Molecular Biology and Evolution*, *34*(5): 1183–1193.

Ball, S.L. (2001). Tychoparthenogenesis and mixed mating in natural populations of the mayfly *Stenonema femoratum*. *Heredity*, *87*(3): 373–380.

Ball, S.L. (2002). Population variation and ecological correlates of tychoparthenogenesis in the mayfly, *Stenonema femoratum*. *Biological Journal of the Linnean Society*, *75*(1): 101–123.

Ballard, J.W.O. (2000). Comparative genomics of mitochondrial DNA in *Drosophila simulans*. *Journal of Molecular Evolution*, *51*(1): 64–75.

Barbosa, P., Krischik, V. & Lance, D. (1989). Life-history traits of forest-inhabiting flightless Lepidoptera. *American Midland Naturalist*, *122*: 262–274.

Bast, J., Parker, D.J., Dumas, Z., Jalvingh, K.M., Tran Van, P., Jaron, K.S., Figuet, E., Brandt, A., Galtier, N. & Schwander, T. (2018). Consequences of asexuality in natural populations: insights from stick insects. *Molecular Biology and Evolution*, *35*(7): 1668–1677.

Bast, J., Schaefer, I., Schwander, T., Maraun, M., Scheu, S. & Kraaijeveld, K. (2015). No accumulation of transposable elements in asexual arthropods. *Molecular Biology and Evolution*, *33*(3): 697–706.

Baxendale, F.P. & Teetes, G.L. (1981). Production of unisexual progenies by the sorghum midge, *Contarinia sorghicola*. *Annals of the Entomological Society of America*, *74*(4): 412–413.

Beatty, R.A. (1967). Parthenogenesis in vertebrates. In: *Fertilization*, *vol. 1* (Eds. Metz, C.B. & Monroy, A.). New York, London: Academic Press, pp. 413–440.

Bell, G. (1982). *The Masterpiece of Nature: The Evolution and Genetics of Sexuality*. Berkeley and Los Angeles: California Uiversity Press.

Bergman, E.A. & Hilsenhoff, W.L. (1978). Parthenogenesis in the mayfly genus Baetis (Ephemeroptera: Baetidae). *Annals of the Entomological Society of America*, *71*: 167–168.

Berçot, F. F. & Prestes, M.E.B. (2017). O enigma do pulgão. Histórias das ciências, epistemologia, gênero e arte: Ensaios para a formação de professores, 11–35.

Beukeboom, L.W., Kamping, A., Louter, M., Pijnacker, L.P., Katju, V., Ferree, P.M. & Werren, J.H. (2007). Haploid females in the parasitic wasp *Nasonia vitripennis*. *Science*, *315*(5809): 206.

Beukeboom, L.W. & Vrijenhoek, R.C. (1998). Evolutionary genetics and ecology of sperm-dependent parthenogenesis. *Journal of Evolutionary Biology*, *11*(6): 755–782.

Beukeboom, L.W. & Werren, J.H. (2000). The paternal-sex-ratio (PSR) chromosome in natural populations of *Nasonia* (Hymenoptera: Chalcidoidea). *Journal of Evolutionary Biology*, *13*(6): 967–975.

Beutel, R.G., Yavorskaya, M.I., Mashimo, Y., Fukui, M. & Meusemann, K. (2017). The phylogeny of Hexapoda (Arthropoda) and the evolution of megadiversity. *Proceedings of the Arthropodan Embryological Society of Japan*, *51*: 1–15.

Bezzi, M. (1922). On the dipterous Passeromyia and Ornithomusca, with notes and bibliography on the non-pupiparous Myiodaria parasitic on birds. *Parasitology*, *14*: 29–46.

Bicknell, R.A. & Koltunow, A.M. (2004). Understanding apomixis: recent advances and remaining conundrums. *The Plant Cell*, *16*(1): S228–S245.

Bierzychudek, P. (1985). Patterns in plant parthenogenesis. *Experientia*, *41*(10): 1255–1264.

Birky, C.W. & Barraclough, T.G. (2009). Asexual speciation. In: *Lost Sex* (Ed. Schön, I., Schön, Isa, Martens, K. & Dijk, P.). Dordrecht: Springer, pp. 201–216.

Blackman, R.L. (1972). The inheritance of life-cycle differences in *Myzus persicae* (Sulz.)(Hem., Aphididae). *Bulletin of Entomological Research*, *62*(2): 281–294.

Borsa, P. & Kjellberg, F. (1996). Experimental evidence for pseudo-arrhenotoky in Hypothenemus hampei (Coleoptera: Scolytidae). *Heredity*, *76*: 130–135.

Braig, H.R., Turner, B.D., Normark, B.B. & Stouthamer, R. (2002). Microorganism-induced parthenogenesis. In: *Reproductive Biology of Invertebrates. Progress in Asexual Reproduction, vol. 11* (Ed. Hughes, R.N.). New Delhi: Oxford & IBH Publishing, pp. 1–62.

Brandeis, M. (2018). New-age ideas about age-old sex: separating meiosis from mating could solve a century-old conundrum. *Biological Reviews*, *93*(2): 801–810.

Brandt, A., Bast, J., Scheu, S., Meusemann, K., Donath, A., Schütte, K., Machida, R. & Kraaijeveld, K. (2019). No signal of deleterious mutation accumulation in conserved gene sequences of extant asexual hexapods. *Scientific Reports*, *9*(1): 5338.

Brandt, A., Schaefer, I., Glanz, J., Schwander, T., Maraun, M., Scheu, S. & Bast, J. (2017). Effective purifying selection in ancient asexual oribatid mites. *Nature Communications*, *8*(1): 873.

Brucker, R.M. & Bordenstein, S.R. (2012). Speciation by symbiosis. *Trends in Ecology & Evolution*, *27*(8): 443–451.

Cariou, M., Duret, L. & Charlat, S. (2017). The global impact of *Wolbachia* on mitochondrial diversity and evolution. *Journal of Evolutionary Biology*, *30*(12): 2204–2210.

Carson, H.L. (1967). Selection for parthenogenesis in *Drosophila mercatorum. Genetics*, *55*(1): 157–171.

Catanach, A.S., Erasmuson, S.K., Podivinsky, E., Jordan, B.R. & Bicknell, R. (2006). Deletion mapping of genetic regions associated with apomixis in Hieracium. *Proceedings of National Academy of Sciences of USA*, *103*(49): 18650–18655.

Chapman, N.C., Beekman, M., Allsopp, M.H., Rinderer, T.E., Lim, J., Oxley, P.R. & Oldroyd, B.P. (2015). Inheritance of thelytoky in the honey bee Apis mellifera capensis. *Heredity*, *1146*: 584–592.

Charlesworth, B., Morgan, M.T. & Charlesworth, D. (1993). The effect of deleterious mutations on neutral molecular variation. *Genetics*, *134*(4): 1289–1303.

Chen, S.J., Lu, F., Cheng, J.A., Jiang, M.X. & Way, M.O. (2012). Identification and biological role of the endosymbionts *Wolbachia* in rice water weevil (Coleoptera: Curculionidae). *Environmental Entomology*, *41*(3): 469–477.

Chen, Z.J. & Ni, Z. (2006). Mechanisms of genomic rearrangements and gene expression changes in plant polyploids. *Bioessays*, *28*(3): 240–252.

Chernova, N.M., Potapov, M.B., Savenkova, Y.Y. & Bokova, A.I. (2010). Ecological significance of parthenogenesis in Collembola. *Entomological Review*, *90*(1): 23–38.

Choudhury, M.N., Uddin, A. & Chakraborty, S. (2017). Codon usage bias and its influencing factors for Y-linked genes in human. *Computational Biology and Chemistry*, *69*: 77–86.

Cole-Clark, M.P., Barton, D.A., Allsopp, M.H., Beekman, M., Gloag, R.S., Wossler, T.C., Ronai, I., Smith, N., Reid, R.J. & Oldroyd, B.P. (2017). Cytogenetic basis of thelytoky in Apis mellifera capensis. *Apidologie*, *48*(5): 623–634.

Comai, L. (2005). The advantages and disadvantages of being polyploid. *Nature Reviews Genetics*, *6*(11): 836–846.

Crow, J.F. (1994). Advantages of sexual reproduction. *Developmental Genetics*, *15*(3): 205–213.

Darwin, C.R. (1862). On the two forms, or dimorphic condition, in the species of *Primula*, and on their remarkable sexual relations. *Journal of the Linnean Society of London Botany*, *6*: 77–96.

Dedryver, C.A., Le Gallic, J.F., Maheo, F., Simon, J.C. & Dedryver, F. (2013). The genetics of obligate parthenogenesis in an aphid species and its consequences for the maintenance of alternative reproductive modes. *Heredity*, *110*(1): 39–45.

Degrange, C. (1960). Recherches sur la reproduction des Ephemeropteres. *Travaux du Laboratoire d'Hydrobiologie et de Pisciculture de l'Université de Grenoble*, *51*: 7–193.

Delmotte, F., Leterme, N., Bonhomme, J., Rispe, C. & Simon, J.C. (2001). Multiple routes to asexuality in an aphid species. *Proceedings of the Royal Society of London, Series B: Biological Sciences*, *268*(1483): 2291–2299.

Delmotte, F., Sabater-Muñoz, B., Prunier-Leterme, N., Latorre, A., Sunnucks, P., Rispe, C. & Simon, J.C. (2003). Phylogenetic evidence for hybrid origins of asexual lineages in an aphid species. *Evolution*, *57*(6): 1291–1303.

Dixon, A.F.G. (1998). *Aphid Ecology*, 2nd Ed. London: Chapman & Hall.

Duron, O., Bouchon, D., Boutin, S., Bellamy, L., Zhou, L., Engelstädter, J. & Hurst, G.D. (2008). The diversity of reproductive parasites among arthropods: *Wolbachia* do not walk alone. *BMC Biology*, *6*(1): 27.

Dzierzon, J. (1845). Gutachten über die von Herrn Direktor Stöhr im ersten und zweiten Kapitel des General-Gutachtens aufgestellten Fragen. *Eichstädter Bienenzeitung*, *1* (109-113): 119–121.

Elias-Costa, A.J., Confalonieri, V.A., Lanteri, A.A. & Rodriguero, M.S. (2019). Game of clones: is *Wolbachia* inducing speciation in a weevil with a mixed reproductive mode? *Molecular Phylogenetics and Evolution*, *133*: 42–53.

Elzinga, J.A., Chevasco, V., Mappes, J. & Grapputo, A. (2012). Low parasitism rates in parthenogenetic bagworm moths do not support the parasitoid hypothesis for sex. *Journal of Evolutionary Biology*, *25*(12): 2547–2558.

Engelstädter, J. (2008). Constraints on the evolution of asexual reproduction. *BioEssays*, *30*(11–12): 1138–1150.

Engelstädter, J. (2017). Asexual but not clonal: evolutionary processes in automictic populations. *Genetics*, *206*(2): 993–1009.

Erlandsson, R., Wilson, J.F. & Pääbo, S. (2000). Sex chromosomal transposable element accumulation and male-driven substitutional evolution in humans. *Molecular Biology and Evolution*, *17*(5): 804–812.

Felsenstein, J. (1974). The evolutionary advantage of recombination. *Genetics*, *78*(2): 737–756.

Fischer, O. & Schmid-Hempel, P. (2005). Selection by parasites may increase host recombination frequency. *Biology Letters*, *1*(2): 193–195.

Fisher, R.A. (1930). *The Genetical Theory of Natural Selection*. Oxford: Clarendon Press.

Flot, J.F., Hespeels, B., Li, X., Noel, B., Arkhipova, I., Danchin, E.G., Hejnol, A., Henrissat, B., Koszul, R., Aury, J.M. & Barbe, V. (2013). Genomic evidence for ameiotic evolution in the bdelloid rotifer *Adineta vaga*. *Nature*, *500*(7463): 453–457.

Fontaneto, D., Herniou, E.A., Boschetti, C., Caprioli, M., Melone, G., Ricci, C. & Barraclough, T.G. (2007). Independently evolving species in asexual bdelloid rotifers. *PLoS Biology*, *5*(4): e87.

Fournier, D., Estoup, A., Orivel, J., Foucaud, J., Jourdan, H., Le Breton, J. & Keller, L. (2005). Clonal reproduction by males and females in the little fire ant. *Nature*, *435*(7046): 1230–1234.

Frantz, A., Plantegenest, M., Bonhomme, J., Prunier-Leterme, N. & Simon, J.C. (2005). Strong biases in the transmission of sex chromosomes in the aphid *Rhopalosiphum padi*. *Genetics Research*, *85*(2): 111–117.

Frati, F., Negri, I., Fanciulli, P.P., Pellecchia, M., De Paola, V., Scali, V., & Dallai, R. (2004). High levels of genetic differentiation between *Wolbachia*-infected and non-infected populations of *Folsomia candida* (Collembola, Isotomidae). *Pedobiologia*, *48*(5–6): 461–468.

Funk, D.H., Sweeney, B.W. & Jackson, J.K. (2010). Why stream mayflies can reproduce without males but remain bisexual: a case of lost genetic variation. *Journal of the North American Benthological Society*, *29*: 1258–1266.

Fuyama, Y. (1986). Genetics of parthenogenesis in *Drosophila melanogaster*. II Characterization of a gynogenetically reproducing strain. *Genetics*, *114*(2): 495–509.
Gade, B. & Parker Jr., E.D. (1997). The effect of life cycle stage and genotype on desiccation tolerance in the colonizing parthenogenetic cockroach *Pycnoscelus surinamensis* and its sexual ancestor *P. indicus*. *Journal of Evolutionary Biology*, *10*(4): 479–493.
Galtier, N., Nabholz, B., Glémin, S. & Hurst, G.D.D. (2009). Mitochondrial DNA as a marker of molecular diversity: a reappraisal. *Molecular Ecology*, *18*(22): 4541–4550.
Galtier, N., Roux, C., Rousselle, M., Romiguier, J., Figuet, E., Glémin, S., Bierne, N. & Duret, L. 2018. Codon usage bias in animals: disentangling the effects of natural selection, effective population size, and GC-biased gene conversion. *Molecular Biology and Evolution*, *35*(5): 1092–1103.
Gavrilov-Zimin, I.A. (2012). A contribution to the taxonomy, cytogenetics and reproductive biology of the genus Aclerda Signoret (Homoptera, Coccinea, Aclerdidae). *Comparative Cytogenetics*, *6*: 389–395.
Gerth, M., Geißler, A. & Bleidorn, C. (2011). *Wolbachia* infections in bees (*Anthophila*) and possible implications for DNA barcoding. *Systematics and Biodiversity*, *9*(4): 319–327.
Glémin, S., François, C.M. & Galtier, N. (2019). Genome evolution in outcrossing vs. selfing vs. asexual species. In: *Evolutionary Genomics. Methods in Molecular Biology, vol. 1910*. (Ed. Anisimova, M.). New York: Humana, pp. 331–369.
Goday, C. & Esteban, M.R. (2001). Chromosome elimination in sciarid flies. *Bioessays*, *23*(3): 242–250.
Gómez-Zurita, J., Funk, D.J., & Vogler, A.P. (2006). The evolution of unisexuality in Calligrapha leaf beetles: molecular and ecological insights on multiple origins via interspecific hybridization. *Evolution*, *60*(2): 328–347.
Gokhman, V.E. & Kuznetsova, V.G. (2018). Parthenogenesis in Hexapoda: holometabolous insects. *Journal of Zoological Systematics and Evolutionary Research*, *56*(1): 23–34.
Gottlieb, Y. & Zchori-Fein, E. (2001). Irreversible thelytokous reproduction in *Muscidifurax uniraptor*. *Entomologia Experimentalis et Applicata*, *100*(3): 271–278.
Gottlieb, Y., Zchori-Fein, E., Werren, J.H. & Karr, T.L. (2002). Diploidy restoration in *Wolbachia*-infected *Muscidifurax uniraptor* (Hymenoptera: Pteromalidae). *Journal of Invertebrate Pathology*, *81*: 166–174.
Grbic, M. (2003). Polyembryony in parasitic wasps: evolution of a novel mode of development. *International Journal of Developmental Biology*, *47*(7-8): 633–642.
Greeff, J.M. & Villet, M.H. (1993). Deducing the coefficient of relationship by the amount of recombination produced during automictic parthenogenesis. *Heredity*, *70*(5): 499–502.
Gregory, T.R. & Mable, B.K. (2005). Polyploidy in animals. In: *The evolution of the Genome* (Ed. Gregory, T.R.). New York: Academic Press, pp. 427–517.
Grossniklaus, U., Nogler, G.A. & van Dijk, P.J. (2001). How to avoid sex: the genetic control of gametophytic apomixis. *The Plant Cell*, *13*(7): 1491–1498.
Gustafsson, A. (1947). Mutations in agricultural plants. *Hereditas*, *33*(1–2): 1–100.
Haddrill, P.R., Halligan, D.L., Tomaras, D. & Charlesworth, B. (2007). Reduced efficacy of selection in regions of the *Drosophila* genome that lack crossing over. *Genome Biology*, *8*(2): R18.
Hamilton, W.D. (1967). Extraordinary sex ratios. *Science*, *156*(3774): 477–488.
Hardy, I.C., Ode, P.J., & Strand, M.R. (1993). Factors influencing brood sex ratios in polyembryonic Hymenoptera. *Oecologia*, *93*(3): 343–348.
Hauschteck, E. (1962). Die Zytologie der Paedogenese und der Geschlechtsbestimmung einer heterogonen Gallmücke. *Chromosoma*, *13*(1962): 163–182.
Hebert, P.D.N. (1987). Genotypic charactersitics of cyclic parthenogens and their obligately asexual derivatives. In: *The Evolution of Sex and Its Consequences* (Ed. Stearns, S.C.). Basel: Birkhauser Verlag, pp. 175–218.
Heimpel, G.E. & De Boer, J.G. (2008). Sex determination in the Hymenoptera. *Annual Review of Entomology*, *53*: 209–230.
Henry, L., Schwander, T. & Crespi, B.J. (2011). Deleterious mutation accumulation in asexual *Timema* stick insects. *Molecular Biology and Evolution*, *29*(1): 401–408.
Hebert, P.D., Stoeckle, M.Y., Zemlak, T.S. & Francis, C.M. (2004). Identification of birds through DNA barcodes. *PLoS Biology*, *2*(10): e312.

Herrick, G. & Seger, J. (1999). Imprinting and paternal genome elimination in insects. In: *Genomic Imprinting. Results and Problems in Cell Differentiation, vol. 25*. Berlin, Heidelberg: Springer, pp. 41–71.
Hey, J. (2001). The mind of the species problem. *Trends in Ecology & Evolution, 16*: 326–329.
Hill, W.G. & Robertson, A. (1966). The effect of linkage on limits to artificial selection. *Genetics Research, 8*(3): 269–294.
Hodson, C.N., Hamilton, P.T., Dilworth, D., Nelson, C.J., Curtis, C.I. & Perlman, S.J. (2017). Paternal genome elimination in *Liposcelis* booklice (Insecta: Psocodea). *Genetics, 206*(2): 1091–1100.
Hollister, J.D., Greiner, S., Wang, W., Wang, J., Zhang, Y., Wong, G.K.S., Wright, S.I. & Johnson, M.T. (2015). Recurrent loss of sex is associated with accumulation of deleterious mutations in *Oenothera*. *Molecular Biology and Evolution, 32*(4): 896–905.
Hoffmann, A.A., Tracy Reynolds, K., Nash, M.A. & Weeks, A.R. (2008). A high incidence of parthenogenesis in agricultural pests. *Proceedings of the Royal Society B: Biological Sciences, 275*(1650): 2473–2481.
Horner, V.L. & Wolfner, M.F. (2008). Transitioning from egg to embryo: triggers and mechanisms of egg activation. *Developmental Dynamics, 237*(3): 527–544.
Hosokawa, T., Koga, R., Kikuchi, Y., Meng, X.Y. & Fukatsu, T. (2010). *Wolbachia* as a bacteriocyte-associated nutritional mutualist. *Proceedings of the National Academy of Sciences USA, 107*(2): 769–774.
Huigens, M.E., Luck, R.F., Klaassen, R.H.G., Maas, M.F.P.M., Timmermans, M.J.T.N. & Stouthamer, R. (2000). Infectious parthenogenesis. *Nature, 405*(6783): 178–179.
Hurst, G.D.D. & Jiggins, F.M. (2005). Problems with mitochondrial DNA as a marker in population, phylogeographic and phylogenetic studies: the effects of inherited symbionts. *Proceedings of the Royal Society B: Biological Sciences, 272*(1572): 1525–1534.
Igic, B. & Kohn, J.R. (2006). The distribution of plant mating systems: study bias against obligately outcrossing species. *Evolution, 60*(5): 1098–1103.
Ibrahim, I. A. & Gad, A. M. (1975). The occurrence of paedogenesis in Eristalis larvae (Diptera: Syrphidae). *Journal of Medical Entomology, 12*:268.
Jaquiéry, J., Stoeckel, S., Larose, C., Nouhaud, P., Rispe, C., Mieuzet, L., Bonhomme, J., Mahéo, F., Legeai, F., Gauthier, J.P. & Prunier-Leterme, N. (2014). Genetic control of contagious asexuality in the pea aphid. *PLoS Genetics, 10*(12): e1004838.
Jarosch, A., Stolle, E., Crewe, R. M. , & Moritz, R. F. (2011). Alternative splicing of a single transcription factor drives selfish reproductive behavior in honeybee workers (Apis mellifera). *Proceedings of the National Academy of Sciences, 108*(37): 15282–15287.
Jensen, L.H., Enghoff, H., Frydenberg, J., Parker Jr., E.D. (2002). Genetic diversity and the phylogeography of parthenogenesis: comparing bisexual and thelytokous populations of *Nemasoma varicorne* (Diplopoda: Nemasomatidae) in Denmark. *Hereditas, 136*: 184–194.
Johannsen, O.A. (1910). Paedogenesis in *Tanytarsus*. *Science, 32*: 768.
Kawecki, T.J. (2008). Adaptation to marginal habitats. *Annual Review of Ecology, Evolution, and Systematics, 39*: 321–342.
Keegan-Rogers, V. (1984). Unfamiliar-female advantage among clones of unisexual fish (Poeciliopsis, Poeciliidae). *Copeia, 1984*: 169–1174.
Kellner, K. & Heinze, J. (2011). Mechanism of facultative parthenogenesis in the ant *Platythyrea punctata*. *Evolutionary Ecology, 25*(1): 77–89.
King, P.E. & Rafai, J. (1970). A possible mechanism for initiating the parthenogenetic development of eggs in a parasitoid Hymenopteran, *Nasonia vitripennis* (Walker) (Pteromalidae).*The Entomologist, 106*: 118–120.
Kirkendall, L.R. (1990). Sperm is a limiting resource in the pseudogamous bark beetle *lps acuninatus* (Scolytidae). *Oikos, 57*: 80–87.
Kirkendall, L.R. & Stenseth, N.C. (1990). Ecological and evolutionary stability of sperm-dependent parthenogenesis: effects of partial niche overlap between sexual and asexual females. *Evolution, 44*: 698–714.
Kondrashov, A.S. (1993). Classification of hypotheses on the advantage of amphimixis. *Journal of Heredity, 84*(5): 372–387.
Kotásková, N., Kolasa, M. & Kajtoch, Ł. (2018). Contrasting patterns of molecular diversity and *Wolbachia* infection in bisexual and parthenogenetic *Strophosoma* weevils (Coleoptera: Curculionidae). *Entomological Science, 21*(4): 385–395.

Kover, P.X. & Caicedo, A.L. (2001). The genetic architecture of disease resistance in plants and the maintenance of recombination by parasites. *Molecular Ecology*, *10*(1): 1–16.

Kraaijeveld, K., Anvar, S.Y., Frank, J., Schmitz, A., Bast, J., Wilbrandt, J., Petersen, M., Ziesmann, T., Niehuis, O., De Knijff, P. & Den Dunnen, J.T. (2016). Decay of sexual trait genes in an asexual parasitoid wasp. *Genome Biology and Evolution*, *8*(12): 3685–3695.

Kraaijeveld, K. & Bast, J. (2012). Transposable element proliferation as a possible side effect of endosymbiont manipulations. *Mobile Genetic Elements*, *2*(5): 253–256.

Kraaijeveld, K., Franco, P., De Knijff, P., Stouthamer, R. & Van Alphen, J.J. (2011). Clonal genetic variation in a *Wolbachia*-infected asexual wasp: horizontal transmission or historical sex? *Molecular Ecology*, *20*(17): 3644–3652.

Kumpulainen, T., Grapputo, A. & Mappes, J. (2004). Parasites and sexual reproduction in psychid moths. *Evolution*, *58*(7): 1511–1520.

Lamb, R.Y. & Willey, R.B. (1987). Cytological mechanisms of thelytokous parthenogenesis in insects. *Genome*, *29*(2): 367–369.

Lamb, R.Y. & Willey, R.B. (1989). Parthenogenetic mechanism and its evolutionary potential in the cave cricket *Euhadenoecus insolitus* (Orthoptera: Rhaphidophoridae). *Annals of the Entomological Society of America*, *82*(1): 101–108.

Lanteri, A.A. & Normark, B.B. (1995). Parthenogenesis in the tribe Naupactini (Coleoptera: Curculionidae). *Annals of the Entomological Society of America*, *88*(6): 722–731.

Lattorff, H. M. G., Moritz, R. F., Crewe, R. M., & Solignac, M. (2007). Control of reproductive dominance by the thelytoky gene in honeybees. *Biology Letters*, *3*(3):292–295.

Laven, H. (1956). Cytoplasmic inheritance in *Culex*. *Nature*, *177*(4499): 141–142.

Lehmann, G.U., Strauß, J. & Lakes-Harlan, R. (2007). Listening when there is no sexual signalling? Maintenance of hearing in the asexual bushcricket *Poecilimon intermedius*. *Journal of Comparative Physiology A*, *193*(5): 537–545.

Lehtonen, J., Parker, G.A. & Schärer, L. (2016). Why anisogamy drives ancestral sex roles. *Evolution*, *70*(5): 1129–1135.

Leniaud, L., Darras, H., Boulay, R. & Aron, S. (2012). Social hybridogenesis in the clonal ant *Cataglyphis hispanica*. *Current Biology*, *22*: 1188–1193.

Levin, D.A. 1975. Pest pressure and recombination systems in plants. *The American Naturalist*, *109*: 437–451.

Levin, D.A. (1983). Polyploidy and novelty in flowering plants. *The American Naturalist*, *122*(1): 1–25.

Lin, Y.P., Edwards, R.D., Kondo, T., Semple, T.L. & Cook, L.G. (2017). Species delimitation in asexual insects of economic importance: the case of black scale (*Parasaissetia nigra*), a cosmopolitan parthenogenetic pest scale insect. *PLoS One*, *12*(5): e0175889.

Lindsey, A.R., Kelkar, Y.D., Wu, X., Sun, D., Martinson, E.O., Yan, Z., Rugman-Jones, P.F., Hughes, D.S., Murali, S.C., Qu, J. & Dugan, S. (2018). Comparative genomics of the miniature wasp and pest control agent *Trichogramma pretiosum*. *BMC Biology*, *16*(1): 54.

Lokki, J. & Saura, A. (1980). Polyploidy in insect evolution. In: *Polyploidy: Biological Relevance* (Ed. Lewis, W.M.). New York: Plenum Press, pp. 277–312.

Lorenzo-Carballa, M.O., Beatty, C.D., Haitlinger, R., Valdecasas, A.G., Utzeri, C., Vieira, V. & Cordero-Rivera, A. (2011). Larval aquatic and terrestrial mites infesting parthenogenetic *Ischnura hastata* (Odonata: Coenagrionidae) from the Azores islands. *Experimental and Applied Acarology*, *54*(3): 225–241.

Lorenzo-Carballa, M.O., Hadrys, H., Cordero-Rivera, A. & Andrés, J.A. (2012). Population genetic structure of sexual and parthenogenetic damselflies inferred from mitochondrial and nuclear markers. *Heredity*, *108*(4): 386–395.

Lu, H. L., Tanguy, S., Rispe, C., Gauthier, J. P., Walsh, T., Gordon, K., & Jaubert-Possamai, S. (2011). Expansion of genes encoding piRNA-associated argonaute proteins in the pea aphid: diversification of expression profiles in different plastic morphs. *PLoS one*, *6*(12): e28051.

Luck, R. F. , Stouthamer, R ., & Nunney, L. (1992). Sex determination and sex ratio patterns in parasitic hymenoptera. InEvolution andDiversity of Sex Ratio in Haplodiploid Insects and Mites (Eds. Wrench D. L. & Ebbert M.A.). New York: Chapman & Hall, pp. 442–476.

Lushai, G., Loxdale, H.D. & Allen, J.A. (2003). The dynamic clonal genome and its adaptive potential. *Biological Journal of the Linnean Society*, *79*(1): 193–208.

Lynch, M. (1984). Destabilizing hybridization, general-purpose genotypes and geographic parthenogenesis. *The Quarterly Review of Biology*, *59*(3): 257–290.
Lynch, M., Bürger, R., Butcher, D. & Gabriel, W. (1993). The mutational meltdown in asexual populations. *Journal of Heredity*, *84*(5): 339–344.
Ma, W.J., Pannebakker, B.A., Beukeboom, L.W., Schwander, T. & van de Zande, L. (2014). Genetics of decayed sexual traits in a parasitoid wasp with endosymbiont-induced asexuality. *Heredity*, *113*(5): 424–431.
Ma, W.J. & Schwander, T. (2017). Patterns and mechanisms in instances of endosymbiont-induced parthenogenesis. *Journal of Evolutionary Biology*, *30*(5): 868–888.
Mable, B.K. (2004). 'Why polyploidy is rarer in animals than in plants': myths and mechanisms. *Biological Journal of the Linnean Society*, *82*(4): 453–466.
Magro, A., Lecompte, E., Hemptinne, J.L., Soares, A.O., Dutrillaux, A.M., Murienne, J., Fürsch, H. & Dutrillaux, B. (2020). First case of parthenogenesis in ladybirds (Coleoptera: Coccinellidae) suggests new mechanisms for the evolution of asexual reproduction. *Journal of Zoological Systematics and Evolutionary Research*, *58*(1): 194–208.
Manandhar, G., Schatten, H. & Sutovsky, P. (2005). Centrosome reduction during gametogenesis and its significance. *Biology of Reproduction*, *72*(1): 2–13.
Martin, P. (2015). Reproductive biology: parthenogenesis: mechanisms, evolution, and its relevance to the role of marbled crayfish as model organism and potential invader. In: *Freshwater Crayfish: A Global Overview* (Ed. Kawai, T., Faulkes, Z. & Scholtz, G.). Boca Raton: CRC Press, pp. 63–82.
Matsuura, K., Fujimoto, M. & Goka, K. (2004). Sexual and asexual colony foundation and the mechanism of facultative parthenogenesis in the termite *Reticulitermes speratus* (Isoptera, Rhinotermitidae). *Insectes Sociaux*, *51*(4): 325–332.
Matzk, F., Prodanovic, S., Bäumlein, H. & Schubert, I. (2005). The inheritance of apomixis in Poa pratensis confirms a five locus model with differences in gene expressivity and penetrance. *The Plant Cell*, *17*(1): 13–24.
Maynard Smith, J. (1971). What use is sex? *Journal of Theoretical Biology*, *30*: 319–335.
Maynard Smith, J. (1978). *The Evolution of Sex*. Cambridge: Cambridge University Press.
McMeniman, C.J. & Barker, S.C. (2006). Transmission ratio distortion in the human body louse, *Pediculus humanus* (Insecta: Phthiraptera). *Heredity*, *96*(1): 63–68.
Meyer, R.E., Delaage, M., Rosset, R., Capri, M. & Aït-Ahmed, O. (2010). A single mutation results in diploid gamete formation and parthenogenesis in a *Drosophila yemanuclein-alpha* meiosis I defective mutant. *BMC Genetics*, *11*: 104.
Milani, L., Ghiselli, F., Pellecchia, M., Scali, V. & Passamonti, M. (2010). Reticulate evolution in stick insects: the case of *Clonopsis* (Insecta: Phasmida). *BMC Evolutionary Biology*, *10*(1): 258.
Mittwoch, U. (1978). Parthenogenesis. *Journal of Medical Genetics*, *15*(3): 165–181.
Mogie, M. (1986). Automixis: its distribution and status. *Biological Journal of the Linnean Society*, *28*(3): 321–329.
Mogie, M., Britton, N.F. & Stewart-Cox, J.A. (2007). Asexuality, polyploidy and the male function. In: *XVII International Botanical Congress*. Gantner Verlag, pp. 195–214.
Montelongo, T. & Gomez-Zurita, J. (2014). Nonrandom patterns of genetic admixture expose the complex historical hybrid origin of unisexual leaf beetle species in the genus *Calligrapha*. *The American Naturalist*, *185*(1): 113–134.
Moore, W.S. (1975). Stability of unisexual-bisexual populations of *Poeciliopsis* (Pisces: Poeciliidae). *Ecology*, *56*: 791–808.
Moore, W.S. (1976). Components of fitness in the unisexual fish *Poeciliopsis monacha-occidentalis*. *Evolution*, *30*: 5644578.
Morgan-Richards, M., Hills, S.F., Biggs, P.J. & Trewick, S.A. (2016). Sticky genomes: using NGS evidence to test hybrid speciation hypotheses. *PLoS One*, *11*(5): e0154911.
Morgan-Richards, M. & Trewick, S.A. (2005). Hybrid origin of a parthenogenetic genus? *Molecular Ecology*, *14*(7): 2133–2142.
Mugal, C.F., Weber, C.C., Ellegren, H. (2015). GC-biased gene conversion links the recombination landscape and demography to genomic base composition. *Bioessays*, *3712*: 1317–1326.
Muller, H.J. (1932). Some genetic aspects of sex. *The American Naturalist*, *66*: 118–138.
Muller, H.J. (1964). The relation of recombination to mutational advance. *Mutation Research*, *1*(1): 2–9.

Murdy, W.H. and Carson, H.L. (1959). Parthenogenesis in *Drosophila mangabeirai* Malog. *The American Naturalist*, *93*(873): 355–363.
Neaves, W.B. & Baumann, P. (2011). Unisexual reproduction among vertebrates. *Trends in Genetics*, *27*(3): 81–88.
Neiman, M. & Koskella, B. (2009). Sex and the red queen. In: *Lost Sex* (Eds. Schön, I., Schön, Isa, Martens, K. & Dijk, P.). Dordrecht: Springer, pp. 133–159.
Neiman, M., Sharbel, T.F. & Schwander, T. (2014). Genetic causes of transitions from sexual reproduction to asexuality in plants and animals. *Journal of Evolutionary Biology*, *27*(7): 1346–1359.
Nokkala, C., Kuznetsova, V.G. & Nokkala, S. (2015). Rare diploid females coexist with rare males: a novel finding in triploid parthenogenetic populations in the psyllid *Cacopsylla myrtilli* (W. Wagner, 1947) (Hemiptera, Psylloidea) in northern Europe. *Genetica*, *143*(5): 589–595.
Normark, B.B. (1994). *Phylogeny and Evolution of Parthenogenesis in the Aramigus tessellatus Complex (Coleoptera: Curculionidae)*. PhD Dissertation, Ithaca, USA: Cornell University.
Normark, B.B. (2003). The evolution of alternative genetic systems in insects. *Annual Review of Entomology*, *48*(1): 397–423.
Normark, B.B. (2014). Modes of reproduction. In: *The Evolution of Insect Mating Systems* (Eds. Shuker, D.M. & Simmons, L.W.). Oxford: Oxford University Press, pp. 1–19.
Normark, B.B. & Kirkendall, L.R. (2009). Parthenogenesis in insects and mites. In: *Encyclopedia of Insects* (Eds. Resh, V.H. & Cardé, R.). 2nd Ed. New York: Academic Press, pp. 753–757.
Normark, B.B. & Lanteri, A.A. (1998). Incongruence between morphological and mitochondrial-DNA characters suggests hybrid origins of parthenogenetic weevil lineages (genus *Aramigus*). *Systematic Biology*, *47*(3): 475–494.
Normark, B.B. & Johnson, N.A. (2011). Niche explosion. *Genetica*, *139*(5): 551–564.
Normark, B.B. & Moran, N.A. (2000). Testing for the accumulation of deleterious mutations in asexual eukaryote genomes using molecular sequences. *Journal of Natural History*, *34*(9): 1719–1729.
Norton, R.A., Kethley, J.B., Johnston, D.E. & O'Connor, B.M. (1993). Phylogenetic perspectives on genetic systems and reproductive modes of mites. In: *Evolution and Diversity of Sex Ratio in Insects and Mites* (Eds. Wrensch, D.L. & Ebbert, M.A.). New York: Chapman & Hall, pp. 8–99.
Noskiewicz, J., & Poluszynski, G. (1935). Embryologische Untersuchungen an Strepsipteren. II. Teil; Polyembryonie. *Zoologica Poloniae*, *1*(1): 53–94.
Nothnagle, P.J. & Schultz J.C. (1987). What is a forest pest? In: *Insect Outbreaks* (Eds. Barbosa, P. & Schultz, J.). New York: Academic Press, pp. 59–80.
Noyes, R.D., Baker, R. & Mai, B. (2007). Mendelian segregation for two-factor apomixis in *Erigeron annuus* (Asteraceae). *Heredity*, *98*(2): 92–98.
Nur, U. (1971). Parthenogenesis in coccids (Homoptera). *American Zoologist*, *11*(2): 301–308.
Nur, U. (1979). Gonoid thelytoky in soft scale insects (Coccidae: Homoptera). *Chromosoma*, *72*: 89–104
Ohkawara, K., Nakayama, M., Satoh, A., Trindl, A. & Heinze, J. (2006). Clonal reproduction and genetic caste differences in a queen-polymorphic ant, Vollenhovia emeryi. *Biology Letters*, *2*(3): 359–363.
Oldroyd, B.P., Allsopp, M.H., Gloag, R.S., Lim, J., Jordan, L.A. & Beekman, M. (2008). Thelytokous parthenogenesis in unmated queen honeybees (*Apis mellifera capensis*): central fusion and high recombination rates. *Genetics*, *180*(1): 359–366.
Ollivier, M., Gabaldón, T., Poulain, J., Gavory, F., Leterme, N., Gauthier, J.P., Legeai, F., Tagu, D., Simon, J.C. & Rispe, C. (2012). Comparison of gene repertoires and patterns of evolutionary rates in eight aphid species that differ by reproductive mode. *Genome Biology and Evolution*, *4*(2): 155–167.
Orr, H.A. & Unckless, R.L. (2008). Population extinction and the genetics of adaptation. *The American Naturalist*, *172*(2): 160–169.
Otto, S.P. (2009). The evolutionary enigma of sex. *The American Naturalist*, *174*(S1): S1–S14.
Owen, R. (1842). Report on British fossil reptiles. *Part II. Reports of the British Association for the Advancement of Science*, *11*: 60–204.
Owen, R. (1849). *On Parthenogenesis*. London: John van Voorst.
Painter, T.S. (1966). The role of the E-chromosomes in Cecidomyiidae. *Proceedings of the National Academy of Sciences USA*, *56*: 853–855.
Pannebakker, B.A., Pijnacker, L.P., Zwaan, B.J. & Beukeboom, L.W. (2004). Cytology of *Wolbachia*-induced parthenogenesis in *Leptopilina clavipes* (Hymenoptera: Figitidae). *Genome*, *47*(2): 299–303.

Parker, D.J., Bast, J., Jalvingh, K., Dumas, Z., Robinson-Rechavi, M. & Schwander, T. (2018). Repeated evolution of asexuality involves convergent gene expression changes. *Molecular Biology and Evolution, 36*(2): 350–364.

Parker Jr., E.D., and Niklasson, M. (1995). Dessication resistance among clones in the invading parthenogenetic cockroach, *Pycnoscelus surinamen*: a search for the general-purpose genotype. *Journal of Evolutionary Biology, 8*: 331–337

Parker Jr., E.D. & Niklasson, M. (2000). Genetic structure and evolution in parthenogenetic animals. In: *Evolutionary Genetics: From Molecules to Morphology* (Eds. Singh, R.S. & Krimbas, C.B.). Cambridge: Cambridge University Press, pp. 456–474.

Pearcy, M., Hardy, O. & Aron, S. (2006): Thelytokous parthenogenesis and its consequences on inbreeding in an ant. *Heredity, 96*(5): 377–382.

Pearcy, M., Hardy, O. & Aron, S. (2011). Automictic parthenogenesis and rate of transition to homozygosity. *Heredity, 107*: 187–188.

Perotti, M.A., Young, D.K. & Braig, H.R. (2016). The ghost sex-life of the paedogenetic beetle *Micromalthus debilis. Scientific Reports, 6*: 27364.

Phillips, E.F. (1903). A review of parthenogenesis. *Proceedings of the American Philosophical Society, 42*: 275–345.

Pigneur, L.M., Hedtke, S.M., Etoundi, E. & Van Doninck, K. (2012). Androgenesis: a review through the study of the selfish shellfish Corbicula spp. *Heredity, 108*(6): 581–591.

Pijnacker, L.P. (1969). Automictic parthenogenesis in the stick insect *Bacillus rossius* Rossi (Cheleutoptera, phasmidae). *Genetica, 40*(1): 393–399.

Pijnacker, L.P. & Ferwerda, M.A. (1978). Additional chromosome duplication in female meiotic prophase of *Sipyloidea sipylus* Westwood (Insecta, Phasmida) and its absence in male meiosis. *Experientia, 34*: 1558–1560.

Plateau, F. (1868). XXIX.-On the production of the Sexes in bees: to the editors of Annals and Magazine of Natural History. *Journal of Natural History, 2*(10): 252–255.

Pollock, D.A. & Normark, B.B. (2002). The life cycle of *Micromalthus debilis* LeConte (1878) (Coleoptera: Archostemata: Micromalthidae): historical review and evolutionary perspective. *Journal of Zoological Systematics and Evolutionary Research, 40*(2): 105–112.

Poon, A. & Chao, L. (2004). Drift increases the advantage of sex in RNA bacteriophage Φ6. *Genetics, 166*(1): 19–24.

Prevost, Y.H. & McFarlane, J.E. (1979). Polyembryony in the house cricket (Orthoptera: Gryllidae) 1. *The Canadian Entomologist, 111*(10): 1199–1200.

Rabeling, C. & Kronauer, D.J. (2013). Thelytokous parthenogenesis in eusocial Hymenoptera. *Annual Review of Entomology, 58*: 273–292.

Rafael, J.A., Marques, D.W.A. & Engel, M.S. (2017). The first gynandromorph of a zorapteran and potential thelytokous parthenogenesis in a population of *Zorotypus brasiliensis* Silvestri (Zoraptera: Zorotypidae). *Revista Brasileira de Entomologia, 61*(4): 318–322.

de Réaumur, R.A.F. (1734–1742). Mémoires pour servir à l'histoire des insectes. Six volumes. Paris: Académie Royale des Sciences.

Reumer, B.M., van Alphen, J.J. & Kraaijeveld, K. (2012). Occasional males in parthenogenetic populations of *Asobara japonica* (Hymenoptera: Braconidae): low *Wolbachia* titer or incomplete coadaptation? *Heredity, 108*(3): 341–346.

Rey, O., Loiseau, A., Facon, B., Foucaud, J., Orivel, J., Cornuet, J.M., Robert, S., Dobigny, G., Delabie, J.H.C., Mariano, C.D.S.F. & Estoup, A. (2011). Meiotic recombination dramatically decreased in thelytokous queens of the little fire ant and their sexually produced workers. *Molecular Biology and Evolution, 28*(9): 2591–2601.

Rice, W.R. & Friberg, U. (2009). A graphical approach to lineage selection between clonals and sexuals. In: *Lost Sex* (Eds. Schön, I., Schön, Isa, Martens, K. & Dijk, P.). Dordrecht: Springer, pp. 75–97.

Riparbelli, M. G., Giordano, R. , & Callaini, G. (2006). *Centrosome inheritance in the parthenogenetic egg of the collembolan Folsomia candida. Cell and tissue research, 326*(3): 861–872.

Riparbelli, M.G., Dallai, R. & Callaini, G. (2010). The insect centriole: a land of discovery. *Tissue and Cell, 42*(2): 69–80.

Rodriguero, M.S. (2009). *Origen y consecuencias de la reproducción asexual en una especie de gorgojo de importancia agronómica*. PhD Dissertation. Buenos Aires, Argentina: University of Buenos Aires.
Riparbelli, M.G., Gottardo, M. & Callaini, G. (2017). Parthenogenesis in insects: the centriole renaissance. In: *Oocytes. Results and Problems in Cell Differentiation, vol 63*. (Ed. Kloc, M.). Cham: Springer, pp. 435–479.
Rodriguero, M.S., Lanteri, A.A. & Confalonieri, V.A. (2010). Mito-nuclear genetic comparison in a *Wolbachia* infected weevil: insights on reproductive mode, infection age and evolutionary forces shaping genetic variation. *BMC Evolutionary Biology, 10*(1): 340.
Rodriguero, M.S., Lanteri, A.A. & Confalonieri, V.A. (2013). Speciation in the asexual realm: is the parthenogenetic weevil *Naupactus cervinus* a complex of species in *statu nascendi*? *Molecular Phylogenetics and Evolution, 68*(3): 644–656.
Rodriguero, M.S., Scannapieco, A.C., Chifflet, L., Monti, D.S., Elias-Costa, A.J., Lanteri, A.A., Confalonieri, V.A. (2021). Dependence of egg hatching on Wolbachia density in a parthenogenetic weevil revealed by an antibiotic treatment. *Entomologia Experimentalis et Applicata* 169(4): 384-392.
Rodriguero, M.S., Wirth, S.A., Alberghina, J.S., Lanteri, A.A. & Confalonieri, V.A. (2018). A tale of swinger insects: signatures of past sexuality between divergent lineages of a parthenogenetic weevil revealed by ribosomal intraindividual variation. *PLoS One, 13*(5): e0195551.
Roff, D.A. (1990). The evolution of flightlessness in insects. *Ecological Monographs, 60*(4): 389–421.
Ross, L., Hardy, N.B., Okusu, A. & Normark, B.B. (2013). Large population size predicts the distribution of asexuality in scale insects. *Evolution, 67*(1): 196–206.
Ross, L. & Normark, B.B. (2015). Evolutionary problems in centrosome and centriole biology. *Journal of Evolutionary Biology, 28*(5): 995–1004.
Ross, L., Pen, I. & Shuker, D.M. (2010). Genomic conflict in scale insects: the causes and consequences of bizarre genetic systems. *Biological Review, 85*:807- 828.
Russell, J.E. & Stouthamer, R. (2011). The genetics and evolution of obligate reproductive parasitism in *Trichogramma pretiosum* infected with parthenogenesis-inducing *Wolbachia. Heredity, 106*(1): 58–67.
Ruttner, F. (1988). *Breeding Techniques and Selection for Breeding of the Honeybee*. Derby: British Isles Bee Breeders' Association.
Salathé, M., Kouyos, R.D. & Bonhoeffer, S. (2008). The state of affairs in the kingdom of the Red Queen. *Trends in Ecology & Evolution, 23*(8): 439–445.
Sanderson, A.R. & Jacob, J. (1957). Artilicial activation of the egg in a gynogenetic spider beetle. *Nature, 179*: 1300.
Sandrock, C., Razmjou, J. & Vorburger, C. (2011). Climate effects on life cycle variation and population genetic architecture of the black bean aphid, *Aphis fabae. Molecular Ecology, 20*(19): 4165–4181.
Sandrock, C. & Vorburger, C. (2011). Single-locus recessive inheritance of asexual reproduction in a parasitoid wasp. *Current Biology, 21*(5): 433–437.
Saura, A., Lokki, J. & Suomalainen, E. (1993). Origin of polyploidy in parthenogenetic weevils. *Journal of Theoretical Biology, 163*(4): 449–456.
Scali, V., Passamonti, M., Marescalchi, O. & Mantovani, B. (2003). Linkage between sexual and asexual lineages: genome evolution in Bacillus stick insects. *Biological Journal of the Linnean Society, 79*(1): 137–150.
Schlupp, I. (2005). The evolutionary ecology of gynogenesis. *Annual Review of Ecology, Evolution and Systematics, 36*: 399–417.
van Alphen, J.J.M. (2002). Geographical distribution and genetic relatedness of sympatrical thelytokous and arrhenotokous populations of the parasitoid *Venturia canescens* (Hymenoptera). *Journal of Evolutionary Biology, 15*: 191–200.
Schneider, M.C. & Cella, D.M. (2010). Karyotype conservation in 2 populations of the parthenogenetic scorpion *Tityus serrulatus* (Buthidae): rDNA and its associated heterochromatin are concentrated on only one chromosome. *Journal of Heredity, 101*(4): 491–496.
Schön, I., Gandolfi, A., Di Masso, E., Rossi, V., Griffiths, H.I., Martens, K. & Butlin, R.K. (2000). Persistence of asexuality through mixed reproduction in *Eucypris virens* (Crustacea, Ostracoda). *Heredity, 84*(2): 161–169.

Schwander, T., Crespi, B.J., Gries, R. & Gries, G. (2013). Neutral and selection-driven decay of sexual traits in asexual stick insects. *Proceedings of the Royal Society of London, Series B: Biological Sciences*, *280*(1764): 0823.

Scott, A.C. (1938). Paedogenesis in the Coleoptera. *Zeitschrift für Morphologie und Ökologie der Tiere*, *33*(4): 633–653.

Seiler, J. (1960). Studies on the origin of parthenogenesis in *Solenobia triquetrella* FR (Lepidoptera, Psychidae). Part II. Analysis of diploid parthenogenetic *S. triquetrella* behavior, cultivation results and cytology. *Chromosoma*, *11*: 29–102.

Shoemaker, D.D., Dyer, K.A., Ahrens, M., McAbee, K. & Jaenike, J. (2004). Decreased diversity but increased substitution rate in host mtDNA as a consequence of *Wolbachia* endosymbiont infection. *Genetics*, *168*(4): 2049–2058.

Shropshire, J.D. & Bordenstein, S.R. (2016). Speciation by symbiosis: the microbiome and behavior. *mBio*, *7*(2): e01785-15.

Simon, J.C., Baumann, S., Sunnucks, P., Hebert, P.D., Pierre, J.S., Le Gallic, J.F. & Dedryver, C.A. (1999). Reproductive mode and population genetic structure of the cereal aphid *Sitobion avenae* studied using phenotypic and microsatellite markers. *Molecular Ecology*, *8*: 531–545.

Simon, J.C., Delmotte, F., Rispe, C. & Crease, T. (2003). Phylogenetic relationships between parthenogens and their sexual relatives: the possible routes to parthenogenesis in animals. *Biological Journal of the Linnean Society*, *79*(1): 151–163.

Slifer, E.H., & Shulow, A. (1947). Sporadic polyembryony in grasshopper eggs. *Annals of the Entomological Society of America*, *40*(4): 652–655.

Smith, M.A., Bertrand, C., Crosby, K., Eveleigh, E.S., Fernandez-Triana, J., Fisher, B.L., Gibbs, J., Hajibabaei, M., Hallwachs, W., Hind, K. & Hrcek, J. (2012). *Wolbachia* and DNA barcoding insects: patterns, potential, and problems. *PLoS One*, *7*(5): e36514.

Soriano, M., Li, H. & Boutilier, K. (2013). Microspore embryogenesis: establishment of embryo identity and pattern in culture. *Plant Reproduction*, *206*:181–196.

Srdic, Z. & Jacobs-Lorena, M. (1978). *Drosophila* egg chambers develop to mature eggs when cultured in vivo. *Science*, *202*: 641–643.

Srdic, Z., Reinhardt, C., Beck, H. & Gloor, H. (1979). Autonomous yolk protein synthesis in ovaries of *Drosophila* cultured in vivo. *Wilhelm Roux's Archives*, *187*: 255–266.

Stahlhut, J.K., Gibbs, J., Sheffield, C.S., Alex Smith, M. & Packer, L. (2012). *Wolbachia* (Rickettsiales) infections and bee (Apoidea) barcoding: a response to Gerth *et al. Systematics and Biodiversity*, *10*(4): 395–401.

Staiber, W. & Schiffkowski, C. (2000). Structural evolution of the germ line-limited chromosomes in *Acricotopus*. *Chromosoma*, *109*: 343–349.

Stalker, H.D. (1954). Parthenogenesis in *Drosophila*. *Genetics*, *39*: 4–34.

Stenberg, P., Lundmark, M., Knutelski, S. & Saura, A. (2003). Evolution of clonality and polyploidy in a weevil system. *Molecular Biology and Evolution*, *20*(10): 1626–1632.

Stenberg, P., Terhivuo, J., Lokki, J. & Saura, A. (2000). Clone diversity in the polyploid weevil *Otiorhynchus scaber*. *Hereditas*, *132*(2): 137–142.

Stenberg, P. &, Saura A. (2009). Cytology of asexual animals. In: *Lost Sex* (Eds. Schön, I., Isa, Schön, Martens, K. & Dijk, P.). Dordrecht: Springer, pp. 63–74.

Stouthamer, R., Russell, J.E., Vavre, F. & Nunney, L. (2010). Intragenomic conflict in populations infected by parthenogenesis inducing *Wolbachia* ends with irreversible loss of sexual reproduction. *BMC Evolutionary Biology*, *10*(1): 229.

Sun, S. & Heitman, J. (2012). Should Y stay or should Y go: the evolution of non-recombining sex chromosomes. *BioEssays*, *34*(11): 938–942.

Suomalainen, E. (1950). Parthenogenesis in animals. *Advances in Genetics*, *3*: 193–253.

Suomalainen, E. (1962). Significance of parthenogenesis in the evolution of insects. *Annual Review of Entomology*, *7*(1): 349–366.

Suomalainen, E., Saura, A. & Lokki, J. (1976). Evolution of parthenogenetic insects. *Evolutionary Biology*, *9*: 209–257.

Suomalainen, E., Saura, A. & Lokki, J. (1987). *Cytology and Evolution in Parthenogenesis*. Boca Raton: CRC Press.

Tada, Y., Nishida, C. & Katakura, H. (2014). Effects of low temperature on ploidy levels in parthenogenetic and sexual forms of the weevil *Scepticus insularis* (Coleoptera: Curculionidae). *Zoological Science, 31*(2): 71–77.
Takenouchi, Y. (1980). Experimental study on the evolution of parthenogenetic weevils (Coleoptera: Curculionidae). *Journal of Hokkaido University of Education, Section II B, 31*(1): 1–12.
Takenouchi, Y. (1983). The occurrence of a decaploid embryo in the pentaploid parthenogenetic weevil race as a result of a low temperature treatment (Curculionidae: Coleoptera). *La Kromosomo*, 2(30–31): 935–936.
Takenouchi, Y. (1984). The chromosomes of 8 species of the Curculionidae (Coleoptera). *Journal of Hokkaido University of Education, Section II B, 34*: 1–6.
Takenouchi, Y. (1985). A study of the influence of low temperatures on the eggs of the pentaploid parthenogenetic *Catapionus gracilicornis* Roelofs (Curculionidae: Coleoptera). *Journal of Hokkaido University of Education, Section II B, 36*: 1–6.
Takenouchi, Y. (1986a). Origin of parthenogenetic weevils. *Iden, 40*(50–56): 84–89.
Takenouchi, Y. (1986b). A chromosome study of a parthenogenetic weevil *Scepticus insularis* Roelofs found in Namioka-machi, Aomori Prefecture, Japan (Curculionidae: Coleoptera). *Journal of Hokkaido University of Education, Section II B, 36*: 71–73.
Takenouchi, Y., Okamoto, H. & Sugawara, H. (1981). A study on the influence of low temperatures on the eggs of the tetraploid parthenogenetic *Catapionus gracilicornis* Roelofs (Curculionidae: Coleoptera). *Journal of Hokkaido University of Education, Section II B, 32*: 1–15.
Templeton, A.R. (1982). The prophecies of parthenogenesis. In: *Evolution and Genetics of Life Histories* (Eds. Dingle, H. & Hegmann, J.P.). New York: Springer-Verlag, pp. 75–101.
Templeton, A.R. (1989). The meaning of species and speciation: a genetic perspective. In: *The Units of Evolution: Essays on the Nature of Species* (Eds. Otte, D. & Endler, J.A.). Sunderland: Sinauer, pp. 159–183.
Tilquin, A. & Kokko, H. (2016). What does the geography of parthenogenesis teach us about sex?. *Philosophical Transactions of the Royal Society, Series B: Biological Sciences, 371*(1706): 0538.
Tinti, E. & Scali, V. (1995). Hemiclonal, all-paternal and mosaic offspring of the hybridogenetic stick insect *Bacillus rossius-grundii grundii*: allozymic and cytological evidence. *Journal of Experimental Zoology, 273*: 149–159.
Tinti, F. & Scali, V. (1996). Androgenetics and triploids from an interacting parthenogenetic hybrid and its ancestors in stick insects. *Evolution, 50*(3): 1251–1258.
Tomiuk, J., Loeschcke, V. & Schneider, M. (1994). On the origin of polyploid parthenogenetic races in the weevil *Polydrusus mollis* (Coleoptera: Curculionidae). *Journal of Theoretical Biology, 167*: 89–92.
Tsutsui, Y., Maeto, K., Hamaguchi, K., Isaki, Y., Takami, Y., Naito, T. & Miura, K. (2014). Apomictic parthenogenesis in a parasitoid wasp *Meteorus pulchricornis*, uncommon in the haplodiploid order Hymenoptera. *Bulletin of Entomological Research, 104*(3): 307–313.
Tucker, A.E., Ackerman, M.S., Eads, B.D., Xu, S. & Lynch, M. (2013). Population-genomic insights into the evolutionary origin and fate of obligately asexual *Daphnia pulex*. *Proceedings of the National Academy of Sciences USA, 110*(39): 15740–15745.
Turelli, M. (1994). Evolution of incompatibility-inducing microbes and their hosts. *Evolution, 48*(5): 1500–1513.
Tvedte, E.S., Logsdon Jr., J.M. & Forbes, A.A. (2019). Sex loss in insects: causes of asexuality and consequences for genomes. *Current Opinion in Insect Science, 31*: 77–83.
Van de Peer, Y., Mizrachi, E. & Marchal, K. (2017). The evolutionary significance of polyploidy. *Nature Reviews Genetics, 18*(7): 411–424.
Van Leeuwenhoek, A. (1695). *Arcana naturae detecta ab Antonio van Leeuwenhoek*. Delphis Batavorum, apud Henricum a Krooneveld.
Van Leeuwenhoek, A. (1700). V. Part of two letters from Mr Anthony Van Leewenhoek, FRS van concerning worms pretended to be taken from the teeth. *Philosophical Transactions of the Royal Society of London, 22*(265): 635–642.
Van Valen, L. (1973). A new evolutionary law. *Evolutionary Theory, 1*: 1–30.
Van Vugt, J.F.A., de Jong, H. & Stouthamer, R. (2009). The origin of a selfish B chromosome triggering paternal sex ratio in the parasitoid wasp *Trichogramma kaykai*. *Proceedings of the Royal Society of London, Series B: Biological Sciences, 276*(1676): 4149–4154.

Van Vugt, J.F.A., Salverda, M., de Jong, J.H. & Stouthamer, R. (2003). The paternal sex ratio chromosome in the parasitic wasp *Trichogramma kaykai* condenses the paternal chromosomes into a dense chromatin mass. *Genome*, *46*(4): 580–587.

Vandel, A. (1928). La parthénogénèse geographique. Contribution à l'étude biologique et cytologique de la parthénogénèse naturelle. *Bulletin Biologique de France et Belge*, *62*: 164–281.

Vandel A. 1931. *La parthenogénèse*. Paris: Doin.

Vavre, F., De Jong, J.H. & Stouthamer, R. (2004). Cytogenetic mechanism and genetic consequences of thelytoky in the wasp *Trichogramma cacoeciae*. *Heredity*, *93*(6): 592–596.

Vershinina, A.O. & Kuznetsova, V.G. (2016). Parthenogenesis in Hexapoda: Entognatha and non-holometabolous insects. *Journal of Zoological Systematics and Evolutionary Research*, *54*(4): 257–268.

Von Siebold, C.T. (1856). Die Drohneneier sind nicht befruchtet. *Bienenzeitung*, *12*: 181–184.

Vrijenhoek, R.C. (1989). Genetic and ecological constraints on the origins and establishment of unisexual vertebrates. In: *Evolution and Ecology of Unisexual Vertebrates, bull. 466* (Eds. Dawley, R.M. & Bogart, J.P.). Albany: New York State Museum, pp. 24–31.

Vrijenhoek, R.C. (1998). Animal clones and diversity. *Bioscience*, *48*(8): 617–628.

Went, D.F. (1971). In vitro culture of eggs and embryos of the viviparous paedogenetic gall midge *Heteropeza pygmaea*. *Journal of Experimental Zoology*, *177*: 301–312.

Went, D.F. (1982). Egg activation and parthenogenetic reproduction in insects. *Biological Reviews*, *57*(2): 319–344.

Went, D.F. & Krause, G. (1974). Alteration of egg architecture and egg activation in an endoparasitic Hymenopteran as a result of natural or imitated oviposition. *Wilhelm Roux'Archiv für Entwicklungsmechanik der Organismen*, *175*(3): 173–184.

Werren, J.H., Baldo, L. & Clark, M.E. (2008). *Wolbachia*: master manipulators of invertebrate biology. *Nature Reviews Microbiology*, *6*(10): 741.

Werren, J.H., Windsor, D. & Guo, L.R. (1995). Distribution of *Wolbachia* among neotropical arthropods. *Proceedings of the Royal Society of London, Series B: Biological Sciences*, *262*(1364): 197–204.

Wertheim, B., Beukeboom, L.W. & Van de Zande, L. (2013). Polyploidy in animals: effects of gene expression on sex determination, evolution and ecology. *Cytogenetic and Genome Research*, *140*(2-4): 256–269.

White, M.J.D., Cheney, J. & Ley, K.H.L. (1963). A parthenogenetic species of grasshopper with complex structural heterozygosity (Orthoptera: Acridoidea). *Australian Journal of Zoology*, *11*(1): 1–19.

Williams, G.C. (1975). *Sex and Evolution*. Princeton University Press.

Wood Jr., E.A. & Starks, K.J. (1975). Incidence of paedogenesis in the greenbug. *Environmental Entomology*, *4*(6): 1001–1002.

Wyatt, I.J. (1961). Pupal paedogenesis in the Cecidomyiidae (Diptera).-I. *Proceedings of the Royal Entomological Society of London, Series A: General Entomology*, *36*(10-12): 133–143.

Wyatt, I.J. (1963). Pupal paedogenesis in the Cecidomyiidae (Diptera).-II. *Proceedings of the Royal Entomological Society of London, Series A: General Entomology*, *38*: 136–144.

Yagound, B., Dogantzis, K.A., Zayed, A., Lim, J., Broekhuyse, P., Remnant, E.J., Beekman, M., Allsopp, M.H., Aamidor, S.E., Dim, O., Buchmann, G. & Buchmann, G. (2020). A Single Gene Causes Thelytokous Parthenogenesis, the Defining Feature of the Cape Honeybee *Apis mellifera capensis*. *Current Biology*, *30*(12), 2248–2259.

Yang, Z. & Nielsen, R. (2008). Mutation-selection models of codon substitution and their use to estimate selective strengths on codon usage. *Molecular Biology and Evolution*, *25*(3): 568–579.

Zchori-Fein, E., Faktor, O., Zeidan, M., Gottlieb, Y., Czosnek, H. & Rosen, D. (1995). Parthenogenesis-inducing microorganisms in *Aphytis* (Hymenoptera: Aphelinidae). *Insect Molecular Biology*, *4*(3): 173–178.

Zug, R. & Hammerstein, P. (2015). Bad guys turned nice? A critical assessment of *Wolbachia* mutualisms in arthropod hosts. *Biological Reviews*, *90*(1): 89–111.

Zug, R. & Hammerstein, P. (2018). Evolution of reproductive parasites with direct fitness benefits. *Heredity*, *120*(3): 266–281.

4

Mating Systems

Shriza Rai, Omkar, and Geetanjali Mishra
Ladybird Research Laboratory, Department of Zoology, University of Lucknow, Lucknow India

CONTENTS

4.1 Introduction

Animal mating systems and their dazzling diversity have been a much preferred and fascinating area of research. Charles Darwin in his book, the Descent of Man (1871) was the first to actively discuss variety of mating systems and their evolutionary aspects.

Other than understanding existence of a diversity of mating systems, it is their influence on sexual selection and role in maximizing individual reproductive success (Emlen & Oring, 1977), which makes investigating them desirable. Also, their dynamic nature and modulations in response to environment make investigations both fascinating and challenging.

DOI: 10.1201/9781003043195-4

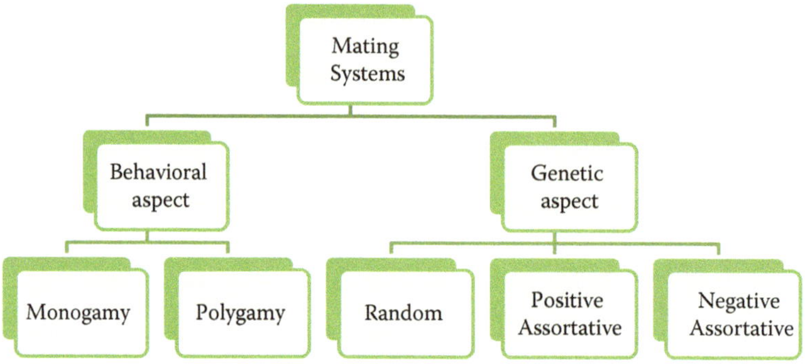

FIGURE 4.1 Classification of mating systems (Modified and adapted from Shuster & Wade, 2003).

Shuster & Wade (2003) suggested that mating systems can be described via behavioural ecology or genetic basis (Figure 4.1). Behavioural ecology defines mating system as number of mating partners, like monogyny (one female mating partner), monandry (one male mating partner), polygyny (multiple female mating partners), polyandry (multiple male mating partners), promiscuity (no restriction of mating partners for either sex), *etc*. The genetic categorization is established on genetic relationships between the two sexes; random mating, negative assortative mating and positive assortative mating. Assortative mating involves mating with individuals with more similar (positive) or dissimilar (negative) phenotype characteristics than would be expected by chance. Positive assortative mating may increase homozygosity in the population, but should not be confused with inbreeding; inbreeding affects all loci, while assortative mating affects only those that play a role in determining similar phenotype characteristics.

Evolutionary ecologists suggest that mating systems are both causes and consequences of sexually selected adaptations. The term "mating system" therefore, in addition to describing the number of individuals involved in, also incorporates how and by whom mating and fertilization are achieved through time in a population (Shuster & Wade, 2003; Klug, 2011; Kokko *et al*., 2014), leading to considerable diversity in mating patterns.

This chapter presents an overview of mating systems, how evolution of sex differences results in varied mating systems in order to gain understanding of persistent patterns across populations and species. We will also discuss various types of mating systems, their resulting reproductive success and the selection pressures that both sexes are subjected to.

4.2 Classification

Mating systems are a critical part of life history of insects and variation in mating success is a key feature. Several factors individually and collectively contribute in shaping each mating system (Karageorge & Wilson 2017; Figure 4.2). The current classification and terminologies with relation to mating systems were suggested prior to advent of paternity testing. and further information and advancement in tools may further evolve this classification.

The basic categorization uses number of mates that each sex has within a defined time period (a breeding season or a lifetime) (Emlen & Oring, 1977), viz. monogamy and polygamy (Davies, 1991). In monogamy, each individual has a single mate for a defined time period, while polygamy describes a system in which both sexes mate multiple times with multiple partners (n). Polygamy can be further subdivided into polygyny (1♀:1♂ but 1♂:n♀), polyandry(1♀:n♂ but 1♂:1♀), polygynandry (2–3♀:2–3♂) and promiscuity (n♀:n♂). (Shuster & Wade, 2003; Breed & Moore, 2016; Table 4.1A).

Some authors have expanded simple mating system classifications to account for how individuals acquire mates (Emlen & Oring, 1977; Shuster & Wade, 2003), pair bond duration between males and females (Davies, 1991; Shuster & Wade, 2003), details of parental care (Reynolds, 1996; Shuster & Wade, 2003), and other specific details of mating behaviour, such as coercion (Shuster & Wade, 2003).

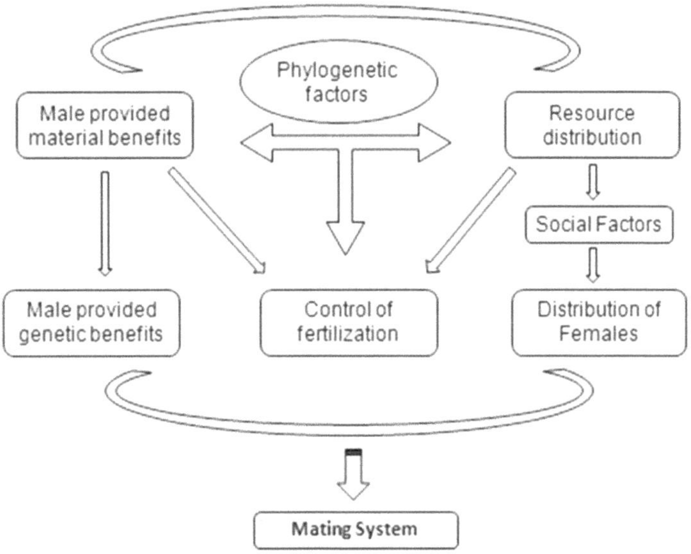

FIGURE 4.2 Factors governing mating system. Phylogenetic factors refer to the presence or absence of adaptations to control fertilization. Social factors include distribution of mates (particularly females) and distribution of resources etc. Adapted and modified from Brown *et al.* (1997).

4.3 Behavioural Ecology Approach to Mating Systems

4.3.1 Monogamy

In monogamy, males and females typically mate and reproduce with only one individual of the opposite sex, for one breeding season or longer (Kvarnemo, 2018). Both partners forgo any mating opportunities voluntarily or by force. Few terms related to monogamy are explained in Table 4.1B.

4.3.1.1 Enforced versus voluntary monogamy

Understanding why some insects mate only with one partner is intriguing, as choosing an inappropriate mate could have catastrophic fitness costs for them. Thus, monogamy is more apparent, or social, than real.

Monogamous animals use two methods to hedge their bets: serial monogamy and extra pair copulations (EPCs) (Breed & Moore, 2015). In *serial monogamy*, individuals form pair bonds that last for one mating season and choose a different mate in the subsequent season. Paternity testing shows thar monogamous pairings are often subject to infidelities, or EPCs and extra pair paternity (EPP) (Parker & Birkhead, 2013). In most monogamous populations at least a few offspring in each generation result from mating with partners other than a pair member. In horned passalus (*Odontotaenius disjunctus*), a socially monogamous beetle with biparental care, 54.8% of offspring were sired by extra-pair males and 70% of nests contained extra-pair young.

Given the prevalence of EPCs, a socially monogamous pairing might have evolved to attempt to enforce genetic monogamy. When paternity is uncertain, mate guarding (Figure 4.3), copulatory plugs, prolonged copulation, aggressive behavior, withholding parental care, infanticide and transfer of compounds (e.g. sex peptides) that control re-mating (Petrie & Kempenaers, 1998; Hosken *et al.,* 2009) could lead to male enforced monandry (Alonzo & Warner, 2000; Zamudio & Sinervo, 2000) by preventing/ reducing female remating (Baer *et al.,* 2001). Such, male-enforced monogyny is probably a means to ensure paternity.

Conversely, females can also force males into monogyny (Andersson, 1994; Petrie & Kempenaers, 1998; Hosken *et al.*, 2009). In burying beetles (*Nicrophorus defodiens*), male–female pairs bury a

TABLE 4.1

(A) Mating systems and their types. (B) Terms related to Monogamy.

A	
Monogamy	• Mating is restricted to one male and one female for one breeding season or longer. • It can be mutual with both partners forgoing mating opportunities by choice. • It can be enforced with one partner restricting the chances of other partner to minimum by keeping other potential mates at bay. This type of monogamy is known as Enforced Monogamy.
Polygamy	• A broader term. • One individual of either sex involved with more than one mate.
Polyandry	• One female having multiple males as mating partners in a breeding season. • Males do most of the parental care.
Polygyny	• One male having multiple females as mating partners in a breeding season. • Common in male dominant social structure. • Males can defend females directly as in Harem or Female defense polygyny. • Males can also defend the group of females by defending resources that females need as in Resource defense polygyny • Or males may attract females to a display site as in leks or may win the widely dispersed mates by roaming as in Scramble competition polygyny. • Females do most of the parental care.
Promiscuity	• Both sexes have multiple mating partners without showing any pair bonds.
Polygynandry	• Both sexes have multiple mating partners. • Form pair bonds.
B	
Enforced monogamy	• Male enforced: Male enforces female to not mate with any other male. • It can be via mate-guarding, repetitive copulations, mating plugs and transfer of compounds (e.g. sex peptides) etc. • It occurs to ensure secured paternity. • Female enforced: Female enforces male to not mate with any other male. • It can occur via sexual cannibalism (praying mantis), killing offspring. • Arises to secure resources or avoid sharing them.
Obligate monogamy	• Monogamy that does not vary in relation to density of mates or other resources.
Facultative monogamy	• Monogamy that varies, within species or even within individuals. • For example in relation to density of mates or some other resource, such as habitat suitable for breeding or feeding.
Social monogamy	• When both sexes voluntarily mate with single partners for a breeding season or for life. • Conceals a polygamous genetic mating system with mating infidelities by one or both sexes. • Both parents contribute in parental care.
Genetic monogamy	• Monogamy that occurs outside pair bond. • Result of extra pair copulations. • Can be analyzed via genetic analysis.

carcass for their young to feed on. Large sized carcasses may induce males to "call" a second female for mating; sharing of carcass will lead to fitness loss of offspring of first females To prevent this, males are bitten or pushed off the perch by female (Eggert & Sakaluk, 1995).

Mating with another partner may result in high variation in fecundity, parental care behaviour or territoriality, leading to a fitness cost to the individual. Thus, monopolizing one high-quality mate can be expected to yield higher fitness than mating promiscuously (Andersson, 1994). Consequently, when both sexes vary in quality, voluntary monogamy is expected to occur (Whiteman & Côté, 2004) and is assumed to benefit both sexes. In most eusocial insects, monogamous relationship persists until they die (Boomsma, 2009). Voluntary lifetime monogamy persists in termites (Nalepa & Jones, 1991; Boomsma *et al.*, 2005; Boomsma, 2009). It is, however, unclear what stops males and females from seeking extra-pair copulations.

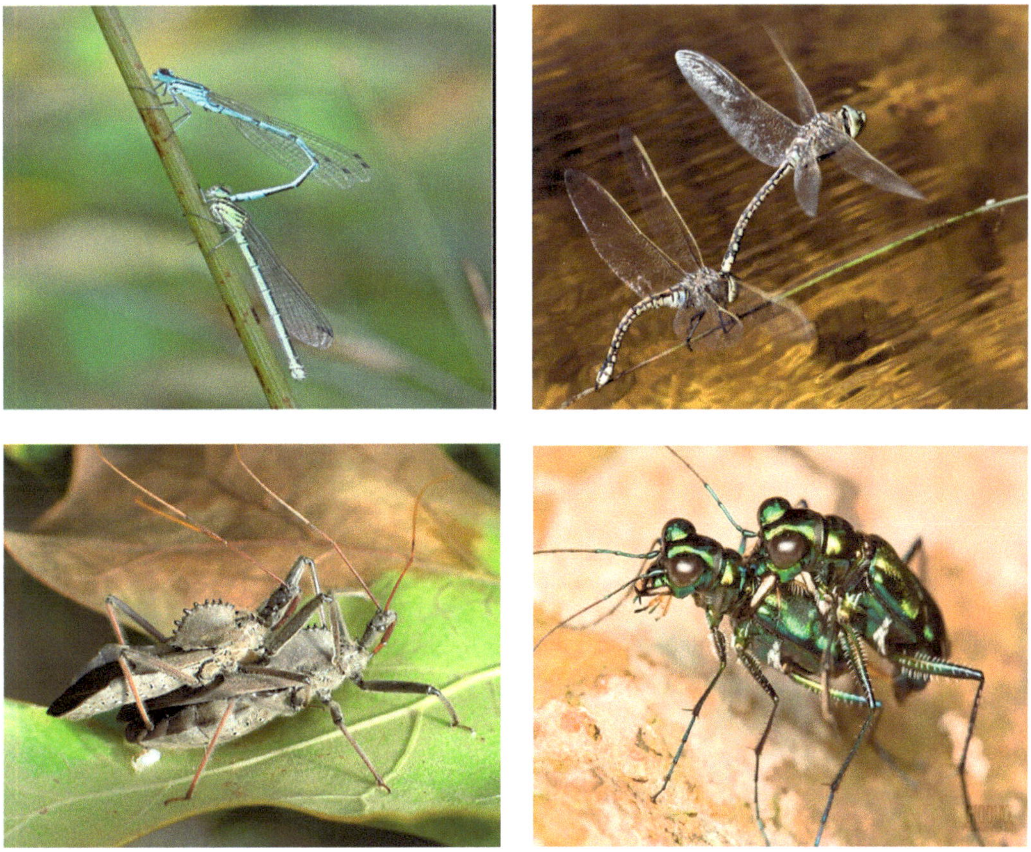

FIGURE 4.3 Mate guarding in insects. Upper left, dragon flies (Paul West); upper right, Dragon flies (Scott Fisher); lower left, Wheel bugs (Lynette Elliott); lower right, Emerald green tiger beetles (Rio Diaz https://askentomologists.com/2016/03/15/insect_coitus/).

4.3.1.2 Causes and consequences of monogamy

Insect mating systems are likely to evolve as a result of varying selection pressures. Theoretically, a monogamous pairing makes certain that both mates contribute to parental care and mutual defense. The evolution of monogamy may, therefore, depend upon high cost of acquiring mates, female's potential to restrict male behaviour, or intensive parental care for offspring survival.

4.3.1.2.1 Spatial constraints

If females are scattered over a large space and are breeding in synchrony, defending more than one female can be a challenge for the male, thereby promoting monogyny (Kvarnemo, 2018); spatially clumped females potentially allow polygyny (Shuster & Wade, 2003).

Habitat limitation and mate availability are two potential aspects of spatial constraints. If guarding sex is benefited more from guarding a single mate than from searching for additional mates, monogamy is expected. When females have restricted home ranges, male mate-guarding of multiple females becomes impractical. This may result in both social and reproductive monogamy. Therefore, mate-guarding could also be responsible for evolution of monogamy (Kvarnemo, 2018). Biased adult sex ratio or long search time to find a mate may lead to the evolution of mate-guarding (Harts & Kokko, 2013; Schacht & Bell, 2016) and monogyny could be consequence of a male-biased adult sex (Fromhage *et al.*, 2005). However, mate-guarding will not evolve towards monogyny unless guarding is efficient and there is last-male sperm precedence (Harts & Kokko, 2013).

4.3.1.2.2 Time constraints

Time constraints can have a considerable influence on mating patterns. High temporal dispersal of breeding by opposite sex promotes polyandry/polygyny (Emlen & Oring, 1977; Shuster & Wade, 2003); breeding synchrony, promotes monogamy by restricting mate availability.

Mating system can also genetically determined, as in green-veined white butterfly, *Pieris napi* (Wedell *et al.*, 2002). Monoandrous females of *P. napi* lay their egg in less than a week, whereas polyandrous females lay fewer eggs per day. Since polyandrous females can keep laying eggs overlong, they can reach a 50% higher life time reproductive success (Välimäki & Kaitala, 2006). However, polyandrous females can only realize their reproductive potential if the weather is favorable for oviposition for a longer duration. Time constraints can thus generate different selection pressures on mating pattern with polyandrous females suffering more from unpredictable weather than monoandrous females (Välimäki & Kaitala, 2006).

4.3.1.2.3 Parental care

Parental care requires allocation of time and energy to parental duties that may reduce energy and time available for self-maintenance or mating (Alonso-Alvarez & Velando 2012). Burying beetles males that have bred and cared for brood attract three times more females than males who had not raised a brood; this was mediated by the former producing a higher amount of the chemical signal (Figure 4.4) (Chemnitz *et al.,* 2017). Monogamy therefore, could also be a consequence of parental care (Emlen & Oring, 1977; Clutton-Brock, 1991).

Monogamy affects evolution of parental care, primarily for two reasons: (*i*) biparental care is best option to improve fitness, when both sexes are bound to mate monogamously because of limitations on offspring survival (Emlen & Oring, 1977), and/or (*ii*) monogamy ensures high paternity assurance (Trivers, 1972; Fromhage & Jennions, 2016).

4.3.1.2.4 Genetic compatibility and mate choice

In insects, differences in compatibility between maternal and paternal genomes contributes to offspring fitness (Tregenza & Wedell, 2000; Puurtinen *et al.,* 2009). Such, consideration during mate choice is rare (Tregenza & Wedell, 2000; Puurtinen *et al.*, 2005), though it helps avoid poor matches. Females seem to determine male genetic quality (‘*good genes*’) pre- or post-mating (Colegrave *et al.,* 2002), resulting in choosy females mating monandrously with high-quality males., leading to higher genetic quality of offspring (Colegrave *et al.,* 2002).

The costs of fecundity of multiple matings (Jennions, 1997) also promotes selection of monoandry (Colegrave *et al.,* 2002) over other mating systems. Mating pattern also affects both relatedness and genetic diversity within broods of young. Increased genetic diversity among half-siblings, compared to full-siblings, also increases offspring survival in some cases (Aguirre & Marshall, 2012; Slatyer *et al.,*2012). High relatedness and thus low genetic diversity among offspring stemming from monogamous mating can also be advantageous. The evolution of eusociality, altruism and cooperation due to kin selection, infanticide, effective population size, mate choice before mating, sexual selection and sexual conflict could be consequences of monogamy (Hamilton, 1964; Peck & Feldman, 1988; Kvarnemo, 2018).

Mating with multiple partners may lead to a compromised immune function (Jennions & Petrie, 2000; Altizer *et al.,* 2003; McLeod & Day, 2014), increased sexual conflict and decreased chances of sexually transmitted diseases are possible constraints of polygamy that favour monogamy. Mate guarding also poses a great challenge for monogamous males and is resource consuming (Ranta & Kaitala, 1999; Clutton-Brock & Isvaran, 2006). Monogamy can be maintained by both sexual and natural selection if polygynous males produce fewer offspring than do monogamous males.

The factors acting against monogamy and hence benefiting multiple matings are also many, such as access to more resources, receiving multiple nuptial gifts (Arnqvist & Nilsson, 2000), increased protection against infanticide through paternity dilution (Klemme & Ylönen, 2010; Lukas & Huchard, 2014). These have been discussed in detail further under multiple matings.

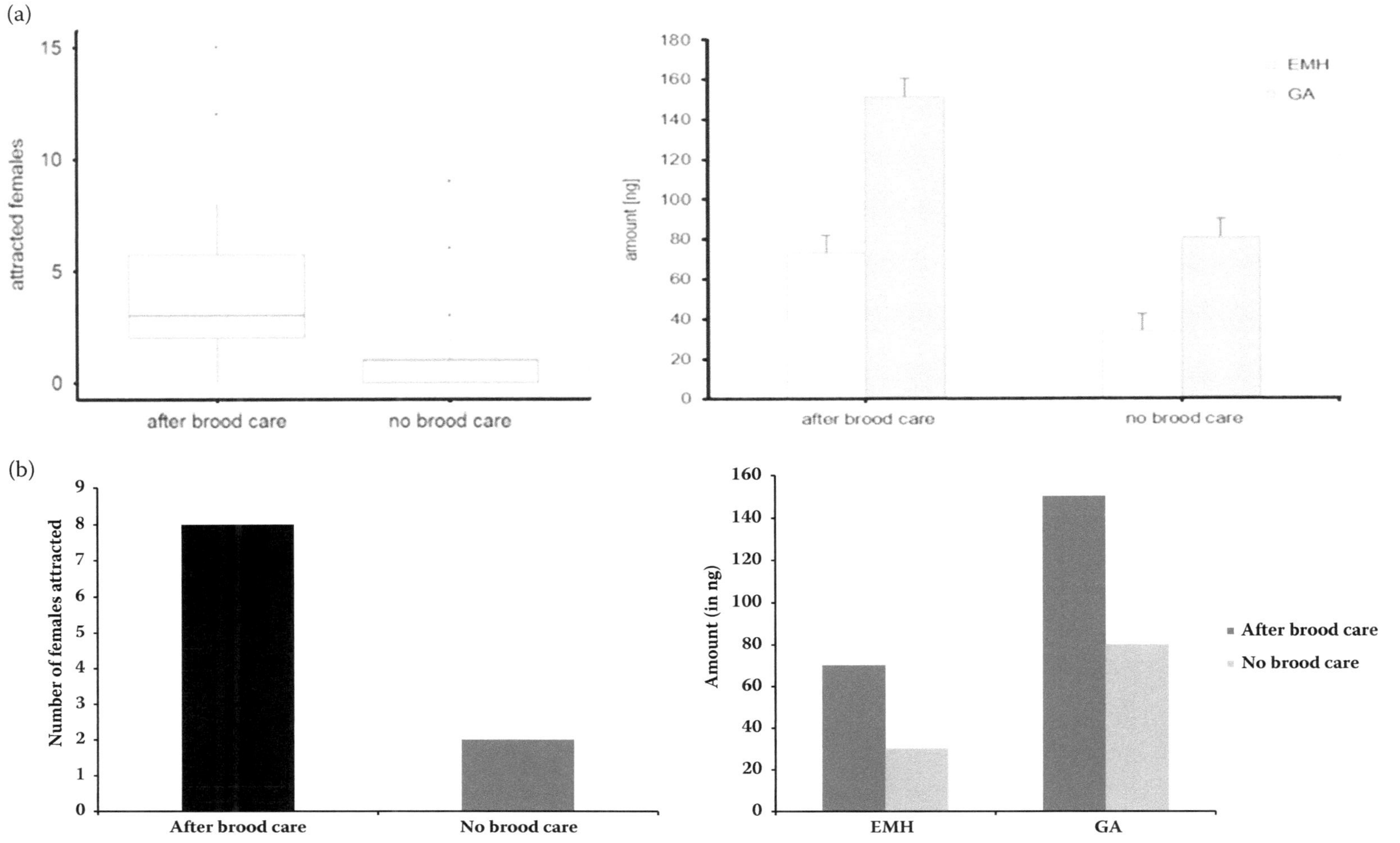

FIGURE 4.4 (A) Females are attracted to males that had provided brood in *Nicrophorus vespilloides* (B) The males that had cared for brood produce more sex pheromone.White bars represent the amount of ethyl 4-methyl heptanoate(EMH), gray bars the amount of (E)- Gerany lacetone (GA) (Adapted from Chemnitz *et al.*, 2017).

4.3.2 Polygamy

This is a mating system where males and females typically mate and reproduce with more than one partner (Table-2).

4.3.3 Polygyny

Polygyny, is association of one male with multiple females. Polygyny is observed in habitats with variation in quality and distribution of resources and is more adaptive when resources (e.g. water, food, female) are fragmentary or clumped. Intrasexual conflict arises among sexually active males to attain mates (Breed & Moore, 2016).

4.3.3.1 Causes and consequences of polygyny

The *polygyny threshold model* suggesting evolution of polygyny from monogamy is valid in a population if fitness of polygynous females is greater than that of monogamous females under poor breeding situations (Orians, 1969).

Success in polygynous mating strategies is related to availability, distribution, and receptivity of females in a given area along with competition among other males for mates, and female or resource defense tactics. To mate with multiple females, males are required to actively search for females. This involves energy and increases the risk of predation, along with reduction in foraging time (Alavi *et al.,* 2016).

4.3.3.2 Types of polygyny and their causes and consequences

4.3.3.2.1 Female defense polygyny: Harem

A harem is a mating system where a single, dominant male defends and mates with a group of females, that may have grouped together for defense or been herded together by a male. Males compete for control of harem. Though harem polygyny is best known amongst mammals, a few insects (Choe, 1994; Kelly, 2006bGriffin *et al.,* 2019) show the same.

There are four aspects of harem-forming species: exclusive maternal care, monandry, male-biased sexual dimorphism and temporal continuity of harem composition (Rowell 1987). However, these characteristics may differ across taxa. For instance, Wellington tree wētā, *Hemideina crassidens* does not show parental care (Kelly, 2006a, 2006b, 2008a). On the contrary, in zorapteran, *Zorotypus gurneyi*, sexual dimorphism (Choe, 1994), and parental care (Griffin *et al.,* 2019) are absent. In *Ips* bark beetles, maternal care is exhibited but no sexual dimorphism. However, females are monandrous and harems do stay together for a prolonged period. Thus, insect species described as harem polygynous do not display all four of these aspects and represent temporary marriages of convenience, maximising their reproductive potential by adjusting their mating behaviour to fit the current situation (Griffin *et al.,* 2019).

4.3.3.2.2 Resource defense polygyny

Here, males control a resource or territory essential for successful female reproduction; within which male will mate with multiple partners during a breeding season, in exchange of access to these resources (e.g. rhinoceros beetle, *Trypoxylus dichotomus;* McCullough & Emlen, 2013; Johns *et al.,* 2014; McCullough, 2014). As this leads to high reproductive skew, sexual selection comes into play and drives evolution of sexual dimorphism and intrasexual conflicts (Emlen, 2014). Widespread resources often results in a rapid breakdown of this mating system.

4.3.3.2.3 Lek polygyny

When females cannot be herded or guarded in one place, males have to find them, or attract them from far away. A *lek* is an aggregation of males, each seeking to attract a mate, and a zeal to do so (Breed & Moore, 2016). Lek polygyny exhibits four features:

1. Males provide no parental care and supply only gametes,
2. Males are spatially aggregated at mating arenas (or leks),
3. Males do not control access to key resources to females,
4. Females are free to select mates at the arena (Shelly, 2018).

In leks, males may aggregate around an attractive male (hotshots), in a "stage in a landscape easily accessible by females. There is a high reproductive skew, with few males achieving high mating success, while most other males have low or no mating success (Herberstein *et al.*, 2017). In comparison to isolated males, lek attracts more females and thereby males sacrifice their needs for food, water, and shelter by being near their competitors (Aspbury & Gibson, 2004). Lek mating systems have been found in several insect species, *viz.* moths (Alem *et al.*, 2011), fruit flies (Shelly 2018), paper wasps (De Souza *et al.*, 2017), orchid bees (Kimsey, 1980), butterflies (Wickman & Jansson, 1997) and tarantula hawk wasps (Alcock, 1983).

A number of studies have focused on dipteran lekking species. Fruit flies of families Tephritidae and Drosophilidae display lek behavior that closely resembles 'classic' lek polygyny. Lek forming flies may aggregate on a substrate linked to a resource (fruit flies, or males may swarm(Diptera: Culicidae) (Shelly & Whittier, 1997).

Males of oriental fruit fly, *Bactrocera dorsalis* form leks on leaves of host trees with each male defending a leaf. In addition, wing-fanning produces a distinctive sound and disperses a sex pheromone; latter providing location of food or ovipositional sites. Attracted by the pheromones, females attend lek and choose mate from among males present (Shelly, 2000).

Among swarming species, males aggregate in non-resource defined areas such as hilltops (Alcock, 1987; Sivinski & Petersson, 1997) with females attend swarms to mate. Swarms are most likely aggregations of competing males, not linked to any resource,. Females do not actively choose in swarms;, males may be selected for their placement in the swarm. In dipteran love bugs, *Pleciane arctica* Hardy, larger males that are positioned low in the swarm have optimal female encounter rate (Thornhill & Alcock, 1983), while in buzzer midge, *Chironomus plumosus*, small males move with great agility and capture females (Crompton *et al.*, 2003).

4.3.3.2.4 Scramble competition polygyny

When females are dispersed widely and neither female-attracting resources nor females themselves are economically defendable (Emlen & Oring, 1977), or when potential mates are abundant for a discrete and limited reproductive time period (Thornhill & Alcock, 1983), scramble competition polygyny is predicted to occur.

Thornhill & Alcock (1983) specifically described and separated two key forms of scramble competition polygyny:

1. Explosive mating assemblages where "receptive females are abundant during a very brief mating period".
2. Prolonged searching polygyny where "receptive females cannot be economically monopolised because of their even distribution or the high rate of competition from other males".

This polygyny involves competition between males for mating opportunities based on differences in their ability to move about and locate females. Males attempt to secure matings with maximum females in a given area at a given time through competitive searching. Therefore, scramble competition represents extremes of a continuum rather than strict category (Herberstein *et al.*, 2017).

4.3.4 Polyandry

Polyandry is the reverse of polygyny: one female, many males (Thornhill & Alcock 1983). This however raises the question that despite "an undiscriminating eagerness in males and a discriminating passivity in females", why do females mate multiply and males do not? (Andersson 1994; Arnqvist & Rowe 2005).

In polyandrous females, multiple mating often increases reproductive success of females (Taylor *et al.,* 2008). Studies have shown that females prefer males that produce ejaculates with more sperm (Harris & Moore, 2005; Scarponi *et al.,* 2015). However, this decreases expected fitness gain per mating for males (Parker & Birkhead, 2013), influencing male ejaculate size and male mating rate (Parker & Pizzari, 2010; Abe & Kamimura, 2015; Bocedi & Reid, 2016); thereby restricting male mating frequency. This restriction on number of successful copulations that a male can perform can lead to evolution of prudent male mate choice and strategic sperm allocation (Wedell *et al.,* 2002). Thus, females tend to avoid mating with low-quality males and prefer males that have had fewer consorts (Harris & Moore, 2005).

4.3.4.1 Causes and consequences of polyandry

For evolutionary ecologists, understanding the causes and consequences of polyandry remains a central aim (Pizzari & Wedell, 2013; Shuster *et al.,* 2013), as it can determine evolution of traits underlying sexual selection, sexual conflict, population viability and disease dynamics.

Polyandry promotes competition between sperms of different males in female reproductive tract leading to adaptations to enhance paternity share (Thornhill & Alcock, 1983). Polyandry is widely hypothesized to have evolved as an adaptation against inbreeding (Michalczyk *et al.,* 2011; Reid *et al.,* 2015; Duthie *et al.,* 2016). In insects that causes galls during larval development (Price, 2005; Shorthouse *et al.,* 2005), or are parasitoids (Werren & Simbolotti, 1989; Martel *et al.,* 2010), females may initially mate within gall or host and later post-dispersal, may mate again in wider and diverse population (West & Herre, 1998; Debout *et al.,* 2002).

Inbreeding depression may have led to various adaptations to avoid inbreeding (Geffen *et al.,* 2011; Szulkin *et al.,* 2013; Tennenhouse, 2014). However, under very constrained conditions, polyandry may evolve to adjust inbreeding. This depends upon the strength of inbreeding depression (β) and is unaffected by costs of polyandry (Duthie & Reid, 2015). Inbreeding avoidance readily evolves given sufficiently strong inbreeding depression ($\beta \geq 1$), while inbreeding preference is more likely to evolve if inbreeding depression is weak ($\beta \leq 0.2$) (Figure 4.5). Consequently, evolution of polyandry to avoid inbreeding is compromised. Inbreeding may benefit inclusive fitness of the parents by strengthening representation of parental genes in future generations (Kokko & Ots, 2006). Females might even mate multiply to increase the degree to which their offspring are inbred (Lehtonen & Kokko, 2015). A female, when socially paired with an unrelated male, can increase her inclusive fitness by mating with a related extra-pair male (Lehtonen & Kokko, 2015).

Polyandry effectively maximizes opportunities to increase total number of female reproductive attempts. Resources and energy availability are key determinants of female reproductive success at each life history stage (Emlen & Oring, 1977; Hosken & Stockley 2003). Consequently, the factors that

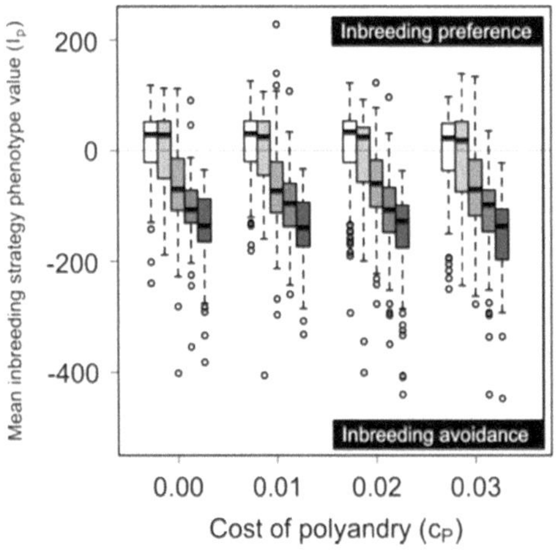

FIGURE 4.5 Inbreeding avoidance readily evolves given sufficiently strong inbreeding depression ($\beta \geq 1$), while inbreeding preference is more likely to evolve if inbreeding depression is weak ($\beta \leq 0.2$). (Source: (Duthie *et al.,* 2016). https://onlinelibrary.wiley.com/doi/full/10.1111/evo.13005.

restrict or enhance rates of energy gain, also determine female reproductive success. Harassment by courting males drains energy, if females continue rejecting male advances. Therefore, if accepting additional copulations than resisting male advances conserve energy and time, cost minimization may drive polyandry (Parker, 1979). Nutritional benefits in form of nuptial gifts increases female longevity and by extension, lifetime reproductive successes; this has likely facilitated evolution of polyandry (Fox, 1993; Brown 1997; Karlsson, 1998).

Polyandry also reduces predation risk. In moth, *Utetheisa ornatrix*, males transfer defensive alkaloids to females during copula, which provide protection against predators to the female and eggs (Gonzalez *et al.*, 1999). Therefore, preference of female towards males with high alkaloid content drives polyandry.

Polyandry increases number of sperm stored by females and mating with more than one male is probably needed to maintain fertility (Boomsma *et al.*, 1999). In Panamanian leaf-cutter ant, number of sperms stored by the queen, increase by approximately 30 million with each subsequent mating. Mating with multiple partners increases female reproductive success as there is a strong positive correlation between number of fathers and overall clutch size (McCullough *et al.*, 2017).

Whenever genetic compatibility is determined post-mating, mating with multiple mates is recommended (Simmons, 2005; Price *et al.*, 2008; Michalczyk *et al.*, 2011). Polyandry also has positive effects on offspring viability through bet-hedging (Garcia-Gonzalez *et al.*, 2015), via greater fertilization rates when gametes from different males are mixed (Henshaw & Holman, 2015).

The evolutionary circumstances favoring polyandry are still controversial, owing to considerable costs despite the benefits. The costs include, time/energy expenditure, increased predation risks, pathogen transmission or harmful ejaculate accessory substances (Okada *et al.*, 2017).

There are considerable variations among individuals within populations, among populations, and across taxa that depends upon degree to which females mate with multiple males within one reproductive bout, and hence the extent of polyandry, (Pannell & Labouche, 2013; Parker & Birkhead, 2013; Taylor *et al.*, 2014). In cigarette beetle, *Lasioderma serricorne*, multiple matings reduce female adult lifespan with no difference in lifetime fecundity and offspring viability in comparison to once mated females (Okada *et al.*, 2017). The direct cost of courting by second more attractive mate was greater. Thus, polyandry carries fitness costs but not benefits to females in *L. serricorne* (Okada *et al.*, 2017). To understand polyandry is theoretically challenging as increased mating does not necessarily increase female reproductive success (Bateman, 1948). Therefore, the evolution of polyandry is probably dependent on the outweighing of benefits to females.

4.3.5 Polygynandry

Polygynandry occurs when two or more females form pair bonds with two or more males in a breeding season. Females mate with several males and the males care for the broods of several females (Breed & Moore, 2015). Polygynandry is a form of polygamy that encompasses both polygyny and polyandry within the same species (Szala & Shackelford, 2019).

4.3.5.1 Causes and consequences of polygynandry

Polygynandry may provide benefits for both females and males with benefits differing between sexes. Females can enhance genetic diversity and improve genetic quality of offspring via sperm competition and also cryptic female choice by mating with multiple males in a brief time period. By increasing the number of female partners, male fitness is directly increased as males can fertilize the egg(s) of multiple females and, therefore, maximize the number of offspring they produce. In polygynandry, there is paternal uncertainty, which decreases infanticide or willingness to harm an offspring and thus, multiple mating partners contribute materially to the well-being and safety of offspring.

Polygynandry encompasses the benefits of both polyandry and polygyny. Multiple mates reduce the risk of absolute reproductive failure as in case of exclusive mating with a sterile mate or a mate of poor genetic quality (Bellamy & Pomiankowski, 2011; Davies *et al.*, 2012). The possibility of 'trading-up' when experiencing direct benefits from multiple higher quality mates and producing offspring that are more viable, sexually attractive or genetically more variable (Slatyer *et al.*, 2012; Culina *et al.*, 2015). According to Fisher's (1930) "sexy son" hypothesis, multiple mating by females increases probability of producing a greater quantity of sons that inherit their father's sexual attractiveness. Furthermore, if genetic compatibility between mates is accessed after mating, mating with multiple mates is expected to evolve, as long as reproduction can be biased towards more compatible mates, hence, increasing the probability of genetically compatible offspring (Tregenza & Wedell, 2000).

During mating, non-self agents (e.g., sperm and other ejaculate components) are transferred to female reproductive tract, where self and non-self agents interact. This may result in immune system compromise along with the transfer of various sexually transmitted infections or parasites (Oku *et al.*, 2019). Male genitalia may also sever female genitalia and reproductive tract (Kamimura, 2007; Lemaître *et al.*, 2020). Thus, multiple mating or polygynandry poses challenges for the female's immune system.

The costs associated with multiple mating in males are less frequently studied. In black scavenger flies, development costs related to the elaborate sexual dimorphisms has been reported (Ang *et al.*, 2008; Rohner *et al.*, 2018) Males engage in energy- and time-consuming behavior to obtain copulations (Scharf *et al.*, 2013; Harwood *et al.*, 2015). In addition, they have to bear viability costs mediated by reproductive investment as well. The reproductive investment depends on both male and female mating status (Boorman & Parker 1976; Reinhardt 2007) as well as on age and fecundity (Lüpold *et al.*, 2011) as sperm viability decreases with male age (Dowling & Simmons 2012).

4.3.6 Promiscuity

Promiscuity occurs when males and females in a population seem to mate randomly without pair bonds (Breed & Moore, 2016). Therefore, promiscuity does not involve mate choice. However, the opportunity to choose mate is advantageous to one or both of the sexes and offers large fitness benefits by choosing the highest quality mates. Hence, promiscuity should be rare or absent in animals.

Promiscuity is likely to have evolved in species that live in completely unpredictable environment; where it is impossible to assess which mate might carry genes that would be good in the next generation. Conditions where choice of mates become unlikely or impossible, like the inability to hold a territory and monopolize access to critical resources; promiscuity is the only option to increase reproductive success (Breed & Moore, 2016).

If males in a population, sire offspring with multiple females and each of the males are not equally successful in fathering young, then the mating system is truly promiscuous. However, mere observation of seemingly random mating does not establish that a mating system is, in fact, genetically promiscuous. For that one needs to go into the genetics of the system.

4.4 Genetic Basis of Mating Systems

Choosing an optimal partner is a task. The quest is to find the best mate(s) among competing mates. However, mating is often costly' therefore, males/females exhibit preferences or differential investment in mates (Bonduriansky, 2001; Scharf *et al.*, 2013). This may pave the way for mutual choice, particularly when both parents invest in caring for the offspring (Trivers, 1972). Mutual mate choice often results in preferences established on genetic relationship among the mating adults. Based on this, there are two types of mating systems: Random mating and assortative (positive and negative) mating.

In positive assortative mating pattern, individuals with similar phenotypes (body size, skin coloration or pigmentation, *etc.*), mate with one another more frequently than would be expected under a random mating pattern. Male insects undergoing multiple mating often showsize-assortative mating. Body size is directly correlated to reproductive success; larger or heavier females may contain more mature eggs (Bonduriansky, 2001; Katvala & Kaitala, 2001). Larger males are more successful in courting or defending larger females (Clutton-Brock, 2007; Jiang *et al.*, 2013).

Size-assortative mating can be achieved by several behavioral mechanisms, viz., intrasexual competition in males for larger females, greater readiness/receptivity of females to mate with more competitive males, physical barriers that interfere with mating when size differences are extreme, or a spatial/temporal correlation between female and male sizes (Harari *et al.*, 1999; Jiang *et al.*, 2013). Since, positive correlation between a coupling pair in any genetically determined phenotypic trait may result in homozygosity, assortative mating may increase genetic relatedness within the family (Jiang *et al.*, 2013).

On the contrary, in negative assortative mating (or heterogamy), individuals with dissimilar genotypes or phenotypes mate with one another more frequently. It reduces the genetic similarities within the family. More instances of positive assortative mating are reported than negative assortative mating (Jiang *et al.*, 2013). Both result in a deviation from the Hardy–Weinberg principle (which states that genotype frequencies in a population will remain constant from generation to generation in the absence of other evolutionary influences, such as "mate choice" in this case). Either strategy may be employed by the individuals to maximize fitness and maximally pass on their genes to the next generation.

4.5 Mating System Theories

4.5.1 Bateman's principle

Bateman's principle postulates the variance in mating success in both the sexes (Figure 4.6). According to this principle, intrasexual selection is more intense in males and males' mate promiscously. Conversely, females are sexually passive and highly selective, as there is a limit to the number of eggs a female can produce. The reproductive success of females increases less prominently or in cases even flattens after a single mating as one mating is adequate to fertilize all the eggs produced by the females. However, reproductive success of males increases with successive mating to fertilize millions of sperms. However, it does not explain the factors determining distribution of mates acquired by each sex.

Based on the idea of Bateman, the evolution of mating systems, sexual dimorphism, and male–female differences in sexual behavior have been explained remarkably. For instance, males are subjected to intrasexual selection to ensure high mating frequency, leading to elaborate behaviour and evolution of

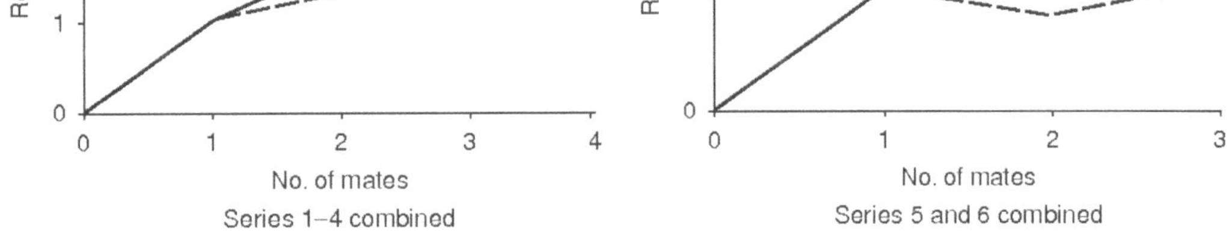

FIGURE 4.6 The relationship between number of mates and relative fertility in *Drosophila melanogaster*, where relative fertility is the number of offspring. The solid line represents male and dashed represents female. These two graphs are the original Bateman's graph produced in his two series of experiments, where graph 1a represents data from Series 1 through 4; Graph 1b represents data from Series 5 and 6. Reprinted from Bateman (1948). http://www2.nau.edu/~shuster/shustercourses/BIO%20698/Literature/Bateman1948.pdf.

structures for attracting mates in males (Emlen, 2008). Thereby, making males more 'showy' and investing in ornaments to showcase their fitness and attract mates.

4.5.2 Parental Investment Theory

The expenditure of time, energy or other resources increasing the fitness of offspring while reducing the ability of a parent to invest in other components of fitness is referred as Parental Investment Theory (Trivers, 1972; Klug 2016). In cypress bark beetle, *Phloeosinus armatus,* biparental investment leads to increased number of broods (Baruch *et al.,* 2017) (Figure 4.7). Though parental care offers clear benefits for offspring, it bears substantial costs for parents (Alonzo, 2012; Scharf *et al.*, 2013). For instance, in earwigs and tree hoppers, parental care increases survival rate of offspring while reducing the future reproduction in parents (Zink, 2003; Kölliker, 2007).

Since there are differences between males and females, these differences can affect parental investment, sexual selection and ultimately mating system (Kokko & Jennions, 2008; Alonzo, 2010; Klug, 2016). The variations in parental investment modulate the probability of mate availability of potential mates. For say, the sex involved in post zygotic care/parental care will be unavailable as potential mates till they care for their offspring. Therefore, sex with greater parental investment is a limiting resource for the counter sex. There is also stronger sexual selection in mate limited sex.

Trivers (1972) pointed that females invest more in parental care per breeding event than males because it will be costly to abandon the brood (Herridge *et al.,* 2016). Therefore, sexual selection is stronger on males. There are evidences of male parental care as well. However, the benefit to the fitness of offspring should be greater than the cost invested by giving up mating opportunities (Clutton-Brock, 1991). Male care is often in species, like water bug, *Abedus herberti*, where high paternity confidence is assured (Smith, 1979). Females oviposit on the back of males and males' brood the eggs (Herridge *et al.,* 2016). The relationship between parental care, sexual selection, and mating dynamics is complicated.

When intrasexual mate competition results in high mortality, the sex that provides less parental care and competes more for mates, will become relatively rare in the population due to greater mortality. In such scenario, whichever sex is rare, will be expected to favour earlier desertion and counter sex investing in parental care (Kokko & Jennions, 2008). If providing parental care is more dangerous, individuals who invest less will experience reduced mortality, and will become more common in the population. In this scenario, individuals of this sex will have difficulty finding mates. Therefore, increased paternal investment by the deserting sex will be expected and the selection would be expected to favor increased parental investment by both sexes (Kokko & Jennions, 2008). Similarly, if the choosier sex prefers mates that provide parental care, the more competitive and mate-limited sex would be expected to also provide parental care in some cases (Alonzo, 2012).

The presence of social interactions and coevolutionary dynamics within and between the sexes are interlinked with the mating behaviour and parental investment (Alonzo, 2010). Hence, individual reproductive traits and the social interactions indelibly shape the mating systems.

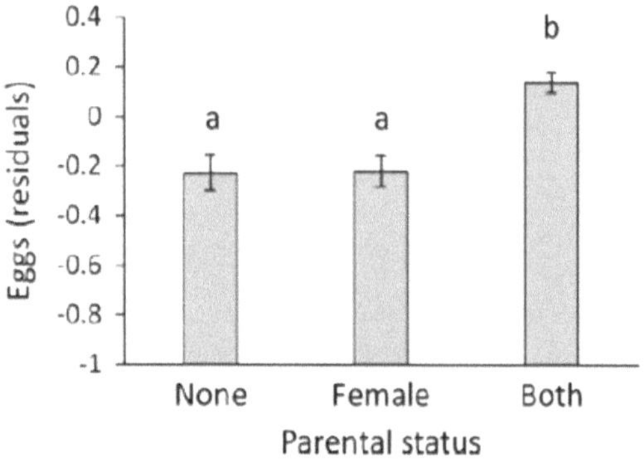

FIGURE 4.7 In *Phloeosinus armatus,* biparental investment leads to increased number of broods. The effect of parental status (both, only female, or no adults were present when opening the gallery) on the number of eggs detected (Means +1SE) (Baruch *et al.,* 2017). https://www.researchgate.net/profile/Ally_Harari2/publication/315813534_Mating_system_mate_choice_and_parental_care_in_a_bark_beetle/links/5e9d670492851c2f52b290bb/Mating-system-mate-choice-and-parental-care-in-a-bark-beetle.pdf.

4.5.3 Operative Sex Ratio (OSR)

OSR is the ratio of males to females prepared to mate at a given time and in a given location (Kvarnemo & Ahnesjo, 1996). It is anticipated that the individuals of mate-limiting sex hold the key to monopolizing the more common sex when OSR is biased. A male-biased OSR will lead to polygyny, and a female-biased OSR will lead to polyandry. When mate monopolization is high, traits that allow a small number of individuals to monopolize mates are strongly selected for. However, this view was challenged and criticized by many (Klug *et al.,* 2010; Jennions *et al.,* 2012).

There is no strong evidentiary support that strength of sexual selection and polyandry or polygyny will increase as OSR becomes more biased toward one sex. Therefore, OSR is largely not considered as a reliable predictor of mate monopolization, mating system, or sexual selection across many biologically-realistic scenarios (Klug *et al.,* 2010; Jennions *et al.,* 2012). Conversely, mate monopolization can be a definitive proportional predictor of the strength of sexual selection, if one or more phenotypic traits are under sexual selection (Emlen & Oring, 1977; Klug *et al.,* 2010).

Mating behavior, parental care, and sexual selection are closely associated and have interactive effects on mating systems (Klug, 2016). There is mate competition and/or attraction, when one sex tends to be the choosier and the other sex is mate limited (Darwin, 1871; Andersson, 1994). As aforementioned, the mate limited sex will experience strong sexual selection if a small number of individuals possess traits, allowing them to monopolize many opposite-sex mates (Klug *et al.,* 2010). Therefore, sexual selection strongly influences mating dynamics and the resulting mating system. For instance, fighting ability, increased body size, traits preferred by the opposite sex ("good genes"), and traits that increase fertilization success (reviewed in Andersson, 1994) are favored by sexual selection as they increase mating success and mate monopolization.

4.5.4 Environmental Potential for Polygyny (EPP)

Temporal availability of mates and spatial distribution of breeding resources determine the degree to which mates can be monopolized and thus define the environmental potential for polygamy (EPP) (Emlen & Oring, 1977).

Asynchrony in spatiotemporal dispersion of resources or receptive females result in great biases in OSR and thereby, increase the opportunity for sexual selection (Clutton-Brock & Parker 1992; Wade 1995; Reynolds 1996). When females are uniformly distributed in space, the opportunity for the monopolization of mates is low, resulting in low EPP. Whereas when females are aggregated, EPP is greater. Thus, EPP seems to offer greater flexibility in mating systems even within a species and thus the strength of sexual selection and degree of polygyny displayed.

4.6 Evolution of mating systems

Different mating systems are natural outcomes of different ecological contexts. The key population-specific differences in ecologies of females and males ultimately determine the mating system (Herberstein *et al.,* 2017). Thus, across a wide range of female ecologies, male fitness is optimized by investing in sequential mate searching and acquisition. Since selection on individual level depends on the behavioural adaptations, by understanding selection on individuals, we can know the evolution of a certain mating system in a given population. For example, with increase in polyandry, postcopulatory sexual selection strengthens to ensure maximum paternity. Spatial and temporal distribution of mates and resources also assists in shaping mate systems. Clumped female distribution favours female defense polygyny whereas clumped resources favours resource defense polygyny.

Mating systems are influenced by populations exposed to different environmental or demographic conditions (Shuster & Wade, 2003). Availability and distribution of critical resources and the spatial and temporal distribution of sexually receptive mates (Emlen & Oring, 1977; Shuster & Wade, 2003) are a few such influential factors. For instance, with increasing variance in habitat quality, there exists a greater potential for resource monopolization and thus a greater environmental potential for polygamy and sexual selection is expected to amplify (Emlen & Oring, 1977; Shuster & Wade, 2003). Subsequently, the ability of individuals to take advantage of the environmental potential for polygamy depends upon changes in the energetic cost of resource or mate monopolization resulting from changes in population density and the length of breeding season are expected to influence.

Courtship or coercion may subsequently influence mating success. However, neither of them is apparent in some systems, where females appear indifferent to the efforts males are putting. This differs greatly from systems where male effort is focused on defending a single female (monogamy), multiple females (female defense polygyny) or resources of limiting value to females (resource defense polygyny).

With reduction in the number of mates, probability of coercive traits increases that restricts a mate from acquiring a new partner. To decrease the propensity to remate, various strategies have evolved with one being the release of seminal chemicals by males into the female reproductive tract. Similarly, another such adaptation gave way to female enforced monogamy. These adaptable strategies are indeed shaping the mating systems, along with several other ecological and environmental factors.

4.6.1 Evolution of sex

Sex differences in selection for mate acquisition and rejection are decisive in predicting mating systems and are fundamental in understanding mating patterns, as it determines mate encounter rate (Kokko *et al.*, 2014). Once a potential mate encounter has happened, there are several critical factors to determine the success of mating. Mate choice brings about deviations from a null model in which a mate encounter always leads to a mating.

A key to explain the evolution of mating systems is explaining the evolutionary maintenance of sex and sexual reproduction. One of the potential benefits of reproducing sexually is that it allows more rapid genetic variations that are advantageous to changing environment. These variations often increase the efficiency with which selection can fix beneficial mutations.

For many evolutionary biologists, the Red Queen's hypothesis is relevant to the evolution of sex. It explains that the value of sexual reproduction fluctuates over time. In the presence of low pressure of disease and parasite, the cost of meiosis and the advantages of proven genetic combinations favor asexual reproduction. Conversely, with surge in diseases and parasites, sexual reproduction is preferred, resulting in the maintenance of sex. The genetic recombination gained from sexual reproduction not only produces new defenses against diseases and parasites but also aids in developing new mating strategies and shaping mating systems.

4.7 Mating systems are "not" what they seem

The standard paradigm till late last century was "*mating systems are what they seem*" (Avise, 2014). In the 1960s, protein-electrophoretic methods evolutionary biologists to assess genetic parentage (maternity and paternity) and thereby characterize a population's actual genetic mating system.

Shift to a new paradigm was gradual and is still ongoing. With the degree to which considerable complex mating systems vary (both within and among populations), it is exceedingly difficult to precisely identify mating patterns without genetic data (Klug, 2011). With the development of genetic markers, the study of mating systems and social structure have revealed a higher degree of promiscuity, which was not possible by behavioral observations alone (DeWoody & Avise, 2001; Gibbs & Weatherhead, 2001; Griffith *et al.*, 2002). For say, a mating system may behaviourally appear socially monogamous, but genetic analysis may reveal that the male and/or the female had rather several mating partners. There are many such other genetic evidences of the social (observed) mating system not complying with the genetic mating system (Turjeman *et al.*, 2016; Janes *et al.*, 2016; Miño *et al.*, 2017,). Therefore, for the characterization of the actual mating system of a species, genetic studies are essential to be carried out as a complement to field observations on social reproductive behavior.

4.8 Conclusions

Mating systems are shaped by mate choice and the cost of mating differs between sexes. The sex with higher cost of mating is reluctant to accept mating and is often choosy, while the other sex accepts all mating. The number of mates and mating frequency are the outcomes of interacting male and female mating strategies. In order to increase our understanding of how mating systems evolve requires characterizing optimal mating strategies in both sexes and how they co-evolve between the sexes, Sexual conflict arises in the race to win mates leading to various adaptations and counter adaptations in both the sexes and shape the evolution of mating systems.

Mating systems can be classified behaviourally and genetically. In terms of behavioral ecology, mating systems are broadly of two types: Monogamy (monandry, monogyny) and Polygamy (Polyandry, Polygyny, Polygynandry, and Promiscuity). Each class has its respective causes and consequences and has evolved to increase the reproductive success of the organism. Based on genetic correlation across mating pairs, mating systems can be random or assortative. These two approaches (behavioral and genetic) when combined capture the covariances between male and female promiscuity, between male guarding and tendency of females to aggregate, or between male sperm plugs and female detergents. These two classification systems can collectively explain the phenotypic variance in sexual dimorphism within and among species.

Experimental evolution along with new molecular technology could be a great tool for investigating the evolutionary consequences of variation in mate numbers and paternity share. Previously studied various monogamous systems have revealed to be polygamous on genetic level. Therefore, it can be inferred that polygamy is the rule while monogamy may be the exception. With recent advances in evolutionary thinking, and the integrated use of currently available molecular technologies, like microsatellite molecular markers, mating systems could be studied more intensely and genetic characterization of mating systems could be conducted.

REFERENCES

Abe J., & Kamimura Y. (2015). Sperm economy between female mating frequency and male ejaculate allocation. *The American Naturalist*, 185, 406–416. 10.1086/679586

Aguirre, J.D., and Marshall, D.J. (2012). Does genetic diversity reduce sibling competition?. *Evolution: International Journal of Organic Evolution*, *66*(1), 94–102.

Alavi Y., Elgar M.A., Jones T.M. (2016) Male mating success and the effect of mating history on ejaculate traits in a facultativelyparthenogenic insect (Extatosomatiaratum). *Ethology*, 122, 523–530. 10.1111/eth.12497.

Alcock, J. (1983). Consistency in the relative attractiveness of a set of landmark territorial sites to two generations of male tarantula hawk wasps (Hymenoptera: Pompilidae). *Animal Behaviour*, 31, 74–80. (1983).

Alcock, J. (1987). Leks and hilltopping in insects. *Journal of Natural History*, 21(2), 319e328.

Alem, S., Koselj, K., Siemers, B.M. and Greenfield, M.D. (2011). Bat predation and the evolution of leks in acoustic moths. *Behavioral Ecology and Sociobiology*, 65, 2105–2116.

Alonso-Alvarez, C., and Velando, A. (2012). Benefits and costs of parental care. In N. J. Royle, P. T. Smiseth & M. Kolliker (Eds.). *The Evolution of Parental Care*. (pp. 40–61). Oxford: Oxford University Press.

Alonzo, S.H. (2010). Social and coevolutionary feedbacks between mating and parental investment. *Trends in Ecology and Evolution*, *25*(2), 99–108.

Alonzo, S.H. (2012). Sexualselectionfavoursmaleparentalcare,whenfemalescan choose. *Proceedings of the Royal Society B: Biological Sciences*, 279, 1784–1790.

Alonzo, S.H., & Warner, R.R. (2000). Allocation to mate guarding or increased sperm production in a Mediterranean wrasse. *The American Naturalist*, 56(3), 266–275.

Altizer, S., Nunn, C.L., Thrall, P.H., Gittleman, J.L., Antonovics, J., Cunningham, A.A., and Poss, M. (2003). Social organization and parasite risk in mammals: integrating theory and empirical studies. *Annual Review of Ecology, Evolution, and Systematics*, *34*(1), 517–547.

Andersson, M. (1994). *Sexual Selection*. Princeton: Princeton University Press.
Ang, Y., Puniamoorthy, N., & Meier, R. (2008). Secondarily reduced foreleg armature in *Perochaeta dikowi* sp. n.(Diptera: Cyclorrhapha: Sepsidae) due to a novel mounting technique. *Systematic Entomology*, *33*(3), 552–559.
Arnqvist G., & Rowe L. (2005). *Sexual Conflict*. Princeton: Princeton University Press.
Arnqvist, G., & Nilsson, T. (2000). The evolution of polyandry: multiple mating and female fitness in insects. *Animal Behaviour*, *60*(2), 145–164.
Aspbury, A.S., & Gibson, R.M. (2004). Long-range visibility of greater sage grouse leks: a GIS-based analysis. *Animal Behaviour*, 67, 1127–1132.
Avise, J.C. (2014). *Conceptual breakthroughs in evolutionary genetics: a brief history of shifting paradigms*. Academic Press.
Baer, B., Morgan, E.D., & Schmid-Hempel, P. (2001). A nonspecific fatty acid within the bumblebee mating plug prevents females from remating. *Proceedings of the National Academy of Sciences*, *98*(7), 3926–3928.
Baruch, O., Mendel, Z., Scharf, I., & Harari, A.R. (2017). Mating system, mate choice and parental care in a bark beetle. *Bulletin of Entomological Research*, *107*(5), 611–619.
Bateman, A.J. (1948). Intra-sexual selection in *Drosophila. Heredity* 2, 349–368.
Bellamy, L., & Pomiankowski, A. (2011). Why promiscuity pays. *Nature*, *479*(7372), 184–185.
Berger-Tal R., & Lubin Y. (2011). High male mate search costs and a femalebiased sex ratio shape the male mating strategy in a desert spider. *Animal Behaviour*, 82, 853–859.
Bocedi G., & Reid J.M. (2016). Coevolutionary feedbacks between female mating interval and male allocation to competing sperm traits can drive evolution of costly polyandry. *The American Naturalist*, 187, 334–350. 10.1086/684746.
Bonduriansky, R. (2001). The evolution of male mate choice in insects: a synthesis of ideas and evidence. *Biological Reviews*, *76*(3), 305–339.
Boomsma, J.J. (2009). Lifetime monogamy and the evolution of eusociality. *Philosophical Transactions of the Royal Society B: Biological Sciences*, *364*(1533), 3191–3207.
Boomsma, J.J., Baer, B. & Heinze, J. (2005). The evolution of male traits in social insects. *Ann. Rev. Entomol.* 50, 395–420. 10.1146/annurev.ento.50.071803.130416.
Boomsma, J.J., Fjerdingstad, E.J., & Frydenberg, J. (1999). Multiple paternity, relatedness and genetic diversity in Acromyrmex leaf–cutter ants. *Proceedings of the Royal Society of London. Series B: Biological Sciences*, *266*(1416), 249–254.
Boorman, E., & G.A. Parker. (1976). Sperm (ejaculate) competition in *Drosophila melanogaster*, and reproductive value of females to males in relation to female age and mating Status. *Ecological Entomology*, 1, 145–155.
Breed, M. D., & Moore, J. (2015). Polygynandry in Nature.
Breed, M.D., & Moore, J. (2016). *Animal behavior*. Academic Press.
Breed M.D., & Moore J., (2016). Chapter 11 - Mating Systems, Editor(s): Michael D. Breed, Janice Moore, *Animal Behavior* (Second Edition), Academic Press, Pages 357–393.
Brown, J.L. (1997) A theory of mate choice based on heterozygosity. *Behavioral Ecology*, 8, 60–65.
Brown, W.D., Crespi, B. J., & Choe, J. C. (1997). Insects and Arachnids. In J. C. Choe & B. J. Crespi (Eds.). (pp. 352–637). Cambridge: Cambridge University Press.
Chemnitz, J., Bagrii, N., Ayasse, M., & Steiger, S. (2017). Staying with the young enhances the fathers' attractiveness in burying beetles. *Evolution*, *71*(4), 985–994. 10.1111/evo.13194
Choe J.C. (1994) Sexual selection and mating system in *Zorotypus gurneyi* Choe (Insecta: Zoraptera): I. Dominance hierarchy and mating success. *Behavioral Ecology and Sociobiology*, 34, 87–93.
Clutton-Brock, T.H. (1991). *The evolution of parental care* (Vol. 64). Princeton University Press.
Clutton-Brock, T.H., & Isvaran, K. (2006). Paternity loss in contrasting mammalian societies. *Biology Letters*, *2*(4), 513–516.
Clutton-Brock, T.H. & Parker, G.A. (1992). Potential reproductive rates and the operation of sexual selection. *Quarterly Review of Biology*, 67, 437–456.
Clutton-Brock, T., (2007). Sexual selection in males and females. *Science(NewYork, NY)*, 318(5858), 1882–1885.
Colegrave, N., Kotiaho, J.S. & Tomkins, J.L. (2002). Mate choice or polyandry: reconciling genetic compatibility and good genes sexual selection. *Evolutionary Ecology Research*, 4, 911–917.

Crompton, B., Thomason, J.C., & McLachlan, A. (2003). Mating in a viscous universe: The race is to the agile, not to the swift. *Proceedings of the Royal Society of London B: Biological Sciences*, 270, 1991e1995.
Culina, A., Radersma, R., & Sheldon, B.C. (2015). Trading up: the fitness consequences of divorce in monogamous birds. *Biological Reviews*, *90*(4), 1015–1034.
Darwin C. (1871). *The Descent of Man, and Selection in Relation to Sex*. London: Murray.
Davies, N.B., Krebs, J.R., & West, S.A. (2012). *An introduction to behavioural ecology*. John Wiley and Sons.
Davies, N.B. (1991). Mating systems. In: Krebs, J.R., Davies, N.B. (Eds.), *Behavioural Ecology: An Evolutionary Approach*, 3rd ed. (pp. 263–294). Oxford: Blackwell.
De Souza, A.R., Lino-Neto, J., Tibbetts, E.A., Turillazzi, S., & Beani, L. (2017). The leks of *Polistes dominula* paper wasps: tiny abdominal spots play a critical role in male attacks toward potential rivals. *Ethology Ecology & Evolution*, 29, 410–419.
Debout, G., Fauvergue X., & Fleury F. (2002). The effect of foundress number on sex ratio under partial local mate competition. *Ecological Entomology*, 27, 242–246.
DeWoody, J.A., & Avise, J.C. (2001). Genetic perspectives on the natural history of fish mating systems. *Journal of Heredity*, *92*(2), 167–172.
Dewsbury, D.A. (1982). Ejaculate cost and male choice. *The American Naturalist*, 119, 601–610. 10.1086/283938
Dowling, D.K., & Simmons, L.W. (2012). Ejaculate economics: testing the effects of male sexual history on the trade-off between sperm and immune function in Australian crickets. *PLoS One*, 7, e30172.
Duthie, A.B., & Reid, J.M. (2015). What happens after inbreeding avoidance? Inbreeding by rejected relatives and the inclusive fitness benefit of inbreeding avoidance. *PLoS One*, 10, e0125140.
Duthie, A.B., Bocedi, G., & Reid, J.M. (2016). When does female multiple mating evolve to adjust inbreeding? Effects of inbreeding depression, direct costs, mating constraints, and polyandry as a threshold trait. *Evolution*, *70*(9), 1927–1943.
Eggert, A.K., & Sakaluk, S.K. (1995). Female-coerced monogamy in burying beetles. *Behavioral Ecology and Sociobiology*, *37*(3), 147–153.
Emlen, D.J. (2008). The evolution of animal weapons. *Annual Review of Ecology, Evolution, and Systematics*, *39*, 387–413.
Emlen, D.J. (2014). Reproductive contests and the evolution of extreme weaponry. In D. Shuker, and L. Simmons (Eds.), *The evolution of insect mating systems* (pp. 92e105). Oxford: Oxford University Press.
Emlen, S., & Oring, L. (1977). Ecology, sexual selection, and the evolution of mating systems. *Science*, 197(4300), 215–223.
Fisher, R. (1930). *The Genetical Theory of Natural Selection*. New York, NY: Dover, 1957.
Fox, C.W. (1993). Multiple mating, lifetime fecundity and female mortality of the bruchid beetle, *Callosobruchus maculatus* (Coleoptera: Bruchidae). *Functional Ecology*, *7*, 203–208.
Fromhage, L., & Jennions, M.D. (2016). Coevolution of parental investment and sexually selected traits drives sex-role divergence. *Nature Communications*, *7*(1), 1–11.
Fromhage, L., Elgar, M.A., & Schneider, J.M. (2005). Faithful without care: the evolution of monogyny. *Evolution*, *59*(7), 1400–1405.
Garcia-Gonzalez, F., Yasui, Y., & Evans, J.P. (2015). Mating portfolios: bet-hedging, sexual selection and female multiple mating. *Proceedings of the Royal Society B: Biological Sciences*, *282*(1798), 20141525.
Geffen, E., Kam M., Hefner R., Hersteinsson P., Angerbjörn A., Dalèn L., Fuglei E., Norèn K., Adams J.R., Vucetich J., Meier T.J., Mech L.D., Vonholdt B.M., Stahler D.R., & Wayne R.K. (2011). Kin encounter rate and inbreeding avoidance in canids. *Molecular Ecology*, 20, 5348–5358.
Gibbs, H.L., & Weatherhead, P.J. (2001). Insights into population ecology and sexual selection in snakes through the application of DNA-based genetic markers. *Journal of Heredity*, *92*(2), 173–179.
Gonzalez, A., Rossini, C., Eisner, M., & Eisner, T. (1999). Sexually transmitted chemical defense in a moth (Utetheisaornatrix). *Proceedings of the National Academy of Sciences*, *96*(10), 5570–5574.
Griffin, M.J., Holwell, G.I., & Symonds, M.R. (2019). Insect harem polygyny: when is a harem not a harem?. *Behavioral Ecology and Sociobiology*, *73*(4), 40.
Griffith, S.C., Owens, I.P., & Thuman, K.A. (2002). Extra pair paternity in birds: a review of interspecific variation and adaptive function. *Molecular Ecology*, *11*(11), 2195–2212.

Hamilton, W.D. (1964). The genetical evolution of social behaviour. II. *Journal of Theoretical Biology*, *7*(1), 17–52.
Harari, A.R., Handler, A.M., & Landolt, P.J. (1999). Size-assortative mating, male choice and female choice in the curculionid beetle *Diaprepes abbreviatus*. *Animal Behaviour*, 58, 1191–1200.
Harris W.E., & Moore P.J. (2005). Female mate preference and sexual conflict: females prefer males that have had fewer consorts. *The American Naturalist*, 165, S64–S71.
Harts, A.M., & Kokko, H. (2013). Understanding promiscuity: when is seeking additional mates better than guarding an already found one?. *Evolution*, *67*(10), 2838–2848.
Harwood, J.F., K. Chen, P. Liedo, H.G. Muller, J.L. Wang, A.E. Morice, & J.R. Carey. (2015). Female access and diet affect insemination success, senescence, and the cost of reproduction in male mexican fruit flies *Anastrepha ludens*. *Physiological Entomology*, 40, 65–71.
Henshaw, J.M., & Holman, L. (2015). Bet-hedging via polyandry: a comment on 'Mating portfolios: bet-hedging, sexual selection and female multiple mating'. *Proceedings of the Royal Society B: Biological Sciences*, *282*(1809), 20150346.
Herberstein, M.E., Painting, C.J., & Holwell, G.I. (2017). Scramble competition polygyny in terrestrial arthropods. In *Advances in the Study of Behavior* (pp. 237–295). Elsevier.
Herridge, E.J., Murray, R.L., Gwynne, D.T., & Bussiere, L. (2016). Diversity in mating and parental sex roles. In R. Kliman (Ed.). *Encyclopedia of Evolutionary Biology*. (pp. 453–458). Oxford: Elsevier.
Hosken, D.J., & Stockley, P. (2003). *Benefits of Polyandry: A Life History Perspective. Evolutionary Biology, 173–194.*
Hosken, D.J., Stockley, P., Tregenza, T., & Wedell, N. (2009). Monogamy and the battle of the sexes. *Annual Review of Entomology*, *54*, 361–378.
Janes, J.K., Roe, A.D., Rice, A.V., Gorrell, J.C., Coltman, D.W., *et al*., (2016). Polygamy and an absence of fine-scale structure in *Dendroctonus ponderosae* (Hopk.) (Coleoptera: Curcilionidae) confirmed using molecular markers. *Heredity*, 116, 68–74.
Jennions, M.D. (1997). Female promiscuity and genetic incompatibility. *Trends in Ecology and Evolution*, *12*(7), 251–253.
Jennions, M.D., & Petrie, M. (2000). Why do females mate multiply? A review of the genetic benefits. *Biological Reviews*, *75*(1), 21–64.
Jennions, M.D., Kokko, H., & Klug, H. (2012). The opportunity to be misled in studies of sexual selection. *Journal of Evolutionary Biology*, 25(3), 591–598.
Jiang, Y., Bolnick, D.I., & Kirkpatrick, M. (2013). Assortative mating in animals. *The American Naturalist*, *181*(6), E125–E138.
Johns, A., Gotoh, H., McCullough, E.L., Emlen, D.J., & Lavine, L.C. (2014). Heightened condition-dependent growth of sexually selected weapons in the rhinoceros beetle, *Trypoxylus dichotomus* (Coleoptera: Scarabaeidae). *Integrative and Comparative Biology*, 54, 614e621.
Kamimura, Y. (2007). Twin intromittent organs of *Drosophila* for traumatic insemination. *Biology Letters*, *3*(4), 401–404.
Karageorge, K.W., & Wilson, R.R. (2017). An integrative mating system assessment of a nonmodel, economically important Pacific rockfish (*Sebastes melanops*) reveals nonterritorial polygamy and conservation implications for a large species flock. *Ecology and Evolution*, *7*(24), 11277–11291.
Karlsson, B. (1998). Nuptial gifts, resource budgets, and reproductive output in a polyandrous butterfly. *Ecology*, *79*(8), 2931–2940.
Katvala, M., & Kaitala, A. (2001). Male choice for current female fecundity in a polyandrous egg-carrying bug. *Animal Behaviour*, *62*(1), 133–137.
Kelly, C.D. (2006a). Movement patterns and gallery use by the sexually dimorphic Wellington tree weta. *New Zealand Journal of Ecology*, *30*, 273–278.
Kelly, C.D. (2006b). Resource quality or harem size: what influences male tenure at refuge sites in tree weta (Orthoptera: Anostostomatidae)?. *Behavioral Ecology and Sociobiology*, *60*(2), 175–183.
Kelly, C.D. (2008). Why do male tree weta aggressively evict females from galleries after mating?. *Ethology*, *114*(2), 203–208.
Kimsey, L.S. The behaviour of male orchid bees (Apidae, Hymenoptera, Insecta) and the question of leks (1980). *Animal Behaviour*, 28, 996–1004.
Klemme, I., & Ylönen, H. (2010). Polyandry enhances offspring survival in an infanticidal species. *Biology Letters*, *6*(1), 24–26.

Klug, H. (2011). *Animal mating systems.Encyclopedia of Life Sciences*. Chichester: John Wiley and Sons. 10.1002/9780470015902.a0022553.

Klug, H. (2016). *Mating Systems, A Brief History of. Encyclopedia of Evolutionary Biology*, 459–464.

Klug, H., Heuschele, J., Jennions, M.D., & Kokko, H. (2010). The mismeasurement of sexual selection. *Journal of Evolutionary Biology*, 23(3), 447–462.

Kokko, H., & Jennions, M.D. (2008). Parental investment, sexual selection and sex ratios. *Journal of Evolutionary Biology, 21*(4), 919–948.

Kokko, H., & Ots I. (2006). When not to avoid inbreeding. *Evolution*, 60, 467–475.

Kokko, H., Klug, H., & Jennions, M.D. (2014). Mating systems. *The evolution of insect mating systems*, 42–58.

Kölliker, M. (2007). Benefits and costs of earwig (*Forficula auricularia*) family life. *Behavioral Ecology and Sociobiology, 61*(9), 1489–1497.

Kvarnemo, C. (2018). Why do some animals mate with one partner rather than many? A review of causes and consequences of monogamy. *Biological Reviews, 93*(4), 1795–1812.

Kvarnemo, C., & Ahnesjo, I. (1996). The dynamics of operational sex ratios and competition for mates. *Trends in Ecology & Evolution, 11*(10), 404–408.

Lehtonen, J., & Kokko H. (2015). Why inclusive fitness can make it adaptive to produce less fit extra-pair offspring. *Proceedings of the Royal Society B: Biological Sciences*, 282, 20142716.

Lemaître, J.F., Gaillard, J.M., & Ramm, S.A. (2020). The hidden ageing costs of sperm competition. *Ecology Letters, 23*(11), 1573–1588.

Lukas, D., & Huchard, E. (2014). The evolution of infanticide by males in mammalian societies. *Science, 346*(6211), 841–844.

Lüpold, S., Manier, M.K., Ala-Honkola, O., Belote, J.M., & Pitnick, S. (2011). Male *Drosophila melanogaster* adjust ejaculate size based on female mating status, fecundity, and age. *Behavioral Ecology*, 22, 184–191.

Martel, V., Doyon J., & Boivin G. (2010). Partial local mate competition in the wasp *Trichogramma euproctidis*: the role of emergence sex ratio on female mating behaviour. *Ecological Entomology*, 35, 698–703.

McCullough, E.L. (2014). Mechanical limits to maximum weapon size in a giant rhinoceros beetle. *Proceedings of the Royal Society of London B: Biological Sciences*, 281(1786). 20140565.

McCullough, E.L., & Emlen, D.J. (2013). Evaluating the costs of a sexually selected weapon: Big horns at a small price. *Animal Behaviour*, 86, 977e985.

McCullough, E.L., Buzatto, B.A., & Simmons, L.W. (2017). Benefits of polyandry: Molecular evidence from field-caught dung beetles. *Molecular Ecology, 26*(13), 3546–3555.

McLeod, D.V., & Day, T. (2014). Sexually transmitted infection and the evolution of serial monogamy. *Proceedings of the Royal Society B: Biological Sciences, 281*(1796), 20141726.

Michalczyk, Ł., Millard A.L., Martin O.Y., Lumley A.J., Emerson B.C., Chapman T., & Gage M.J.G. (2011). Inbreeding promotes female promiscuity. *Science*, 333, 1739–1742.

Miño, C.I., de Souza, E.D., Moralez-Silva, E., Valdes, T.A., CortiçoCorrêa Rodrigues, V.L., & Del Lama, S.N. (2017). Use of noninvasive 'bug-eggs' to enable comparative inferences on genetic mating system with and without parental information: A study in a cattle egret colony. *PLoS one, 12*(8), e0183153.

Nalepa, C.A., & Jones, S.C. (1991). Evolution of monogamy in termites. *Biological Reviews, 66*(1), 83–97.

Okada, K., Suzaki, Y., Sasaki, R., and Katsuki, M. (2017). Fitness costs of polyandry to female cigarette beetle Lasiodermaserricorne. *Behavioral Ecology and Sociobiology, 71*(5), 86.

Oku, K., Price, T.A., & Wedell, N. (2019). Does mating negatively affect female immune defences in insects? *Animal Biology, 69*(1), 117–136.

Orians, G.H. (1969). On the evolution of mating systems in birds and mammals. *The American Naturalist, 103*(934), 589–603.

Pannell, J.R., & Labouche, A.M. (2013). The incidence and selection of multiple mating in plants. *Philosophical Transactions of the Royal Society B: Biological Sciences, 368*(1613), 20120051.

Parker G.A., & Birkhead (2013). Polyandry: the history of a revolution. *Philosophical Transactions of the Royal Society B: Biological Sciences, 368*(1613), 20120335.

Parker G.A., & Pizzari T. (2010). Sperm competition and ejaculate economics. *Biological Reviews*, 85, 897–934.

Parker, G.A. (1979). Sexual selection and sexual conflict Pp. 123–166in Blum M.S. and Blum N.A., eds. *Sexual selection and reproductive competition in insects*. New York: Academic Press, Inc.

Parker, G.A., & Birkhead, T.R., (2013). Polyandry: the history of a revolution. *Philosophical Transactions of the Royal Society B: Biological Sciences*, 368, 20120335.

Peck, J.R., & Feldman, M.W. (1988). Kin selection and the evolution of monogamy. *Science, 240*(4859), 1672–1674.

Petrie, M., & Kempenaers, B. (1998). Extra-pair paternity in birds: explaining variation between species and populations. *Trends in Ecology and Evolution, 13*(2), 52–58.

Pizzari, T., & Wedell, N. (2013). The polyandry revolution. *Philosophical Transactions of the Royal Society B: Biological Sciences*, 368, 20120041.

Price, P.W. (2005). Adaptive radiation of gall-inducing insects. *Basic and Applied Ecology*, 6:413–421.

Price, T.A.R., Hodgson, D.J., Lewis, Z., Hurst, G.D.D., & Wedell, N. (2008). Selfish genetic elements promote polyandry in a fly. *Science, 322*(5905), 1241–1243.

Puurtinen, M., Ketola, T., & Kotiaho, J.S. (2005). Genetic compatibility and sexual selection. *Trends in Ecology and Evolution, 20*(4), 157–158.

Puurtinen, M., Ketola, T., & Kotiaho, J.S. (2009). The good-genes and compatible-genes benefits of mate choice. *The American Naturalist, 174*(5), 741–752.

Ranta, E., & Kaitala, V. (1999). Punishment of polygyny. *Proceedings of the Royal Society of London. Series B: Biological Sciences, 266*(1435), 2337–2341.

Reid, J.M., Arcese P., Keller L.F., Germain R.R., Duthie A.B., Losdat S., Wolak M.E., & Nietlisbach P. (2015). Quantifying inbreeding avoidance through extra-pair reproduction. *Evolution* 69, 59–74.

Reinhardt, K. (2007). Ejaculate size varies with remating interval in the grasshopper *Chorthippus parallelus erythropus* (Caelifera:Acrididae). *European Journal of Entomology*, 104, 725–729.

Reynolds, J.D. (1996). Animal breeding systems. *Trends in Ecology & Evolution*, 11, 68–72.

Rohner, P.T., Yoong, K.S., Tuan, M.J., & Meier, R. (2018). Evolution of male costs of copulation in sepsid flies (Diptera: Sepsidae). *BioRxiv*, 265439.

Rowell, T.E. (1987). On the significance of the concept of the harem when applied to animals. *Social Science Information, 26*(3), 649–669.

Scarponi V., Chowdhury D., & Godin J.-G.J. (2015). Male mating history influences female mate choice in the Trinidadian guppy (*Poeciliareticulata*). *Ethology*, 121, 1091–1103.

Schacht, R., & Bell, A.V. (2016). The evolution of monogamy in response to partner scarcity. *Scientific Reports, 6*, 32472.

Scharf, I., F. Peter, & O.Y. Martin. (2013). Reproductive trade-offs and direct costs for males in arthropods. *Evolutionary Biology*, 40, 169–184.

Shelly, & Whittier. (1997). Lek behavior of insects. In J. Choe, and B. Crespi (Eds.), *The evolution of mating systems in insects and arachnids* (pp. 273e293). Cambridge: Cambridge University Press.

Shelly, T.E. (2000). Flower-feeding affects mating performance in male oriental fruit flies *Bactrocera dorsalis*. *Ecological Entomology*, 25(1), 109e114.

Shelly, T.E. (2018). Sexual selection on leks: a fruit fly primer. *Journal of Insect Science, 18*(3), 9.

Shorthouse, J.D., Wool D., & Raman A. (2005). Gall-inducing insects – Nature's most sophisticated herbivores. *Basic and Applied Ecology*, 6, 407–411.

Shuster, S.M., & Wade, M.J. (2003). *Mating systems and strategies* (Vol. 61). Princeton University Press.

Shuster, S.M., Briggs, W.R., & Dennis, P.A. (2013). How multiple mating by females affects sexual selection. *Philosophical Transactions of the Royal Society B: Biological Sciences, 368*(1613), 20120046.

Simmons, L.W. (2005) The evolution of polyandry: sperm competition, sperm selection, and offspring viability. *Annual Review of Ecology, Evolution, and Systematics* 36, 125–146.

Sivinski, J.M., & Petersson, E. (1997). Mate choice and species isolation in swarming insects. In J. Choe, and B. Crespi (Eds.), *The evolution of mating systems in insects and arachnids* (pp. 294e309). Cambridge: Cambridge University Press.

Slatyer, R.A., Mautz, B.S., Backwell, P.R., & Jennions, M.D. (2012). Estimating genetic benefits of polyandry from experimental studies: a meta-analysis. *Biological Reviews, 87*(1), 1–33.

Smith, R.L. (1979). Repeated copulation and sperm precedence: paternity assurance for a male brooding water bug. *Science, 205*(4410), 1029–1031.

Szulkin, M., Stopher, K.V., Pemberton, J.M., & Reid, J.M. (2013). Inbreeding avoidance, tolerance, or preference in animals? *Trends in Ecology & Evolution*, 28, 205–211.

Taylor M.L., Wigmore C., Hodgson D.J., Wedell N., & Hosken D.J. (2008). Multiple mating increases female fitness in *Drosophila simulans*. *Animal Behaviour*, 76, 963–970.

Taylor, M.L., Price, T.A., & Wedell, N. (2014). Polyandry in nature: a global analysis. *Trends in Ecology and Evolution*, *29*(7), 376–383.

Tennenhouse, E.M. (2014). Inbreeding avoidance in male primates: a response to female mate choice? *Ethology* 120, 111–119.

Thornhill, R., & Alcock, J. (1983). *The evolution of insect mating systems*. Cambridge: Harvard University Press.

Tregenza, T. & Wedell, N. (2000). Genetic compatibility, mate choice and patterns of parentage: invited review. *Molecular Ecology*, *9*(8), 1013–1027.

Trivers, R.L. (1972). Parental investment and sexual selection. in Campbell, B. (Ed.) *Sexual Selection and the Descent of Man, 1871–1971* (pp. 136–179). Chicago: Aldine.

Turjeman S.F., Centeno-Cuadros A., Eggers U., Rotics S., Blas J., *et al.*, (2016). Extra-pair paternity in the socially monogamous white stork (*Ciconia ciconia*) is fairly common and independent of local density. *Scientific Reports*, 6, 27976.pmid:27328982.

Välimäki, P., & Kaitala, A. (2006). Does a lack of mating opportunities explain monandry in the green-veined white butterfly (*Pieris napi*)?. *Oikos*, *115*(1), 110–116.

Wade, M.J. (1995). The ecology of sexual selection: mean crowding of females and resource-defence polygyny. *Evolutionary Ecology*, 9, 118–124.

Wedell N., Gage M.J.G., & Parker G.A. (2002). Sperm competition, male prudence and sperm-limited females. *Trends in Ecology & Evolution*, 17, 313–320.

Werren, J.H., & Simbolotti G. (1989). Combined effects of host quality and local mate competition on sex allocation in *Lariophagus distinguendus*. *Evolutionary Ecology*, 3, 203–213.

West, S.A., & Herre E.A. (1998). Partial local mate competition and the sex ratio: A study on non-pollinating fig wasps. *Journal of Evolutionary Biology*, 11, 531–548.

Whiteman, E.A., & Côté, I.M. (2004). Monogamy in marine fishes. *Biological Reviews*, *79*(2), 351–375.

Wickman, P.O. & Jansson, P. (1997). An estimate of female mate searching costs in the lekking butterfly *Coenonympha pamphilus*. *Behavioral Ecology and Sociobiology*, 40(5), 321–328.

Zamudio, K.R., & Sinervo, E., (2000). Polygyny, mateguarding, and posthumous fertilization as alternative male mating strategies. *Proceedings of the National Academy of Sciences*, 97 (26), 14427–14432.

Zink, A.G. (2003). Quantifying the costs and benefits of parental care in female treehoppers. *Behavioral Ecology*, *14*(5), 687–693.

5

Modifications of Copulatory Organs

Fabiano Stefanello[1] **and José Ricardo Inacio Ribeiro**[2]
[1]*Laboratório de Biologia Comparada e Abelhas (LBCA), Departamento de Biologia, Faculdade de Filosofia, Ciências e Letras deRibeirão Preto (FFCLRP), Universidade deSão Paulo, Ribeirão Preto, Brazil*
[2]*Laboratório de Estudos da Biodiversidade do Pampa (LEBIP), Universidade Federal do Pampa, Campus São Gabriel, Avenida Antônio Trilha 1847, 97300-162, São Gabriel, RS, Brazil*

CONTENTS

DOI: 10.1201/9781003043195-5

5.1 Introduction

Before discussing genitalia of insects, it is important to understand some concepts and basic structure of insects and their components. Serial homology is an important concept in understanding of studies of zoological classification and comparative biology. Serial homology is a type of homology representing similarity between parts of same organism with a common genetic origin. A good and didactic example of serial homology are legs of centipede: such legs are produced as a repetition of parts, presenting same shape, function, and embryonic origin. However, it is sometimes not so easy to identify a serial homology. The body of insects, for example, is composed of a series of repeated units with different shapes and different functions, but all have same genetic origin. The appendages of head, thorax, and abdomen of insects are serially homologous among segments of body and may have been modified through reductions, fusions, and sub-segmentations (Machida, 2000; Tomoyasu *et al.*, 2017). Therefore, using the above criteria, maxillae, palps, and genitalia of insects, for example, are modified thoracic legs. Thus, we can discuss elements of genitalia of insects in each order, based on premise that all appendages of insects are composed of arthropods' ground plan leg with some reductions or fusions.

5.1.1 Abdomen

Insect abdomen is composed of 11 segments without legs and a post-metameric segment called telson or periproct (Snodgrass, 1935; Klass & Matushkina, 2018). It is difficult to identify all discernable segments, but all insects show a reduction of at least one segment. The first to seventh abdominal segments are pregenital, and eighth and ninth segments form genital segments. Usually, the tenth abdominal segment is reduced in size, and the 11th segment may be identified by presence of a dorsal lobe, called epiproct, and two latero-ventral lobes, called paraprocts. Despite these aspects, insect abdomen can be divided into two plates: tergum (dorsal plate) and sternum (ventral plate). Each plate consists of basic secondary segmentation that can be identified by presence of small sclerites on surface. The sclerites are hypothesized to be homologous with coxae of ancestral legs, being called coxopodites. Between tergum and sternum is pleural membrane. The genital segments of male apterygote ectognaths are on eighth and ninth segments, as in females, but they differ from females by formation of pregenital segments, which are sclerite-like, and in skeleto-musculature details (Bitsch, 1974; Klass & Matushkina, 2018; Boudinot, 2018). The cerci are appendages on the 11th segment, typically with a sensory function, although these have developed into forceps in Dermaptera and some Diplura and provide a grasping function. Since 11th segment is typically lost, fused, or highly reduced, they typically appear to articulate with tenth tergum or other segments.

5.1.2 Genitalia

In males, ninth-segmental coxopods are antero-posteriorly elongated, relative to previous segments, and bear multiannulate gonapophyses, as in eighth segment, unlike coxal vesicles of preceding segments (Klass & Matushkina, 2018). Due to modification of ninth-segmental coxopods and styli for sexual purposes, these are termed *gonopods* and *gonostyli,* respectively. Situated dorso-medially to ninth-segmental gonapophyses is unpaired *penis*, which bears gonopore and is divided into proximal and distal penial sclerites; known as *phallobase* and *aedeagus*, respectively (Snodgrass, 1935; Klass & Matushkina, 2018).

Penis of Hexapoda is inferred to be derived from medially fused primary gonopods (gonopore-bearing limbs), while genitalia of Ectognatha include both tenth-segmental penis and ninth-segmental secondary gonopods, similar to genitalia of female insects, which comprise gonopods of eighth and ninth segments. According to Boudinot (2018), several competing hypotheses have been proposed to explain evolution of penis: (1) *de novo* origin without homology to other orders (Snodgrass, 1957), (2) homology with medially fused gonapophyses of eighth segment (Verhoeff, 1902), (3) homology with medially fused secondary subdivisions of ninth segmental gonapophyses (Bitsch, 1974), and (4) homology with appendages of tenth abdominal segment (Heymons, 1897). Hypothesis 4, known as coxopod hypothesis, is supported by developmental and innervation patterns. Qadri (1940) observed that penis of Archaeognatha and Ephemeroptera develops from paired outgrowths of genital chamber, which is posterior to ninth segment, contrary to Snodgrass (1935, 1957), who reported but a single anlage. However, apterygote penis is innervated by neurons of tenth segment (Niven *et al.*, 2008), as in Ephemeroptera (Brinck, 1957).

The correspondence between muscles strongly indicates that penis of Archaeognatha is homologous with phallic complex of Protura which, are homologous with gonopore-bearing appendages of Remipedia (Boudinot, 2018). Hypothesis 4 for origin of penis, relating to derivation from modified appendages, is therefore supported, although ancestral segmental identity is uncertain. Two structures are consistently observed across most ectognathan orders: a median unpaired 'penis' and outer paired appendages, which function as claspers or bear single-segmented apical styli. This observation is basis for coxopod hypothesis, which homologizes intromittent organ and claspers with penis and coxopods, respectively, of Archaeognatha and Zygentoma, without assumptions of origin (Boudinot, 2018).

The ancestral hexapod plausibly had a single 'genital segment' bearing multisegmented gonopods. The basal segments of gonopods, protopods, are inferred to have fused medially in ancestor of extant Hexapoda, forming a phallic apparatus that was sunken into a genital chamber. Inferred to be borne on this apparatus were unisegmental (i.e. reduced) exopods, forming styli ('side pieces'), as well as multisegmented endopods ('basistylus', 'acrostyli'), which were situated at apex of distal protopodal segment (Boudinot, 2018). The so-called genital segment of hexapod ground plan is uncertain due to differences in development between Protura and Ectognatha, but it is clear that tenth segment bore medially fused primary gonopod (= penis) in ancestral ectognathan. The female of this ancestral hexapod may have been derived similarly to male, given structural similarity of male and female Protura (Tuxen, 1970a).

This ground plan inference aligns with female ectognaths, namely, that female genitalia of insects comprise paired appendages of eighth and ninth abdominal segments, which together form ovipositor complex (Snodgrass, 1933, 1935; Hünefeld *et al.*, 2012). The ovipositor complex comprises paired gonapophyses of eighth segment, also called anterior, ventral, or *first valvulae*. During development, gonapophyses of ninth segment shift their positions forward and come to lie above anterior valvulae. Therefore, they are called dorsal, posterior, or *second valvulae* (Matsuda, 1976). The ancestral condition of insemination in Hexapoda is indirect through spermatophore, without copula.

In Pterygota, the group with most species in Hexapoda, males transfer sperm directly into females. Male genitalia form on ninth abdominal segment. The aedeagus is often weakly sclerotized or entirely membranous and is flanked by *parameres*, which form clasping structures. The parameres bear gonostyli at their apices although these are given alternate names in many groups, because homology with true gonostyli is not always evident. Basally, parameres articulate with phallobase or gonobasis. An *endophallus* is typically invaginated within aedeagus, through which ejaculatory duct runs. The endophallus is often everted, turning inside out so that gonopore is extruded (Grimaldi *et al.*, 2005). Some subsequent modifications also occur in ovipositor of insects.

In female Zygentoma and Pterygota, anterolatero coxites and posterolatero coxites may fuse to form *gonangulum*, which is a triangular sclerite that improves mechanical efficiency of ovipositor (Scudder, 1957; Klass *et al.*, 2012). Also, a series of modifications occur in ovipositor of Pterygota: (1) gonostylus is apparently lost; (2) a new structure appears, called *gonoplac* or *third valvula* (Scudder, 1961); (3) inner surface of gonapophyses are lined with scales or ridges; and (4) gonocoxae of eighth tergum shifts forward and forms a second articulation with ninth tergum.

5.1.3 Gonopore

Among Hexapoda, gonopore (Snodgrass, 1936) may be internal or external, single or paired. The gonopore(s) of Hexapoda open(s) on fifth abdominal segment in both sexes of Collembola; on eighth segment in both sexes of Diplura; on ninth or tenth segments for female and male apterygote ectognaths, respectively; and on ninth segment or more-anterior segments in pterygotes. Segmental identity of genitalia opening is evolutionarily unstable, particularly near root of hexapod phylogeny (Boudinot, 2018), with the process unknown (Klass & Matushkina, 2018), but possibly *Hox*-delimitation of segmentation (Maderspacher, 2016; Smith *et al.*, 2016). It is plausible that deletion of pregonopodial abdominal segments via mutation of *Hox* genes could render a variable segmental count, while primary and secondary gonopods remain homologous (Boudinot, 2018).

5.2 Modifications of Copulatory Organs across Hexapoda

The groups defined here represent clades recovered by Misof *et al.* (2014), most recent and comprehensive phylogenetic study in Hexapoda.

5.2.1 Ellipura and Diplura

Both sexes of these two orders may be considered as highly modified but simple, represented by a single gonopore situated on a genital papilla, which is of abdominal fifth segment in Collembola (Tuxen, 1970b; Matsuda, 1976) and just posterior to eighth segment in Diplura (Tuxen, 1970c; Matsuda, 1976). Although appendages are absent from fifth segment in Collembola, stumpy 'appendices genitales' occur on eighth segment of japygoid Diplura (Tuxen, 1970c). In Protura, possibly intromittent male genital structures (Pass & Szucsich, 2011) of phallic apparatus comprise two primary elements: (1) a ring-like sclerite, termed periphallus, which bears a pair of long anterior apodemes (basal apodemes); and (2) a phallus situated within ring of periphallus, divided into a basal ring-like stipes and a distal pair of styli (Boudinot, 2018). The styli are divided into cone-like basistyli and distal acrostyli, with gonopores opening in acrostyli (Tuxen, 1963; Matsuda, 1976) (Figure 5.2a). In females of Protura, genital apparatus is undivided and is called perigynium. The paired styli are borne by perigynium, and common oviduct opens between bases of styli (Tuxen, 1963; Matsuda, 1976).

5.2.2 Archaeognatha and Zygentoma

In these two orders, external genitalia structures are absent altogether or are present but not functioning as intromittent organs. The external male genitalia are identical to female, except that gonapophyses are smaller and frequently missing first pair (Grimaldi *et al.*, 2005). In males, ninth-segmental coxopods are antero-posteriorly elongated relative to previous segments and bear multiannulate gonapophyses, as in eighth segment, unlike coxal vesicles of preceding segments (Klass & Matushkina, 2018). The adult penis is a simple tube, borne on simple carina situated dorsomedially to ninth-segmental gonapophyses, which bears gonopore and is divided into proximal and distal penial sclerites (Klass & Matushkina, 2018). In summary, male genitalia of Archaeognatha and Zygentoma have retained several ancestral conditions of Hexapoda (Figure 5.1a). On other hand, there is a fundamental difference with regard to female genitalia: while Archaeognatha retain ancestral conditions of Hexapoda, gonangulum, a homologous structure shared with Pterygota, appears in Zygentoma (Scudder, 1961) (Figure 5.1b).

5.2.3 Ephemeroptera and Odonata

The male external genitalia in Ephemeroptera consist of median-paired penis and lateral clasping organs (Brinck, 1957). The lateral clasping organs are multiannulate gonostyli situated on a transverse sclerite, termed *styliger plate*; which corresponds to medially fused ninth gonocoxite (Kluge, 2004). The penis varies in relative size between species and may be completely fused or absent. When present, gonopores open in each penis lobe (Matsuda, 1976) (Figure 5.2b). Interestingly, male genitalia seem to undergo metamorphosis, from subconical hollow evaginations of nymphs to a well-defined penis and lateral clasping through a rapid differentiation during subimaginal stage (Matsuda, 1976). The oviducts open separately in most mayflies but may vary in degree of fusion or still, they may open to exterior through a single gonopore. Also, in Caenidae, gonopore is absent, and eggs are expelled by rupture of abdomen (Matsuda, 1976). In general, oviducts are tubiform, situated laterally, and extend posteromesdially in posterior part of seventh segment. The modifications of area surrounding openings of oviducts involve

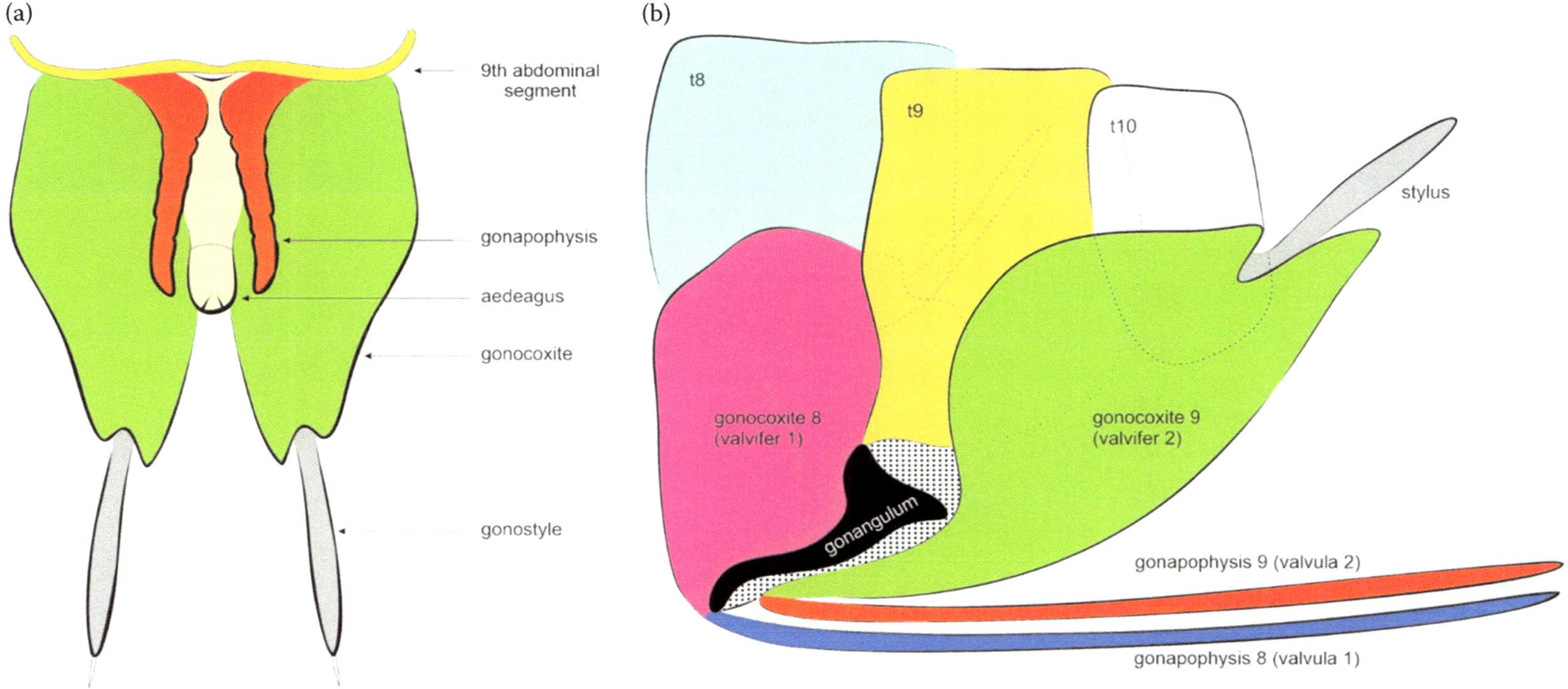

FIGURE 5.1 (a) Generalized male genitalic structures of a bristletail, modified from Grimaldi *et al.* (2005). Male genitalia of apterygotes insects is relatively simple retaining several conditions of ancestral genitalia of Hexapoda. (b) Sagittal view of skeletal parts of female genitalia of silverfish, modified from Scudder (1957). The female genitalia of silverfish represent a transition between ancestral condition of Hexapoda and ancestral condition of Pterygotes, once gonangulum is present but gonoplac is absent. The shared colors represent homology of structure between sexes.

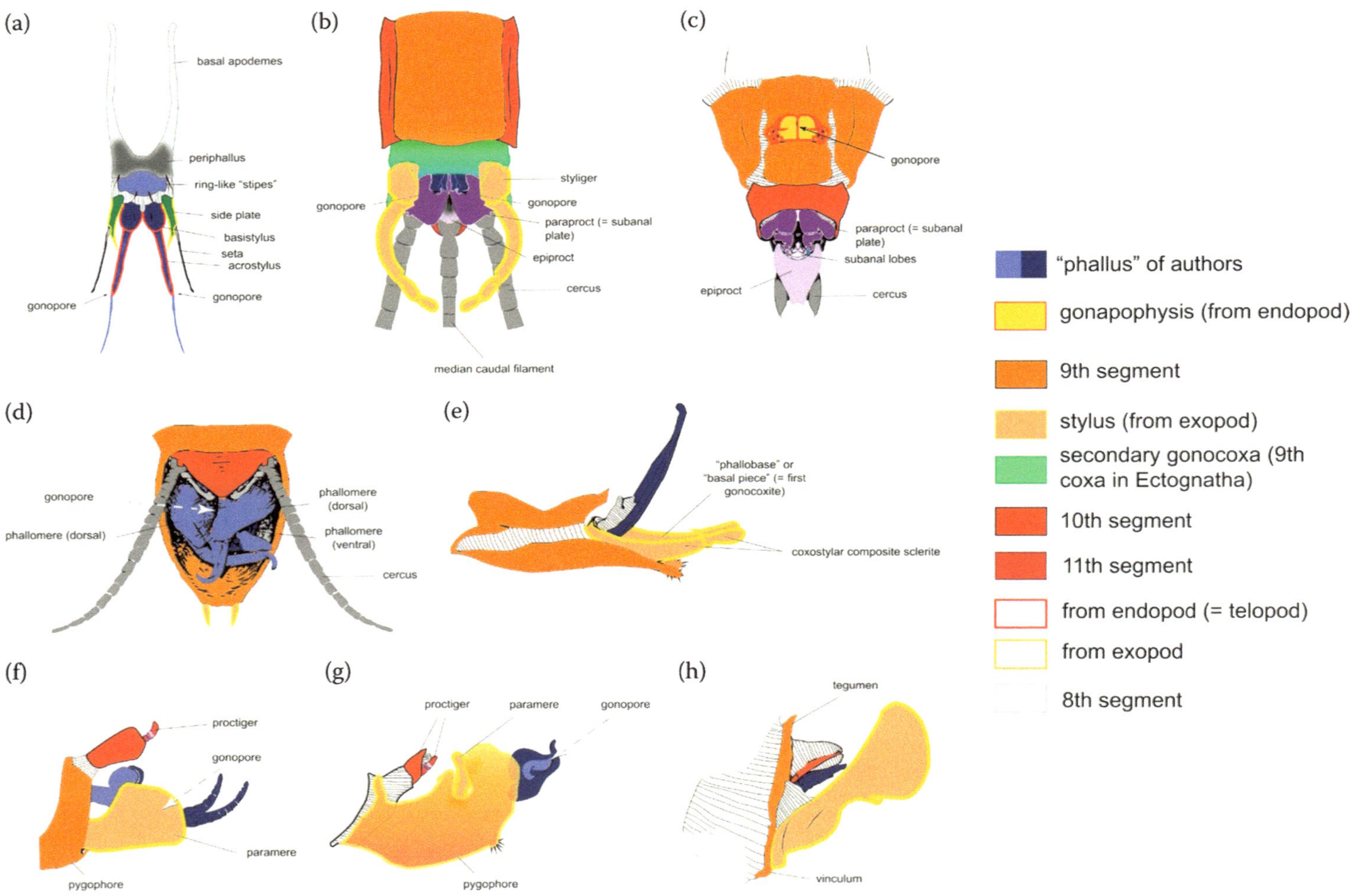

FIGURE 5.2 Schematic representation of male genilia for some order across Hexapoda, modified from Matsuda (1976) and Snodgrass (1957). a = Protura, b = Ephemeroptera, c = Odonata, d = Blattodea, e = Coleoptera, f = Hemiptera (Auchenorryncha), g = Hemiptera (Heteroptera), h = Lepidoptera. The shared colors represent homology among parts of genitalia.

development of a shallow cavity, *vestibulum,* formed between seventh and eighth sternum by extension of subgenital plate (Brinck, 1957). The gonapophyseal ovipositor does not occur in Ephemeroptera.

The primary genitalia of odonates have been observed to consist of a pair of lobes covering genital opening on ninth segment (Fraser & Asahina, 1970). It has been postulated that these lobes are modified gonapophyses, based on developmental observations (Snodgrass, 1935) and paleontological evidence (Bechly *et al.*, 2001). The primary external genitalia in ninth segment are almost obsolete (Figure 5.2c). The copulatory apparatus of male odonates has developed from second abdominal sternite and consists of a 'penis' (*ligula*), a seminal vesicle, and a series of accessory structures (hamuli anteriores and hamuli posteriores) (Matsuda, 1976). The ligula is intromittent organ and shows a variable morphology (a pair of lobes or flagella), closely related to sperm displacement. After removal of rival's sperm, sperm is conducted in a furrow of ligula during insemination (Cordero-Rivera & Córdoba-Aguilar, 2010). The presence of copulatory organ in anterior abdominal segment is a highly autapomorphic feature of Odonata and represents a conspicuous case of production of new structures in evolution (Matsuda, 1976). In addition, copulatory organs of male odonates are not homologous with any animal organs known (Tillyard, 1917). In general, when fully developed, ovipositor consists of paired first and second valvulae and gonoplac. Each first valvula or gonapophysis is provided with a short saw-like edge near its distal end and is basally fixed to eighth gonocoxite. On each side, eighth gonocoxite is connected with gonangulum, which is, in turn, connected with third valvula and lies along lateral margin of ninth tergum (Matsuda, 1976).

5.2.4 Polyneoptera

In hemimetabolous insects, especially in Polyneoptera, spermatophores are more in evidence as a means of transferring male gametes to females, since intromittent organs are less elaborate. In Dictyoptera, Phasmatodea, Dermaptera, Embioptera, some Ensifera, and probably Grylloblatodea, spermatophore is inserted more or less completely inside a female's body (Alexander & Otte, 1967). Gryllidae, Tettigonidae, and some Gryllacrididae are more elaborate, the ampulla either hangs outside a female's body, with a tube long enough to reach spermathecal tube, or the tube of ampulla is inserted into female's genital opening and held in place by male during insemination, as occurs in some Gryllidae and Caelifera (Alexander & Otte, 1967). The ancestral condition of external female genitalia is absence of subgenital plate, as in Zygentoma (Matsuda, 1976).

5.2.4.1 Dermaptera and Zoraptera

In Dermaptera, male genitalia consist of ninth sternum (lacking styli throughout development), a pair of penes with individual ducts or a single organ containing both ducts or only one duct, and a pair of parameres (Qadri, 1940). The paired penes are similar to those of Ephemeroptera and are considered to be plesiomorphic for order (Popham, 1965). On the other hand, development of parameres at lateral branches of phallic rudiments appears to relate penis rudiments of Dermaptera to primary phallic lobes of Eumetabola (Snodgrass, 1957). On ninth coxosternum, at the base of penis, there is an anterior elongation, *manubrium,* where ejaculatory duct passes over and is attached to the latter by connective tissue. The degree of development of manubrium varies greatly and is of great taxonomic importance (Burr, 1916). In some Protodermaptera, ovipositor is present (Giles, 1963). In *Cranopygia* Burr, paired gonapophyses of eighth and ninth segments, though reduced, are clearly recognizable, and are most probably homologous with those in other orders (Matsuda, 1976). However, in *Forficula,* a representative genus of Eudermaptera, subgenital plate is a prolongation of seventh sternum and forms the floor of genital chamber. The dorsal wall of genital chamber in *Forficula* is eighth and ninth membranized sternum, without gonapophyseal ovipositor (Qadri, 1940).

The male genitalia in Zoraptera is poorly studied but seems highly modified, showing complex structures (Walker, 1922). In *Zorotypus hubbardi* Caudell, ninth sternum (also called *hypandrium*) is apparently represented by an extension of eighth sternum (Crampton, 1920a). The copulatory organs consist of an unpaired asymmetrical aedeagus, doubtlessly a derivate of ninth segment, and left outer paramere is strongly sclerotized and elongated, with hooks at tip, which is referred to as penis valve

(Hünefeld, 2007). Crampton (1920b) and Walker (1922) pointed out that asymmetrically developed parameres of zorapterans resemble those of mantids. The aedeagus seems markedly different among species: *Z. caudelli* Karny consists of a long cuticular spiral, whereas that of *Z. hubbardi* is formed by a short cuticular structure (Dallai *et al.*, 2011). In females, eighth sternum forms subgenital plate, and external genitalia in *Z. brasiliensis* consist of several pairs of vestigial lobes that surround genital opening which represents basal portion of reduced ovipositor (Delamare-Deboutteville, 1970). An ovipositor gonapophyseal is completely absent in zorapterans females (Engel, 2004).

5.2.4.2 Plecoptera

The Plecoptera have no styli or appendages of any kind on the ninth sternum. The penis is represented by an endophallus, a genital cavity, where ejaculatory duct discharges through a long opening, either directly to exterior or onto a small penis (Brinck, 1956; Snodgrass, 1957). In *Chloroperla* Newman, penis is formed from a pair of small outgrowths at anterior end of genital chamber, between ninth and tenth sternum (Qadri, 1940). The degree of development of endophallus varies considerably within Plecoptera, as accessory genitalic parts, cerci, paraprocts, and epiproct, with considerable reduction in some cases, are directly involved in copulation and display striking variation (Brinck, 1956; Nelson & Hanson, 1971). The ovipositor is entirely lost in stonefly females.

5.2.4.3 Embioptera and Phasmatodea

As in Plecoptera, penis is presumably of endophallus type in Embioptera. Males have no prominent, true phallic structures and posterior abdominal segments have been modified into copulatory organs (Hünefeld, 2007). The male genitalia consist of ninth sternum plus cerci and tenth tergum, all of which are asymmetrical and modified for clasping (Ross, 1940). Styli are absent. The ovipositor is entirely lost in webspinner females.

In Phasmatodea, male genitalia consist of a subgenital plate (ninth coxosternum) covering penis from beneath (Helm *et al.*, 2011). The penis may have a different number of membranous or sclerotized valves. The tenth segment bears a sclerotized process of ventral surface called *vomer*, which extends horizontally backward from anterior margin of tenth sternum (Helm *et al.*, 2011). The vomer intrudes into preopercular organ of female during copulation (Matsuda, 1976). The cerci may also assist in copulation. Styli are absent. The female genitalia of *Timema* Scudder consist of a subgenital plate (eighth sternum) or operculum, broadly triangular and partly covering valves of ovipositor. The first gonapophyses are curved upward between gonoplacs, covering second gonapophyses ventrally while second gonapophysis and gonoplac are distinct blades, with gonoplac overlapping smaller second gonapophysis ventrolaterally (Tilgner *et al.*, 1999). The gonangulum is apparently not formed. The ovipositor in *Timema* is remarkably similar to nymphal ovipositor in *Grylloblatta* Walker (Matsuda, 1976). In other phasmids, well-developed subgenital plate covers ovipositor from beneath. The most peculiar feature of female external genitalia in Phasmida is preopercular organ arising from posterior margin of seventh sternum in many phasmids. This organ is a kind of pouch often surrounded by a sclerotized wall and is covered anteriorly by a single or double sclerotized process, which holds vomer of male during copulation (Matsuda, 1976).

5.2.4.4 Grylloblatodea and Mantophasmatodea

In Grylloblatodea, ninth coxosternum of male nymphs comprises a broad basal area demarcated by a line from paired coxites bearing styli, left coxite being larger than right (Walker, 1919). In adult male, asymmetry of left and right coxites is more pronounced. The penis also consists of two asymmetrical lobes, or phallomeres, which are different in shape and size. The left phallomere bears an eversible tube, and right one bears several sclerites of irregular shapes. The gonopore opens on inner margin of right phallomere (Walker, 1922) or between paired phallomeres near their base (Scudder, 1970). In females, three pairs of valvulae are well-developed, with the third pair (gonoplac) being longest. The gonangulum is a small triangular plate through where each anterior valvula is connected with lower extremity of lateral apodeme of ninth tergum and with dorsal gonoplac (Walker, 1919). The gonopore opens on medially emarginated posterior margin of eighth sternum, where it meets the base of first valvulae (Walker, 1919). The gonoplac is greatly developed and incorporated into ovipositor, representing an intermediate stage among orthopteroids, where function of reduced, second gonapophysis is assumed by gonoplac (Grimaldi *et al.*, 2005).

In Mantophasmatodea, ninth coxosternum is undivided, styli are absent, and phallic organ comprises up to eight individual sclerites (Klass *et al.*, 2003), which are probably penial sclerites, although not confirmed according to Boudinot (2018). In most Mantophasmatodea, genitalia are asymmetrical, with one sclerotic possibly corresponding to right phallomere of Dictyoptera. Although not mentioned by Klass *et al.* (2003), this right phallomere may be homologous with those phallomeres observed by Walker (1919) for Grylloblattodea. Moreover, males have a sclerotized process, composed of a posteriorly projecting vomeroid body in middle and a pair of slender lateral arms, each of which articulates with anteroventral corner of tenth tergum of same side, resembling vomer in Phasmatodea (Klass *et al.*, 2003). Styli are absent. In females, posterior part of eighth segment (eighth coxosternum) bears first valvifers articulating upon two others paired with sclerites, gonangulum, and first valvulae, which are conspicuously blunt, distally, in lateral view (Klass *et al.*, 2003). On venter of ninth segment, anterior part comprises gonangulum, which extends along ventrolateral margins of ninth tergum. Ninth coxosternum bears second valvifer, and posterior part of ninth sternum is mainly constituted by a bulky projection that includes two large, stout gonoplacs with their sclerites, two small second valvulae (Klass *et al.*, 2003).

5.2.4.5 Orthoptera

In Ensifera, subgenital plate is formed by modification of ninth sternum; in Caelifera embryonic ninth pair of appendages and ninth sternum together form rudiment of subgenital plate (Walker, 1922; Matsuda, 1976). According to Snodgrass (1935), external genital organs arise from a single pair of primary rudiments in orthopteroid insects. The primitive phallic lobes, also called *phallomeres*, generally seen in nymph, lie behind ninth abdominal sternum, and may vary widely in number and size across Orthoptera (Matsuda, 1976; Eades, 2000). In general, these phallomeres are hollow and swollen, perhaps also glandular or, more likely, functioning as inflatable claspers that operate through hydraulic pressure to hold female's copulatory papilla during sperm transfer (Mello, 2007). In Caelifera, penis is derived from abdominal appendages of embryonic tenth segment with profound modifications of genitalic structures, which obscure homology of genitalic parts with those of Ensifera, occurring during last moult into adult (Walker, 1922). In Acrididae, penis is enclosed within a genital cavity, formed below subanal lobes, and enclosed posteriorly by vertically erected distal end of subgenital plate (Snodgrass, 1937). The styli of ninth segment occur in some Tettigonioidea, although they are not functionally accessory copulatory organs. In Grylloidea and Califera, styli never occur (Snodgrass, 1935).

Orthoptera females are always cited for morphology of ovipositor because of well-developed three pairs of valvulae and valvifers of eighth and ninth segments. In some families, there is a reduction of some valvulae, as in Gryllidae, in which second valvulae is reduced and membranous, or in Tettigoniidae where gonangulum is not clear (Matsuda, 1976). The reduction of ovipositor has occurred independently in some groups of Ensifera, as in Stenopelmatidae, or as Schizodactyloidea and Gryllotalpoidea, in which ovipositor gonapophyseal is absent (Matsuda, 1976).

5.2.4.6 Dictyoptera

Male roaches and mantises have genitalia hung to left, and are so asymmetrical that their large and irregular phallomeres on left side have barely recognizable corresponding parts on vestigial right side (Grimaldi *et al.*, 2005; Klass, 1997) (Figure 5.2d). Yet, phallomeres in Blattaria have various sclerites, hooks, lobes, and spines concealed by subgenital plate (ninth sternum), which usually bears paired styli (Roth, 1970). Phallomeres in Isoptera are almost impossible to identify because male genitalia are reduced to minute knobs, virtually lost in all but a few termites, such as *Stolotermes*

Hagen (Klass *et al.*, 2000). The homologous relationship between phallomeres of Dictyoptera and elements of external genitalia of other ectognathan taxa are completely unclear.

The gonapophyseal ovipositor in Dictyoptera is much reduced. It belongs to abdominal segments eighth and ninth, and is comprised of paired appendages. These include two pairs of valvifers on segments eighth and ninth, so-called because of apparent serial homology with coxae of legs; two pairs of gonapophyses (first and second valves or valvifers), one each on segments eighth and ninth; a pair of gonoplacs on segment ninth (third valvifers); and a pair of gonangulum, also on segment ninth. The longest appendages appear in Dictyoptera, gonapophyses eighth, and gonoplacs, others often being rudimentary knobs. Mantises have the best-developed ovipositor, which is slightly protruding; roaches and termite *Mastotermes* Froggatt have an ovipositor entirely concealed within an internal pouch, vestibulum, and suspended from its roof. All other termites have an ovipositor that is virtually lost (Grimaldi *et al.*, 2005; Marks &Lawson, 1962) (Figure 5.2).

5.2.5 Condylognatha

5.2.5.1 Hemiptera

In Heteroptera, the ninth sternum is developed in most taxa into a distinct genital capsule, also called a *pygophore,* containing aedeagus (Schuh & Slater, 1995). The subgenital plate (the *hypandrium*) is extension of ninth coxites, forming genital cavity above, and phallic organs lie within cavity. In many adult Heteroptera, subgenital plate is apically bilobed, also called *paratergites* (Scudder, 1959), indicating its paired origin (Matsuda, 1976). The phallic organs consist of median penis, phallobase, and paired parameres (Matsuda, 1976). The parameres are highly variable and often asymmetrical, developed from each lateral side of ninth sternum in fifth-stage nymph, with various degrees separated from aedeagus (Qadri, 1949). Qadri homologized parameres with styli in Ephemeroptera, arguing that parameres developed quite independently of phallic lobes. Heteropteran claspers organs are displaced phallic parameres, not styli, since its apodemes are attached on a ridge connecting parameres with base of aedeagus, like those in Auchenorrhyncha (Dupuis, 1955). The intromittent organ consists of sclerotized aedeagus, also called *phallosoma*, forming a tubular outer wall of penis, and eversible, much less sclerotized, endophallus (Schuh & Slater, 1995) (Figure 5.2g).

Males of Micronectinae (Corixidae) show a very uncommon case of stridulation. The sound is produced by rotating pygophore and rubbing a pars stridens (structure with a well-ridge) of right paramere against a well-sclerotized ridge on left lobe of eighth segment (Jansson & Meyer-rochow, 1990). The intensity of signal can reach 100 dB, which is far below that estimated for large mammals but, when scaled to body size, *Micronecta scholtzi* (Fieber) has highest decibel/body size ratio (Sueur *et al.*, 2011). Only males produce sounds. A second weird modification in genitalia occurs in Cimicomorpha with traumatic insemination. Males of Prostemmatinae (Nabidae) have a sharp genitalic spicule, also called *acus*, in aedeagus, which is used to pierce wall of bursa copulatrix and deposit semen in haemocoel instead of in female's genital cavity (Tatarnic *et al.*, 2006). In Plokiophilidae and Lyctocoridae, male directly pierces female's abdominal wall using acus, while in Anthocoridae, Polyctenidae, and Cimicidae, male's left paramere is coupled with aedeagus, forming a scythe-like intromittent organ (Schuh & Štys, 1991).

The Heteroptera ovipositor consists of first and second valvulae (gonapophyses), usually fused with first and second valvifers (gonocoxae), respectively (Scudder, 1959; Schuh & Slater, 1995). The ninth segment may also bear a pair of third valvulae (gonoplacs), firmly attached to apex of second gonocoxa, which form as ovipositor sheaths, although these are frequently absent (Schuh & Slater, 1995). In Pentatomorpha and Cimicomorpha, two types of ovipositors can be distinguished, depending on oviposition site and habitat: lacinate type, with lanceolate blades and a heavily sclerotized anterior part of gonangulum; and plate-shaped type, with a flap-like ovipositor valvulae (Scudder, 1959). According to Scudder (1959), lacinate type occurs in those that oviposit in plant tissues, and plate-like type occurs in those Heteroptera that oviposit eggs on surface of leaves, trunks, etc.

In Auchenorrhyncha and Coleorrhyncha, copulatory organ consists of median penis and a pair of parameres (Matsuda, 1976; Burckhardt, 2009). The penis is borne by a relatively simple phallobase with parameres typically also hinged to phallobase (Figure 2f). In general, penis is a simple sclerotized aedeagus that contains eversible endophallus without formation of phallosoma. The parameres are two-paired in some Auchenorrhyncha, and are absent in most Cicadidae (Helms, 1968). In females, ovipositor gonapophyseal is quite similar to those in Heteroptera. In Sternorrhyncha, aedeagus and parameres are widely separated basally and are sometimes reduced. In some Psyllidae, both parameres and aedeagus are rotated during moult into adult stage, or aedeagus may be two-segmented with a peculiar elbow mechanism enabling penis to stretch and lie on ventral side of female's ninth segment during copulation (Matsuda, 1976). The ovipositor gonapophyseal may be reduced or absent.

5.2.5.2 Thysanoptera

The subgenital plate is an extension of ninth sternum, forming a genital cavity above, which bears penis. It consists of phallobase, phallosoma, endophallus, *periandrium*, and *pseudovirga* (Priesner, 1960). In some thrips, two pairs of parameres (dorsal and ventral) occur, and they may be homologous with two pairs of parameres in Psocodea (Matsuda, 1976). The parameres arise as outgrowths from lateral walls of phallobase during pupal stage (Priesner, 1970). The gonapophyseal ovipositor is quite similar to those of Hemiptera. As mentioned by Grimaldi *et al.* (2005), ovipositor of terebrants is saw-like or reduced, while ovipositor in tubiliferans is developed into a chute.

5.2.6 Psocodea

The subgenital plate (hypandrium) is the modified ninth sternum and may vary among groups, including being represented by paired coxopodites bearing styli (Sommerman, 1946; Badonnel, 1970). The copulatory organ in Psocodea consists of basal plate (phallobase), external parameres, internal parameres (*mesomeres*), and ventral plates, which together comprise phallosoma, abutted ventrally by a hypandrium (Yoshizawa, 2005; Yoshizawa & Johnson, 2006). Thus, phallosoma of psocodean may be homologized with penis of other Ectognatha. However, homologies of parameres and mesomeres are hard to establish since development of organs has not been studied. In some psocodean, male epiproct, paraproct, clunial arms play a principal role in holding female subgenital plate during copulation (Cheng & Yoshizawa, 2019). In general, male genitalia of bark lice and true lice is quite similar, except for phallobase, which is longer in true lice than in bark lice. The terminology may vary, but structures are considered homologous (Matsuda, 1976). The male genitalia in Ischnocera are generally more specialized than those in Amblycera, and specializations include prolongation of phallobase and frequent loss of *endoparameres* (mesomeres) (Qadri, 1936). In bark lice females, subgenital plate forms a genital chamber that contains gonapophyses of ovipositor. The ovipositor consists of three pairs of valvulae, which include ventral valvulae of eighth segment, and dorsal and lateral valvulae of ninth segment, as in other orders. Often, however, one or two pairs become reduced or completely lost. The gonapophyseal ovipositor of true lice is completely lost.

One of most striking variations occurs in genus *Neotrogla* Lienhard, where females have a large penis-like structure, *gynosome*, which acts as an intromittent organ to receive voluminous spermatophores from male (Yoshizawa *et al.*, 2014). The apical sclerotized part of gynosome, bearing opening of spermathecal duct, deeply penetrates male, and its tip fits into opening of seminal duct. The membranous part inflates within male genital chamber, numerous spines on membrane internally anchor female to male, and female can then hold male tightly using gynosome and paraprocts (Yoshizawa *et al.*, 2014).

5.2.7 Endopterygota

The penis of Endopterygota is developmentally and mechanically integrated with inner surfaces of coxopods of ninth segment, differing from apterygote Ectognatha, Palaeoptera, and Polyneoptera, where it is positioned on the ninth sternum (Boudinot, 2018). Once integrated into gonocoxae, penial promotor and remotor muscles duplicate laterally and dorsally, with sclerite of insertion for lateral duplicates splitting off laterally, resulting in *lateropenite* (the *parandrite* of Bechly *et al.*, 2001; *volsella* of

Michener, 1944; and other, numerous terms) (Boudinot, 2018). In general, coxopods of Endopterygota function as claspers. Thus, Endopterygote male genitalia comprise paired coxopods of ninth segment (secondary gonopods), divided into basal coxae (gonocoxae) and distal styli (gonostyli); medial to gonopods are paired lateropenites and median penis (Snodgrass, 1957; Boudinot, 2018).

5.2.7.1 Hymenoptera

In Hymenoptera, male subgenital plate is the ninth sternum that can be prolonged. It forms genital cavity above, in which external genitalia lie (Matsuda, 1976). The male genitalia comprises a sclerotic ring (the *cupula* of Schulmeister, 2001; *gonobase* of Michener, 1944), which is situated at anterior apex of paired appendages; paired appendages comprise clasping structures (the *latimeres* of Schulmeister, 2001), which themselves are often divided into a basal lamina (*gonostipes* = coxae), which is petal-shaped in profile view and bears dorsomedian processes with their long axes oriented anteroposteriorly, and an apical lobate structure (*harpe* = stylus); medial to latimeres are two sets of paired sclerites: a smaller clasping structure (*volsella* = *parossiculus*, minor part of ninth gonocoxa, plus lateropenite), which is lateromedially narrow and has its length oriented along anteroposterior axis; medial most to volsellae are a pair of sclerites (*penisvalva*) with a conspicuous posterior lobe and anterior, longitudinally oriented apodeme. Michener (1956) interpreted gonocoxal origin of parossiculus, stating that volsella is a fragmentum of gonocoxa. Conversely, parossiculus, rather than volsellar complex itself, may be considered a uniquely derived apomorphy in ground plan of Hymenoptera (Sharkey, 2007). The endophallus is typically supported on each side by penis valve, and valves extend cephalad into body cavity as paired apodemes (Schulmeister, 2001). Three components of gonocoxopodite (gonocoxite and stylus) of hymenopterans (gonostipes, harpe, and volsella) are completely homologous to three parts of gonocoxopodite in Zygentoma (Matsuda, 1976).

In females, structural plan of ovipositor and associated structures in Hymenoptera are remarkably uniform (Snodgrass, 1933). The subgenital plate is seventh sternum, which covers the base of ovipositor formed by two valvifers and three valvulae (Matsuda, 1976). The gonangulum (first valvifer of many authors) is entirely dissociated from other parts of eighth segment and forms important basal mechanism of ovipositor. Each gonangulum is a small plate continuous with ramus of anterior first valvula on its anterior end. In Aculeata (except Sapygidae, Chrysididae, and Drynidae), the ovipositor functions as a sting, eggs being laid at base of sting apparatus; in some, organ is used both as an ovipositor and as a sting (Matsuda, 1976). The sting of Aculeata is composed of first (ventral and anterior) and second (dorsal, inner) pairs of valvulae, and third (lateral) valvulae ensheath distal part of shaft of ovipositor (Matsuda, 1976). In some Ichneumonoidea, first and second valvulae of ovipositor may be three times longer than body of female, such a long ovipositor allows eggs to be deposited inside fruits, for example.

5.2.7.2 Neuroptera, Megaloptera, and Raphidioptera (Neuropterida)

Generalities about male genitalia in Neuropterida are difficult to make since many independent structures present in various families, probably more than in any other insect order (New, 1989). In Neuroptera, male genitalia comprise a free, undivided ninth sternum, with coxae migrated dorsally and fragmented as (1) anterior part of gonocoxa fused with tergum, forming a tergocoxal composite; and (2) posterior fragments fused with styli, forming acoxostylar composite (Acker, 1960; Boudinot, 2018). Long, trunk-like styli are fused with gonocoxites, and articulated with tergocoxal composite, anteriorly. The lateropenites (also called parameres by some authors) are tightly integrated with penial sclerite (*gonarcus*) and fused medially, forming dorsoventrally tall laminae, which bear ventrolateral lobes and an acutely pointed V-shaped process, dorsomedially. A disassociated penial sclerite with gonopore fused medially and migrated dorsally, forming inverted, U-shaped sclerite is associated with lateropenites (Aspöck, 2002). In Megaloptera, the ninth sternum is medially divided by longitudinal sulcus or is large and tongue-like with coxae, comprising dorsal tergocoxal composite, with a lateral fragment fused to and bearing stylus, forming a coxostylar composite with ventrolateral lobe. Short and stout or club-like styli, each bearing a single, sharp spine, are fused to lateral gonocoxite. Lateropenites digitiform are fused or partially fused medially, and tightly integrated with penial sclerite (gonarcus), forming penial-lateropenital complex, which is migrated dorsally. Penial sclerites dissociated from gonopore are fused medially, forming gonarcus, and are tightly associated with lateropenites (Boudinot, 2018). In Raphidioptera, the ninth sternum is fused to base of gonocoxae. The coxae are unmodified except for fusion with abdominal sternum and tenth tergum. The styli articulate freely with gonocoxae posterodorsally via a single anterolateral condyle; gonocoxal-gonostylar condyle is supported by a longitudinal apodeme connecting to anterior gonocoxal margin. The lateropenites are small and lobate. Penial sclerites are fused anteroventrally with media margins of gonocoxae (Boudinot, 2018). The sclerotized intromittent organ or aedeagus occurs only in Coniopterygidae (Neuroptera); in others, penis is represented by eversible endophallus, which carries gonopore (Tjeder, 1970a).

The strictly gonapophyseal ovipositor does not occur in Neuroptera. In Raphidioptera, unpaired ventral valvula is a prolongation of eighth sternum, although it may well be paired with anterior gonapophyses that have become completely fused (Matsuda, 1976; Mickoleit, 1973). The gonangulum, with which anterior valvula (gonapophysis) usually articulates, is absent. The paired dorsolateral valvulae of ninth segment is ninth coxites bearing apical styli. The inner valvulae (posterior valvulae, posterior gonapophyses) are not differentiated. A peculiar sliding articulation between dorsal lateral valvulae and ventral valvula in *Raphidia* is therefore different in terms of components involved from ovipositor in other orders (Matsuda, 1976). In *Osmylus*, sclerotized eighth sternum is present, but anterior (ventral) valvula is absent. Therefore, ovipositor is formed exclusively by approximation of ninth valvifers, which are well developed (Mickoleit, 1973).

5.2.7.3 Coleoptera and Strepsiptera (Coleopterida)

The subgenital plate does not occur, most probably as a result of secondary loss (Matsuda, 1976). The male genitalia of beetles are characterized by a tripartite structure composed of aedeagus and parameres, supported on a proximal plate or phallobase (Snodgrass, 1957). The phallobase is often called *basal piece*, phallobase and parameres together are referred to as *tegmen*, and aedeagus is known as *median lobe* or penis by coleopterists (Figure 2e). However, according to current theory of genital homologies, phallobase is gonocoxite, and parameres are gonostyli. The sclerotized aedeagus contains internal sac (endophallus), which is dilated distal portion of ejaculatory duct; during copulation, sac is everted into female genitalia (Matsuda, 1976). The endophallus sometimes bears multiple spine-like or jaw-like structures, which are involved in postcopulatory selection, as in seed beetles (Hotzy *et al.*, 2012; Van Haren *et al.*, 2017). The male genitalia of Coleoptera display astonishing diversity, and a variety of morphological forms, or types, have been recognized (Crowson, 1955, 1981). One of most common types is phytophagan type, occurring in Curculionoidea and Chrysomeloidea, with coxae fragmented into two portions: (1) first fragment forming a thin ring around second gonocoxites and penis, and bearing rod-like dorsomedial apodeme; (2) second fragments fused with penis (forming 'coxopenis'), with styli and lateropenites absent, and penial sclerite indistinguishably integrated with second gonocoxites (Boudinot, 2018). The gonapophyseal ovipositor has become completely lost in Coleoptera. For oviposition, posterior segments (seventh to ninth) have often become prolonged to form an ovipositor (Matsuda, 1976). In some species, females have a completely developed cloaca, as for those of Lepidoptera (see section *2.7.4*), which have been formed by convergence (Matsuda, 1976).

In Strepsiptera, external male genitalia are highly simplified, being reduced to intromittent organ, penis (Hünefeld *et al.*, 2011). This organ consists solely of a tube- or scythe-like structure connected to ninth sternum by thinly walled phallobase (Matsuda, 1976; Beani *et al.*, 2005; Pohl & Beutel, 2008). This reduction is likely associated with modifications of female genitalic systems, at least for ancestor of crown Strepsiptera (Pohl & Beutel, 2008). Given that penis muscles originate on abdominal sternum and ninth tergum, and close association of penis with second gonocoxites in Coleoptera, it is here postulated that penis of Strepsiptera is a coxopenis, with complete loss of gonostyli, lateropenites, and intrinsic musculature (Hünefeld *et al.*, 2011; Boudinot, 2018). The gonapophyseal ovipositor is completely lost in Strepsiptera. In endoparasitic females, disuse of female genitalia could then drive extreme reduction of these external and internal organs, with females lacking a vagina, genital chamber, bursa copulatrix, and receptaculum seminis, and eggs floating freely in hemolymph (Peinert *et al.*, 2016). Traumatic insemination occurs in endoparasitic strepsipterans (Beani *et al.*, 2005; Peinert *et al.*, 2016).

5.2.7.4 Lepidoptera and Trichoptera (Amphiesmenoptera)

In most Lepidoptera, the ninth segment ring is divided into a dorsal 'tegumen' and a ventral 'vinculum' (Kristensen, 1984). The coxae of ninth sternum bearing one- or two-segmented gonopods are sometimes fused ventromedially by a sclerotized median plate, *juxta*. The gonostyle is fused with gonocoxa and lateropenites are fused ventromedially, forming a vase- or ring-like structure, which may be closely associated with gonopods or ninth tergum (Kristensen, 1984) (Figure 5.2h). The penial sclerite forms a tube quite similar to those from Trichoptera (Kristensen, 1984). The juxta of Lepidoptera is homologizable with lateropenites, as for parameres of Trichoptera (Nielsen, 1957; Boudinot, 2018). In females, gonapophyseal ovipositor is completely lost. In many groups, whether common oviduct is dorsal or ventral to bursa, genital and rectal systems combine posteriorly to form a terminal cloaca. There are two types of cloaca in Lepidoptera: a complete cloaca, where chamber exit serves as anus, ovipore, and a copulatory pore, as in Micropterigidae; and an incomplete cloaca, serving as anus and ovipore only, as in some Ditrysia (Dugdale, 1974). The cloaca appears to be a characteristic of lepidopterans' development.

In Trichoptera, ninth tergum and sternum are partially fused, with ninth sternum bearing one- or two-segmented gonopods, sometimes fused ventromedially (Wojtowicz, 1982). The two-segmented condition consists of a gonocoxite and apical gonostyle, usually immovably, while one-segmented condition results from loss of gonostyle or from its fusion to coxopodite; sometimes coxopodite itself is fused to ninth segment (Holzenthal *et al.*, 2007). The lateropenites, also called *titilators* or parameres (Betten, 1950; Holzenthal *et al.*, 2007), present in form of long, thin, and flexible rods, situated dorsally to penial sclerite, within invaginated membranous sheath. The penial sclerite is tube-shaped, with a thickened basal section, also called phallobase (Holzenthal *et al.*, 2007), is situated ventromedially between gonocoxae in an invaginated membranous sheath, along with lateropenites (Boudinot, 2018). As in Lepidoptera, female of trichopterans lack gonapophyseal ovipositor but instead have terminal abdominal segments (eighth to tenth and eleventh segments) either elongated into a protrusible oviscapt for placing eggs on substrate or modified into a shorter apparatus for forming and holding egg mass (Holzenthal *et al.*, 2007). External female genitalia are considerably simplified compared to those of male, but internal structures that engage apex of phallus during copulation are often complex and species-specific (Nielsen, 1980).

5.2.7.5 Mecoptera, Siphonaptera, and Diptera (Antliophora)

In all of these three orders, genital segment consists of epandrium (ninth tergum) and hypandrium (ninth sternum), with genital apparatus positioned in-between them (Matsuda, 1976). In Mecoptera, male genitalia are comprised of laterally paired gonopods and median aedeagus complex. The gonopods are two-segmented, being composed of basal gonocoxites and distal gonostyli. The lateropenites are absent. The aedeagus consists of paired ventral and dorsal valves, with phallotrema situated centrally between two paired valves. The parameres are Y-shaped with a basal stalk and two apical branches (ventral and dorsal parameres), each of which terminates into a flattened lobe (Ma *et al.*, 2010; Boudinot, 2018). In Diptera, copulatory apparatus comprises epandrium, hypandrium, genital appendages, and proctiger, all are substantially modified (Boudinot, 2018). Gonocoxa are situated at posterior apices of ninth sternum, about a third as long anteroposteriorly as ninth sternum, bearing styli bifid, with horn-shaped ventrolateral tine and phylliform dorsomedial. The lateropenites are small and lobate. The aedeagus is divided into two distinct units: an anterior penial sclerite consisting of a bulb, bearing a long needle dorsomedially; and a posterior penial sclerite with a single disc- or plate-shaped apodeme nestled anteroventrally between lateropenites, and bearing paired horn-like processes, which guide penial needle (Boudinot, 2018). The male genitalia of Siphonaptera include several sclerites: ninth tergum fused with a set of gonocoxal fragments; ninth sternum more-or-less T-shaped, with paired anterior arms and a lateromedially narrow posteromedial lobe; gonocoxa divided into several fragments, with first subfragment bearing styli that vary in shape, and second subfragment bearing lateropenites (Boudinot, 2018). The penis consists of external aedeagus, divided several times, that bears aedeagal apodeme internally and internal endophallus, which is invaginated from aedeagus (Matsuda, 1976).

In most females of Mecoptera and Cyclorrapha (Diptera), terminal segments are retractable into each other in a telescope-like manner (Hünefeld & Beutel, 2012). In Nannochoristidae, Meropidae, and Panorpidae, an internal structure called medigynium lies above subgenital plate (Tjeder, 1970b). The genital valvulae of eighth and ninth segments are homologizable with gonapophyses in other orders (Matsuda, 1976; Hünefeld & Beutel, 2012). In Siphonaptera, common oviduct opens posteriorly through vulva between eighth and ninth sternum, and rudiments of gonapophyses may be found in form of ectodermal thickenings on eighth abdominal segment; they are present in pupa but disappear in adult (Matsuda, 1976). In Diptera, functional visceral region is reduced to five segments by conversion of posterior segments into a tubular organ of oviposition without appendices (valvulae) (Snodgrass, 1931).

5.3 Mechanism for Evolution of Genitalia

5.3.1 Sexual Selection Driving Genital Morphology

Males and females have divergent interests in reproduction that are rooted in anisogamy and lead to distinct roles, where traits favoured by one sex might be costly to other (Chapman *et al.*, 2003). The result is sexually antagonistic coevolution (SAC) between interacting traits in males and females. Some examples are mentioned below:

Perhaps most striking case of SAC occurs in some Heteroptera with traumatic insemination. Such insemination is known to impose a cost on female in form of physical damage and by increasing risk of infection (Morrow & Arnqvist, 2003; Reinhardt *et al.*, 2003). To mitigate these costs, females of many species have evolved unique internal and external paragenital counter-adaptation for traumatic insemination (Morrow & Arnqvist, 2003; Reinhardt *et al.*, 2003), also known as *spermalege* (Carayon, 1959).

In bruchid beetle *Callosobruchus maculatus* Fabricius, expression of some male genital traits (notably, spines of male intromittent organ) has been shown to affect male competitive fertilization success (Hotzy *et al.*, 2012) and to impose substantial fitness costs on females (Rönn *et al.*, 2007). Likewise, *C. maculatus* females possess a complex genital tract that also includes spines in bursa copulatrix. Although function of bursal spines is unclear, their marked enlargement in another *Callosobruchus* species suggests a potential sexually antagonistic role (e.g. prevention or limitation of intromission) (Cayetano & Bonduriansky, 2015).

In water striders, *Gerris*, male genital segments are elongated and shaped for grasping females. During premating struggles and subsequent mating, inflatable genitalia emerge from pre-genital segment that houses them and grasp female's genital segments (Perry & Rowe, 2012). After controlling for body size, it was found that shape of sclerite, influenced fertilization success: males that had stout ventral sclerites and narrow and inwardly curved dorsal sclerites sired significantly more offspring (Arnqvist & Danielsson, 1999). On the other hand, females possess traits that help them resist male mating attempts as enlarged and upturned connexival spines above their genitalia, which are angled downward, make it easier to repel copulation attempts by males (Arnqvist, 1997; Arnqvist & Rowe, 2002).

Additionally, gynosome of psocodean *Neotrogla* could represents a novel genital organ that has probably been driven by reversed sexual selection on copulatory function (Andersson, 1994; Yoshizawa *et al.*, 2014), but it remains unclear.

However, sperm competition (when ejaculates of two or more males compete to fertilize an ovum) could also cause genital divergence (Parker, 1984). Some examples are: (1) in odonates and other insects, selection has favoured males with genitalia that remove rival sperm from female. In certain damselflies, males displace sperm of other precedent males, which is stored inside female's spermatheca or bursa copulatrix using pair of lobes or flagella of ligula (penis) (Waage, 1979, 1986); (2) several species of *Drosophila* Fallén exhibit what is called insemination reaction. This phenomenon is an enlargement of vagina produced by male's ejaculate preventing female from re-mating for a certain period and thus ensuring him paternity of offspring (Patterson, 1946); (3) during copulation, male beetles of genus *Tenebrio* Linnaeus mark females with an anti-aphrodisiacal pheromone, which is transferred via genital apparatus, repelling other males (Happ, 1969); (4) mating plugs, or copulatory plugs, are male-produced substances or structures that are transferred to female upon copulation and promote sealing of her genital opening, which will, ultimately, prevent her from copulating with other males, at least for a certain period of time. The spermatophore itself, in some orthopterans, may constitute mating plug once it is inside female's genital duct (Gregory, 1965; Mello, 2007).

5.3.2 Natural Selection as a Predictor of Coevolution between Male and Female Genitalia

Cryptic female choice (CFC), as well as SAC are plausible hypotheses, which have been erected by many authors to explain evolution of genitalia. In fact, both mechanisms are known to predict coevolution between male and female genitalia; for years, many studies have focused only on sexual selection to explain diversity of genitalia. However, natural selection is also important and can also be put forward to consistently explain faster morphological diversification of parts of male and female genitalia. More specifically, 'lock-and-key' mechanism (a natural selection mechanism) can be evoked as being driver of genital diversity, since only male genitalia parts with a specific morphology are expected to 'key' corresponding female genitalia part, a mechanism against hybridization. Ishikawa (1987) demonstrated that female genitalia have coevolved with male genitalia in *Ohomopterus* (Coleoptera), both in size and shape. In addition, Sota and Kubota (1998) also indicated such a mechanism to be an important cause of covariation between male and female genital morphology, by keeping a barrier against hybridization in two species of *Carabus* (Coleoptera).

It is admissible to accept idea that we cannot easily discriminate underlying evolutionary responsible mechanisms causing covariation between male and female genitalia, especially in studies of a single species. The problem is to assume that 'females use sensory mechanisms' to discriminate superior males from inferior ones ('cooperative evolution' according to Yassin & Orgogozo, 2013). However, females do not necessarily use their neural system. Sometimes, just having an appropriate genital morphology is enough for this purpose (see Peretti & Aisenberg, 2015). So, just as 'lock-and-key' is a natural selection mechanism, sexual selection via CFC can pose strong stabilizing selection pressure not only on shape of male genitalia, but also on those of female genitalia. On other hand, 'lock-and-key' natural selection to discriminate conspecific males from heterospecific ones can also be mediated by sensory mechanisms (*sensory lock-and-key*; see Simmons, 2014). A recent study with *Belostoma angustum* (Heteroptera) has stressed apparent lack of evidence in disentangling isolation from CFC mechanisms (Ribeiro *et al.*, 2019). Apparently, male genitalia in this species covary with dimensions of female genitalia. Also, a significant correlation between shape of some intromittent male genitalic parts and genital chamber size of females was found, suggesting a coevolution of these traits.

5.4 Conclusions

Despite many striking variations in male and female genitalia and difficulty in homologizing genitalic elements, much progress was made in the past century. For example, studies of phylogenetic relationship between taxa of Hexapoda allowed comparison between morphological variations and proposed some good hypotheses of homology between genitalic components. Also, study of skeletal musculature considering phylogenetic relationship allowed refinement of those homologies and added clarity origin of genitalic components (e.g. Boudinot, 2018; Klass & Matushkina, 2018). For example, coxopod hypothesis for origin of penis has been supported nowadays. Additionally, study of genetics of development integrated with qualitative or quantitative observations of fossils could also help elucidate open questions about genital homologies in hexapods and make estimates of tempo and mode of evolution of male and female genitalia.

Although knowledge about homologies and variations of male and female genitalia has made impressive progress, mechanisms that have driven such variation remains unclear. Some alternative sources involving sexual selection were postulated and massively discussed in the past decades as intersexual sources (SAC and CFC) and intrasexual as sperm competition. However, natural selection should not be ignored, since only male genitalic parts with a specific morphology are expected to 'key' corresponding female genitalic part as postulated by lock-and-key hypothesis. It is expected that controlled observational experiments and genetics approaches with many insect taxa could contribute to discriminate underlying evolutionary responsible mechanisms driving genital variation.

REFERENCES

Acker, T.S. (1960). The comparative morphology of male genitalia of Neuroptera (Insecta). *Microentomology*, *24*(2), 25–84.

Alexander, R.D., & Otte, D.D. (1967). *The Evolution of Genitalia and Mating Behavior in Crickets (Gryllidae) and other Orthoptera*. Miscellaneous Publications, Museum of Zoology, University of Michigan. *133*, 1–62.

Andersson, M. (1994). *Sexual Selection* (Vol. 72). Princeton University Press.

Arnqvist, G., & Danielsson, I. (1999). Postmating sexual selection: effects of male body size and recovery period on paternity and egg production rate in a water strider. *Behavioral Ecology*, *10*(4), 358–365.

Arnqvist, G., & Rowe, L. (2002). Correlated evolution of male and female morphologies in water striders. *Evolution*, *56*(5), 936–947.

Arnqvist, G. (1997). The evolution of animal genitalia: distinguishing between hypotheses by single species studies. *Biological Journal of Linnean Society*, *60*(3), 365–379.

Aspöck, U. (2002). Male genital sclerites of Neuropterida: an attempt at homologisation (Insecta: Holometabola). *Zoologischer Anzeiger-A Journal of Comparative Zoology*, *241*(2), 161–171.

Badonnel, A. (1970). Psocoptera. In: Tuxen, S.L. (Ed.), *Taxonomist's Glossary of Genitalia in Insects*, second ed. (pp. 172–175). Munksgaard, Copenhagen.

Beani, L., Giusti, F., Mercati, D., Lupetti, P., Paccagnini, E., Turillazzi, S., & Dallai, R. (2005). Mating of Xenos vesparum (Rossi) (Strepsiptera, Insecta) revisited. *Journal of Morphology*, *265*(3), 291–303.

Bechly, G., Brauckmann, C., Zessin, W., & Gröning, E. (2001). New results concerning morphology of most ancient dragonflies (Insecta: Odonatoptera) from Namurian of Hagen-Vorhalle (Germany). *Journal of zoological Systematics and Evolutionary Research*, *39*(4), 209–226.

Betten, C. (1950). The genus Pycnopsyche (Trichoptera). *Annals of Entomological Society of America*, *43*(4), 508–522.

Bitsch, J. (1974). Morphologie abdominale des machilides (thysanura)—III. Squelette et musculature des segments génitaux mâles et des segments postgénitaux. *International Journal of Insect Morphology and Embryology*, *3*(2), 203–224.

Boudinot, B.E. (2018). A general theory of genital homologies for Hexapoda (Pancrustacea) derived from skeletomuscular correspondences, with emphasis on Endopterygota. *Arthropod Structure & Development*, *47*(6), 563–613.

Brinck, P. (1956). Reproductive system and mating in Plecoptera. *Opuscula Entomologica*, *21*, 57–127.

Brinck, P. (1957). Reproductive system and mating in Ephemeroptera. *Opuscula Entomologica*, *22*, 1–37.

Burckhardt, D. (2009). Taxonomy and phylogeny of Gondwanan moss bugs or Peloridiidae (Hemiptera, Coleorrhyncha). *Deutsche Entomologische Zeitschrift*, *56*(2), 173–235.

Burr, M. (1916). IX. Note on Manubrium of Ninth Sternite in Male Earwig. *Transactions of Royal Entomological Society of London*, *63*(3-4), 269–272.

Carayon, J. (1959). Insémination par "spermalege" et cordon conducteur de spermatozoıdes chez *Stricticimex brevispinosus* Usinger (Heteroptera, Cimicidae). *Revue de zoologie et de botanique africaines*, *60*, 81–104.

Cayetano, L., & Bonduriansky, R. (2015). Condition dependence of male and female genital structures in seed beetle *Callosobruchus maculatus* (Coleoptera: Bruchidae). *Journal of Evolutionary Biology*, *28*(7), 1364–1372.

Chapman, T., Arnqvist, G., Bangham, J., & Rowe, L. (2003). Sexual conflict. *Trends in Ecology & Evolution*, *18*(1), 41–47.

Cheng, Z., & Yoshizawa, K. (2019). Functional morphology of Trichadenotecnum male and female genitalia analyzed using µCT (Insecta: Psocodea: Psocomorpha). *Journal of Morphology*, *280*(4), 555–567.
Cordero-Rivera, A., & Córdoba-Aguilar, A. (2010). Selective forces propelling genitalic evolution in Odonata. *The Evolution of Primary Sexual Characters in Animals*, *332*, 352.
Crampton, G.C. (1920a). Some anatomical details of remarkable winged zorapteron, *Zorotypus hubbardis* [sic] Caudell, with notes on its relationships. *Proceedings of Entomological Society of Washington*, *22*(5), 98–106.
Crampton, G.C. (1920b). A comparison of genitalia of Male Hymenoptera, Mecoptera, Neuroptera, Diptera, Trichoptera, Lepidoptera, Homoptera, and Strepsiptera, with those of lower insect. *Psyche*, *27*(2–3), 34–45.
Crowson, R.A. (1955). *The Natural Classification of Families of Coleoptera*. London: Nathanial Lloyd & Co., Ltd..
Crowson, R.A. (1981). *The Biology of Coleoptera*. London: Academic Press.
Dallai, R., Mercati, D., Gottardo, M., Machida, R., Mashimo, Y., & Beutel, R.G. (2011). The male reproductive system of *Zorotypus caudelli* Karny (Zoraptera): sperm structure and spermiogenesis. *Arthropod Structure & Development*, *40*(6), 531–547.
Delamare-Deboutteville, C. (1970). Zoraptera. In: Tuxen, S.L. (Ed.), *Taxonomist's Glossary of Genitalia in Insects*, second ed. (pp. 46–49) Munksgaard, Copenhagen.
Dugdale, J.S. (1974). Female genital configuration in classification of Lepidoptera. *New Zealand Journal of Zoology*, *1*(2), 127–146.
Dupuis, C. (1955) Les genitalia des Hemipteres Heteropteres (Genitalia externes des deux sexes; voies ectodermiques femelles). *Memoires du Museum National d'Histoire Naturelle*, *6*(4), 183–278.
Eades, D.C. (2000). Evolutionary relationships of phallic structures of Acridomorpha (Orthoptera). *Journal of Orthoptera Research*, *9*, 181–210.
Engel, M.S. (2004). Zoraptera. In: Bousquets, J.L., Morrone, J.J., Ordóñez, O.Y., Fernández, I.V. (Eds.), *Biodiversidad, Taxonomía y Biogeografía de Artrópodos de México*, (vol. IV, pp. 637–640). Universidad Nacional Autónoma de México.
Fraser, F.G., & Asahina, S. (1970). Odonata. In: Tuxen, S.L. (Ed.), *Taxonomist's Glossary of Genitalia in Insects*, second ed. (pp. 32–37). Munksgaard, Copenhagen.
Giles, E.T. (1963). The comparative external morphology and affinities of Dermaptera. *Transactions of Royal Entomological Society of London*, *115*(4), 95–164.
Gregory, G.E. (1965). The formation and fate of spermatophore in African migratory locust, Locusta migratoria migratorioides Reiche and Fairmaire. *Transactions of Royal Entomological Society of London*, *117*(2), 33–66.
Grimaldi, D., Engel, M.S., & Engel, M.S. (2005). *Evolution of Insects*. Cambridge University Press.
Happ, G.M. (1969). Multiple sex pheromones of mealworm beetle, Tenebrio molitor L. *Nature*, *222*(5189), 180–181.
Helm, C., Treulieb, S., Werler, K., Bradler, S., & Klass, K.D. (2011). The male genitalia of *Oxyartes lamellatus*–phasmatodeans do have complex phallic organs (Insecta: Phasmatodea). *Zoologischer Anzeiger*, *250*(3), 223–245.
Helms, T.J. (1968). Postembryonic reproductive systems development in *Empoasca fabae*. *Annals of Entomological Society of America*, *61*, 316–332.
Heymons, R. (1897). Entwicklungsgeschichte Untersuchungen an *Lepisma saccharina* L. *Zeitschrift für wissenschaftliche Zoologie*, *62*, 582–631.
Holzenthal, R.W., Blahnik, R.J., Prather, A.L., & Kjer, K.M. (2007). Order Trichoptera Kirby, 1813 (Insecta), Caddisflies. In: Zhang, Z.-Q., & Shear, W.A. (Eds), *Linnaeus Tercentenary: Progress in Invertebrate Taxonomy. Zootaxa*. (Vol. 1668, pp. 1–766).
Hotzy, C., Polak, M., Rönn, J.L., & Arnqvist, G. (2012). Phenotypic engineering unveils function of genital morphology. *Current Biology*, *22*(23), 2258–2261.
Hünefeld, F. (2007). The genital morphology of *Zorotypus hubbardi* Caudell, 1918 (Insecta: Zoraptera: Zorotypidae). *Zoomorphology*, *126*(3), 135–151.
Hünefeld, F., & Beutel, R.G. (2012). The female postabdomen of enigmatic Nannochoristidae (Insecta: Mecopterida) and its phylogenetic significance. *Acta Zoologica*, *2*(93), 231–238.

Hünefeld, F., Mißbach, C., & Beutel, R.G. (2012). The morphology and evolution of female postabdomen of Holometabola (Insecta). *Arthropod Structure & Development*, *41*(4), 361–371.
Hünefeld, F., Pohl, H., Wipfler, B., Beckman, F., Beutel, R.G. (2011). The male postabdomen and genital apparatus of †*Mengea tertiaria*, a strepsipteran amber fossil (Insecta). *Journal of Zoological Systematics and Evolutionary Research*, *49*, 298–308.
Ishikawa, R. (1987). On function of copulatory organs of *Ohomopterus* (Coleoptera, Carabidae, Genus *Carabus*). *Kontyu*, *55*, 202–206.
Jansson, A., & Meyer-rochow, V.B. (1990). Sound production in *Synaptonecta issa* (Heteroptera: Corizidae, Micronectinae): an Asian bug that turned up in a New Zealand aquarium. *Applied Entomology and Zoology*, *25*(2), 328–331.
Klass, K.D. (1997). The external male genitalia and phylogeny of Blattaria and Mantodea. *Bonner Zoologische Monographien*, *42*, 1–341.
Klass, K.D., & Matushkina, N.A. (2018). The exoskeleton of male genitalic region in archaeognatha with hypotheses on early evolution and morphological interpretation of genitalia in insects. *Arthropod Systematics and Phylogeny*, *76*(2), 235–294.
Klass, K.D., Matushkina, N.A., & Kaidel, J. (2012). The gonangulum: a reassessment of its morphology, homology, and phylogenetic significance. *Arthropod Structure & Development*, *41*(4), 373–394.
Klass, K.D., Picker, M.D., Damgaard, J., van Noort, S.I.M.O.N., & Tojo, K. (2003). The taxonomy, genitalic morphology, and phylogenetic relationships of Southern African Mantophasmatodea (Insecta). *Entomologische Abhandlungen*, *61*(1), 3–67.
Klass, K.D., Thorne, B.L., & Lenz, M. (2000). The male postabdomen of *Stolotermes inopinus*: a termite with unusually well-developed external genitalia (Dictyoptera: Isoptera: Stolotermitinae). *Acta Zoologica*, *81*(2), 121–130.
Kluge, N. (2004). *The Phylogenetic System of Ephemeroptera*. Springer Science & Business Media.
Kristensen, N.P. (1984). The male genitalia of Agathiphaga (Lepidoptera: Agathiphagidae) and lepidopteran ground plan. *Insect Systematics & Evolution*, *15*(2), 151–178.
Ma, N., Zhong, W., & Hua, B. (2010). Genitalic morphology and copulatory mechanism of scorpionfly Panorpa jilinensis (Mecoptera: Panorpidae). *Micron*, *41*(8), 931–938.
Machida, R. (2000). Serial homology of mandible and maxilla in jumping bristletail Pedetontus unimaculatus Machida, based on external embryology (Hexapoda: Archaeognatha, Machilidae). *Journal of Morphology*, *245*(1), 19–28.
Maderspacher, F. (2016). Zoology: walking heads. *Current Biology*, *26*(5), R194–R197.
Matsuda, R. (1976). *Morphology and Evolution of Insect Abdomen: With Special Reference to Developmental Patterns and Their Bearing upon Systematics*. Oxford: Pergamon Press.
Mello, F.D.A. (2007). Female monopolization and paternity assurance in South American crickets (Orthoptera, Grylloidea): mating plugs, extra claspers and forced copulation. *Papéis avulsos de zoologia*, *47*(20), 245–257.
Michener, C.D. (1944). A comparative study of appendages of eighth and ninth abdominal segments of insects. *Annals of Entomological Society of America*, *37*(3), 336–351.
Michener, C.D. (1956). Hymenoptera. In: Tuxen, S.L. (Ed.), *Taxonomist's Glossary of Genitalia in Insects*, first ed. (pp. 131–140). Eljnar Munksgaard, Copenhagen.
Mickoleit, G. (1973). Über den ovipositor der neuropteroidea und coleoptera und seine phylogenetische bedeutung (Insecta, Holometabola). *Zeitschrift für Morphologie der Tiere*, *74*(1), 37–64.
Misof, B., Liu, S., Meusemann, K., Peters, R.S., Donath, A., Mayer, C., ... & Niehuis, O. (2014). Phylogenomics resolves timing and pattern of insect evolution. *Science*, *346*(6210), 763–767.
Morrow, E.H., & Arnqvist, G. (2003). Costly traumatic insemination and a female counter-adaptation in bed bugs. *Proceedings of Royal Society of London. Series B: Biological Sciences*, *270*(1531), 2377–2381.
New, T.R. (1989). Planipennia, lacewings. *Handbuch der Zoologie (Berlin)*, *4*, 1–132.
Nielsen, A. (1957). On evolution of genitalia in male insects. *Entomologiske Meddelelser*, *28*, 27–57.
Nielsen, A. (1980). A comparative study of genital segments and genital chamber in female Trichoptera. *Biologiske Skrifter Danske Videnskabernes Selskab*, *23*, 1–200.
Niven, J.E., Graham, C.M., & Burrows, M. (2008). Diversity and evolution of insect ventral nerve cord. *Annual Review of Entomology*, *53*, 253–271.

Parker, G.A. (1984). Sperm competition and evolution of animal mating strategies. In: Smith R.L. (Ed.), *Sperm Competition and Evolution of Animal Mating Systems* (pp. 1–60). New York: Academic Press.
Pass, G., & Szucsich, N.U. (2011). 100 years of research on Protura: many secrets still retained. *Soil Organisms*, *83*(3), 309–334.
Patterson, J.T. (1946). A new type of isolating mechanism in Drosophila. *Proceedings of National Academy of Sciences of USA*, *32*(7), 202.
Peinert, M., Wipfler, B., Jetschke, G., Kleinteich, T., Gorb, S.N., Beutel, R.G., & Pohl, H. (2016). Traumatic insemination and female counter-adaptation in Strepsiptera (Insecta). *Scientific Reports*, *6*(1), 1–10.
Peretti, A.V., & Aisenberg, A. (2015). Cryptic female choice in arthropods. *Patterns, Mechanisms and Prospects*. Springer.
Perry, J.C. & Rowe, L. (2012). Sexual conflict and antagonistic coevolution across water strider populations. *Evolution*, *66*(2), 544–557.
Pohl, H., & Beutel, R.G. (2008). The evolution of Strepsiptera (Hexapoda). *Zoology*, *111*(4), 318–338.
Popham, E.J. (1965). The functional morphology of the reproductive organs of the Common earwing (Forficula auricularia) and other Dermaptera with reference to the natural classification of the order. Journal of Zoology, *146*, 1–43.
Priesner, H. (1960). A monograph of Thysanoptera of Egyptian deserts. *Publications de la Institute Desert Egypt*, *13*, 1–549.
Priesner, H. (1970). Thysanoptera. In: Tuxen, S.L. (Ed.), *Taxonomist's Glossary of Genitalia in Insects*, second ed. (pp. 209–213). Copenhagen: Munksgaard.
Qadri, M.A.H. (1936). Male genitalia of Mallophaga infesting North-Indian Birds. *Proceedings of Indian Academy of Sciences*,*4*(6), 454–470.
Qadri, M.A.H. (1940). On development of genitalia and their ducts of orthopteroid insects. *Transactions of Royal Entomological Society of London*, *90*(6), 121–175.
Qadri, M.A.H. (1949). On morphology and post-embryonic development of male genitalia and their ducts in Hemiptera (Insecta). *Journal of Zoolofical Society of India*, *1*(2), 129–143.
Reinhardt, K., Naylor, R., & Siva–Jothy, M.T. (2003). Reducing a cost of traumatic insemination: female bedbugs evolve a unique organ. *Proceedings of Royal Society of London. Series B: Biological Sciences*, *270*(1531), 2371–2375.
Ribeiro, J.R.I., Stefanello, F., Bugs, C., Stenert, C., Maltchik, L., & Guilbert, E. (2019). Coevolution between male and female genitalia in *Belostoma angustum* Lauck, 1964 (Insecta, Heteroptera, Belostomatidae): disentangling size and shape. *Zoology*, *137*, 125711.
Rönn, J., Katvala, M., & Arnqvist, G. (2007). Coevolution between harmful male genitalia and female resistance in seed beetles. *Proceedings of National Academy of Sciencesof USA*, *104*(26), 10921–10925.
Ross, E.S. (1940). A revision of Embioptera of North America. *Annals of Entomological Society of America*, *33*(4), 629–676.
Roth, L.M. (1970). Evolution and taxonomic significance of reproduction in Blattaria. *Annual Review of Entomology*, *15*(1), 75–96.
Schuh, R.T., & Slater, J.A. (1995). *True Bugs of World (Hemiptera: Heteroptera): Classification and Natural History*. Cornell University press.
Schuh, R.T., & Štys, P. (1991). Phylogenetic analysis of cimicomorphan family relationships (Heteroptera). *Journal of New York Entomological Society*, *99*(3), 298–350.
Schulmeister, S. (2001). Functional morphology of male genitalia and copulation in lower Hymenoptera, with special emphasis on Tenthredinoidea *s. str.*(Insecta, Hymenoptera,'Symphyta'). *Acta Zoologica*, *82*(4), 331–349.
Scudder, G.G.E. (1957). Reinterpretation of some basal structures in insect ovipositor. *Nature*, *180*(4581), 340–341.
Scudder, G.G.E. (1959). The female genitalia of Heteroptera: morphology and bearing on classification. *Transactions of Royal Entomological Society of London*, *111*(4), 405–467.
Scudder, G.G.E. (1961). The comparative morphology of insect ovipositor. *Transactions of Royal Entomological Society of London*, *113*(2), 25–40.
Scudder, G.G.E. (1970). Grylloblattodea. In: Tuxen, S.L. (Ed.), *Taxonomist's Glossary of Genitalia in Insects*, second ed. (pp. 55–58). Munksgaard, Copenhagen.

Sharkey, M.J. (2007). Phylogeny and classification of Hymenoptera. *Zootaxa, 1668*(1), 521–548.
Simmons, L.W. (2014). Sexual selection and genital evolution. *Austral Entomology, 53*, 1–17.
Smith, F.W., Boothby, T.C., Giovannini, I., Rebecchi, L., Jockusch, E.L., & Goldstein, B. (2016). The compact body plan of tardigrades evolved by loss of a large body region. *Current Biology, 26*(2), 224–229.
Snodgrass, R.E. (1931). Morphology of insect abdomen. Part I. General structure of abdomen and its appendages. *Smithsonian Miscellaneous Collection, 85*, 1–128.
Snodgrass, R.E. (1933). Morphology of insect abdomen. Part II. The genital ducts and ovipositor. *Smithsonian Miscellaneous Collection, 89*, 1–148.
Snodgrass, R.E. (1935). *Principles of Insect Morphology*. London: Comstock Publishing Associates.
Snodgrass, R.E. (1936). Morphology of insect abdomen. Part III. The male genitalia (including arthropods other than insects). *Smithsonian Miscellaneous Collection, 96*, 1–96.
Snodgrass, R.E. (1937). The male genitalia of orthopteroid insects. *Smithsonian Miscellaneous Collections, 96*(5), 1–107.
Snodgrass, R.E. (1957). A revised interpretation of external reproductive organs of male insects. *Smithsonian Miscellaneous Collections, 135*, 1–60.
Sommerman, K.M. (1946). A revision of genus *Lachesilla* north of Mexico (Corrodentia: Caeciliidae). *Annals of Entomological Society of America, 39*(4), 627–661.
Sota, T., & Kubota, K. (1998). Genital lock-and-key as a selective agent against hybridization. *Evolution, 52*, 1507–1513.
Sueur, J., Mackie, D., & Windmill, J.F. (2011). So small, so loud: extremely high sound pressure level from a pygmy aquatic insect (Corixidae, Micronectinae). *PLoS One, 6*(6), e21089.
Tatarnic, N.J., Cassis, G., & Hochuli, D.F. (2006). Traumatic insemination in plant bug genus *Coridromius* Signoret (Heteroptera: Miridae). *Biology Letters, 2*(1), 58–61.
Tilgner, E.H., Kiselyova, T.G., & McHugh, J.V. (1999). A morphological study of *Timema cristinae* Vickery with implications for phylogenetics of Phasmida. *Deutsche Entomologische Zeitschrift, 46*(2), 149–162.
Tillyard, R.J. (1917). *The Biology of Dragonflies:(Odonata or Paraneuroptera)*. CUP Archive.
Tomoyasu, Y., Ohde, T., & Clark-Hachtel, C. (2017). What serial homologs can tell us about the origin of insect wings. F1000Research, 6.
Tjeder, B. (1970a). Neuroptera. In: Tuxen, S.L. (Ed.), *Taxonomist's Glossary of Genitalia in Insects*, second ed. (pp. 89–99). Munksgaard, Copenhagen.
Tjeder, B. (1970b). Mecoptera. In: Tuxen, S.L. (Ed.), *Taxonomist's Glossary of Genitalia in Insects*, second ed. (pp. 99–104). Munksgaard, Copenhagen.
Tuxen, S.L. (1963). Phylogenetical trends in Protura as shown by relationship between recent genera. *Journal of Zoological Systematics and Evolutionary Research, 1*(3-4), 277–310.
Tuxen, S.L. (1970a). Protura. In: Tuxen, S.L. (Ed.), *Taxonomist's Glossary of Genitalia in Insects*, second ed. (pp. 21–24). Munksgaard, Copenhagen.
Tuxen, S.L. (1970b). Collembola. In: Tuxen, S.L. (Ed.), *Taxonomist's Glossary of Genitalia in Insects*, second ed. (pp. 24–25). Munksgaard, Copenhagen.
Tuxen, S.L. (1970c). Diplura. In: Tuxen, S.L. (Ed.), *Taxonomist's Glossary of Genitalia in Insects*, second ed. (pp. 25–27) Munksgaard, Copenhagen.
Van Haren, M.M., Rönn, J.L., Schilthuizen, M., & Arnqvist, G. (2017). Postmating sexual selection and enigmatic jawed genitalia of *Callosobruchus subinnotatus*. *Biology Open, 6*(7), 1008–1012.
Verhoeff, K.W. (1902). Zur vergleichenden Morphologie der Coxalorgane und Genitalanhänge der Tracheaten. *Zoologischer Anzeige, 26*, 60–77.
Waage, J.K. (1979). Dual function of damselfly penis: sperm removal and transfer. *Science, 203*(4383), 916–918.
Waage, J.K. (1986). Evidence for widespread sperm displacement ability among Zygoptera (Odonata) and means for predicting its presence. *Biological Journal of Linnean Society, 28*(3), 285–300.
Walker, E.M. (1919). On male and immature state of *Grylloblatta campodeiformis* Walker. *The Canadian Entomologist, 51*(6), 131–139.
Walker, E.M. (1922). The terminal structures of orthopteroid insects: a phylogenetic study. *Annals of entomological Society of America, 15*(1), 1–76.

Wojtowicz, J.A. (1982). A review of adults and larvae of genus *Pycnopsyche* (Trichoptera: Limnephilidae) with revision of *Pycnopsyche scabripennis* (Rambur) and *Pycnopsyche lepida* (Hagen) complexes. Ph.D. dissertation, University of Tennessee. 292 pp.

Yassin, A., Orgogozo, V. (2013). Coevolution between male and female genitalia in *Drosophila melanogaster* species subgroup. *PLoS One*, *8*, e57158.

Yoshizawa, K. (2005). Morphology of Psocomorpha (Psocodea: 'Psocoptera'). *Insecta matsumuranaNew series: Journal of Faculty of Agriculture Hokkaido University, series entomology.*, *62*, 1–44.

Yoshizawa, K., & Johnson, K.P. (2006). Morphology of male genitalia in lice and their relatives and phylogenetic implications. *Systematic Entomology*, *31*(2), 350–361.

Yoshizawa, K., Ferreira, R.L., Kamimura, Y., & Lienhard, C. (2014). Female penis, male vagina, and their correlated evolution in a cave insect. *Current Biology*, *24*(9), 1006–1010.

6

Courtship

Ahmad Pervez[1] and Omkar[2]
[1]*Biocontrol Laboratory, Department of Zoology, Sri Dev Suman Uttarakhand University, Rishikesh Campus, Uttarakhand, India*
[2]*Ladybird Research Laboratory, Department of Zoology, University of Lucknow, Lucknow - 226 007, India*

CONTENTS

DOI: 10.1201/9781003043195-6

6.1 Introduction

Courtship is an important behavioural activity of sexual partners, particularly males, in majority of insect for fitness gains via selection as mates. Insects from orders Diptera, Coleoptera, Hymenoptera, Orthoptera, Lepidoptera, and Hemiptera, particularly, exhibit conspicuous courtship displays to appease their sexual partners (Table 6.1). It is mediated by different kinds of visual, acoustic, tactile, and other quality indicating signals, which vary considerably between different insect taxa. A few of these courtships occur in high frequencies that may have serious ecological and evolutionary implications.

Sexual selection theory states that courtship displays may be considered as an intrasexual competition among males for their copulation probabilities and paternity (Ritchie, 2007; Miller & Svensson, 2014). Courtship by males allows choosy females to assess their quality prior to mating, as females refrain from indiscriminate mating (Rowe *et al.,* 1994; Hernǻndez & Fabre, 2017). Female mate choice may further be governed by 'sexy sons' hypothesis, which states female directly (mate preference) or indirectly (cryptic mate choice) choose ideal mate for higher offspring fitness (Weatherhead & Robertson, 1979). The Fisherian runaway theory explains the evolution of exaggerated male ornaments and female mate choice as a form of sexual selection mediated by courtship activities (Fisher, 1930).

Proximate causes for courtship are chemical cues, operational sex ratio (OSR), female mating preference, body size, and overcrowding during immature development of males. Similarly, ultimate evolutionary causes are male sexual traits, intrasexual competition, cryptic mate choice, and sexual conflict. This review highlights influence of environmental (temperature and light) and physiological (juvenile and ecdysis hormones, and neuropeptides) factors on courtship behaviour.

We have reviewed the current literature on courtship behaviour in insects to understand: (i) types of courtship, (ii) proximate and ultimate causes, (iii) factors affecting them, and (iv) role in sexual selection.

6.2 Chemical Cues

Chemical signals, particularly pheromones, both volatile and non-volatile, are courtship initiating factors meant for intraspecific communication in almost all insects (Wyatt, 2014). Volatile or air-borne pheromones are detected through sensory neurons of olfactory system via sensillae located on antennae (Hansson & Stensmyr, 2011), while non-volatile signals are perceived by contact chemoreception through neurons of gustatory system via sensilla located on other appendages, viz. proboscis and legs (Kohl *et al.*, 2015; Fleischer *et al.*, 2018).

Sex pheromones induce intense male courtship and aggression towards competitive males (Stamps & Shaw, 2019). Mixing with plant volatiles from locality of mating partner, increases pheromone potency (Markow & O'Grady, 2008; Trona *et al.*, 2013; Xu & Turlings, 2018), even inducing females to release sex pheromones. Sex pheromone glands could be at base of second valvifer (Field & Keller, 1994), Dufour's gland (Ernst *et al.*, 2013), etc. Chemical cues do not work in isolation but are often integrated with other chemical, contact, and visual stimuli (Kurtovic *et al.*, 2007; Markow & O'Grady, 2008; Trona *et al.*, 2013). Research, however, is needed to identify and characterize these sex pheromones and aphrodisiacs.

6.3 Different Types of Courtship

Courting insects use different forms of visual displays, auditory stimuli, tactile stimulations, and nuptial gifts to appease mating partners.

TABLE 6.1

List of visual courtship events by insects of various taxonomic groups

Name of Insect	Order: Family	Courtship Events	References
Hermetia illucens L.	Diptera: Stratiomyidae	Approach, mounting, fanning, tapping abdomen, mounting	Giunti *et al.*, 2018
Drosophila melanogaster	Diptera: Drosophilidae	Male orientation, chasing, tapping, licking, singing, mounting	Yamamoto & Koganezawa, 2013
Anastrepha fraterculus	Diptera: Tephritidae	Contact, Alignment, and Arrowhead 1 in Brazil 1 morphotype Alignment, Arrowhead 1, and Fanning in Brazil 3 morphotype	Roriz *et al.*, 2018
Chrysomya varipes	Diptera: Calliphoridae	Orienting, tapping, waving, arching, wing vibration, and mounting	Jones *et al.*, 2017
Lucilia sericata	Diptera: Calliphoridae	Approach, foreleg tapping, mount	Benelli & Romano, 2019
Phlebotomus papatasi	Diptera: Psychodidae	Stationary wing flapping, abdominal bending, touching, leg rubbing, facing, leg rubbing, touching, approach flapping, wing flapping, copulation attempt, mount	Chelbi *et al.*, 2012
Heliconius erato phyllis	Lepidoptera: Nymphalidae	Inspection, hovering, alighting of wing, slow flapping, androconia exposition, abdomen bending, touching with proboscis, mounting	Klein & de Araujo, 2010
Spalgis epius (Westwood)	Lepidoptera: Lycaenidae	Territory markings, perching, inspection, marking, chasing an intruder by straight flight, territorial flight, flight pursuit, initiation of courtship flight, tracking, alighting on vegetation, crawling, abdominal bending, coupling, couple rotation	Dinesh & Venkatesha, 2013
Coccinella septempunctata	Coleoptera: Coccinellidae	Approach, watch, examine, mount and copulatory attempt	Omkar & Srivastava, 2002
Propylea dissecta	Coleoptera: Coccinellidae	Approach, watch, examine, embrace, mount and attempt	Omkar & Pervez, 2005
Anegleis cardoni	Coleoptera: Coccinellidae	Approach, watch, examine, sharp turn, mount, attempt, mate	Omkar *et al.*, 2013
Mnesarete pudica	Odonata: Calopterygidae	Approach, circling, spinning, grabbing female's wing, mount	Ferreira & Bispo, 2012
Lucihormetica verrucosa	Blattodea: Blaberidae	Palpating with antennae, palpating with palps, pursuing, pushing, climbing, raising wings and mounting	Greven & Zwanzig, 2013
Campoplex capitator	Hymenoptera: Ichneumonidae	Wing fanning, antennal tapping and mounting	Benelli *et al.*, 2019
Calliptamus wattenwylianus Pantel	Orthoptera: Catantopinae	Raise and lower one of hind femora, Raise and lower both hind femora, kicking with extension of tibia, knock, progress, rest, Fore legs kicking, cleanness, Chas, assault	García *et al.*, 2015
Endecous onthophagus	Orthoptera: Grylloidea	Approach, antennation, mount, stridulating sporadically, copulatory courtship	Acosta *et al.*, 2020

6.3.1 Visual Displays

Visual cues are essentially important in locating mates during daytime. These cues could be both static (colours and shapes) and dynamic (locomotor activity, wing displays, and motion) (Greenspan & Ferveur, 2000). Male *D. melanogaster* approaches female, who is of same size, by using visual cues

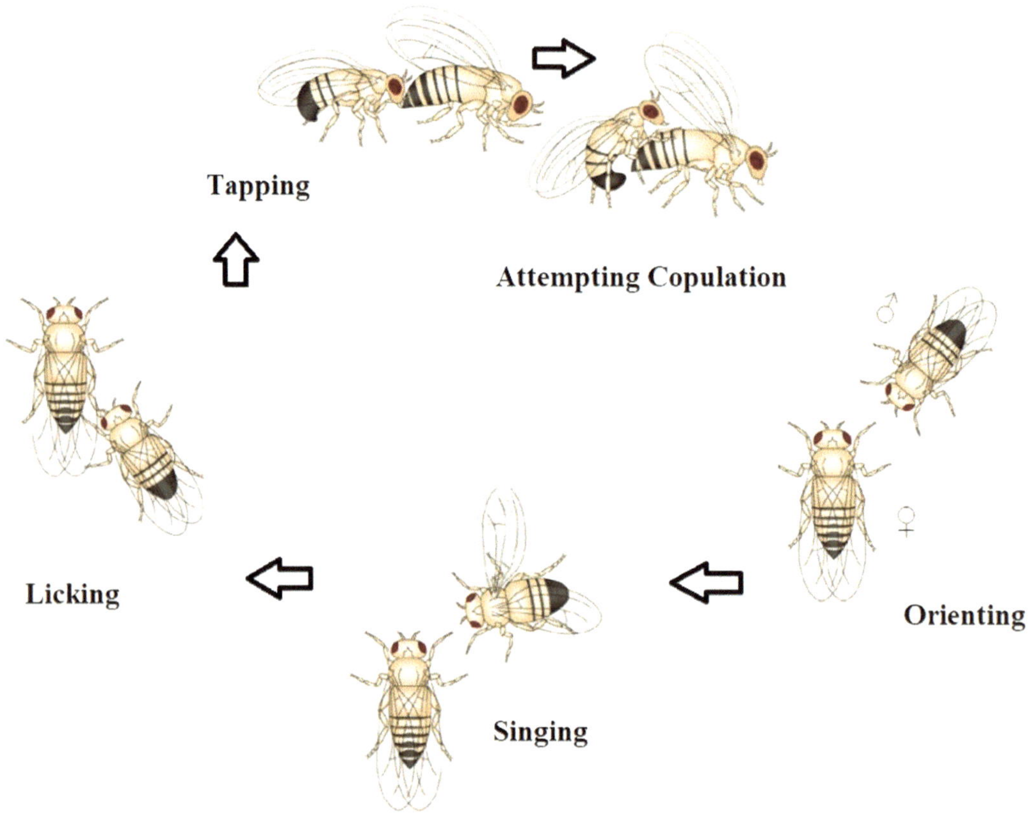

FIGURE 6. 1 Courtship displays by male, ***Drosophila melanogaster*** (Adapted and redrawn following Sokolowski, 2001).

(Agrawal *et al.*, 2014). Males initiate courtship dance and song by extending wings and vibrating them after locating females (Figure 6.1) (Coen *et al.*, 2016). Males of *Drosophila subobscur*a Collin exhibit intricate wing scissoring patterns and rapid side stepping (Mittleman *et al.*, 2017), while those of *Drosophila persimilis* Dobzhansky and Epling perform an intricate postural display (acrobatic contortions with tremulation of males' abdomen, and production of substrate borne vibrations) along with standard courtship (Hernández & Fabre, 2017). Accompanying this display is a nuptial gift in the form of a nutritional droplet. A minimum of 20 neurons coordinate male *Drosophila* courtship behaviour (Kohatsu & Yamamoto, 2015), though in some a single neuron conveys relevant courtship signals from visual system to central brain (Ribeiro *et al.*, 2018). Nojima *et al.* (2018) found visual neurons in male *Drosophila* that identify and pursue mating partners. Males of *D. melanogaster* readily court young, virgin, and unfamiliar females and show no courtship with old and familiar ones (Feld & Valone, 2018). 'Leg vibration', counteracts re-mating suppression in *D. melanogaster* (Minekawa *et al.*, 2018); it is effective only on mated females.

Males of hairy maggot blowfly, *Chrysomya varipes* (Macquart), court females by orienting, tapping, waving, arching, wing vibration, and mounting (Jones *et al.*, 2017). Ladybirds exhibit courtship displays, like approach, watch, examine, embrace, mount, sharp turns, and attempt before mating (Figure 6.2) (Omkar, 2004; Omkar & James, 2005; Omkar & Pervez, 2005; Omkar *et al.*, 2013).

Approximately, one-third females of dancefly (Diptera: Empididae) possess conspicuous sex-specific ornaments, viz. enlarged or darkened wings, pinnate leg scales, and inflatable abdominal airsacs (Cumming, 1994). Male *Lispecons anguinea* Loew performs long pedestrian dances in circular arcs around female fly interrupted by pauses of 0.35 seconds (Frantsevich & Gorb, 2006).

In odonates, males hover over females for 'wing display' (Cordero Rivera & Córdoba Aguilar, 2010) and hold females by head (dragonflies) and prothorax (damselflies). Thereafter, females raise abdominal

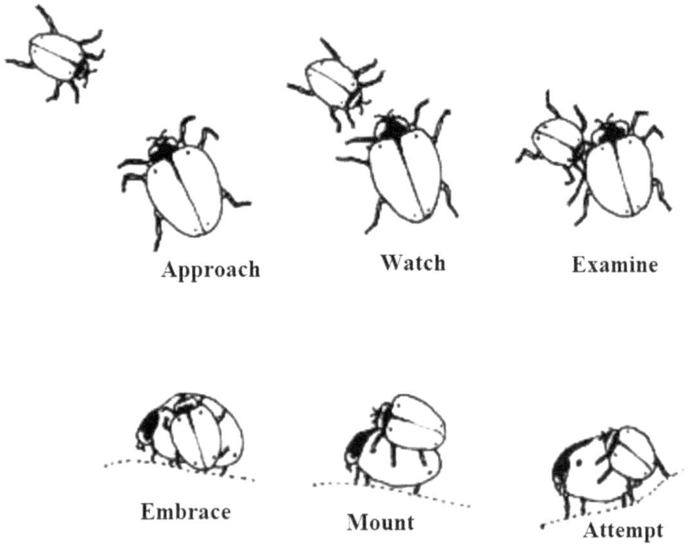

FIGURE 6.2 Courtship display by the male, ladybird, ***Propylea dissecta*** (pale morph - Adapted from Pervez, 2002).

tip to receive male genitalia, forming a typical 'mating wheel' (Wellenreuther & Sánchez Guillén, 2016); males displace sperms of prior males using their genitalia (Córdoba Aguilar *et al.*, 2003). Wing pigmentation influences male–male competition (Guillermo Ferreira *et al.*, 2019) and female mate choice (Tsubaki *et al.*, 2010).

In damselflies, there exist females who look like males (*androchromes*) and normal females (*gynochromes*) (Sanmartín Villar & Cordero Rivera, 2016). These androchromes or male-mimic females are less courted by males and lay fewer eggs (Cordero Rivera & Sanchez Guillen, 2007), primarily because of high aggression. Presence of androcromes reduces sexual harassments and coercive matings (Gosden & Svensson, 2009).

In *Heliconius eratophyllis* (Fabricius) (Lepidoptera: Nymphalidae), courtship involves inspection, hovering, alighting of wing, slow flapping, androconia exposition, abdomen bending, and touching with proboscis (Klein & de Araujo, 2010).

6.3.2 Acoustic Courtship Signals

Courtship songs are generated by stridulation, percussion, tremulation, tymbals, and/ or air expulsion) (Cocroft *et al.*, 2014).

6.3.2.1 Calling and Courtship Songs

Insects produce species-specific '**calling song**' (to find mate) and '**courtship song**' (after mate location and recognition), with former being stereotypic call for species recognition, and latter probably depicting individual male's quality (Zuk *et al.*, 2008). Females usually prefer males with highly intense longer calling songs (Baker *et al.*, 2019).

Generally, calling songs of crickets (*Gryllus bimaculatus* De Geer) have a narrow range (3–6 kHz) pulse frequency with certain species composing continuous pulse trains known as trills (Figure 6.3a) (Baker *et al.*, 2019). However, *Chorthippus biguttulus* (Linn.) uses broadband carrier frequency (5–30 kHz) (Figure 6.3b) for their calling songs (Baker *et al.*, 2019). Calling songs of *G. bimaculatus* vary less among different populations than their morphometric traits (Ferreira & Ferguson, 2002). Courtship songs of crickets and grasshoppers are short ranged with a higher carrier frequency (11–16 kHz) and/or have subtle differential temporal pattern compared to calling songs (Zuk *et al.*, 2008).

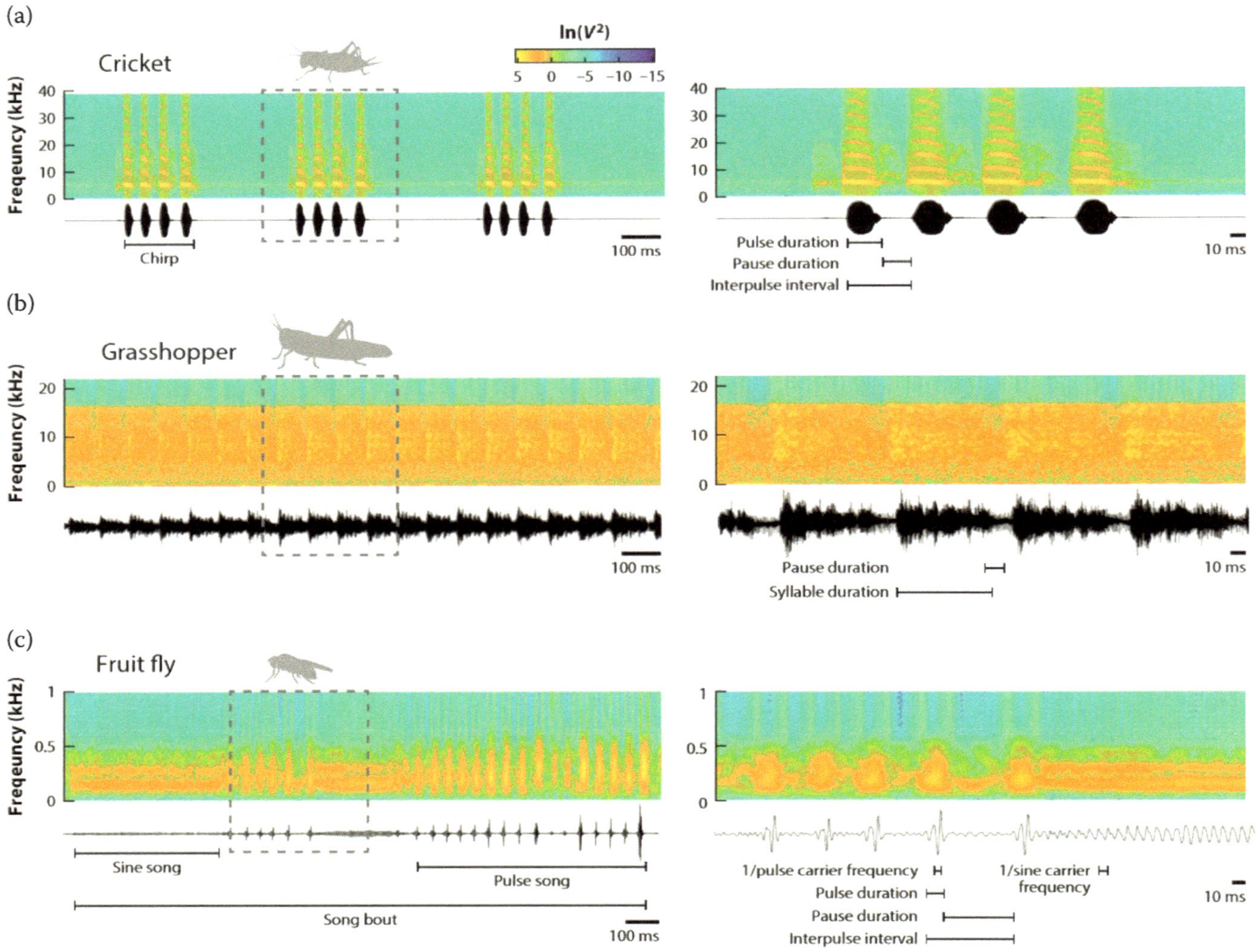

FIGURE 6.3 Details of the calling and courtship songs of the three important insects, (a) cricket, ***Gryllus bimaculatus***, (b) grasshopper, ***chorthippus bigutulus***, and (c) fruitfly, ***Drosophila melanogaster***, which heavily rely on these songs for mate-location, mate-recognition, courtship, and mating (Adapted from Baker *et al.*, 2019).

Courtship behaviour of male, *D. melanogaster,* has both acoustic signals and visual displays (Figure 1) (Tomaru & Yamada, 2011). They do not call but sing short-ranged courtship songs composed of almost pure tone sine song inserted with trains of brief pulses (Figure 3c) (Baker *et al.*, 2019). The carrier frequencies of these songs are exceptionally low (0.1–0.8 KHz (Clemens *et al.*, 2018). Agonistic pulse songs by males greatly differ from usual courtship songs (Jonsson *et al.*, 2011).

Females of *Gryllus lineaticeps* Stål. prefer courtship songs with higher chirp rates (Wagner & Reiser, 2000). Cricket females walk towards calling songs, while grasshopper initiate duets (Finck *et al.*, 2016; Baker *et al.*, 2019). Response of female fruitflies is similar to those of crickets (Bussell *et al.*, 2014). A few male fruitflies might also respond and show same sex courting displays (Vaughan *et al.*, 2014; Zhou *et al.*, 2015). Males of red turpentine beetle, *Dendroctonus valens* LeConte (Coleoptera: Curculionidae) produce proverbial 'passwords' in form of interrupted chirps, which 'females' accept and give males entry to their galleries for mating (Lindeman & Yack, 2015).

The great insect song diversity may reflect genetic drift and displacement of reproductive character. Little evidence exists on adaptation of songs to physical environments (Greenfield, 2016). Song divergence is directly associated with high frequency of population differentiation and speciation. Song analyses in combination of genetic and morphological data can help recompose ancestral songs (Nattier *et al.*, 2011), revise phylogenies (Montealegre & Morris, 2004), and identify cryptic species (Vigoder *et al.*, 2015).

6.3.2.2 Insect Duets

Duets are initiated by males during mate search to which females respond with their own songs. Duets enable pair formation and partner location (Legendre *et al.*, 2012). Duets are characterized by low variance in response latency of female. In a duet, a species-specific reply from female within a particular time duration is mandatory. Female grasshoppers are highly sensitive to fine tunings of calling songs and can easily identify negative callers, particularly heterospecific ones (Schmidt *et al.*, 2008; Clemens *et al.*, 2014).

Males of cicada, *Aphrodes makarovi* Zakhvatkin emit long and complex calling signals composed of several sections (Groot *et al.*, 2012); females reply with a long and continuously overlapping song near end of male call. This duet increases in complexity, with each modifying song in response to other; indicating evolutionary benefit of duets in sexual selection. Males of two cryptic species of phaneropterines, *Amblycorypharo tundifolia* (Scudder) and *Amblycorypha alexanderi* Walker sing duets at almost similar frequencies with their female partners (Forrest *et al.*, 2006), but for minute differences in series of syllables.

6.3.3 Tactile Stimulation

Tactile courting may involve bumping, antennal tapping or drumming on declivity, brushing of antennae or antennal scape setae against elytra, and mandibular gnawing, etc. (Smith & Cognato, 2011). *Hermetia illucens* Linnaeus (Diptera: Stratiomyidae) males, post wing fanning tap female abdomen using tarsi of their second and third pairs of legs (Giunti *et al.*, 2018). *Drosophila* males also court females by chasing, tapping, licking, and singing (Yamamoto & Koganezawa, 2013).

Males use their genital organs as stimulators and encourage females to accept and store their sperm for future progeny (Eberhard, 2011). Parasitic wasp, *Leptomastidea abnormis* (Girault), shows lateralized courtship by tapping antennae on right side of females' head (Romano *et al.*, 2016). Male of ladybirds court their females by antennal tapping (Omkar & Srivastava, 2002; Omkar & Singh, 2010). Tactile stimulation from antennae could be important to nocturnal insects, as it may help in both identifying and courting females in dark.

6.3.4 Nuptial Gifts

Nuptial gifts are materials, viz. food, salivary gland secretions, carrions, plant products, spermatophores, free ejaculates, and even body parts, donated by males to female to ensure mating and to provide

FIGURE 6.4 Photographs showing (a) courtship behaviour by male ***Harmonia dimidiata***, and (b) harmonine offered to female as nuptial gift (Adapted from Monalisa *et al.,* 2020).

benefits to females (Brown, 2010). These may be important, where females depend on males for nourishment, as is the case in dance flies (Cumming, 1994).

Gifts maybe endogenous and/or exogenous and could be taken up orally or seminally by the female (Gwynne, 2008). Male scorpion fly gifts spitballs and dead arthropods; latter may even be stolen back post-mating (Lewis *et al.*, 2011).

Seminal gifts in form of spermatophores are common in Lepidoptera, Orthoptera, and Coleoptera. Male crickets transfer spermatophore comprising of a sperm containing ampulla connected to a structure known as '*spermatophylax*' (Sakaluk *et al.*, 2019), which is consumed by female while sperm is directed into her reproductive tract (Lehmann, 2012). Consumption of spermatophores by females may enhance their fecundity and longevity, while defensive chemicals may protect females and their eggs against predators.

Males of ladybird, *Harmonia dimidiata* (Fabricius) release yellowish sticky reflex blood containing harmonine, from tibio femoral joint of its hindleg and offer it to female as nuptial gift (Monalisa *et al.,* 2020). Female accepts and feeds on the nuptial gift while allowing male to copulate (Figure 6.4a & b).

Nuptial gifts increase chance of mating success as well as increased paternity share; latter may be achieved by reducing females' receptivity to additional mating (Arnqvist & Rowe, 2005; Vahed, 2007). Such gifts are the outcome of sexual conflict and extortion, and addressing them as nuptial gifts may be misleading, as evidence suggests that they may be toxic for female fitness and inhibit further mating probabilities (Sakaluk *et al.*, 2006), thereby decreasing sperm competition (Fedorka & Mousseau, 2002).

6.3.5 Same Sex Courting Events

Widespread incidence of male same sex courtship behaviour in insects is bewildering and is attributed to isolation, high density, and increased prevalence of female pheromones (Dukas, 2010; Sales *et al.*, 2018). Same sex pairs show similar courtship, as found in male courting a female, albeit shorter (Giunti *et al.*, 2018). Similarly, males of *Ptecticus nigrifrons* Enderlei (Diptera: Stratiomyidae) approach both sexes after landing at nearby lekking sites (Chatzimanolis, 2003). Olfactory resemblance of young males with females is suggested as the cause; it may benefit young males by diverting mature males from females and practicing courtship behaviour (Bailey & Zuk, 2009; Benelli *et al.*, 2014). It may also act as a dominance display among males (Lane *et al.*, 2016).

6.3.6 Sexual Cannibalism

Sexual cannibalism is quite common in praying mantids (Wilder *et al.*, 2009; Kolnegari, 2020), where males prefer to court fat females more because risk of sexual cannibalism is more with nutritionally poor females and fat females are more fecund (Scardamaglia *et al.*, 2015; Kadoi *et al.*, 2017).

Evolution of sexual cannibalism could be because of (i) *fecundity advantage theory* (nutrients for egg production) and (ii) *size disparity theory* (as males are highly reduced in size). Wilder *et al.* (2009) pointed out a few ecological (food availability, population density, and operational sex ratio) and phylogenetic factors (mode of feeding, sexual size dimorphism, and genetic constraints) for occurrence of sexual cannibalism-based courtship in mantids.

6.3.7 Heterospecific Courtship

Male insects, occasionally, court closely related heterospecific females due to incomplete species and mate recognition, with latter even becoming receptive probably owing to similar courtship, even at cost of rejection of conspecific males (Friberg *et al.*, 2013; Mair *et al.*, 2018).

Females of wasp *Nasonia longicornis* Darling greatly reject conspecific courters after having been courted by *Nasonia vitripennis* (Walker) males (Mair *et al.*, 2018). Males of sympatric butterflies, *Leptidea sinapis* (Linn.) and *Leptidea reali* Reissinger perform costly and intricate courtships for heterospecific females but are strongly rejected (Berger *et al.*, 2012). Heterospecific courtships have greater costs to females in form of premating opportunity and post mating hybridization costs (Barry & Kokko, 2010). It also forces females to oviposit in non-optimal habitats which triggers habitat niche separationimposing great costs, particularly, to local minority species (Hochkirch *et al.*, 2007). Heterospecific courtship in fruitflies determines supremacy of dominant courters, as weaker species have to leave high quality fruits and occupy niche containing low-quality fruits (Kitano *et al.*, 2018).

6.3.8 Honest and Dishonest Courtship Signals

Insects court their partners with both honest and dishonest signals. Honest signals include true information regarding internal and external biological traits.

Female crickets usually get attracted towards dominant males, with better calling songs and a strong immune system, as choosing 'good genes' provide direct benefits for their offspring (Qvarnstrom & Forsgren, 1998). However, weaker (infectious/injured) or older male crickets might become more aggressive and send dishonest calling songs as a last resort to increase fitness; but females are able to discriminate and remain unresponsive (Copeland & Fedorka, 2012; Pölkki *et al.*, 2013). Male mealworm beetles, *Tenebrio molitor* L. with reduced immunity increase pheromone productions to attract females, who do fall in trap (Sadd *et al.*, 2006).

During female biased OSR, females may also transmit dishonest signals to lure males for food and matings (Funk & Tallamy, 2000). Female danceflies, *Rhamphomyia longicauda* Loew, inflate their abdomen with air and pretend to carry mature eggs to attract males with food-based nuptial gifts (Murray *et al.*, 2018); males are attracted to such females owing to last male sperm precedence (Laturney *et al.*, 2018). Androchrome morphs of some damselflies also exhibit dishonest signals to display their sexual immaturity to avoid mating and sexual harassments (Willink *et al.*, 2019).

6.4 Proximate Causes of Courtship

6.4.1 Chemical Cues and Exposure

Males use chemical cues and signals to distinguish between mated and unmated females (Thomas, 2011). This change in female odour is probably due to transmission of male substances, which might suppress attractiveness to competing males. Female cuticular hydrocarbons (CHCs) induce male courtship behaviour in co-generic grasshoppers, *Chorthippus biguttulus* (Linnaeus) and *Chorthippus mollis* (Charpentier) (Finck *et al.*, 2016).

Certain CHCs function as species-specific signals or sex pheromones that provide both inter- and intraspecific information (Howard & Blomquist, 2005). Long chain CHCs in *D. melanogaster* are perceived by contact or at a short distance (Ferveur, 2005). These even show sexual dimorphism, as only females synthesize CHCs with two double bonds (often 7,11 dienes) that arouse male courtship, while males possess mono-enes (with one double bond, such as 7 tricosene; 7 T), which inhibit courtship (Foley *et al.*, 2007).

6.4.2 Operational Sex Ratio (OSR)

OSR is the ratio of sexually active males and females in a given population. It represents level of competition required to access mates and probability of finding potential mate. Male-biased OSR has a positive correlation with courtship activities in males of broad-horned flour beetle, *Gnatocerus cornutus* (Fabr.) (House *et al.*, 2019). Female-biased OSR may induce courtship of smaller sized females than bigger ones (Dougherty *et al.*, 2020), production of costly nuptial gifts (Yiğit *et al.*, 2019), and male courtship flexibility of but not female mate choice (Villarreal *et al.*, 2018).

6.4.3 Female Mating Preferences

Female mating preferences or mate choice is selection or acceptance of a courting male for copulation and is a proximal cause for initiation of mating. Females of soldier fly, *H. illucens* prefer mates who court them by wing fanning for a longer duration (Giunti *et al.*, 2018). Female crickets, *G. lineaticeps* spend less time on low-quality male songs (Atwell & Wagner, 2014). Female age and perceived male density also trigger female choosiness, as younger females show more choosiness in high male-biased OSR. This shows that insect populations dominated by younger females promote sexual selection. Familiarity with mates may change the female preference with some preferring familiar (Westerman *et al.*, 2012) and some unfamiliar ones (Saxena *et al.*, 2017).

6.4.4 Body Size

Bigger and stronger males have selective advantage over intrasexual competition and female mate choice (Benelli *et al.*, 2016; Ekanayake *et al.*, 2017).

Larger males of solitary bee, *Diadasia rinconis* Cockerell (Hymenoptera: Apidae), successfully court their mates and copulate for shorter durations (Russell *et al.*, 2018). Large male size in *Aedes albopictus* (Skuse) (Diptera: Culicidae) mostly indicates 'good quality' in males and an assurance of quality genes and greater ejaculate (Hatala *et al.*, 2018).

6.4.5 Overcrowding during Development

Overcrowding during early development decreases courtship tendencies in male mosquitoes (Ng'habi *et al.*, 2005). On contrary, *D. melanogaster* males from larval crowding adapted population had higher courtship frequency than those reared in isolation (Shenoi *et al.*, 2016); but was not translated into mating success. Males of lepidopteran, *Ephestia kuehniella* Zeller, reared within highest density population had reduced body mass, smaller forewings, and other body parts, which theoretically (not actually) make them inferior courters (Bhavanam & Trewick, 2017).

6.5 Ultimate Evolutionary Causes

6.5.1 Male Sexual Traits

Cook (2018) stated that an individual's attractiveness is a combination of various morphological traits with each differently influence courtship behaviour. Highly evolved and elaborate male genitalia are used by males during copulatory courtship to arouse female and stabilize mating (Eberhard, 1994, 2011; Kopp, 2011; Rhebergen *et al.*, 2016; Lefèvre *et al.*, 2019).

Males of certain insect species present their distal abdomen to female, such as abdominal flexion displays in Persian boxer mantid, *Holaptilon brevipugilis* Kolnegari (Kolnegari, 2020), genitalia curling in moth *Phlogophora meticulosa* (Linn.) (Conner & Iyenger, 2016), or abdomen curving in cockroach *Oxyhalo adeusta* (Thunberg) (Lizée *et al.*, 2017).

6.5.2 Intrasexual Competition

Male insects frequently compete among themselves before courting females in a bid to attain mates and paternity (Gwynne & Simmons, 1990). Males of *Bolitotherus cornutus* (Panzer) use their thoracic horns and relative body size in an intrasexual struggle which may predict contest outcome that lead to female courtship (Mitchem *et al.*, 2019). Larger males of solitary bee, *Diadasia rinconis* Cockerell, were better mates and continue complex courtship even while mounting over females (Russell *et al.*, 2018. In courting males with non-aggressive and aggressive courtship behaviour of *D. melanogaster*, male aggression was a determinant of male mating success (Baxter *et al.*, 2018).

6.5.3 Cryptic Female Choice

Cryptic female choice allows females to regulate offspring paternity by choosing sperms of males with superior morphological, behavioural, and physiological traits. It includes female control throughout copulation and sperm transfer (Pilastro *et al.*, 2004) and internal sperm manipulation by female (Otronen & Siva Jothy, 1991).

CFC may direct evolution of intricate male genitalia and sperm, non-genital male structures specialized to hold and stimulate females, manipulative effects of semen on female reproduction, and male copulatory courtship (Eberhard, 2009). High frequency of male copulatory courtship is attributed to male attempts to influence cryptic female choice (Eberhard, 1994). Females of flour beetle, *Tribolium castaneum* (Herbst), reject transfer of sperm and spermatophore of males with low phenotypic quality (Fedina & Lewis, 2008). *T. castaneum* males who responded greatly towards female odours and courted well were found to be positively correlated with both successful insemination and paternity share.

6.5.4 Sexual Conflict

Sexual conflict occurs in insects when benefits of a particular behaviour to each sex are diametrically opposite leading to the coevolution of sexually antagonistic traits in both sexes (Arnqvist & Rowe, 2005). Cost-inducing or coercive male traits are complemented by evolutionary counter adaptations in females (Marchini *et al.*, 2010). Sexual conflicts often result in traumatic insemination, toxic semen, sexual harassment, coercive mating, mating resistance, mating refusals, etc.

Sexual conflict may even arise over mating, as usually male optimal mating rate outnumbers that of females (Chapman *et al.*, 2003) incurring increasing costs in females due to physical struggle or impaired immunity (Rönn *et al.*, 2007), or in female resource acquisition (Reinhardt *et al.*, 2009). This may result in evolution of deceptive mating suitability signals as a means to avoid premating sexual harassment and male mating attempts (Rooker & Gavrilets, 2018) and evolve anti-aphrodisiacs to control male sexual desires (Engel *et al.*, 2016).

6.6 Benefits and Costs of Courtship Signals

Courtship signals in insects have both ecological and evolutionary benefits and costs. Courted females select better, which yield more attractive offspring (*'sexy sons' hypothesis*). These honest and inexpensive signals are beneficial to both mating partners (Smith, 1994). Nuptial gifts provide additional nutrients to females and may also affect female immune defences (Worthington & Kelly, 2016).

Courtship signals also come with costs which prevent Fisherian runaway selection of these signals (Smith & Harper, 2003). Signal costs cause immediate deterioration in signallers' fitness through parasitism/predation or increased energy expenditure (Woods *et al.*, 2007). There are certain costs, viz. risk of predation (Zuk & Kolluru, 1998), disease transmission (Thrall *et al.*, 1997), injury, or physical damage (Blanckenhorn *et al.*, 2002). Certain females avoid costly and premature mating (Stoltz *et al.*, 2007) via delayed sexual maturity (Teder, 2014) and long re-mating latencies (Schärer *et al.*, 2012). Certain courtship costs are directly related to metabolism, fitness parameters, and immunity (Table 6.2).

TABLE 6.2

Courtship costs, currency with cost examples in various insect species

Type of Cost	Insect Species	Currency	Cost Example	References
Longevity	*D. melanogaster*	Longevity	25% decrease in longevity	Cordts & Partridge (1996)
	C. capitata		Upto 85% decrease	Ferkau & Fischer (2006)
	Onthophagus binodis (Thunberg)			Simmons & Kvarnemo (2006)
	P. napi			Wedell (2010)
	C. capitata		50-75% decrease in longevity	Papadopoulos *et al.* (2010)
Immunity	*Tenebrio molitor* L.	Phenoloxidase activity	30% reduction in immunity due to JH, which down regulates enzyme activity involved in humoral immune system	Rolff & Siva-Jothy (2002)
	Allonemobius socius (Scudder)	Lytic activity	60% reduction in immunity by decrease in circulating hemocytes	Fedorka *et al.* (2004)
	D. melanogaster	Juvenile Hormone	Due to JH, decrease in resistance against bacterial infection	Schwenke & Lazzaro (2017)
Predation	*Orthodera novaezealandiae* (Colenso)	Life	Females of *Miomantis caffra* attract and prey upon courting males of *Orthodera novaezealandiae*	Fea *et al.* (2013)
	Photinus pyralis (L.)		Firefly, *Photinus pyralis* is preyed upon by another firefly, *Photuris versicolor* Fabricius	Woods *et al.* (2007)
Parasitism	*Teleogryllus oceanicus* (Le Guillou)	Life	Male cricket's song also attracts female parasitoid flies (*Ormia ochracea* (Bigot)) who parasitize and kill their host	Zuk *et al.* (1993)
Energy	*Photinus greeni* Lloyd *Photinus obscurellus* LeConte	ATP	Flashing fireflies consumed 20 40 μW more energy than non flashing ones during courtship	Woods *et al.* (2007)
	Achroia grisella Fabricius		Singing increases metabolic rate by up to 4 x Standard Metabolic Rate and results in daily loss of 5% of body mass	Reinhold *et al.* (1998)
	Gryllotalpa monanka Otte & Alexander		Singing increases metabolic rate by 13.5 times Standard Metabolic Rate	White *et al.* (2008)

6.7 Factors Affecting Courtship Behaviour

6.7.1 Environmental Factors

6.7.1.1 Temperature

High temperatures can be detrimental to courtship and mating, with deviation from optimal temperature range leading to physiological constraints that may limit such activities (Kingsolver & Huey, 2008). Shifts in body temperatures of courting partners enhance mismatches in mating activity in army ants (Baudier *et al.*, 2015). Tree hopper, *Enchenopa binotata* (Say), differs in thermal requirements at gender level during mating season, which may become a barrier for courtship activities (Macchiano *et al.*, 2019). Females of damselflies, *Mnais costalis* Selys, show more intensive courtship and are preferred by females colder males (Tsubaki *et al.*, 2010). Similarly, male fruitflies, *D. melanogaster,* with thermo-genetically activated P1 neurons have elevated levels of courtship compared to those which are not thermo-activated (Agrawal & Dickinson, 2019). Male butterflies, *Bicyclus anynana*, reared on cool temperature (17°C), court females less frequently than those raised at 27°C (Bear & Monteiro, 2013). Temperature variations can alter male thoracic vibrations and odour bouquets of antennae of mason bees, *Osmia bicornis* (Linn.) (Conrad *et al.*, 2017).

6.7.1.2 Light

Light is an important environmental factor affecting courtship behaviour of certain male insects, especially ones who search for their mates during night. Emblematic species, such as fireflies and glow worms, use bioluminescent flashes as courtship displays to attract females. Increased light pollution reduces flashing activities in nocturnal firefly, *Photuris versicolor* (Fabr.) and courtship activities in a twilight active, *Photinus pyralis* (Linn.) (Firebaugh & Haynes, 2016). Similarly, artificial light at night (ALAN) negatively affects insect communication processes, viz. signal production, transmission, and detection (Metzger *et al.*, 2010) and can impede courtship (Desouhant *et al.*, 2019). Dim light also might reduce frequency of male flash signals in Taiwanese firefly, *Aquatica ficta* (Olivier) (Owens & Lewis, 2018). Acoustic signals in sexual communication are also affected by ALAN (Botha *et al.*, 2017) as is chemical communication via sex pheromones (van Geffen *et al.*, 2015).

6.7.2 Physiological and Biochemical Factors

6.7.2.1 Juvenile and Ecdysis Hormones

Juvenile Hormone (JH) seems to regulate courtship oriented male mating behaviour and reduction in levels of JH may reduce this behaviour (Wijesekera *et al.*, 2016). Juvenile-hormone-acid-O-methyltransferase (JHAMT) regulates final step of JH biosynthesis in corpora allata. The courtship behaviour is greatly reduced in males of *D. melanogaster* who have reduced levels of JHAMT (Wijesekera *et al.*, 2016). Reducing JHAMT conditionally in mature flies lead to courtship defects that are rescuable by methoprene (Argue *et al.*, 2013). JH in adult insects is directly related to reproduction by regulating cuticular pheromone synthesis in females and initiating courtship in male insects (Bilen *et al.*, 2013). Ecdysis triggering hormone receptor (ETHR) signalling regulates courtship behaviour in *Drosophila* through chemosensory processing in its antennal lobe. In addition, silencing of ETHR signalling in *fruitless* neurons and/or local interneurons triggers male–male courtship (Deshpande *et al.*, 2019).

6.7.2.2 Neuropeptides

Male insects transmit certain sex peptides (SP) through sperm into females after first courting, which inhibits further courtship and mating in females (Wigby & Chapman, 2005). Injection of *Drosophila* SP in *Helicoverpa armigera* (Hubner) suppresses courtship behaviour by inhibiting biosynthesis of sex pheromone and expression of myo-inhibitory peptide receptor (Hanin *et al.*, 2012).

Pheromone biosynthesis-activating neuropeptide (PBAN) initiates production of pheromones in females who have undergone knockdown due to unattractive males, to attract mates (Lee *et al.*, 2011). Natalisin and Neuro Peptide F (NPF) are also important neuropeptides, which are involved in both courtship and mating behaviour (Lee *et al.*, 2006; Jiang *et al.*, 2013).

6.8 Role of Courtship in Sexual Selection

Sexual selection is, precisely, selection and propagation of certain sexual traits and behaviour of an individual, which may lead to speciation (see reviews, Ritchie, 2007; Miller & Svensson, 2014). It involves male–male competition, cryptic mate choice, and female mate choice based on intraspecific differences in characteristics of courting males.

Sexual selection through female mating preference drives evolution of diversified male courtship songs across closely associated species, which triggers speciation. Thus, neurons involved in song production are susceptible to rapid evolution (von Philipsborn *et al.*, 2011). Explosive speciation of *Laupala* crickets in Hawaii is directly linked with diversity of calling and courtship songs (Oh *et al.*, 2012). It is argued that such sexual selection overpowers ecological selection or reinforces selection and becomes the primary cause of speciation in such group of insects (Mendelson & Shaw, 2005).

Ecological radiation in Hawaiian *Drosophila* with elaborate and intricate secondary sexual traits, diversified courtship songs, and painted wings (Kang *et al.*, 2016; Nie & Kaneshiro, 2016) is courtship driven. Thus, sexual selection seems to be facilitated by various coadapted gene complexes and evolution of intricate and unusual courtship displays (Castillo & Barbash, 2017). Sexual selection of *Drosophila* in Hawaii may also be credited to divergence of CHCs, as 138 flies belonging to 27 Hawaiian *Drosophila* species had significant differences in CHC patterns (Alves *et al.*, 2010).

6.9 Conclusions

We conclude that insects perform highly intricate, elaborate, and myriad courtship displays, which is key to their reproductive activities and success. Insects rely on all possible cues and utilize their maximum resources to court females, as certain insects, particularly *Drosophila,* use chemicals cues followed by visual displays, audio calls, and tactile stimulations. *Drosophila melanogaster* is most studied insect in light of courtship and mating behaviour, ecology, and evolution. Evidences from *D. melanogaster* along with other well-studied insects provide ecological and evolutionary explanations on diverse and myriad courtship behaviour in insects. Courtship displays have evolved gradually. It is intriguing that various insect groups use different sexually selected traits, viz. odours, songs, colours, horns, genitalia, etc., to court their partners. Furthermore, one of the selected traits mostly plays a crucial role in courtship and that trait is further divided into various types and forms. For instance, visual displays may further be divided into various types: colours, dances, head ornaments, morphological traits, etc. This leads to obvious questions that why do some insect groups lack all these selected traits and some of clades have excessive forms of these traits, e.g. *Drosophila.* Insect groups lacking these sexually selected traits and courtship signals are poorly investigated and understood. We reviewed that courtship signals carry costs, like decline in fitness, energy, metabolism, immunity, and threats against natural enemies, in a few insect groups. The costs of these signals as compared to their benefits in light of selection need to be addressed. The implications of selected traits and dishonest signals on insect ecology and courtship patterns also need to be further studied.

Acknowledgement

AP thanks financial support by the Science and Engineering Research Board, Department of Science and Technology, Government of India in the form of Major Research Project (EMR/2016/006296).

REFERENCES

Acosta, R.C., Timm, V.F., Szinwelski, N., da Costa, M.K.M., & Zefa, E. (2020). Mating behavior and acoustic communication of long legged cricket *Endecous (Notendecous) onthophagus* (Berg, 1891) from Southern Brazil (Orthoptera: Grylloidea: Phalangopsidae). *Zootaxa*, *4743*(3), 427–437. https://doi.org/10.11646/zootaxa.4743.3.10.

Agrawal, S., Safarik, S., & Dickinson, M. (2014). The relative roles of vision and chemosensation in mate recognition of *Drosophila melanogaster*. *Journal of Experimental Biology*, *217*, 2796–2805. https://doi.org/10.1242/jeb.105817.

Agrawal, S., & Dickinson, M.H. (2019). The effects of target contrast on *Drosophila* courtship. *Journal of Experimental Biology*, *222*, jeb203414. 10.1242/jeb.203414.

Alves, H., Rouault, J.D., Kondoh, Y., Nakano, Y., Yamamoto, D., *et al.* (2010). Evolution of cuticular hydrocarbons of Hawaiian Drosophilidae. *Behavioural Genetics*, *40*, 694–705. 10.1007/s105190109364y.

Argue, K.J., Yun, A.J., & Neckameyer, W.S. (2013). Early manipulation of juvenile hormone has sexually dimorphic effects on mature adult behaviour in *Drosophila melanogaster*. *Hormonal Behaviour*, *64*(4), 589–597. https://doi.org/10.1016/j.yhbeh.2013.08.018.

Arnqvist, G., & Rowe, L. (2005). *Sexual conflict*. Princeton University Press, Princeton, NJ.

Atwell, A., & Wagner, W.E. Jr. (2014). Female mate choice plasticity is affected by interaction between male density and female age in a field cricket. *Animal Behaviour*, *98*, 177–183. https://doi.org/10.1016/j.anbehav.2014.10.007.

Bailey, N.W., & Zuk, M. (2009). Same sex sexual behavior and evolution. *Trends in Ecology & Evolution 24*, 439–446. https://doi.org/10.1016/j.tree.2009.03.014.

Baker, C.A., Clemens, J., & Murthy, M. (2019). Acoustic pattern recognition and courtship songs: Insights from insects. *Annual Review of Neuroscience*, *42*, 129–147. https://doi.org/10.1146/annurev-neuro-080317-061839.

Barry, K.L., & Kokko, H. (2010). Male mate choice: Why sequential choice can make its evolution difficult. *Animal Behaviour*, *80*, 163–169. https://doi.org/10.1016/j.anbehav.2010.04.020.

Baudier, K.M., Mudd, A.E., Erickson, S.C., & O'Donnell, S. (2015). Microhabitat and body size effects on heat tolerance: Implications for responses to climate change (army ants: Formicidae, Ecitoninae). *Journal of Animal Ecology*, *84*, 1322–1330. https://doi.org/10.1111/1365-2656.12388.

Baxter, C., Mentlik, J., Shams, I., & Dukas, R. (2018). Mating success in fruit flies: Courtship interference versus female choice. *Animal Behaviour*, *138*, 101–108. https://doi.org/10.1016/j.anbehav.2018.02.010.

Bear, A., & Monteiro, A. (2013). Male courtship rate plasticity in butterfly *Bicyclusanynana* is controlled by temperature experienced during pupal and adult stages. *PLoS One*, *8*(5), e64061. https://doi.org/10.1371/journal.pone.0064061.

Benelli, G., Daane, K.M., Canale, A., Niu, C.Y., Messing, R.H., & Vargas, R.I. (2014). Sexual communication and related behaviours in Tephritidae: Current knowledge and potential applications for integrated pest management. *Journal of Pest Science*, *87*, 385–405. https://doi.org/10.1007/s10340-014-0577-3.

Benelli, G., Doanti, E., Romano, D., Ragni, G., Bonsignori, G., Stefanini, C., & Canale, A. (2016). Is bigger better? Male body size affects wing-borne courtship signals and mating success in olive fruit fly, *Bactroceraoleae* (Diptera: Tephritidae). *Insect Science*, *23*(6), 869–880. https://doi.org/10.1111/1744-7917.12253.

Benelli, G., Ricciardi, R., Romano, D., Cosci, F., Stefanini, C., & Lucchi, A. (2019). Wing-fanning frequency as a releaser boosting male mating success—High-speed video analysis of courtship behavior in *Campoplexcapitator*, a parasitoid of *Lobesiabotrana*. *Insect Science*, 26. 10.1111/17447917.12740.

Benelli, G., & Romano, D. (2019). Looking for right mate—What do we really know on courtship and mating of *Lucilia sericata* (Meigen)? *Acta Tropica*, *189*, 145–153. https://doi.org/10.1016/j.actatropica.2018.08.013.

Berger, D., Olofsson, M., Friberg, M., Karlsson, B., Wiklund, C., & Gotthard, K. (2012). Intraspecific variation in body size and rate of reproduction in female insects: Adaptive allometry or biophysical constraint? *Journal of Animal Ecology*, *81*, 1244–1258. https://doi.org/10.1111/j.1365-2656.2012.02010.x.

Bhavanam, S., & Trewick, S. (2017). Effects of larval crowding and nutrient limitation on male phenotype, reproductive investment and strategy in *Ephestia kuehniella* Zeller (Insecta: Lepidoptera). *Journal of Stored Products Research*, *71*, 64–71. https://doi.org/10.1016/j.jspr.2017.01.004.

Bilen, J., Atallah, J., Azanchi, R., Levine, J.D., & Riddiford, L.M. (2013). Regulation of onset of female mating and sex pheromone production by juvenile hormone in *Drosophila melanogaster. Proceedings of National Academy of Sciences of USA. 110*(45), 18321–18326. 10.1073/pnas.1318119110.

Blanckenhorn, W.U., Hosken, D.J., Martin, O.Y., Reim, C., Teuschl, Y., & Ward, P.I. (2002). The costs of copulating in dung fly *Sepsis cynipsea. Behavioral Ecology, 13*, 353–358. https://doi.org/10.1093/beheco/13.3.353.

Botha, L.M., Jones, T.M., & Hopkins, G.R. (2017). Effects of lifetime exposure to artificial light at night on cricket (*Teleogryllus commodus*) courtship and mating behaviour. *Animal Behaviour, 129*, 181–188. https://doi.org/10.1016/j.anbehav.2017.05.020.

Brown, W. D. (2010). Allocation of nuptial gifts in tree crickets changes with both male and female diet. *Behavioral Ecology and Sociobiology*, 65, 1007–101410.1007/s00265-010-1105-y.

Bussell, J.J., Yapici, N., Zhang, S.X., Dickson, B.J., & Vosshall, L.B. (2014). Abdominal B neurons control *Drosophila* virgin female receptivity. *Current Biology, 24*(14), 1584–1595. https://doi.org/10.1016/j.cub.2014.06.011.

Castillo, D., & Barbash, D.A. (2017). Moving speciation genetics forward: Modern techniques build on foundational studies in *Drosophila. Genetics, 207*(3), 825–842. https://doi.org/10.1534/genetics.116.187120.

Chapman, T., Arnqvist, G., Bangham, J., & Rowe, L. (2003). Sexual conflict. *Trends in Ecology and Evolution 18*, 41–47.

Chatzimanolis, S. (2003). Notes on mating behavior of *Ptecticus nigrifrons* (Diptera: Stratiomyidae). *The Journal of Kansas Entomological Society, 76*, 663–666.

Chelbi, I., Bray, D., & Hamilton, J. (2012). Courtship behaviour of *Phlebotomus papatasi* sand fly vector of cutaneous leishmaniasis. *Parasites Vectors, 5*, 179. 10.1186/175633055179.

Clemens, J., Krämer, S., & Ronacher, B. (2014). Asymmetrical integration of sensory information during mating decisions in grasshoppers. *Proceedings of the National Academy of Sciences of USA, 111*(46), 16562–16567. https://doi.org/10.1073/pnas.1412741111.

Clemens, J., Coen, P., Roemschied, F., Pereira, T., Mazumder, D., *et al.* (2018). Discovery of a new song mode in *Drosophila* reveals hidden structure in sensory and neural drivers of behavior. *Current Biology, 28*, 2400–2412. https://doi.org/10.1016/j.cub.2018.06.011.

Cocroft, R.B., Gogala, M., Hill, P.S.M., & Wessel, A. (2014). *Studying vibrational communication.* New York: Springer.

Coen, P., Xie, M., Clemens, J., & Murthy M. (2016). Sensorimotor transformations underlying variability in song intensity during *Drosophila* courtship. *Neuron, 89*, 629–644. https://doi.org/10.1016/j.neuron.2015.12.035.

Conner, W.E., & Iyenger, V.K. (2016). Male pheromones in Moth: Reproductive Isolation, Sexy sons, and good genes. In: Allison, J.D., & CardŽ, R.T. (eds.), *Pheromone communication in moths: Evolution, behavior, and application* (pp. 191–210). Berkeley, CA: University of California Press.

Conrad, T., Stöcker, C., & Ayasse, M. (2017). The effect of temperature on male mating signals and female choice in red mason bee, *Osmia bicornis* (L.). *Ecology and Evolution, 7*, 8966–8975. https://doi.org/10.1002/ece3.3331.

Cook, L. (2018) How does individual quality and mate attractiveness affect courtship and mating behaviour? *The Plymouth Student Scientist, 11*(1), 24–38.

Copeland, E.K., & Fedorka, K.M. (2012). The influence of male age and simulated pathogenic infection on producing a dishonest sexual signal. *Proceedings of Royal Society B, 279*, 4740–4746. https://doi.org/10.1098/rspb.2012.1914

Cordero Rivera, A., & Córdoba Aguilar A. (2010). Selective forces propelling genitalic evolution in Odonata. In: Leonard, J., & Cordoba, A. (eds.), *The evolution of primary sexual characters in animals* (pp. 332–352). Oxford: Oxford University Press.

Cordero Rivera, A., & Sanchez Guillen, R.A. (2007). Male like females of a damselfly are not preferred by males even if they are majority morph. *Animal Behaviour, 274*, 247–252. https://doi.org/10.1016/j.anbehav.2006.06.023.

Córdoba Aguilar, A., Uhía Castro, E., & Cordero Rivera, A. (2003). Sperm competition in Odonata (Insecta): The evolution of female multiple mating and rivals' sperm displacement. *Journal of Zoology, 261*, 381–398. https://doi.org/10.1017/S0952836903004357.

Cordts, R., & Partridge, L. (1996). Courtship reduces longevity of male *Drosophila* melanogaster. *Animal Behaviour*, *52*, 269–278.
Cumming, J.M. (1994). Sexual selection and evolution of dance fly mating systems (Diptera: Empididae; Empidinae). *The Canadian Entomologist*, *126*, 907–920.
Deshpande, S.A., Meiselman, M., Hice, R.H., Arensburger, P., Pereze, R.C., Kim, D.H., Croft, R.L., Noriega, F.G., & Adams, M.E. (2019). Ecdysis triggering hormone receptors regulate male courtship behavior via antennal lobe interneurons in *Drosophila*. *General and Comparative Endocrinology*, *278*, 79–88. https://doi.org/10.1016/j.ygcen.2018.12.003.
Desouhant, Emmanuel, Gomes, Elisa, Mondy, Nathalie, & Amat, Isabelle (2019). Mechanistic, ecological, and evolutionary consequences of artificial light at night for insects: review and prospective. *Entomologia Experimentalis et Applicata*, 167, 37–5810.1111/eea.12754.
Dinesh, A.S., & Venkatesha, M.G. (2013). Analysis of Territorial, Courtship and Coupling Behavior of Hemipterophagous Butterfly, *Spalgisepius* (Westwood) (Lepidoptera: Lycaenidae). *Journal of Insect Behaviour*, *26*, 149–164.
Dougherty, L.R., Dewhurst, A.D., & Lewis, Z. (2020). Operational sex ratio does not influence evolution of male mate choice in Indian meal moth. *Journal of Ethology*, *38*, 21–28. https://doi.org/10.1007/s10164-019-00616-3.
Dukas, R. (2010). Causes and consequences of male–male courtship in fruit flies. *Animal Behaviour*, *80*(5), 913–919. https://doi.org/10.1016/j.anbehav.2010.08.017.
Eberhard, W.G. (1994). Evidence for widespread courtship during copulation in 131 species of insects and spiders, and implications for cryptic female choice. *Evolution*, *48*(3), 711–733.
Eberhard, W.G. (2009). Postcopulatory sexual selection: Darwin's omission and its consequences. *Proceedings of the National Academy of Sciences of USA*, *106*(1), 10025–10032. https://doi.org/10.1073/pnas.0901217106.
Eberhard, W.G. (2011). Experiments with genitalia: A commentary. *Trends in Ecology and Evolution*, *26*, 17–21. https://doi.org/10.1016/j.tree.2010.10.009.
Ekanayake, E.W.M.T.D., Clarke, A.R., & Schutze, M.K. (2017). Effect of body size, age, and premating experience on male mating success in *Bactroceratryoni* (Diptera: Tephritidae), *Journal of Economic Entomology*, *110*(5), 2278–2281. https://doi.org/10.1093/jee/tox186.
Engel, K.C., Stökl, J., Schweizer, R., Vogel, H., Ayasse, M., Ruther, J., & Steiger S., (2016). A hormone related female anti aphrodisiac signals temporary infertility and causes sexual abstinence to synchronize parental care. *Nature Communications*, *7*, 11035. https://doi.org/10.1038/ncomms11035.
Ernst, A.F., Mikó, I., & Deans, A.R. (2013). Morphology and function of ovipositor mechanism in Ceraphronoidea (Hymenoptera, Apocrita). *Journal of Hymenoptera Research*, *33*, 25–61. https://doi.org/10.3897/JHR.33.5204.
Fea, M.P., Stanley, M.C., & Holwell, G.I. (2013). Fatal attraction: Sexually cannibalistic invaders attract naive native mantids. *Biology Letters*, *9*, 20130746. 10.1098/rsbl.2013.0746.
Fedina, T.Y., & Lewis, S.M. (2008). An integrative view of sexual selection in *Tribolium* flour beetles. *Biological Reviews*, *83*(2), 151–171. https://doi.org/10.1111/j.1469-185X.2008.00037.x.
Fedorka, K.M., Zuk, M., & Mousseau, T.A. (2004). Immune suppression and cost of reproduction in ground cricket, *Allonemobius socius*. *Evolution*, *58*, 2478–2485. https://doi.org/10.1111/j.0014-3820.2004.tb00877.x.
Feld, B.J., & Valone, T.J. (2018). Changes in courtship behaviour following rejection: The influence of female phenotype in *Drosophila melanogaster*. *Ethology*, *124*(3), 149–154. https://doi.org/10.1111/eth.12715.
Ferkau, C., & Fischer, K. (2006). Costs of reproduction in male *Bicyclus anynana* and *Pieris napi* butterflies: Effects of mating history and food limitation. *Ethology*, *112*, 1117–1127. https://doi.org/10.1111/j.1439-0310.2006.01266.x.
Ferreira, M., & Ferguson, J. W. H. (2002). Geographic variation in the calling song of the field cricket *Gryllus bimaculatus* (Orthoptera: Gryllidae) and its relevance to mate recognition and mate choice. *Journal of Zoology*, 257, 163–17010.1017/s0952836902000766.
Ferreira, R.G., & Bispo, P.C. (2012). Male and female interactions during courtship of Neotropical damselfly *Mnesarete pudica* (Odonata: Calopterygidae). *Acta Ethologica*, *15*, 173–178. https://doi.org/10.1007/s10211-012-0122-4.

Ferveur, J.F. (2005). Cuticular hydrocarbons: Their evolution and roles in *Drosophila* pheromonal communication. *Behavioural Genetics*, *35*(3), 279–295. https://doi.org/10.1007/s10519-005-3220-5.
Field, S.A., & Keller, M.A. (1994). Localization of female sex pheromone gland in *Cotesia rubecola* Marshall (Hymenoptera: Braconidae). *Journal of Hymenoptera Research*, *3*, 151156.
Finck, J., Kuntze, J., & Ronacher, B. (2016). Chemical cues from females trigger male courtship behaviour in grasshoppers. *Journal of Comparative Physiology A*, *202*, 337–345. https://doi.org/10.1007/s00359-016-1081-4.
Firebaugh, A., & Haynes, K.J. (2016). Experimental tests of light pollution impacts on nocturnal insect courtship and dispersal. *Oecologia*, *182*, 1203–1211. https://doi.org/10.1007/s00442-016-3723-1.
Fisher, R.A. (1930). *The genetical theory of natural selection*. Oxford, U.K: Clarendon Press.
Fleischer, J., Pregitzer, P., Breer, H., & Krieger, J. (2018). Access to odor world: Olfactory receptors and their role for signal transduction in insects. *Cellular and Molecular Life Sciences*, *75*, 485–508. https://doi.org/10.1007/s00018-017-2627-5.
Foley, B., Chenoweth, S.F., Nuzhdin, S.V., & Blows, M.W. (2007). Natural genetic variation in cuticular hydrocarbon expression in male and female *Drosophila melanogaster*. *Genetics*, *175*(3), 1465–1477. https://doi.org/10.1534/genetics.106.065771.
Forrest, T.G., Lajoie, D.R., & Cusick, D. (2006). Calling songs, duets, and auditory tuning in two cryptic katydids (Tettigoniidae: Phaneropterinae: Amblycorypha). *Annals of Entomological Society of America*, *99*(5), 978–987.
Frantsevich, L., & Gorb, S. (2006). Courtship dances in flies of genus Lispe (Diptera: Muscidae): From fly's viewpoint. *Archives of Insect Biochemistry and Physiology*, *62*, 26–42.
Friberg, M., Leimar, O., & Wiklund C. (2013). Heterospecific courtship, minority effects and niche separation between cryptic butterfly species. *Journal of Evolutionary Biology*, *26*, 971–979. https://doi.org/10.1111/jeb.12106.
Funk, D. H., & Tallamy, D. W. (2000). Courtship role reversal and deceptive signals in the long-tailed dance fly, Rhamphomyia longicauda. *Animal Behaviour*, 59, 411–42110.1006/anbe.1999.1310.
García, M.D., Clemente, M.E., & Presa, J.J. (2015). The courtship behaviour of *Calliptamus wattenwylianus* Pantel, 1896 (Orthoptera, Catantopinae). *Articulata*, *30*, 33–52.
Giunti, G., Campolo, O., Laudani, F., & Palmeri, V. (2018). Male courtship behaviour and potential for female mate choice in black soldier fly *Hermetia illucens* L. (Diptera: Stratiomyidae). *Entomologia Generalis*, *38*(1), 29–46. DOI: 10.1127/entomologia/2018/0657.
Gosden, T.P., & Svensson, E.I. (2009). Density dependent male mating harassment, female resistance, and male mimicry. *The American Naturalist*, *173*(6), 709 721. https://doi.org/10.1086/598491.
Greenfield, M.D. (2016). Evolution of Acoustic Communication in Insects. In: Pollack, G., Mason, A., Popper, A., & Fay, R. (eds.), *Insect hearing. Springer handbook of auditory research*, (*Vol. 55*, pp. 17–47). Cham: Springer.
Greenspan, R.J., & Ferveur, J.F. (2000). Courtship in *Drosophila*. *Annual Review of Genetics*, *34*, 205–232. https://doi.org/10.1146/annurev.genet.34.1.205.
Greven, H., & Zwanzig, N. (2013). Courtship, mating, and organisation of pronotum in glowspot cockroach *Lucihormetica verrucose* (Brunner von Wattenwyl, 1865) (Blattodea: Blaberidae). *Entomologie heute*, *25*, 77–97.
Groot, M.D., Derlink, M., Pavlovčič, P., JanezPrešern, J., Čokl, A., & Doberlet, M.V. (2012). Duetting Behaviour in Leafhopper *Aphrodes makarovi* (Hemiptera: Cicadellidae). *Journal of Insect Behaviour*, *25*, 419–440. https://doi.org/10.1007/s10905-011-9304-6.
Guillermo Ferreira, R., Bispo, P.C., Appel, E., Kovalev, A., & Gorb, S.N. (2019). Structural coloration predicts outcome of male contests in Amazonian damselfly *Chalcopteryx scintillans* (Odonata: Polythoridae). *Arthropod Structure & Development*, *53*, 100884. https://doi.org/10.1016/j.asd.2019.100884
Gwynne, D.T. (2008). Sexual conflict over nuptial gifts in insects. *Annual Reviews of Entomology*, *53*, 83–101. https://doi.org/10.1146/annurev.ento.53.103106.093423.
Gwynne, D.T., & Simmons, L.W. (1990). Experimental reversal of courtship roles in an insect. *Nature 346*, 172–174. https://doi.org/10.1038/346172a0.
Hanin, O., Azrielli, A., Applebaum, S.W., & Rafaeli, A. (2012). Functional impact of silencing *Helicoverpa armigera* sex peptide receptor on female reproductive behaviour. *Insect Molecular Biology*, *21*, 161–167. https://doi.org/10.1111/j.1365-2583.2011.01122.x.

Hansson, B.S., & Stensmyr, M.C. (2011). Evolution of insect olfaction. *Neuron*, *72*, 698–711. https://doi.org/10.1016/j.neuron.2011.11.003.
Hatala, A., Harrington, L., & Degner, E. (2018). Age and body size influence sperm quantity in male *Aedes albopictus* (Diptera: Culicidae) mosquitoes. *Journal of Medical Entomology*, *55*, 1051 1054. https://doi.org/10.1093/jme/tjy040.
Hernández, M.V., & Fabre, C.C.G. (2017). Triggers of postural display of courtship in *Drosophila persimilis* flies. *Journal of Insect Behaviour*, *30*, 582–594. https://doi.org/10.1007/s10905-017-9641-1.
Hochkirch, A., Groning, J., & Bucker, A. (2007). Sympatry with devil: Reproductive interference could hamper species coexistence. *Journal of Animal Ecology*, *76*, 633–662. https://doi.org/10.1111/j.1365-2656.2007.01241.x.
House, M.C., Rapkin, J., Hunt, J., & Hosken, D.J. (2019). Operational sex ratio and density predict potential for sexual selection in broad horned beetle. *Animal Behaviour*, *152*, 63–69. https://doi.org/10.1016/j.anbehav.2019.03.019.
Howard, R.W., & Blomquist, G.J. (2005). Ecological, behavioral, and biochemical aspects of insect hydrocarbons. *Annual Reviews of Entomology*, *50*, 371–393. https://doi.org/10.1146/annurev.ento.50.071803.130359.
Jiang, H., Lkhagva, A., Daubnerova, I., Chae, H.S., Simo, L., *et al.* (2013). Natalisin, a tachykinin like signaling system, regulates sexual activity and fecundity in insects. *Proceedings of the National Academy of Sciences of USA*, *110*, E 3526–E 3534. https://doi.org/10.1073/pnas.1310676110.
Jones, S.D., Byrne, P.G., & Wallman, J.F. (2017). Exploring influence of individual courtship behaviors on male mating success in a blow fly. *Journal of Insect Behaviour*, *30*, 528–543. https://doi.org/10.1007/s10905-017-9633-1.
Jonsson, T., Kravitz, E.A., & Heinrich, R. (2011). Sound production during agonistic behavior of male *Drosophila melanogaster*. *Fly*, *5*(1), 29–38. https://doi.org/10.4161/fly.5.1.13713.
Kadoi, M., Morimoto, K., & Takami, Y. (2017). Male mate choice in a sexually cannibalistic species: Male escapes from hungry females in praying mantid *Tenodera angustipennis*. *Journal of Ethology*, *35*, 177–185. doi: 10.1007/s10164-017-0506-z.
Kang, L., Settlage, R., McMahon, W., Michalak, K., Tae, H., *et al.* (2016). Genomic signatures of speciation in sympatric and allopatric Hawaiian picture winged *Drosophila*. *Genome Biology and Evolution*, *8*(5), 1482–1488. https://doi.org/10.1093/gbe/evw095.
Kingsolver, J.G., & Huey, R.B. (2008). Size, temperature, and fitness: Three rules. *Evolutionary Ecology Research*, *10*, 251–268.
Kitano, D., Fujii, N., Yamaue, S., Kitamura, T., Honma, A., *et al.* (2018). Reproductive interference between two serious pests, oriental fruit flies *Bactrocera carambolae* and *B. dorsalis* (Diptera: Tephritidae), with very wide but partially overlapping host ranges. *Applied Entomology and Zoology*, *53*, 525–533. https://doi.org/10.1007/s13355-018-0584-6.
Klein, A.L., & de Araujo, A.M. (2010). Courtship behavior of *Heliconius erato Phyllis* (Lepidoptera, Nymphalidae) towards virgin and mated females: Conflict between attraction and repulsion signals? *Journal of Ethology*, *28*, 409–420. https://doi.org/10.1007/s10164-010-0209-1.
Kohatsu, S., & Yamamoto, D. (2015). Visually induced initiation of *Drosophila* innate courtship like following pursuit is mediated by central excitatory state. *Nature Communications*, *6*, 6457. https://doi.org/10.1038/ncomms7457.
Kohl, J., Huoviala, P., & Jefferis, G.S. (2015). Pheromone processing in *Drosophila*. *Current Opinion on Neurobiology*, *34*, 149–157. https://doi.org/10.1016/j.conb.2015.06.009.
Kolnegari, M. (2020). Mating behavior of Persian boxer mantid, *Holaptilon brevipugilis* (Mantodea: Mantidae). *Journal of Orthoptera Research*, *29*(1), 35–39.
Kopp, A. (2011). *Drosophila* sex combs as a model of evolutionary innovations. *Evolution and Development*, *13*, 504–522. https://doi.org/10.1111/j.1525-142X.2011.00507.x.
Kurtovic, A., Widmer, A., & Dickson, B.J. (2007). A single class of olfactory neurons mediates behavioural responses to a *Drosophila* sex pheromone. *Nature*, *446*, 542–546.https://doi.org/10.1038/nature05672.
Lane, S.M., Haughan, A.E., Evans, D., Tregenza, T., & House, C.M. (2016). Same sex sexual behaviour as a dominance display. *Animal Behaviour*, *114*, 113–118. https://doi.org/10.1016/j.anbehav.2016.01.005.
Laturney, M., van Eijk, R., & Billeter, J.C. (2018). Last male sperm precedence is modulated by female remating rate in *Drosophila melanogaster*. *Evolution Letters*, *2*(3), 180–189. https://doi.org/10.1002/evl3.50.

Lee, G., Bahn, J.H., & Park, J.H. (2006). Sex and clock controlled expression of neuropeptide F gene in *Drosophila. Proceedings of the National Academy of Sciences USA 103*, 12580–12585.

Lee, D.W., Shrestha, S., Kim, A.Y., Park, S.J., Yang, C.Y., *et al.* (2011). RNA interference of pheromone biosynthesis activating neuropeptide receptor suppresses mating behavior by inhibiting sex pheromone production in *Plutella xylostella* (L.). *Insect Biochemistry and Molecular Biology, 41*, 236–243. https://doi.org/10.1016/j.ibmb.2011.01.001.

Lefèvre, B.M., Catté, D., Orgogozo, V.C., & Lang, M. (2019). A primary sexual trait involved in courtship in insects: Male genital lobe morphology affects chance to copulate in *Drosophila pachea. BioRxiv, The Preprint server of Biology.* 10.1101/816538.

Legendre, F., Marting, P.R., & Cocroft, R.B. (2012). Competitive masking of vibrational signals during mate searching in a treehopper. *Animal Behaviour, 83*, 361–368. https://doi.org/10.1016/j.anbehav.2011.11.003.

Lehmann, G.U.C. (2012). Weighing costs and benefits of mating in bushcrickets (Insecta: Orthoptera: Tettigoniidae), with an emphasis on nuptial gifts, protandry and mate density. *Frontiers in Zoology, 9*, 19. https://doi.org/10.1186/1742-9994-9-19.

Lewis, S., South A., Burns, R., & Al Wathiqui, N. (2011). Nuptial gifts. *Current Biology, 21*(17), 644.

Lindeman, A.A., & Yack, J.E. (2015). What is password? Female bark beetles (Scolytinae) grant males access to their galleries based on courtship song. *Behavioural Processes, 115*, 123–131. https://doi.org/10.1016/j.beproc.2015.03.009.

Lizée, M.H., Barascud, B., Cornec, J.P., & Sreng, L. (2017). Courtship and mating behavior of cockroach *Oxyhaloa deusta* [Thunberg, 1784] (Blaberidae, Oxyhaloinae): Attraction bioassays and morphology of pheromone sources. *Journal of Insect Behavior, 30*, 674–694. https://doi.org/10.1007/s10905-017-9648-7.

Macchiano, A., Sasson, D.A., Leith, N.T., & Fowler Finn, K.D. (2019). Patterns of thermal sensitivity and sex specificity of courtship behavior differs between two sympatric species of *Enchenopa* treehopper. *Frontiers in Ecology and Evolution, 7*, 361. 10.3389/fevo.2019.00361.

Mair, M.M., Seifert, N., & Ruther, J. (2018). Previous interspecific courtship impairs female receptivity to conspecifics in parasitoid wasp *Nasonia longicornis* but not in *N. vitripennis. Insects, 9*, 112. 10.3390/insects9030112.

Marchini, D., Bene, G.D., & Dallai, R. (2010). Functional morphology of female reproductive apparatus of *Stephanitis pyrioides* (Heteroptera, Tingidae): A novel role for pseudospermathecae. *Journal of Morphology 271*, 473–482. https://doi.org/10.1002/jmor.10811.

Markow, T.A., & O'Grady, P. (2008). Reproductive ecology of *Drosophila. Functional Ecology, 22*, 747–759. https://doi.org/10.1111/j.1365-2435.2008.01457.x.

Mendelson, T.C., & Shaw, K.L. (2005). Rapid speciation in an arthropod. *Nature, 433*, 375–376. https://doi.org/10.1038/433375a.

Metzger, M., Fischbein, D., Auguste, A., Fauvergue, X., Bernstein, C., & Desouhant, E. (2010). Synergy in information use for mate finding: Demonstration in a parasitoid wasp. *Animal Behaviour, 79*, 1307–1315. https://doi.org/10.1016/j.anbehav.2010.03.003.

Miller, C.W., & Svensson, E.I. (2014). Sexual selection in complex environments. *Annual Review of Entomology, 59*, 427–445. https://doi.org/10.1146/annurev-ento-112408-085500.

Minekawa, K., Miyatake, T., Ishikawa, Y., & Matsuo, T. (2018). The adaptive role of a species specific courtship behaviour in coping with remating suppression of mated females. *Animal Behaviour, 140*, 29–37. https://doi.org/10.1016/j.anbehav.2018.04.002.

Mitchem, L.D., Debray, R., Formica, V.A., & Brodie III, E.D. (2019). Contest interactions and outcomes: Relative body size and aggression independently predict contest status. *Animal Behaviour, 157*, 43–49. https://doi.org/10.1016/j.anbehav.2019.06.031.

Mittleman, B.E., Winkler, B.M., Hall, J.B., Korunes, K.L., & Noor, M.A.F. (2017). The large X effect on secondary sexual characters and genetics of variation in sex comb tooth number in *Drosophila subobscura. Ecology and Evolution, 7*, 533–540. https://doi.org/10.1002/ece3.2634.

Monalisa, Pervez A., & Jahan, M. (2020). Intricacies in mating behaviour of a predaceous ladybird, *Harmonia dimidiata. European Journal of Environmental Sciences, 10*(1), 5–8. https://doi.org/10.14712/23361964.2020.1.

Montealegre, Z.F., & Morris, G.K. (2004). The spiny devil katydids, Panacanthus Walker (Orthoptera: Tettigoniidae): An evolutionary study of acoustic behaviour and morphological traits. *Systematic Entomology, 29*, 21–57. https://doi.org/10.1111/j.1365-3113.2004.00223.x.

Murray, R.L., Wheeler, J., Gwynne, D.T., & Bussiere, L.F. (2018). Sexual selection on multiple female ornaments in dance flies. *Proceeding of Royal Society B*, *285*, 20181525. https://doi.org/10.1098/rspb.2018.1525.

Nattier, R., Robillard, T., Amedegnato, C., Couloux, A., Cruaud, C., & Grandcolas, D.L. (2011). Evolution of acoustic communication in Gomphocerinae (Orthoptera: Caelifera: Acrididae). *Zoologica Scripta*, *40*, 479–497. https://doi.org/10.1098/rspb.2018.1525.

Ng'habi, K.R., John, B., Nkwengulila, G., Knols, B.G.J., Killeen, G.F., & Ferguson, H.M. (2005). Effect of larval crowding on mating competitiveness of *Anopheles gambiae* mosquitoes. *Malaria Journal*, *4*, 49. 10.1186/14752875449.

Nie, H., & Kaneshiro, K. (2016). Sexual selection and incipient speciation in Hawaiian *Drosophila*. *Science Bulletin*, *61*(2), 125–131. https://doi.org/10.1007/s11434-015-0976-8.

Nojima, T., Rings, A., & Goodwi, S.F. (2018). *Drosophila* courtship: Love is not blind. *Current Biology*, *28*(15), 840–842. https://doi.org/10.1016/j.cub.2018.06.003.

Oh, K.P., Fergus, D.J., Grace, J.L., & Shaw, K.L. (2012). Interspecific genetics of speciation phenotypes: Song and preference coevolution in Hawaiian crickets. *Journal of Evolutionary Biology*, *25*(8), 1500–1512. https://doi.org/10.1111/j.1420-9101.2012.02531.x.

Omkar (2004). Reproductive behaviour of two aphidophagous ladybeetles, *Cheilomenes sexmaculatus* and *Coccinella transversalis*. *Entomologia Sinica*, *11*(2), 113–124.

Omkar, & James, B.E. (2005). Reproductive behaviour of an aphidophagous ladybeetle *Coccinella transversalis* (Coleoptera: Coccinellidae). *International Journal of Tropical Insect Science*, *25*(2), 96–102.

Omkar, & Pervez, A. (2005). Mating Behaviour of an aphidophagous ladybird beetle, *Propylea dissecta* (Mulsant). *Insect Science*, *12*(1), 37–44. https://doi.org/10.1111/j.1672-9609.2005.00006.x.

Omkar, & Singh, S.K. (2010). Mating behaviour of aphidophagous ladybird beetle *Coelophora saucia* (Coleoptera: Coccinellidae). *International Journal of Tropical Insect Science*, *30*(1), 3–10. https://doi.org/10.1017/S1742758409990403.

Omkar, & Srivastava, S. (2002). The reproductive behaviour of an aphidophagous ladybeetle, *Coccinella septempunctata* (Coleoptera: Coccinellidae). *European Journal of Entomology*, *99*(4), 465–470.

Omkar, Sahu, J., & Kumar, G. (2013). Age specific mating incidence and reproductive behavior of ladybird beetle, *Anegleis cardoni* (Weise) (Coleoptera: Coccinellidae). *Journal of Asia Pacific Entomology*, *16*, 263–268. https://doi.org/10.1016/j.aspen.2013.03.003.

Otronen, M., & Siva Jothy, M.T. (1991). The effect of postcopulatory male behaviour on ejaculate distribution within female sperm storage organs of fly, *Dryomyza anilis* (Diptera: Dryomyzidae). *Behavioral Ecology and Sociobiology 29*, 33–37. https://doi.org/10.1007/BF00164292.

Owens, A.C., & Lewis, S.M. (2018). The impact of artificial light at night on nocturnal insects: A review and synthesis. *Ecology and Evolution*, *8*, 11337–11358. https://doi.org/10.1002/ece3.4557.

Papadopoulos, N.T., Liedo, P., Müller, H.G., Wang, J.L., Molleman, F., & Carey, J.R. (2010). Cost of reproduction in male medflies: The primacy of sexual courting in extreme longevity reduction. *Journal of Insect Physiology*, *56*(3), 283–287. https://doi.org/10.1016/j.jinsphys.2009.10.014.

Pervez, A. (2002). Contribution on Prey predator Relationship and Reproductive Biology of a colour morph of *Propylea dissecta* (Mulsant) (Coccinellidae: Coleoptera). Ph.D. Thesis. University of Lucknow, Lucknow, India, p. 197.

Pilastro, A., Simonato, M., Bisazza, A., & Evans, J.P. (2004). Cryptic female preference for colorful males in guppies. *Evolution*, *58*, 665–669. https://doi.org/10.1111/j.0014-3820.2004.tb01690.x.

Pölkki, M., Kortet, R., Hedrick, A., & Rantala, M.J. (2013). Dominance is not always an honest signal of male quality, but females may be able to detect dishonesty. *Biology Letter*, *9*, 1–5. 10.1098/rsbl.2012.1002.

Qvarnstrom, A., & Forsgren, E. (1998). Should females prefer dominant males? *Trends in Ecology and Evolution*, *13*, 498–501. https://doi.org/10.1016/S0169-5347(98)01513-4.

Reinhold, K., Greenfield, M.D., Jang, Y., & Broce, A. (1998). Energetic cost of sexual attractiveness: Ultrasonic advertisement in wax moths. *Animal Behaviour*, *55*, 905–913. https://doi.org/10.1006/anbe.1997.0594.

Reinhardt, K., Naylor, R.A., & Siva Jothy, M.T. (2009). Situation exploitation: Higher male mating success when female resistance is reduced by feeding. *Evolution 63*, 29–39. https://doi.org/10.1111/j.1558-5646.2008.00502.x.

Rhebergen, F.T., Courtier Orgogozo, V., Dumont, J., Schilthuizen, M., & Lang, M. (2016). *Drosophila pachea* asymmetric lobes are part of a grasping device and stabilize one sided mating. *BMC Evolutionary Biology*, *16*, 176. 10.1186/s1286201607474.

Ribeiro, I.M.A., Drews, M., Bahl, A., Machacek, C., Borst, A., & Dickson, B.J. (2018). Visual projection neurons mediating directed courtship in *Drosophila*. *Cell*, *174*, 607–621. https://doi.org/10.1016/j.cell.2018.06.020.

Ritchie, M.G. (2007). Sexual selection and speciation. *Annual Review of Ecology, Evolution, and Systematics*, *38*, 79–102. https://doi.org/10.1146/annurev.ecolsys.38.091206.095733.

Rolff, J., & Siva-Jothy, M. T. (2002). Copulation corrupts immunity: A mechanism for a cost of mating in insects. *Proceedings of the National Academy of Sciences*, 99, 9916–991810.1073/pnas.152271999.

Romano, D., Donati, E., Canale, A., Messing, R.H., Benelli, G., & Stefanini, C. (2016). Lateralized courtship in a parasitic wasp. *Laterality*, *21*(3), 243–254. https://doi.org/10.1080/1357650X.2016.1150289.

Rönn, J., Katvala, M., & Arnqvist, G. (2007). Coevolution between harmful male genitalia and female resistance in seed beetles. *Proceedings of National Academy of Sciences of USA*, *104*, 10921–10925. https://doi.org/10.1073/pnas.0701170104.

Rooker, K., & Gavrilets, S. (2018). On evolution of visual female sexual signalling. *Proceedings of Royal Society B*, *285*:20172875. https://doi.org/10.1098/rspb.2017.2875.

Roriz, A.K.P., Japyassú, H.F., & Joachim Bravo, I.S. (2018). Courtship in two morphotypes of *Anastrepha fraterculus* (Diptera: Tephritidae) cryptic species complex and their implications for understanding mate recognition. *Journal of Insect Behaviour*, *31*, 535–551. https://doi.org/10.1007/s10905-018-9697-6.

Rowe, L, Arnqvist, G, Sih, A, & Krupa J, J (1994). Sexual conflict and the evolutionary ecology of mating patterns: water striders as a model system. *Trends in Ecology & Evolution*, 9(8), 289–9310.1016/0169-5347(94)90032-9.

Russell, A.L., Buchmann, S.L., de O Sabino, W., & Papaj, D.R. (2018). Brawls bring buzz: Male size influences competition and courtship in *Diadasia rinconis* (Hymenoptera: Apidae). *Journal of Insect Science*, *18*(4). 10.1093/jisesa/iey083.

Sadd, B., Holman, L., Armitage, H., Lock, F., Marland, R., & Siva Jothy, M.T. 2006 Modulation of sexual signalling by immune challenged male mealworm beetles (*Tenebrio molitor* L.): Evidence for terminal investment and dishonesty. *Journal of Evolutionary Biology*, *19*, 321–325. https://doi.org/10.1111/j.1420-9101.2005.01062.x.

Sakaluk, S.K., Avery, R.L., & Weddle, C.B. (2006). Cryptic sexual conflict in gift giving insects: Chasing chase away. *The American Naturalist*, *167*(1), 94–104. https://doi.org/10.1086/498279.

Sakaluk, S.K., Duffield, K.R., Rapkin, J., Sadd, B.N., & Hunt, J. (2019). The troublesome gift: The spermatophylax as a purveyor of sexual conflict and coercion in crickets. In: Naguib, M., Barrett, L., Healy, S.D., Podos, J., Simmons, L.W., & Zuk, M. (eds.), *Advances in study of behaviour* (*Vol. 51*, pp. 1–30). Academic Press.

Sales, K., Trenta, T. Gardner, J., Lumley, A.J., Vasudeva, R., Michal, L., Martin, O.Y., & Gage, M.J.G. (2018). Experimental evolution with an insect model reveals that male homosexual behaviour occurs due to inaccurate mate choice. *Animal Behaviour*, *139*, 51–59. https://doi.org/10.1016/j.anbehav.2018.03.004.

Sanmartín Villar, I., & Cordero Rivera, A. (2016). Female colour polymorphism and unique reproductive behaviour in Polythore damselflies (Zygoptera: Polythoridae). *Neotropical Entomology*, *45*, 658–664. https://doi.org/10.1007/s13744-016-0417-7.

Saxena, S., Mishra, G., & Omkar (2017). Familiarity dominates relatedness in mate selection in ladybirds. *Journal of Zoology*, 304, 55–64. https://doi.org/10.1111/jzo.12497.

Scardamaglia, R.C., Fosacheca, S., & Pompilio, L. (2015). Sexual conflict in a sexually cannibalistic praying mantid: Males prefer low risk over high risk females. *Animal Behaviour*, *99*, 9–14. https://doi.org/10.1016/j.anbehav.2014.10.013.

Schärer, L., Rowe, L., & Arnqvist, G. (2012). Anisogamy, chance and evolution of sex roles. *Trends in Ecology and Evolution*, *27*, 260–264. https://doi.org/10.1016/j.tree.2011.12.006.

Schwenke, R.A., & Lazzaro, B.P. (2017). Juvenile hormone suppresses resistance to infection in mated female *Drosophila melanogaster*. *Current Biology*, *27*, 596–601. https://doi.org/10.1016/j.cub.2017.01.004.

Schmidt, A., Ronacher, B., & Hennig, R.M. (2008). The role of frequency, phase and time for processing of amplitude modulated signals by grasshoppers. *Journal of Comparative Physiology A*, *194*(3), 221–233. https://doi.org/10.1007/s00359-007-0295-x.

Shenoi, V.N., Banerjee, S.M., Guruswamy, B., Sen, S., Ali, S.Z., & Prasad, N.G. (2016). *Drosophila melanogaster* males evolve increased courtship as a correlated response to larval crowding. *Animal Behaviour*,*120*, 183–193. https://doi.org/10.1016/j.anbehav.2016.08.004.

Simmons, L.W., & Kvarnemo, C. (2006). Costs of breeding and their effects on direction of sexual selection. *Proceedings of Royal Society B, 273*, 465–470. https://doi.org/10.1098/rspb.2005.3309.

Smith, M.J. (1994). Must reliable signals always be costly? *Animal Behavior, 47*, 1115–1120.

Smith, M.J., & Harper, D.C.G. (2003). *Animal signals*. New York: Oxford University Press.

Smith, S.M., & Cognato, A.I., (2011). Observations of biology of *Camptocerus* Dejean (Coleoptera: Curculionidae: Scolytinae) in Peru. *Coleopterists Bulletin, 65*, 27–32. https://doi.org/10.1649/0010-065X-65.1.27.

Sokolowski, MB (2001). Drosophila: genetics meets behaviour. *Nat Rev Genet*, 2(11), 879–9010.1038/35098592.

Stamps, G.F., & Shaw, K.L. (2019). Male use of chemical signals in sex discrimination of Hawaiian swordtail crickets (genus *Laupala*). *Animal Behaviour, 156*, 111–120. https://doi.org/10.1016/j.anbehav.2019.06.033.

Stoltz, J.A., McNeil, J.N., & Andrade, M.C.B. (2007). Males assess chemical signals to discriminate just mated females from virgins in redback spiders. *Animal Behavior, 74*, 1669–1674. https://doi.org/10.1016/j.anbehav.2007.03.011.

Teder, T. (2014). Sexual size dimorphism requires a corresponding sex difference in development time: A meta analysis in insects. *Functional Ecology, 28*, 479–486.https://doi.org/10.1111/1365-2435.12172.

Thomas, M.L. (2011). Detection of female mating status using chemical signals and cues. *Biological Reviews, 86*(1), 1–13. https://doi.org/10.1111/j.1469-185X.2010.00130.x.

Thrall, P.H., Antonovics, J., & Bever, J.D. (1997). Sexual transmission of disease and host mating systems: Within season reproductive success. *The American Naturalist, 149*, 485–506.

Tomaru, M., & Yamada, H. (2011). Courtship of *Drosophila*, with a special interest in courtship songs. *Low Temperature Science, 69*, 61–85.

Trona, F., Anfora, G., Balkenius, A., Bengtsson, M., Tasin, M., *et al.* (2013). Neural coding merges sex and habitat chemosensory signals in an insect herbivore. *Proceeding of Royal Society B, 280*, 20130267 https://doi.org/10.1098/rspb.2013.0267.

Tsubaki, Y., Samejima, Y., & Siva Joth, M.T. (2010). Damselfly females prefer hot males: Higher courtship success in males in sunspots. *Behavioural Ecology and Sociobiology* 64: 1547–1554. 10.1007/s00265 01009682.

Vahed, K. (2007). All that glisters is not gold: Sensory bias, sexual conflict and nuptial feeding in insects and spiders. *Ethology, 113*, 105–127.https://doi.org/10.1111/j.1439-0310.2006.01312.x.

van Geffen, K.G., van Eck, E., de Boer, R.A., van Grunsven, R.H.A., Salis, L. *et al.* (2015). Artificial light at night inhibits mating in a geometrid moth. *Insect Conservation and Diversity, 8*, 282–287. https://doi.org/10.1111/icad.12116.

Vaughan, A.G., Zhou, C., Manoli, D.S., & Baker, B.S. (2014). Neural pathways for detection and discrimination of conspecific song in *D. melanogaster*. *Current Biology, 24*(10), 1039–1049. https://doi.org/10.1016/j.cub.2014.03.048.

Vigoder, F.M., Souza, N.A., Brazil, R.P., Bruno, R.V., Costa, P.L., *et al.* (2015). Phenotypic differentiation in love song traits among sibling species of *Lutzomyia longipalpis* complex in Brazil. *Parasites and Vectors, 8*, 290. 10.1186/s1307101509008.

Villarreal, A.E., Godin, J.G.J., & Bertram, S.M. (2018). Influence of operational sex ratio on mutual mate choice in Jamaican field cricket (*Gryllus assimilis*): Testing predictions of switch point theorem. *Ethology, 124*(11), 816–828. https://doi.org/10.1111/eth.12816.

von Philipsborn, A.C., Liu, T., Yu, J.Y., Masser, C., Bidaye, S.S., & Dickson, B.J. (2011). Neuronal control of *Drosophila* courtship song. *Neuron, 69*, 509–522. https://doi.org/10.1016/j.neuron.2011.01.011.

Wagner, W.E. Jr., & Reiser, M.G. (2000). The importance of calling song and courtship song in female mate choice in variable field cricket. *Animal Behaviour, 59*, 1219–1226. https://doi.org/10.1006/anbe.1999.1428.

Weatherhead, P.J., & Robertson, R.J. (1979). Offspring quality and polygyny threshold: "The sexy son hypothesis". *The American Naturalist, 113* (2), 201–208. https://doi.org/10.1086/283379.

Wedell, N. (2010). Variation in male courtship costs in butterflies. *Behavioral Ecology and Sociobiology, 64*, 1385–1391. https://doi.org/10.1007/s00265-010-0953-9.

Westerman, E.L., Davis, A.H., Dinwiddie, A., & Monteiro, A. (2012). Biased learning affects mate choice in a butterfly. *Proceedings of the National Academy of Sciences of USA, 109*, 10948–10953. https://doi.org/10.1073/pnas.1118378109.

Wellenreuther, M., & Sánchez Guillén, R.A. (2016). Non adaptive radiation in damselflies. *Evolutionary Applications*, *9*, 103–118. https://doi.org/10.1111/eva.12269.
White, C.R., Matthews, P.G.D., & Seymour, R.S. (2008). In situ measurement of calling metabolic rate in an Australian mole cricket, *Gryllotalpa monanka*. *Comparative Biochemistry and Physiology A*, *150*, 217–221. https://doi.org/10.1016/j.cbpa.2006.08.030.
Wigby, S., & Chapman, T. (2005). Sex peptide causes mating costs in female *Drosophila melanogaster*. *Current Biology*, *15*, 316–321. https://doi.org/10.1016/j.cub.2005.01.051.
Wijesekera, T.P., Saurabh, S., & Dauwalder, B. (2016). Juvenile hormone is required in adult males for *Drosophila* courtship. *PLoS One*, *11*(3), e0151912. 10.1371/journal.pone.0151912.
Wilder, S.M., Rypstra, A.L., & Elgar, M.A. (2009). The importance of ecological and phylogenetic conditions for occurrence and frequency of sexual Cannibalism. *Annual Review of Ecology, Evolution, and Systematics*, *40*, 21–39. https://doi.org/10.1146/annurev.ecolsys.110308.120238.
Willink, B., Duryea, M.C., & Svensson, E.I. (2019). Macroevolutionary origin and adaptive function of a polymorphic female signal involved in sexual conflict. *The American Naturalist*, *194*(5), 707 724. https://doi.org/10.5061/dryad.d0bh91n.
Woods, W.A.J., Hendrickson, H., Mason, J., & Lewis, S.M. (2007). Energy and predation costs of firefly courtship signals. *The American Naturalist*, *170*(5), 702–708. https://doi.org/10.1086/521964.
Worthington, A.M., & Kelly, C.D. (2016). Direct costs and benefits of multiple mating: Are high female mating rates due to ejaculate replenishment? *Behavioural Processes*, *124*, 115–122. https://doi.org/10.1016/j.beproc.2015.12.009.
Wyatt, T.D. (2014). *Pheromones and animal behavior*. Cambridge: Cambridge University Press.
Xu, H., & Turlings, T.C.J. (2018). Plant volatiles as mate finding cues for insects. *Trends in Plant Science*, *23*(2), 100–111. https://doi.org/10.1016/j.tplants.2017.11.004.
Yamamoto, D., & Koganezawa, M. (2013). Genes and circuits of courtship behaviour in *Drosophila* males. *Nature Reviews Neuroscience*, *14*, 681–692. https://doi.org/10.1038/nrn3567.
Yiğit, A., Sevgili, H., & Özdemir, H. (2019). Male bush crickets in female biased environment allocate fewer sperm per ejaculation. *Entomological News*, *128*(4), 393–403. https://doi.org/10.3157/021.128.0412.
Zhou, C., Franconville, R., Vaughan, A.G., Robinett, C.C., Jayaraman, V., & Baker, B.S. (2015). Central neural circuitry mediating courtship song perception in male *Drosophila*. *eLife*, *4*, e08477. https://doi.org/10.7554/eLife.08477.001.
Zuk, M., & Kolluru, G.R. (1998). Exploitation of sexual signals by predators and parasitoids. *Quarterly Review of Biology* *73*, 415–438. https://doi.org/10.1086/420412.
Zuk, M., Rebar, D., & Scott, S.M. (2008). Courtship song is more variable than calling song in field cricket *Teleogryllus oceanicus*. *Animal Behaviour*, *76*, 1065–1071. https://doi.org/10.1016/j.anbehav.2008.02.018.
Zuk, M., Simmons, L.W., & Cupp, L. (1993). Calling characteristics of parasitized and unparasitized populations of field cricket *Teleogryllus oceanicus*. *Behavioural Ecology and Sociobiology*, *33*, 339–343. https://doi.org/10.1007/BF00172933.

7

Precopulatory Sexual Selection

Ankita Dubey, Omkar, and Geetanjali Mishra
Ladybird Research Laboratory, Department of Zoology, University of Lucknow, Lucknow-226007, India

CONTENTS

7.1 Introduction

Charles Darwin proposed and expressed his ideas on a phenomenon that he called as 'sexual selection' (Darwin, 1859), where he suggested inclusion of certain types of non-survival adaptations that seemed misfits under parasol of natural selection (Darwin, 1871). Alcock (2001) defined sexual selection as, *'A form of natural selection that occurs when individuals differ in their ability to compete with others for mates or to attract members of opposite sex'*. Ever since its inception, sexual selection has continued to garner attention of biologists because of its strength to drive evolution of characters that affect fitness of organisms both directly as well as indirectly (Andersson, 1994; Andersson & Simmons, 2006). Sexual selection might also affect speciation via influencing rate of gene flow between diverging taxa (Coyne & Orr, 2004; Ritchie, 2007).

Among the many organisms studied, insects have especially contributed to sexual selection with evidences of both pre- and post-copulatory mechanisms. The premating mechanisms, such as aggressive or appeasement displays, and display of exaggerated characters are quite evident and well-studied in

DOI: 10.1201/9781003043195-7

many organisms (Wheeler *et al.*, 2012), while post-mating mechanisms are more subtle and are being paid attention to only in recent years (Pischedda & Rice, 2012). In this chapter, we discuss about the traits of interest in insects that shape selection of mates at the precopulatory level.

7.2 Causes of Sexual Selection

The underlying resultant of sexual selection arises from competition for scarce and rarer sources, and assessment of fitness of an individual by number of mates it has gained access to (Andersson & Iwasa, 1996). Therefore, in sexual selection it is non-limiting sex that faces stronger selection pressure. In Darwinian model of sexual selection idea of 'nonlimiting sex' was associated with males typically undergoing competition and 'limiting sex' with mostly females exhibiting 'mate choice'. However, this model was not a universal fit and hence, several hypotheses, like parental investment (Bateman, 1948; Trivers, 1972, Lehtonen *et al.*, 2016), mating system (Darwin, 1871), sex ratio/operational sex ratio (Emlen & Oring, 1977), etc. were put forth to explain causes and intensity of sexual selection.

7.3 Operational Level of Sexual Selection and Attributes Targeted

Sexual selection can operate at various stages of sexual reproduction. Though usually associated with phenotypic characters evaluated before copulation (precopulatory selection), strong sexual selection does continue after copulation as well (post-copulatory selection) (Pizzari & Parker, 2009).

Precopulatory sexual selection involves traits/attributes that signal higher fitness both directly and indirectly and help individuals of non-limiting sex win contests that maximize their access to mates. Overall, precopulatory sexual selection focuses on mate selection by exercising choice with respect to overt traits/attributes that include: (1) size: in contests involving mate seeking, fighting, endurance, productivity; (2) colour: in conditions involving morphological appearance, spatio-temporal competence; (3) age: determinant factor in scenarios considering length of reproductively active window, survivorship; (4) mating history/pattern: considered in contests comprising extent of probable mating success and offspring share; (5) nutritional plasticity: as a gateway to external and physiological strength, gametic yield; and (6) chemical signals: as a measure to assess condition of an individual and express mate choice thereof.

7.4 Traits of Interest in Precopulatory Sexual Selection

7.4.1 Body Size

Size is a prime driver in evolution, with no living entity capable of evolving or developing without taking it into consideration. It is not just a by-product of evolution, but a major player (Bonner, 2012) that indicates physiological state and indirectly reflects good genes (Ryder & Siva-Jothy, 2001; Pech-May *et al.*, 2012).

In most species across animal world, sexes differ in size with females usually being larger than males. Such sexual dimorphism, its intensity and elucidation of selective factors regulating its intensity has always been a point of discussion but their importance is still to be understood well. The gain incurred by adults of both sizes across taxa is nothing but outcome of 'who fits better with respect to time and space'.

Darwin (1871) suggested that in species, which grow throughout life, females are larger than males, because egg production increases with body size. Further, in mammals, males are larger than females as it gives them an advantage in contests over mates. However, this may not be a universal scenario, as arms race for escalation or reduction of body size among two sexes depends on selective forces acting on them as well as gains incurred by individuals expressing that size.

7.4.1.1 Body Size as a Criterion of Mate Preference in Insects

In insects, mate choice as well as contest competition are major selective mechanisms responsible for sexual selection on body size, operating either singly or simultaneously.

In insects, large body size provides a competitive edge to males and fecundity advantage to females (Bonduriansky, 2001). Larger females lay large eggs as well as clutch (Dixon, 2007; Salavert *et al.*, 2011), promote faster development (Katvala & Kaitala, 2001) and affect offspring fitness (Lalonde, 2005). Under fluctuating environments large offspring outperform smaller ones (Roff, 2002).

For large males, large body size confers advantages in terms of: (1) better fighting and competitive ability (Filin & Ovadia, 2007) leading to increased chances of selection (Byrne & Rice, 2006; Anjos-Duarte *et al.*, 2011), and/or (2) larger ejaculates with more numbers of sperm (Bissoondath & Wiklund, 1996), nutrients, and accessory gland proteins (Arnqvist *et al.*, 2005; Avila *et al.*, 2011), which influence sperm uptake (Eberhard, 1996; Wedell, 2005), remating propensity (Avila *et al.*, 2011), fecundity, and offspring fitness (Vahed, 1998).

However, being large imposes costs, *viz.* higher risk of parasitism (Zuk & Kolluru, 1998) and predation (Andersson, 1994; Blanckenhorn, 2000) because of longer exposure (Abrams *et al.*, 1996). Large individuals cannot efficiently utilise energy under food stress (Gotthard *et al.*, 1994; Blanckenhorn, 1998) as they require more food to support themselves (and their offspring), which increases mortality risk under resource limitation (Blanckenhorn *et al.*, 1995; but see Millar & Hickling, 1990; Dixon, 2007).

In odonates and orthopterans, larger adults are preferred by both sexes (Moore, 1990) as well as win male–male competition (Jacobs, 1955). In order Thysanoptera, competition between males ends with selection of larger adults as mates (Crespi, 1988). In Coleoptera too, large adults are preferred as mates across genera (Otronen, 1988; Brown, 1990a, b; Omkar & Afaq, 2013; Dubey *et al.*, 2016a). Also in lepidopterans, mate choice is selective force behind selection of large sized mates (Rutowski, 1982; Phelan & Baker, 1986).

The preference for larger adults as mate is however not universal, (Andersson, 1994). Both processes may actually occur in same species (Blanckenhorn *et al.*, 1995), resulting in stabilizing sexual selection (Mason, 1964; Moore, 1990). Small females are able to manoeuvre more resources into reproduction (Wiley, 1974), and also breed earlier in fluctuating environment with *ad libitum* food (Grant & Grant, 1989). Small males also reach maturity and breed earlier at a young age (Wallace, 1987) and are better at scramble competition. Small males also dominate in contests where manipulation is more crucial than strength (Ralls, 1976), short generation time (Singer, 1982), higher agility (Andersson & Norberg, 1981), rapid rate of reproduction and female preference (Steele & Partridge, 1988).

7.4.2 Body Colour

Variation in body colour at organismal level has been one of major evidences exemplifying existence of biodiversity in nature. Among such conspicuous variations lies most simple, common yet enigmatic phenomenon of melanism. It involves occurrence of forms which are either mostly/completely dark, i.e., ***melanic*** or lighter ones called ***typicals*** at both intra- and interspecific levels (True, 2003).

Melanism is common across taxa (Majerus, 1998). Work on melanism has provided crucial models for understanding role of natural selection in expression of polymorphism (Brakefield, 1987; True, 2003).

7.4.2.1 Selective Pressures and Correlated Effects Associated with Insect Melanism

7.4.2.1.1 Visual/Non-visual Mechanisms

In lepidopterans, coleopterans, and dipterans, melanism is often associated with body colour, thermoregulation, UV resistance, desiccation tolerance, and resistance to abrasion. In *Biston betularia* L., melanism confers advantage during predator attack via crypsis, mimicry, and camouflage (Majerus, 1998). Melanic *Manduca sexta* (L.) are less tolerant to desiccation than typicals (Safranek &

Riddiford, 1975), while, in coleopteran, desert beetle melanic cuticle is more resistant to abrasion than nonmelanic cuticle (Majerus, 1998).

7.4.2.1.2 Pleiotropic Effects Associated with Melanism

Several morphological, developmental, physiological, and behavioural costs are also associated with melanism. In *Xylophasia monoglipha* (Lepidoptera), cuticle of melanic pupa is stronger than typical (Brakefield, 1987), while in *Drosophila melanogaster* Meigen, melanic wing cuticle is weaker than its wild type (Jacobs, 1985). The melanic larvae of *M. sexta* are smaller than their typical counterparts (Safranek & Riddiford, 1975). In *Tenebrio molitor* L. (Coleoptera), melanic cuticle darkens faster than does typical. In many lepidopterans, melanic forms are more effective than typicals at encapsulating parasitoids (Verhoog *et al.,* 1996). Also, melanics are more resistant to fungal and viral infections (Wilson *et al.,* 2001). In *Adalia bipunctata* L., melanic males require more copulation time than normal ones (de Jong & Brakefield, 1998).

7.4.2.2 Melanism as a Criterion for Mate Selection

Melanism in Lepidoptera and Diptera is displayed on wings as spots; in Coleoptera complete elytral melanism is seen (Figure 7.1). Usually, a strong hereditary component determines degree of melanism. However, temporal and geographical variations also contribute for fluctuations in frequency of melanic

FIGURE 7.1 Range of polymorphism in *Menochilus sexmaculatus* (Fabricius).

individuals (Wang *et al.,* 2009; Michie *et al.,* 2010, 2011). Various theories viz. 'thermal melanism' (de Jong *et al.,* 1996; Michie *et al.,* 2010), mimicry (Brakefield & Wilmer, 1985), and mating preferences (Majerus *et al.,* 1982a, b) have been put forward to explain selective forces responsible for maintaining colour polymorphisms in different species.

In lepidopterans, pigmentation on wings affects flight performance (Davis *et al.*, 2012), species recognition (Wiernasz & Kingsolver, 1992), and sexual selection (Wiernasz, 1989, 1995; Tuomaala *et al.*, 2012). Even location and choice of mate rely on visual cues present on wings (Chamberlain *et al.*, 2009; Kemp & Rutowski, 2011), with dorsal surface and forewings more likely to bear melanin change via sexual selection forces (Oliver *et al.*, 2009).

In *D. melanogaster*, melanism affects fertility (Fahmy & Fahmy, 1959), desiccation tolerance (Kalmus, 1941), UV resistance (Jacobs, 1985), vision (Borycz *et al.*, 2002; Takahashi *et al.*, 2007), circadian activity (Newby & Jackson, 1991), and mating success (Dev *et al.*, 2013). In female mate choice trials, *D. biarmipes* males bearing an apical dark wing patch achieved higher mating success as compared to males lacking wing patch (Singh & Chatterjee, 1987). Apart from this, under competitive and non-competitive situations, spotted males of *D. biarmipes* displayed enhanced frequency of courtship as compared to non-spotted males (Hegde *et al.*, 2005).

In coleopterans, especially ladybird beetles, melanic polymorphism (Majerus, 1998) has been linked with mating success (Wang *et al.*, 2009; Dubey *et al.*, 2016b). Early studies showed higher mating success is achieved by dark morphs (Brakefield, 1984a, b;Osawa & Nishida, 1992; Srivastava & Omkar, 2005; Wang *et al.,* 2009; Dubey *et al.*, 2016b;). Low temperature facilitates melanic morphs via faster absorption of radiation, thus enhancing foraging activity and reproductive success (Brakefield, 1984b). Other advantages associated with selection of dark morph as mates include better adaptability to local conditions (Proulx, 2001), higher fecundity in melanic females of *C. septempunctata* (Srivastava & Omkar, 2005), faster development due to increased voracity (Soares *et al.*, 2001, 2005), and greater nutritional plasticity (Berkvens *et al.,* 2008). However, melanism is disadvantageous at higher temperature as it leads to overheating, thus reducing daytime foraging and ultimately decreasing individual's fitness (Stewart & Dixon, 1989).

7.4.3 Age

Aging is naturally selected and genetically programmed across all animal phyla (Rose, 1991; Guarente & Kenyon, 2000; Finch & Ruvkun, 2001). Evolutionary biology defines aging as an irreversible, intrinsic decline in survival rate and fertility with age (Rose, 1991). The evolution of senescence is explained as (i) accumulation of deleterious alleles at advanced ages that are subject only to weak counter-selection (Medwar, 1952); (ii) antagonistic pleiotropy, whereby genes beneficial at young ages have lethal effects in later part of age (Williams, 1957); and (iii) trade-offs between somatic maintenance and repair and investment in growth and reproduction (Kirkwood, 2002).

Age is a major determinant of reproductive success and adult fitness in insects (Fox, 1993). Age-related decline in mating and fertilization success have been reported in a number of species (Kidd *et al.*, 2001; Bonduriansky & Brassil, 2002; Bista & Omkar, 2015).

Other than directly affecting adult fitness, transgenerational effects of age are also known. Older senescent parents produce offspring with slower growth rate or lower survival probability (Mohaghegh *et al.,* 1998; Mousseau & Fox, 1998a, b; Hercus & Hoffmann, 2000). However, theory postulates that phenotypic effects of senescence do not accumulate from generation to generation as they will lead to accumulation of deleterious mutations, thereby causing elimination (Kern *et al.,* 2001).

7.4.3.1 Effect of Age on Mate Selection

It has been observed that as females age, their reproductive effort is expected to decline and selection to be weak (Futuyma, 1998). Many have previously studied interaction between age and mating experience (Prosser *et al.,* 1997; Moore & Moore, 2001; Martin & Hosken, 2002). Numerous studies have emphasized age-specific mating incidences (Rose, 1991; Arnqvist & Nilsson, 2000; Srivastava & Omkar, 2004; Omkar & Pervez, 2005; Omkar *et al.,* 2010; Omkar & Singh, 2010).

The good gene model predicts that females should prefer older males to younger males (Trivers, 1972; Andersson, 1994;Avent *et al.*, 2008; Perez-Staples *et al.*, 2010; Prathibha *et al.*, 2011), as old males reflect higher survival ability and thus signal better genes (Kokko & Lindstrom, 1996). Another model predicts that younger males are more likely to be preferred as mates and pairing with older males may result in reduction in progeny output (Hansen & Price, 1995; Price & Hansen, 1998; Xu & Wang, 2009). A few studies document selection of middle-aged males as mates (Beck & Powell, 2000; Proulx *et al.*, 2002; Jones & Elgar, 2004; Pandey & Omkar, 2013).

Empirical studies have found mixed support for both a decline in fertility with male age and age-based female mate preference (Johnson & Gemmell, 2012). Several studies in insects show that females become less selective of potential mates as they get older. In female cockroach, *Nauphoeta cinera* (Olivier), older females were less fertile suggesting that decline in choice might be an outcome of reduced reproductive potential (Moore & Moore, 2001). Under high male densities, young female field crickets, *Gryllus lineaticeps* Stål., were choosier comparative to older females (Atwell & Wagner, 2014). As age-specific changes are irreversible, mate selection will occur considering reproductive value for that age and how relative benefits accrued by mating with them outweighs costs associated with time or energy spent in searching for preferred mates.

7.4.4 Mating History/Pattern

Reproductive success of adults relies on informative signals that can assess reproductive value of a mate and make suitable selection (Smith & Harper, 2003). Males and females often have conflicting interests when mating is correlated with lifetime fitness levels (Arnqvist, 1989; Arnqvist & Rowe, 2002). Although multiple copulations decrease number of sperm, males are still benefitted by further matings, as it increases number of offspring. However, for females, mating with an already mated male can be detrimental in terms of fewer sperm, fewer nutrients from nuptial gifts, prolonged duration of copulation, increased risk of unsuccessful fertilization, or shorter subsequent lifespan (Wedell & Ritchie, 2004; Jones *et al.*, 2006; Lauwers & Van Dyck, 2006).

Since mating history affects allocation and utilization of sperm, it is likely to affect mate selection. Mate choice based on mating status is a method to optimize distribution and utilization of one's gametes because neither ejaculate is cheap to be donated nor ova are inexpensive to be fertilized (Trivers, 1972; Koene *et al.*, 2008). There is increasing evidence that animals are able to assess mating history of their mate, and decide whether to mate or not. Empirical and theoretical studies of mate choice have primarily focused on priority for certain traits and questions whether these preferences are capricious or associated with direct or indirect rewards for males and females (Andersson & Simmons, 2006; Kokko *et al.*, 2006).

7.4.4.1 Costs and Benefits Associated with Varying Mating History

Species in which males and females mate multiply, sexual selection would likely favour evolution of those traits that can be used to distinguish between sexes on basis of their reproductive status to minimize competition (Ridley, 1988, 1989). Mating with a virgin female ensures no competition with another male's sperm, at least until female re-mates (Parker, 2006), while unmated males generally provide more sperm number and accessory proteins in ejaculates to unmated females than previously mated females (Rutowski *et al.*, 1987).

Several studies indicate that females mating with previously mated males have lower lifetime fecundity than ones mating with virgin males (Jimenez-Perez & Wang, 2004; McCartney & Heller, 2008), with exceptions (Unnithan & Paye, 1991; Sadek, 2001). Similarly, males mating with already mated females confer heavy costs in terms of ejaculate expenditure (Wedell *et al.*, 2002; Wedell & Ritchie, 2004) as repeated matings reduce production of ejaculates (Wedell & Ritchie, 2004; McNamara *et al.*, 2007; Dowling & Simmons, 2012).

In species that store sperm, female mating status (virgin vs. mated) is crucial for male reproductive success due to risk of sperm competition that arises when females are already mated (Simmons, 2001;

Wedell *et al.*, 2002). Sperm usage patterns that confer advantage to male predicts direction of preferential mating, i.e. mating with either virgin or mated females, which increases fertilization success and lowers risk of infection by sexually transmitted diseases (Bonduriansky, 2001; Simmons, 2001; Wedell *et al.*, 2002; Knell & Webberley, 2004).

7.4.4.2 Mate Choice in Relation to Mating Status

Variation in quality of mates leads to increased selective mate choice by males as well as females (Byrne & Rice, 2006). However, practical manifestation of direct/indirect benefits pertaining to mate preferences remains exigent despite an intense investigation on it (Kokko *et al.* 2003; Andersson & Simmons, 2006).

Mate choice in relation to mating history has been reported in several insects. Males of *Tribolium castaneum* (Herbst) easily recognize female status and prefer to inseminate previously mated females as it reflects their greater reproductive potential (Arnaud & Haubruge, 1999). Females of *D. melanogaster* do not accept recently mated males because of their reduced fertility as mates (Markow *et al.*, 1978; Nakatsuru & Kramer, 1982). Besides above, a few other examples of mating history dependent mating success have been observed in many other insects (Larsson, 1988; McCartney & Heller, 2008;; Helinski & Harrington, 2011).

Low-quality males turn less choosy when they face strong competition for mates. As a result, incidence of mate choice is reduced and males are allowed to lower time spent in replenishing sperm reserves in order to raise their own mating rate. This, in turn, produces lower levels of sperm at mating and consequently leads to negative effects on female fitness via sperm limitation. Therefore, individuals prefer to mate with virgins than non-virgins ones (Arnaud & Haubruge, 1999), also known as Coolidge effect (Pizzari *et al.*, 2003).

Under certain circumstances where direct benefits outweigh costs, increased preference for experienced mates could be observed. Indirect benefits related to genetically superior partners or inherited male attractiveness might also drive sexual selection (Kirkpatrick & Barton, 1997). In case of pentatomid females, *Euschistus conspersus* Uhler and *Utetheisa ornatrix* (L.) it is indirect fitness gains rather than direct fitness benefits which leads to preferential mating with experienced males (Krupke *et al.*, 2008; Iyengar, 2009). The preference for experienced males has also been observed in *Lucilia cuprina* Wiedmann (Cook, 1995) and a cockroach (Edvardsson *et al.*, 2008).

7.4.5 Nutritional Plasticity

The process of developmental plasticity poses potential effects on individual fitness. Organisms modulate their behaviour, physiology, and morphology in accordance with perpetuating conditions. Experience gained by an organism in early life stages shapes its range of adaptability towards future environmental fluctuations (Monaghan, 2008; Pilakouta *et al.*, 2015).

Individuals experiencing good nutritional conditions in juvenile stages tend to mature early and have a larger body size (Day & Rowe, 2002), and possible higher fitness levels. However, environment-matching hypothesis suggests that fitness is not only limited to juvenile nutrition, but also on continuation of favourable environment in adult stages (Monaghan, 2008).

Food stress in early life stages produces individuals that perform better under energy-limited environments than in energy-abundant environments (Barker *et al.*, 1989; Bateson *et al.*, 2004); however, growth and reproductive performance are adversely affected. Studies have shown that larval and adult stages of beetles are capable of surviving prolonged food stress (Omkar & Pervez, 2003) but their reproductive success is hampered in comparison to those with *ad libitum* food (Koch, 2003; Dmitriew & Rowe, 2007).

Nutritional stress during growth period also affects body composition and allometric relationships (Hahn, 2005; Scharf *et al.*, 2009). It is assumed that natural selection shapes allocation of resources making phenotype adaptive for expected adult environment. For example, costs of maintenance and locomotion are generally lower for small individuals (Blanckenhorn, 2000). Some other characteristics of food restriction include reduced wing loading and high fat content (Boggs & Freeman, 2005; Pellegroms *et al.*, 2009) that increase survival fitness by improving dispersal ability, thus escaping poor local conditions of food restriction (Dingle, 2001). It is thus possible that offspring emerging from fluctuating dietary regimes perform better when subsequently facing poor food conditions as adults than those raised on *ad libitum* larval food supply.

Nutritional condition of both mating partners play an important role in mate choice decisions (Arnqvist & Rowe, 2005; Fisher & Rosenthal, 2006; Garbutt & Little, 2014). Mating decisions of males and females may be influenced by dynamics of food abundance in a particular time window, thus intensifying sexual conflict (Fisher & Rosenthal, 2006).

7.4.5.1 Food and Mate Choice

Mating preferences of both sexes vary during a lifetime (Burley & Foster, 2006; Fisher & Rosenthal, 2006) and changes with an individual's own condition (Cotton *et al.*, 2006; Holveck & Riebel, 2010). Studies have focused on how conditions or resources available to allocate to life history traits (Rowe & Houle, 1996), influence mating behavior.

In black field cricket, *Teleogryllus commodus* (Walker), females reared on high protein diets are more sexually responsive and show stronger preference functions than those reared on low-protein diets (Hunt *et al.*, 2005). In *Cyrtodiopsis dalmanni* (Weidemann), females fed on corn reflect stronger preferences for large-eye span males than those fed on sucrose while female of *Diasemopsis meigenii* (Westwood) are positively associated with female eye span, a condition-dependent trait (Cotton *et al.*, 2006). However, not all studies of diet manipulations have led to observable differences in mate selection.

Previous results from two insect orders, Lepidoptera (Ferkau & Fischer, 2006; Lewis & Wedell, 2007) and Orthoptera (Wedell, 1993), revealed that males in good condition produced larger non-sperm ejaculates than males in poor condition and thus achieved higher mating success than low condition males. In ladybird, *Cryptolaemus montrouzieri* Mulsant, where dietary regimes were varied in both larval and adult stages, female ladybirds preferred mating with males reared under high-nutrition diet regimes postpupal emergence and this was independent of nutritional conditions they experienced during their larval stages, indicating that mate choice is non-random and phenotype-dependent (Xie *et al.*, 2015).

Variation in food availability may select for environment dependent maternal or paternal effects. Individuals that acquire abundant conditions from a resource rich environment may benefit by transferring resources to their offspring, thus enhancing offspring fitness (Mousseau & Fox, 1998a, b). Offspring will do best in environment which holds similarity to one experienced by their parents (Mousseau & Fox, 1998a, b; Gilchrist & Huey, 2001; Holbrook & Schal, 2004). A few studies in insects indicate that parental diet has a carryover or transgenerational effect on offspring even when affecting factors are nullified (Reynolds, 1944; Rossiter, 1991; Stamp, 2001).

7.4.6 Chemical Signals

The process of mate selection implies selection of a mate which is superior to its counterparts in order to achieve higher fitness directly or indirectly. This process of selection is based on 'assessment' of mate with respect to certain signals (physical or chemical) it expresses as they reflect condition of individual bearing them; chemical signals could be condition-dependent (Steiger & Stökl, 2014).

In tomato fruit borer moth *Neoleucinodes elegantalis* (Guenee), males preferred chemicals sent by heavier females over those of light females (Jaffe *et al.*, 2007). Similarly, in mantid, *Stagmomantis limbata* Hahn, chemical cues sent by females fed on high-quality diet attracted more males than those from females fed on a low-quality diet (Maxwell *et al.*, 2010). Olfactory signals of females act as reliable indicators of their fecundity function (Barry *et al.*, 2010). Pheromone production by virgin and mated females in *Heliothis virescens* (Fabricius) depends on hemolymph trehalose concentration and males use as a tactic to assess quality of female for mating (Foster 2009; Foster & Johnson, 2011). In *Tenebrio molitor*, pheromones released by immune-challenged males reduced their attractiveness as well as reproductive success (Worden *et al.*, 2000). *Tenebrio molitor* females found chemical signals produced by inbred males to be less than those produced by outbred males (Pölkki *et al.*, 2012).

Chemical signals can also indicate age of mates and thereby influence mate choice. In European corn borer *Ostrinia nubilalis* (Hubner) and butterfly *Bicyclus anynana* (Butler), females prefer to mate with old males and this decision is based on male sex pheromone (Lassance & Löfstedt, 2009; Nieberding *et al.*, 2012).

7.5 Conclusions

Mate choice is central to sexual selection; however, causes of individual variation in mate choice behaviour and the way it is related to life history trade-offs differs with species. A selection made at one time might not be maintained uniformly as insects inhabit areas that are subjected to rapid changes, so a choice made by an individual is also a reflectance of environmental suitability as well. Factors such as size, colour, age, mating history, nutritional plasticity, and chemical signals influence mate selection, as they reflect both apparent as well as indirect fitness of organism. In general, larger adults, darker morphs, middle-aged individuals, unmated adults, and well-fed individuals emerge as supreme choice for mating across most insect groups. However, under stress lesser preferred outweigh supreme choice owing to their efficient adaptability for those conditions. Besides, precopulatory selection is just a step as more complex mechanisms operate at both pre- and post-copulatory levels that truly decide mating success incurred, which is independent of precopulatory choice.

Acknowledgements

Ankita Dubey acknowledges the Department of Science and Technology, New Delhi, India for awarding the DST INSPIRE Fellowship (IF 120365).

REFERENCES

Abrams, P.A., Leimar, O., Nylin, S. & Wiklund, C. (1996). The effect of flexible growth rates on optimal sizes and development times in a seasonal environment. *The American Naturalist*, *147*, 381–395.

Alcock, J. (2001). *Animal Behavior*. Sunderland, MA: Sinauer. 543 pp.

Andersson, M. (1994). *Sexual Selection*. Princeton, NJ: Princeton University Press.

Andersson, M. & Norberg, R.A. (1981). Evolution of reversed sexual size dimorphism and role partitioning among predatory birds, with a size scaling of flight performance. *Biological Journal of Linnean Society*, *15*, 105–130.

Andersson, M. & Simmons, L.W. (2006). Sexual selection and mate choice. *Trends in Ecology & Evolution*, *21*, 296–302.

Andersson, M. & Iwasa, Y. (1996). Sexual selection. *Trends in Ecology & Evolution*, *11*(2), 53–58.

Anjos-Duarte, C.S., Costa, A.M. & Joachim-Bravo, I.S. (2011). Sexual behaviour of Mediterranean fruit fly (Diptera: Tephritidae): influence of female size on mate choice. *Journal of Applied Entomology*, *135*, 367–373.

Arnaud, L. & Haubruge, E. (1999). Mating behaviour and male mate choice in *Tribolium castaneum* (Coleopteran: Tenebrionidae). *Behaviour*, *136* (1), 67–77.

Arnqvist, G., Nilsson, T. & Katvala, M. (2005). Mating rate and fitness in female bean weevils. *Behavioural Ecology*, *16*, 123–127.

Arnqvist, G. (1989). Multiple matings in a water strider: mutual benefits or intersexual conflict? *Animal Behaviour*, *38*, 749–756.

Arnqvist, G. & Nilsson, T. (2000). The evolution of polyandry: multiple matings and female fitness in insects. *Animal Behaviour*, *60*, 145–164.

Arnqvist, G. & Rowe, L. (2002). Antagonistic coevolution between sexes in a group of insects. *Nature*, *415*, 787–789.

Arnqvist, G. & Rowe, L. (2005). *Sexual Conflict*. Princeton, NJ: Princeton Univ. Press.
Atwell, A. & Wagner, Jr W.E. (2014). Female mate choice plasticity is affected by interaction between male density and female age in a field cricket. *Animal Behaviour*, *98*, 77–83.
Avent, T.D., Price, T.A.R. & Wedell, N. (2008). Age-based female preference in fruit fly *Drosophila pseudoobscura*. *Animal Behaviour*, *75*, 1413–1421.
Avila, F.W., Sirot, L.K., LaFlamme, B.A., Rubinstein, C.D. & Wolfner, M.F. (2011). Insect seminal fluid proteins: identification and function. *Annual Review of Entomology*, *56*, 21–40.
Barry, K.L., Holwell, G.I., & Herberstein, M.E. (2010). Multimodal mate assessment by male praying mantids in a sexually cannibalistic mating system. *Animal Behaviour*, *79*(5), 1165–1172.
Barker, D.J.P., Osmond, C. & Law, N.E. (1989). The intrauterine and early postnatal origins of cardiovascular disease and chronic-bronchitis. *Journal of Epidemiology and Community Health*, *43*, 237–240.
Bateman, A. J. (1948). Intrasexual selection in *Drosophila*. *Heredity*, *2*, 349–368.
Bateson, P., Barker, D., Clutton-Brock, T., Deb, D. & D'Udine, B. (2004). Developmental plasticity and human health. *Nature*, *242*, 419–421.
Beck, C.W. & Powell, L.A. (2000). Evolution of female mate choice based on male age: are older males better mates? *Evolutionary Ecology Research*, *2*, 107–118.
Berkvens, N., Bonte, J., Berkvens, D., Tirry, L. & De Clercq, P. (2008). Influence of diet and photoperiod on development and reproduction of European populations of *Harmonia axyridis* (Pallas) (Coleoptera: Coccinellidae). *Biocontrol*, *53*, 211–221.
Bissoondath, C.J. and Wiklund, C. (1996). Effect of male mating history and body size on ejaculate size and quality in two polyanderous butterflies, *Pieris napi* and *Pieris rapae* (Lepidoptera: Pieridae). *Functional Ecology*, *10*, 457–464.
Bista, M. & Omkar (2015). Age dependent mate choice influences reproductive and progeny attributes in aphidophagous ladybird beetles (Coleoptera: Coccinellidae). *European Journal of Entomology*, *112*, 648–657.
Blanckenhorn, W.U. (1998). Adaptive phenotypic plasticity in growth, development, and body size in yellow dung fly. *Evolution*, *52*, 1394–1407.
Blanckenhorn, W.U. (2000). The evolution of body size: What keeps organisms small? *The Quarterly Review of Biology*, *75*, 385–407.
Blanckenhorn, W.U., Preziosi, R.F. & Fairbairn, D.J. (1995). Time and energy constraints and evolution of sexual size dimorphism-to eat or to mate? *Evolutionary Ecology*, *9*, 369–381.
Boggs, C.L. & Freeman, K.D. (2005). Larval food limitation in butterflies: effects on adult resource allocation & fitness. *Oecologia*, *144*, 353–361.
Bonduriansky, R.A. (2001). The evolution of male mate choice in insects: a synthesis of ideas and evidence. *Biological Reviews*, *76*, 305–339.
Bonduriansky, R.A. & Brassil, C.E. (2002). Rapid and costly ageing in wild male flies. *Nature*, *420*, 377.
Bonner, J.T. (2012). *Why Size Matters: From Bacteria to Blue Whales*. Princeton University Press. p. 176.
Borycz, J., Borycz, J.A., Loubani, M., Meinertzhagen, I.A. (2002). Tan and ebony genes regulate a novel pathway for transmitter metabolism at fly photoreceptor terminals. *Journal of Neuroscience*, *22*, 10549–10557.
Brakefield, P.M. (1984a). Ecological studies on polymorphic ladybird *Adalia bipunctata* in Netherlands. I. Population biology and geographical variation of melanism. *Journal of Animal Ecology*, *53*, 761–774.
Brakefield, P.M. (1984b). Ecological studies on polymorphic ladybird *Adalia bipunctata* in Netherlands. II. Population dynamics, differential timing of reproduction and thermal melanism. *Journal of Animal Ecology*, *53*, 775–790.
Brakefield, P.M. (1987). Industrial melanism: do we have answers? *Trends in Ecology & Evolution*, *2*, 117–122.
Brakefield, P.M. & Wilmer, P.G. (1985). The basis of thermal melanism in ladybird *Adalia bipunctata*. Differences in reflectance and thermal properties between morphs. *Heredity*, *54*, 9–14.
Brown, W.D. (1990a). Constraints on size-assortative mating in blister beetle (*Tegrodera aloga*) (Coleoptera: Meloidae). *Ethology*, *86*, 146–160.
Brown, W.D. (1990b). Size-assortative mating in blister beetle (*Lytta magister*) (Coleoptera: Meloidae) is due to male and female preference for larger mates. *Animal Behaviour*, *40*, 901–909.
Burley, N.T. & Foster, V.S. (2006). Variation in female choice of mates: condition influences selectivity. *Animal Behaviour*, *72*(3), 713–719.

Byrne, P.G. & Rice, W.R. (2006). Evidence for adaptive male mate choice in fruit fly *Drosophila melanogaster*. *Proceedings of Royal Society B Biological Sciences*, *273*, 917–922.

Chamberlain, N.L., Hill, R.I., Kapan, D.D., Gilbert, L.E. & Kronforst, M.R. (2009). Polymorphic butterfly reveals missing link in ecological speciation. *Science*, *326*, 847–850.

Cook, D.F. (1995). Influence of previous mating experience on future mating success in male *Lucilia cuprina* (Diptera: Calliphoridae). *Journal of Insect Behaviour*, *8* (2), 207–217.

Cotton, S., Small, J. & Pomiankowski, A. (2006). Sexual selection and condition-dependent mate preferences. *Current Biology*, *16*, 755–765.

Coyne, J.A. & Orr, H.A. (2004). *Speciation*. Sunderland: Sinauer Associates.

Crespi, B.J. (1988). Risks and benefits of lethal male fighting in colonial, polygynous thrips *Hoplothrips karnyi* (Insecta: Thysanoptera). *Behavioural Ecology and Sociobiology*, *22*, 293–301.

Darwin, C. (1859). *On Origin of Species by Means of Natural Selection, or Preservation of Favoured Races in Struggle for Life*. 1st ed. London: John Murray.

Darwin, C. (1871). *The Descent of Man and Selection in Relation to Sex*. 1st ed. London: John Murray.

Davis, A.K., Chi, J., Bradley, C. & Altizer, S. (2012). The redder better: wing color predicts flight performance in monarch butterflies. *PLoS One*, *7*(7), e41323.

Day, T. & Rowe, L. (2002). Developmental thresholds and evolution of reaction norms for age and size at life-history transitions. *The American Naturalist*, *159*(4), 338–350.

de Jong, P.W. & Brakefield, P.M. (1998). Climate and change in clines for melanism in two-spot ladybird, *Adalia bipunctata* (Coleoptera: Coccinellidae). *Proceedings of Royal Society B Biological Sciences*, *265*, 39–43.

de Jong, P.W., Gussekloo, S.W.S. & Brakefield, P.M. (1996). Differences in thermal balance, body temperature and activity between non-melanic and melanic two-spot ladybird beetles (*Adalia bipunctata*) under controlled conditions. *Journal of Experimental Biology*, *199*, 2655–2666.

Dev, K., Chahal, J. & Parkash, R. (2013). Seasonal variations in mating-related traits of *Drosophila melanogaster*. *Journal of ethology*, *31*(2), 165–174.

Dingle, H. (2001). The evolution of migratory syndromes in insects. *Insect movement: mechanisms and consequences*. *2001*, 159–181.

Dixon, A.F.G. (2007). Body size and resource partitioning in ladybirds. *Population Ecology*, *49*, 45–50.

Dmitriew, C. & Rowe, L. (2007). Effects of early resource limitation and compensatory growth on lifetime fitness in ladybird beetle (*Harmonia axyridis*). *Journal of Evolutionary Biology*, *20*, 1298–1310.

Dowling, D.K. & Simmons, L.W. (2012). Ejaculate economics: testing effects of male sexual history on trade-off between sperm and immune function in australian crickets. *PLoS ONE.*, *7*, e30172.

Dubey, A., Omkar & Mishra, G. (2016a). Adult body size drives sexual selection mutually in ladybird *Menochilus sexmaculatus* (Coleoptera: Coccinellidae). *Acta Entomologica Sinica*, *59*(2):209–218.

Dubey, A., Omkar & Mishra, G. (2016b). Influence of temperature on reproductive biology and phenotype of a ladybird, *Menochilus sexmaculatus* (Fabricius)(Coleoptera: Coccinellidae). *Journal of Thermal Biology*, *58*, 35–42.

Eberhard, W.G. (1996). *Female Control: Sexual Selection by Cryptic Female Choice*. NJ: Princeton University Press.

Edvardsson, M., Hunt, J., Moore, P.J. & Moore, A.J. (2008). Female agreement over male attractiveness is not affected by cost of mating with experienced males. *Behavioural Ecology*, *19*, 854–859.

Emlen, S.T. & Oring, L.W. (1977). Ecology, sexual selection, and evolution of mating systems. *Science*, *197*, 215–223.

Fahmy, O.G. & Fahmy, M. (1959). New mutants report. *Drosophila Information Service*, *33*, 82–94.

Ferkau, C. & Fischer, K. (2006). Costs of reproduction in male *Bicyclus anynana* and *Pieris napi* butterflies: effects of mating history and food limitation. *Ethology*, *112*(11), 1117–1127.

Filin, I. & Ovadia, O. (2007). Individual size variation and population stability in a seasonal environment: a discrete- time model and its calibration using grasshoppers. *American Naturalist*, *170*, 719–733.

Finch, C.E. & Ruvkun, G. (2001). The genetics of aging. *Annual Reviews of Genomics and Human Genetics*, *2*, 435–462.

Fisher, H.S. & Rosenthal, G.G. (2006). Hungry females show stronger mating preferences. *Behavioral Ecology*, *17*(6), 979–981.

Foster, S. (2009). Sugar feeding via trehalose haemolymph concentration affects sex pheromone production in mated *Heliothis virescens* moths. *Journal of Experimental Biology*, *212*(17), 2789–2794.

Foster, S.P. & Johnson, C.P. (2011). Signal honesty through differential quantity in female-produced sex pheromone of moth *Heliothis virescens*. *Journal of Chemical Ecology*, *37*(7), 717–723.
Fox, C.W. (1993). Multiple mating, lifetime fecundity and female mortality of bruchid beetle, *Callosobruchus maculatus* (Coloeptera: Bruchidae). *Functional Ecology*, *7*, 203–208.
Futuyma, D.J. (1998). *Evolutionary Biology*. 3rd ed. Sunderland, MA: Sinauer Associates.
Garbutt, J.S. & Little, T.J. (2014). Maternal food quantity affects offspring feeding rate in *Daphnia magna*. *Biology Letters*, *10*, 856–863.
Gilchrist, G.W. & Huey, R.B. (2001). Parental and developmental temperature effects on thermal dependence of fitness in *Drosophila melanogaster*. *Evolution*, *55*(1), 209–214.
Gotthard, K., Nylin, S. & Wiklund, C. (1994). Adaptive variation in growth rate: life history costs and consequences in speckled wood butterfly, *Parargea egeria*. *Oecologia*, *9*, 281–289.
Grant, B.R. & Grant, P.R. (1989). *Evolutionary Dynamics of a Natural Population: Large Cactus Finch of Galápagos*. Chicago, IL: University of Chicago Press.
Guarente, L. & Kenyon, C. (2000). Genetic pathways that regulate ageing in model organisms. *Nature*, *408*, 255–262.
Hahn, D. (2005). Larval nutrition affects lipid storage and growth, but not protein or carbohydrate storage in newly eclosed adults of grasshopper *Schistocerca americana*. *Journal of Insect Physiology*, *51*, 1210–1219.
Hansen, T.F. & Price, D.K. (1995). Good genes and old age: do old mates provide superior genes? *Journal of Evolutionary Biology*, *8*, 769–778.
Hegde, S.N., Chethan, B.K. & Krishna, M.S. (2005). Mating success of males with and without wing patch in *Drosophila biarmipes*. *Indian Journal of Experimental Biology*, *43*, 902–909.
Helinski, M.E.H. & Harrington, L.C. (2011). Male mating history and body size influence female fecundity and longevity of dengue vector *Aedes aegypti*. *Journal of Medical Entomology*, *48*(2), 202–211.
Hercus, M.J., & Hoffmann, A.A. (2000). Maternal and grandmaternal age influence offspring fitness in *Drosophila*. *Proceedings of Royal Society B Biological Sciences*, *267*, 2105–2110.
Holbrook, G. L. & Schal, C. (2004). Maternal investment affects offspring phenotypic plasticity in a viviparous cockroach. *Proceedings of National Academy of Sciences of USA*, *101*(15), 5595–5597.
Holveck, M.J. & Riebel, K. (2010). Low-quality females prefer low-quality males when choosing a mate. *Proceedings of Royal Society B: Biological Sciences*, *277*(1678), 153–160.
Hunt, J., Brooks, R. & Jennions, M.D. (2005). Female mate choice as a condition-dependent life- history trait. *The American Naturalist*, *166*, 79–92.
Iyengar, V.K. (2009). Experience counts: females favor multiply mated males over chemically endowed virgins in a moth (*Utetheisa ornatrix*). *Behavioural Ecology and Sociobiology*, *63*, 847–855.
Jacobs, M.E. (1955). Studies on territorialism and sexual selection in dragonflies. *Ecology*, *36*, 566-586.
Jacobs, M.E. (1985). Role of beta-alanine in cuticular tanning, sclerotization, and temperature regulation in *Drosophila melanogaster*. *Journal of Insect Physiology*, *31*, 509–515.
Jaffe, K., Mirás, B. & Cabrera, A. (2007). Mate selection in moth *Neoleucinodes elegantalis*: evidence for a supernormal chemical stimulus in sexual attraction. *Animal Behaviour*, *73*(4), 727–734.
Jimenez-Perez, A. & Wang, Q. (2004). Male remating behavior and its effect on female reproductive fitness in *Cnephasia jactatana* Walker (Lepidoptera: Tortricidae). *Journal of Insect Behaviour*, *17*, 685–694.
Johnson, S.L. & Gemmell, N.J. (2012). Are old males still good males and can females tell difference? Do hidden advantages of mating with old males off-set costs related to fertility. *Bioessays*, *34*, 609–619.
Jones, T.M. & Elgar, M.A. (2004). The role of male age, sperm age and mating history on fecundity and fertilization success in hide beetle. *Proceedings of Royal Society B Biological Sciences*, *271*, 1311–1318.
Jones, T.M., McNamara, K.B., Colvin, P., Featherston, R. & Elgar, M.A. (2006). Mating frequency, fecundity and fertilization success in hide beetle, *Dermestes maculatus*. *Journal of Insect Behaviour*, *19*, 357–371.
Kalmus, H. (1941). The resistance to desiccation of *Drosophila* mutants affecting body colour. *Proceedings of Royal Society B Biological Sciences*, *130*, 185–201.
Katvala, M. & Kaitala, A. (2001). Male choice for current female fecundity in a polyandrous egg-carrying bug. *Animal behaviour*, *62*(1), 133–137.
Kemp, D.J. & Rutowski, R.L. (2011). The role of coloration in mate choice and sexual interactions in butterflies. In: Brock- mann H.J., Roper T., Naguib M., Wynne-Edwards K., Barnard C., Mitani J., eds. *Advances in Study of Behavior* (*Vol. 43*, 55–92). Amsterdam: Elsevier.

Kern, S., Ackermann, M., Stearns, S.C. & Kawecki, T.J. (2001). Decline in offspring viability as a manifestation of aging in *Drosophila melanogaster. Evolution, 55*, 1822–1831.

Kidd, S.A., EskenazI, B. & Wyrobek, A.J. (2001). Effects of male age on semen quality and fertility: a review of literature. *Fertility and Sterility, 75*, 237–248.

Kirkpatrick M. & Barton N.H. (1997). The strength of indirect selection on female mating preferences. *Proceedings of National Academy of Sciences of USA, 94*, 1282–1286.

Kirkwood, T.B.L. (2002). Evolution of ageing. *Mechanisms of Ageing and Development, 123*, 737–745.

Knell, R.J. & Webberley, K.M. (2004). Sexually transmitted diseases of insects: distribution, evolution, ecology and host behaviour. *Biological Reviews, 79*, 557–581.

Koch, R.L. (2003). The multicoloured Asian lady beetle *Harmonia axyridis*: a review of its biology, uses in biological control, and non-target impacts. *Journal of Insect Science, 3*, 1–16.

Koene, J.M., Loose, M.J. & Wolters, L. (2008). Mate choice is not affected by mating history in simultaneously hermaphroditic snail, *Lymnaea stagnalis. Journal of Molluscan Studies, 74*, 331–335.

Kokko, H. & Lindstrom, J. (1996). Evolution of female preference for old mates. *Proceedings of Royal Society B Biological Sciences, 263*, 1533–1538.

Kokko, H., Brooks, R., Jennions, M.D. & Morley, J. (2003). The evolution of mate choice and mating biases. *Proceedings of Royal Society B Biological Sciences, 270*, 653–664.

Kokko, H., Jennions, M.D. & Brooks, R. (2006). Unifying and testing models of sexual selection. *Annual Review of Ecology Evolution and Systematics, 37*, 43–66.

Krupke, C.H., Brunner, J.F. & Jones, V.P. (2008). Factors influencing mate choice in *Euschistus conspersus* Uhler (Heteroptera: Pentatomidae). *Environmental Entomology, 37*, 192–197.

Lalonde, R.G. (2005) Egg size variation does not affect offspring performance under intraspecific competition in *Nasonia vitripennis*, a gregarious parasitoid. *Journal of Animal Ecology, 74*, 630–635.

Larsson, F.K. (1988). Experimentally induced density-dependent mating patterns *in Phyllobius maculicornis* Germ (Coleoptera: Curculionidae). *American Midland Naturalist, 119*, 380–384.

Lassance, J.M. & Löfstedt, C. (2009). Concerted evolution of male and female display traits in European corn borer, *Ostrinia nubilalis. BMC Biology, 7*(1), 10.

Lauwers, K. & Van Dyck, H. (2006). The cost of mating with a non-virgin male in a monandrous butterfly: experimental evidence from speckled wood *Parargea egeria. Behavioural Ecology and Sociobiology, 60*, 69–76.

Lehtonen, J., Parker, G.A. & Schärer, L. (2016). Why anisogamy drives ancestral sex roles. *Evolution 70*, 1129–1135.

Lewis, Z. & Wedell, N. (2007). Effect of adult feeding on male mating behaviour in butterfly, *Bicyclus anynana* (Lepidoptera: Nymphalidae). *Journal of Insect Behaviour, 20*(2), 201–213.

Majerus, M.E.N. (1998). *Melanism: Evolution in Action*. Oxford University Press.

Majerus, M.E.N., O'Donald, P. & Weir, J. (1982a). Evidence for preferential mating in *Adalia bipunctata. Heredity, 49*, 37–49.

Majerus, M.E.N., O'Donald, P. & Weir, J. (1982b). Female mating preference is genetic. *Nature. 300*, 521–523.

Markow, T.A., Quaid, M. & Kerr, S. (1978). Male mating experience and competitive courtship success in *Drosophila melanogaster. Nature, 276*, 821–822.

Martin, O.Y. & Hosken, D.J. (2002). Strategic ejaculation in common dung fly *Sepsis cynipsea. Animal Behaviour, 63*, 541–546.

Mason, L.G. (1964). Stabilizing selection for mating fitness in natural populations of Tetraopes. *Evolution, 18*, 492–497.

Maxwell, M.R., Gallego, K.M. & Barry, K.L. (2010). Effects of female feeding regime in a sexually cannibalistic mantid: fecundity, cannibalism, and male response in *Stagmomantis limbata* (Mantodea). *Ecological Entomology, 35*(6), 775–787.

McCartney, J. & Heller, K.G. (2008). A preliminary analysis of mate choice in a bush cricket (*Poecilimon laevissimus:* Tettigoniidae) suggests virginity is more important than body size. *Journal of Orthoptera Research, 17*, 227–230.

McNamara, K.B., Jones, T.M. & Elgar M.A. (2007). No cost of male mating experience on female reproductive success in almond moth, *Cadra cautella. Behavioural Ecology and Sociobiology, 61*, 1177–1184.

Medwar, P.B. (1952). *An Unsolved Problem in Biology*. London: H. K. Lewis.

Michie, L.J., Mallard, F., Majerus, M.E.N. & Jiggins, F.M. (2010). Melanic through nature or nurture: genetic polymorphism and phenotypic plasticity in *Harmonia axyridis*. *Journal of Evolutionary Biology*, *23*, 1699–1707.

Millar, J.S. & Hickling, G.J. (1990). Fasting endurance and evolution of mammalian body size. *Functional Ecology*, *4*, 5–12.

Mohaghegh, J., De Clercq, P. & Tirry, L. (1998). Effects of maternal age and egg weight on developmental time and body weight of offspring of *Podisus maculiventris* (Heteroptera: Pentatomidae). *Annals of Entomological Society of America*, *91*, 315–322.

Monaghan, P. (2008). Early growth conditions, phenotypic development and environmental change. *Philosophical Transactions of Royal Society B*, *363*, 1635–1645.

Moore, A.J. (1990). The evolution of sexual dimorphism by sexual selection: separate effects of intrasexual selection and intersexual selection. *Evolution*, *44*, 315–331.

Moore, P. J. & Moore, A. J. (2001). Reproductive aging and mating: the ticking of the biological clock in female cockroaches.*Proceedings of the National Academy of Sciences*, *98*(16), 9171–9176.

Mousseau, T.A. & Fox, C.W. (1998a). *Maternal Effects as Adaptations*. Oxford, U.K.: Oxford Univ. Press.

Mousseau, T.A. & Fox, C.W. (1998b). The adaptive significance of maternal effects. *Trends in Ecology & Evolution*, *13*, 403–407.

Nakatsuru, K. & Kramer, D.L. (1982). Is sperm cheap? Limited male fertility and female choice in Lemon Tetra (Pisces, Characidae). *Science*, *216*, 753–754.

Newby, L.M. & Jackson, F.R. (1991). *Drosophila* ebony mutants have altered circadian activity rhythms but normal eclosion rhythms. *Journal of Neurogenetics*, *7*, 85–101.

Nieberding, C.M., Fischer, K., Saastamoinen, M., Allen, C.E., Wallin, E.A., Hedenström, E. & Brakefield, P.M. (2012). Cracking olfactory code of a butterfly: scent of ageing. *Ecology Letters*, *15*(5), 415–424.

Oliver, J.C., Robertson, K.A. & Monteiro, A. (2009). Accommodating natural and sexual selection in butterfly wing pattern evolution. *Proceedings of Royal Society B Biological Sciences*, *276*, 2369–2375.

Omkar & Afaq, U. (2013). Evaluation of Darwin's fecundity advantage hypothesis in Parthenium beetle, *Zygogramma bicolorata*. *Insect Science*, *20*, 531–540.

Omkar & Pervez, A. (2005). Mating behaviour of aphidophagous ladybird beetle, *Propylea dissecta*. (Mulsant). *Insect Science*, *12* (1), 37–44.

Omkar & Singh, S.K. (2010). Mating behaviour of aphidophagous ladybird beetle *Coelophora saucia* (Coleoptera: Coccinellidae). *International Journal of Tropical Insect Science*, *30*(1), 3–10.

Omkar, Pandey, P., Rastogi, S. & Mishra, G. (2010). Influence of age at mating on reproductive performance of Parthenium beetle, *Zygogramma bicolorata* Pallister (Coleoptera: Chrysomelidae). *Insect Science*, *17*(2), 112–120.

Omkar & Pervez, A. (2003). Influence of prey deprivation on biological attributes of pale morphs of lady beetle *Propylea dissecta* (Mulsant). *International Journal of Tropical Insect Science*, *23*, 143–148.

Osawa, N. & Nishida, T. (1992). Seasonal variation in elytral colour polymorphism in *Harmonia axyridis* (the ladybird beetle), role of non-random mating. *Heredity*, *69*, 297–307.

Otronen, M. (1988). The effect of body size on outcome of fights in burying beetles (Nicrophorus). *Annales Zoologici Fennici*, *25*, 882–890.

Pandey, P. & Omkar (2013). Age based mate choice improves reproductive performance and offspring attributes in parthenium beetle, *Zygogramma bicolorata* Pallister. *Canadian Entomologist*, *145*, 292–301.

Parker, G.A. (2006). Sexual conflict over mating and fertilization: an overview. *Philosophical Transactions of Royal Society Series B*, *361*, 235–259.

Pech-May, F.G., Medina-Medina, L., May-Itzá, W. de J., Paxton, R.J. & Quezada-Euán, J.J.G. (2012). Colony pollen reserves affect body size, sperm production and sexual development in males of stingless bee *Melipona beecheii*. *Insectes Sociaux*, *59*, 417–424.

Pellegroms, B., Van Dongen, S., Van Dyck, H. & Lens, L. (2009). Larval food stress differentially affects flight morphology in male and female speckled woods (*Parargea egeria*). *Ecological Entomology*, *34*, 387–393.

Perez-Staples, D., Martinez-Hernandez, M.G. & Aluja, M. (2010). Male age and experience increases mating success but not female fitness in Mexican fruit fly. *Ethology*, *116*, 778–786.

Phelan, P.L. & Baker, T.C. (1986). Male –size related courtship success and intersexual selection in tobacco moth, *Ephestia elutella*. *Experientia*, *42*, 1291–1293.
Pilakouta, N., Jamieson, S., Moorad, J.A. & Smiseth, P.T. (2015). Parental care buffers against inbreeding depression in burying beetles. *Proceedings of National Academy of Sciences of USA*, *112*(26), 8031–8035.
Pischedda, A. & Rice, W.R. (2012). Partitioning sexual selection into its mating success and fertilization success components. *Proceedings of National Academy of Sciences of USA*, *109*, 2049–2053.
Pizzari, T. & Parker, G.A. (2009). Sperm competition and sperm phenotype. In: *Sperm Biology* (pp. 207–245). Academic Press.
Pizzari, T., Cornwallis, C.K., Lovlie, H., Jakobsson, S. & Birkhead, T.R. (2003). Sophisticated sperm allocation in male fowl. *Nature*, *426*, 70–74.
Pölkki, M., Krams, I., Kangassalo, K. & Rantala, M.J. (2012). Inbreeding affects sexual signalling in males but not females of *Tenebrio molitor*. *Biology Letters*, *8*(3), 423–425.
Prathibha, M., Krishna, M.S. & Jayaramu, S.C. (2011). Male age influence on male reproductive success in *Drosophila ananassae* (Diptera: Drosophilidae). *Italian Journal of Zoology*, *78*, 168–173.
Price, D.K. & Hansen, T.F. (1998). How does offspring quality change with age in male *Drosophila melanogaster*? *Behavior Genetics*, *28*, 395–402.
Prosser, M.R., Murray, A.E. & Cade, W.H. (1997). The influence of female age on phototoxic during single & multiple song presentations in field cricket, *Gryllus integer* (Orthoptera: Gryllidae). *Journal of Insect Behaviour*, *10*, 437–449.
Proulx, S.R. (2001). Female choice via indicator traits easily evolves in face of recombination and migration. *Evolution*, *55*(12), 2401–2411.
Proulx, S.R., Day, T. & Rowe, L. (2002). Older males signal more reliably. *Proceedings of Royal Society B Biological Sciences*, *269*, 2291–2299.
Ralls, K. (1976). Mammals in which females are larger than males. *Quarterly Reviews of Biology*, *51*, 245–276.
Reynolds, J.M. (1944). The biology of *Tribolium destructor* Uytt. I. Some effects of fertilization and food factors on fecundity and fertility. *Annals of Applied Biology*, *31*(2), 132–142.
Ridley, M. (1988). Mating frequency and fecundity in insects. *Biological Reviews*, *63*, 509–549.
Ritchie, M.G. (2007). Sexual selection and speciation. *Annual Review of Ecology Evolution and Systematics*, *38*, 79–102.
Roff, D.A. (2002). *Life History Evolution*. Sunderland, MA: Sinauer Associates. 527 pp.
Rose, M.R. (1991). *Evolutionary Biology of Aging*. New York: Oxford University Press.
Rossiter, M.C. (1991). Maternal effects generate variation in life history: consequences of egg weight plasticity in Gypsy moth. *Functional Ecology*, *5*(3), 386–393.
Rowe, L. & Houle, D. (1996). The lek paradox and capture of genetic variance by condition dependent traits. *Proceedings of Royal Society B Biological Sciences*, *263*, 1415–1421.
Rutowski, R.L. (1982). Epigamic selection by males as evidenced by courtship partner preferences in checkered white butterfly (*Pieris protodice*). *Animal Behaviour*, *30*, 108–112.
Rutowski, R.L., Gilchrist, G.W. & Terkanian, B. (1987). Female butterflies mated with recently mated males show reduced reproductive output. *Behavioural Ecology and Sociobiology*, *20*, 319–322.
Ryder, J.J. & Siva-Jothy, M.T. (2001). Quantitative genetics of immune function and body size in house cricket, *Acheta domesticus*. *Journal of Evolutionary Biology*, *14*, 646–653.
Sadek, M.M. (2001). Polyandry in field-collected *Spodoptera littoralis* moths and laboratory assessment of effects of male mating history. *Entomologia Experimentalis et Applicata*, *98*, 165–172.
Safranek, L. & Riddiford, L.M. (1975). The biology of black larval mutant of tobacco hornworm. *Manduca sexta*. *Journal of Insect Physiology*, *21*, 1931–1938.
Salavert, V., Zamora-Munoz, C., Ruizrodriguez, M. & Soler, J.J. (2011). Female-biased size dimorphism in a diapausing caddisfly, *Mesophylax aspersus*: effect of fecundity and natural and sexual selection. *Ecological Entomology*, *36*, 389–395.
Scharf, I., Filin, I. & Ovadia, O. (2009). A trade-off between growth and starvation endurance in a pit-building antlion. *Oecologia*, *160*, 453–460.
Simmons, L.W. (2001). *Sperm Competition and its Evolutionary Consequences in Insects*. Princeton, NJ: Princeton University Press.
Singer, M.C. (1982). Sexual selection for small size in female butterflies. *American Naturalist*, *119*, 440–443.

Singh, B.N. & Chatterjee, S. (1987). Greater mating success of *Drosophila biarmipes* males possessing an apical dark black wing patch. *Ethology*, *75*, 81–83.
Smith, J.M. & Harper, D. (2003). *Animal signals*. Oxford University Press.
Soares, A.O., Coderre, D. & Schanderl, H. (2001). Fitness of two phenotypes of *Harmonia axyridis* (Coleoptera, Coccinellidae). *European Journal of Entomology*, *98*, 287–293.
Srivastava, S. & Omkar (2004). Age specific mating and reproductive senescence in seven spotted ladybird, *Coccinella septempunctata*. *Journal of Applied Entomology*, *128*, 452–458.
Srivastava, S. & Omkar (2005). Mate choice and reproductive success of two morphs of seven spotted ladybird, *Coccinella septempunctata* (Coleoptera: Coccinellidae). *European Journal of Entomology*, *102*, 189–194.
Stamp, N.E. (2001). Effects of prey quantity and quality on predatory wasp. *Ecological Entomology*, *26*, 292 301.
Steele, R.H. & Partridge, L. (1988). A courtship advantage for small males in *Drosophila subobscura*. *Animal Behaviour*, *36*, 1190–1197.
Stewart, L.A. & Dixon, A.F.G. (1989). Why big species of ladybird beetles are not melanic. *Functional Ecology*, *3*, 165–177.
Steiger, S. & Stökl, J. (2014). The role of sexual selection in evolution of chemical signals in insects. *Insects*, *5*(2), 423–438.
Takahashi, A., Takahashi, K., Ueda, R. & Takano-Shimizu, T. (2007). Natural variation of ebony gene controlling thoracic pigmentation *in Drosophila melanogaster*. *Genetics*, *177*(2), 1233–1237.
Trivers, R.L. (1972). Parental investment and sexual selection. In: Campbell B., ed., *Sexual selection and Descent of Man, 1871–1971* (pp. 136–179). Chicago: Aldine-Atherton.
True, J.R. (2003). Insect melanism: molecules matter. *Trends in Ecology and Evolution*, *18*, 640–647.
Tuomaala, M., Kaitala, A. & Rutowski, R.L. (2012). Females show greater changes in wing colour with latitude than males in green-veined white butterfly, *Pieris napi* (Lepidoptera: Pieridae). *Biological Journal of Linnean Society*, *107*(4), 899–909.
Unnithan, G.C. & Paye, S.O. (1991). Mating, longevity, fecundity and egg fertility of *Chilo partellus* (Lepidoptera: Pyralidae): effects of delayed or successive matings and their relevance to pheromonal control methods. *Environmental Entomology*, *20*, 150–155.
Vahed, K. (1998). The function of nuptial feeding in insects: review of empirical studies. *Biological Reviews*, *73*, 43–78.
Verhoog, M.D. Van Boven, A., & Brakefield, P.M. (1996) Melanic moths and ability to encapsulate parasitoid eggs and larvae. *Proceedings of Netherland Entomological Society*, *7*, 127–133.
Wallace, B. (1987). Ritualistic combat and allometry. *American Naturalist*, *129*, 775–776.
Wang, S., Michaud, J.P., Zhang, R.Z., Zhang, F. & Liu, S. (2009). Seasonal cycles of assortative mating and reproductive behaviour in polymorphic populations of *Harmonia axyridis* in China. *Ecological Entomology*, *34*, 483–494.
Wedell, N. (1993). Spermatophore size in bushcrickets: comparative evidence for nuptial gifts as a sperm protection device. *Evolution*, *47*, 1203–1212.
Wedell, N. (2005). Female receptivity in butterflies and moths. *Journal of Experimental Biology*, *208*, 3433–3440.
Wedell, N. & Ritchie, M.G. (2004). Male age, mating status and nuptial gift quality in a bush cricket. *Animal Behaviour*, *67*, 1059–1065.
Wedell, N., Gage, M.J.G. & Parker, G.A. (2002). Sperm competition, male prudence and sperm-limited females. *Trends in Ecology and Evolution*, *17*, 313–320.
Wheeler, J., Gwynne, D.T. & Bussier, L.F. (2012). Stabilizing sexual selection for female ornaments in a dance fly. *Journal of Evolutionary Biology*, *25* (7), 1233–1242.
Wiernasz, D.C. & Kingsolver, J.G. (1992). Wing melanin pattern mediates species recognition in Pieris occidentalis. *Animal Behaviour*, *43*(1), 89–94.
Wiernasz, D.C. (1989). Female choice and sexual selection of male wing melanin pattern in *Pieris occidentalis*(Lepidoptera). *Evolution*, *43*, 1672–1682.
Wiley, R.H. (1974). Evolution of social organization and life-history patterns among grouse. *Quarterly Reviews of Biology*, *49*:201–227.
Williams, C.G. (1957). Pleiotropy, natural selection, and evolution of senescence. *Evolution*, *11*, 398–411.

Wilson, K., Cotter, S.C., Reeson, A.F. & Pell, J.K. (2001). Melanism and disease resistance in insects. *Ecology Letters*, *4*, 637–649.

Worden, B.D., Parker, P.G. & Pappas, P.W. (2000). Parasites reduce attractiveness and reproductive success in male grain beetles. *Animal Behaviour*, *59*(3), 543–550.

Xie, J., De Clercq, P., Zhang, Y., Wu, H., Pan, C. & Pang, H. (2015). Nutrition-dependent phenotypes affect sexual selection in a ladybird. *Scientific Reports*, *5*, 13111.

Xu, J. & Wang, Q. (2009). Male moths undertake both pre- and in-copulation mate choice based on female age and weight. *Behavioural Ecology and Sociobiology*, *63*, 801–808.

Zuk, M. & Kolluru, G.R. (1998). Exploitation of sexual signals by predators and parasitoids. *Quarterly Reviews of Biology*, *73*, 415–438.

8

Postcopulatory Sexual Selection

Swati Saxena, Geetanjali Mishra, and Omkar

CONTENTS

8.1 Introduction

Interactions between males and females are asymmetrical across animal world, leading to several conflicts of interests. These conflicting interests can be understood via comprehensive study of sexual selection and its evolution. Darwin attributed sexual selection to pre-copulatory duration because of assumption that females mate monogamously. In 1871, he stated that sexual selection arises from differences in reproductive success caused by competition of mates. However, in past 50 years, it has become apparent that females are far from monogamous and most seek multiple mates (Smith *et al.*,

DOI: 10.1201/9781003043195-8

1990; Birkhead, 1998). Sexual selection can be of different forms and requires accurate understanding prior to interpretation. It also depends on several factors, such as (1) members of different sex, (2) rejection as mates, and (3) coercing against interests of mate (Clutton-Brock & Parker, 1995).

Quantifying contributions of pre- and postcopulatory process has become a major issue for many researchers (Jones & Ratterman, 2009; Pischedda & Rice, 2012). Arnold and Wade (1984a, 1984b) were first to propose an analytical framework to quantify contribution of pre- and post-mating sexual selection. Pischedda and Rice (2012) revealed that by controlling matings in *Drosophila melanogaster* and applying statistical regression techniques, fraction of variance in male's total performance was 27% due to equal contribution of fertilization success and mating success (Figure 8.1).

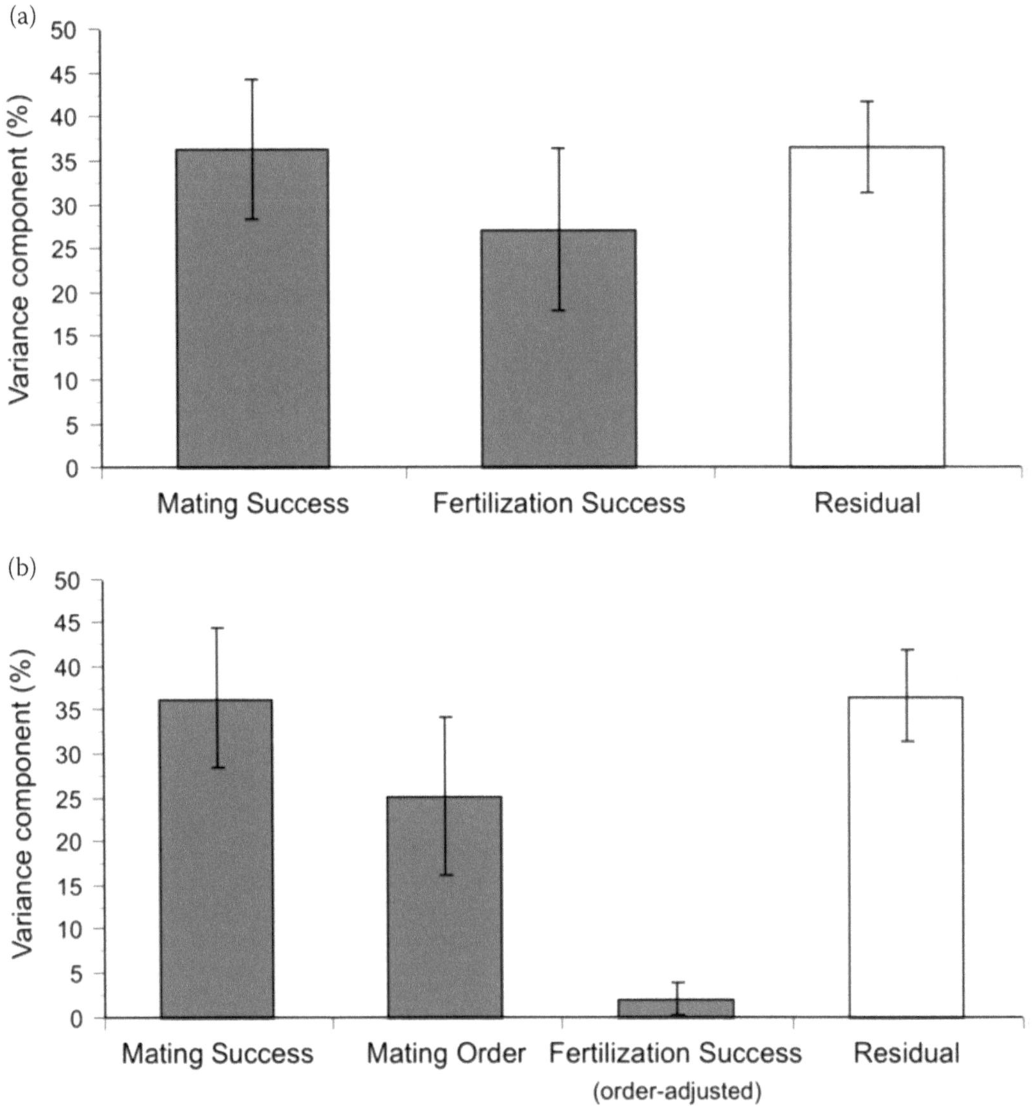

FIGURE 8.1 (a) The variance component estimates from a stepwise regression of male lifetime reproductive success on male mating success and male fertilization success. These values estimate amount of variation in male lifetime reproductive success that can be attributed to variation in mating success or fertilization success. (b) The variance component estimates from a second stepwise regressions of male lifetime reproductive success on male mating success and male fertilization success, but with male fertilization success decomposed into its mating order and fertilization success order-adjusted (after adjusting for mating order) components. (Pischedda & Rice, 2012; https://www.pnas.org/content/109/6/2049.long).

Postcopulatory sexual selection is important to measure strength of selection acting on sexual traits or phenotypes. Usually postcopulatory sexual selection is studied with reference to male reproductive success; however, that acting on females has gained very little attention (Ah-King *et al.*, 2014; Arnqvist, 2014). A frequently used method to measure postcopulatory sexual selection is competitive fertilization when subjected to multiple matings with reproductive success being determined by phenotypic and genetic markers (Simmons, 2001). In many species males sometimes fail to fertilize females (Greenway & Shuker, 2015; Greenway *et al.*, 2015) hence, proving that postcopulatory sexual selection may rise by differential selection of females. Single mating experiments can also be designed to investigate proximate outcomes of postcopulatory sexual selection, such as factors influencing number of sperms in female reproductive tract and their storage (Holwell *et al.*, 2010).

8.1.1 Relation between Pre- and Postcopulatory Sexual Selection

The relationship between pre-and postcopulatory selection is unclear (Andersson & Simmons, 2006). To understand intensity of sexual selection, it is necessary to understand relationship between these two selection processes and how they generate variations in reproductive success. Positive correlation between pre- and postcopulatory selection was studied in *D. melanogaster*, where males gain more reproductive success (Hosken *et al.*, 2008). This was attributed to cryptic female choice (CFC) with fate of sperm depending on female's perception of male quality (Figure 8.2). There is also a negative correlation possibility as in water striders, *Gerris lacutris*; large males perform better in pre-copulatory male-male competition but have reduced fertilization success (Danielsson, 2001; Figure 8.3). Most observations indicate that male reproductive adaptations are functional either in mating success or competitive fertilization. Since each mechanism is energetically costly, trade-off usually occur against each other. In Sometimes male traits are responsible for pre-copulatory male-male competition, and can lead to different patterns of postcopulatory sexual selection.

8.1.2 Multiple Mating and Postcopulatory Sexual Selection

Why do females' mate with more than one male despite cost of matings? A common explanation is that mating with multiple mates is a bet-hedging exercise that ensures highest offspring fitness (Birkhead & Pizzari, 2002). Studies in birds clearly reveal that monogamy was more social and less genetic in nature (Birkhead & Møller, 1995), with molecular studies confirming multiple matings in several mammals and insects (Gromko *et al.*, 1984; Oldroyd *et al.*, 1995). Under polygamous regime, mating frequency is higher for males than females (Bateman, 1948) with male fitness directly correlated with maximum

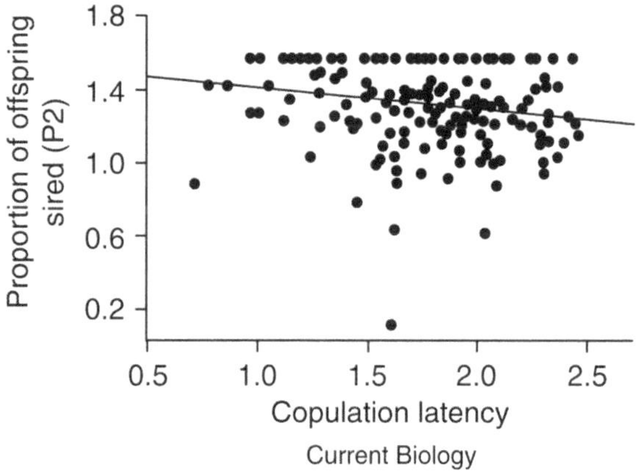

FIGURE 8.2 Female preference and paternity (Hosken *et al.*, 2008; https://www.sciencedirect.com/science/article/pii/S0960982208005137).

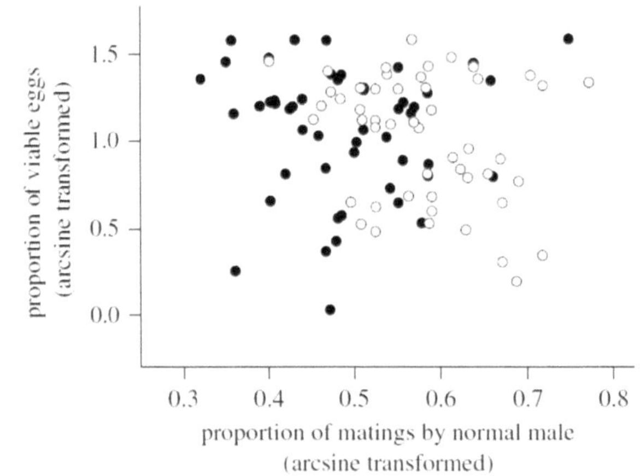

FIGURE 8.3 Relation between matings by small and large males (closed and open circles respectively) ad number of viable eggs. (Danielsson, 2001; https://www.ncbi.nlm.nih.gov/pmc/articles/PMC1087603/pdf/PB010077.pdf).

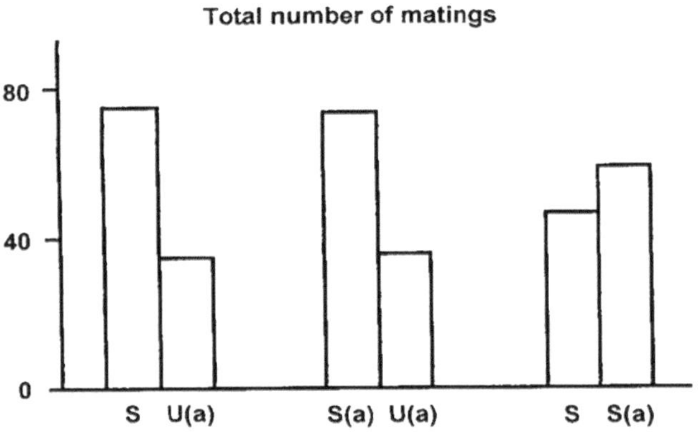

FIGURE 8.4 Offspring sired by successful males (S) has obtained more number of matings than those sired by less successful males (U). (Wedell & Tregenza, 1999; https://onlinelibrary.wiley.com/doi/pdf/10.1111/j.1558-5646.1999.tb03797.x).

paternity share which exceeds cost associated with soliciting multiple males (Figure 8.4). Conversely, for females, reproductive output is constrained by number of ova. In simple words, it can be stated that female only re-mates to sire her eggs with superior males.

The direct benefits of multiple matings include nutrient and fertility success even when fecundity does not increase. Non-additive or indirect genetic benefits of multiple matings include avoidance of inbreeding and other parental genotype incompatibility (Tregenza & Wedell, 2002). Within a natural population, negative impact of homozygosity for deleterious recessive alleles and at loci with heterozygote advantage, means that mating with close relatives is a major source of genetic incompatibility (Thornhill, 1993) (Figure 8.5). Various direct and indirect benefits of multiple matings are listed in Table 8.1.

8.1.3 Mechanisms of Postcopulatory Sexual Selection

Postcopulatory process can be broadly categorized into sperm competition (direct male-male competition) (Eberhard, 2009; Manier *et al.*, 2010) and cryptic female choice (female modulation of male-male competiton) (Arnqvist & Rowe, 2005; Eberhard, 2009).

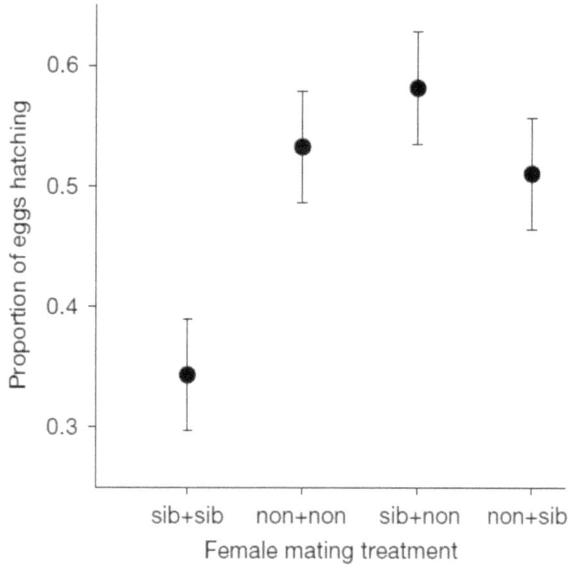

FIGURE 8.5 Eggs from female field cricked who mated with different combinations of males (means and standard errors) (Tregenza & Wedell., 2002; http://eprints.whiterose.ac.uk/77/1/tregenzat1.pdf).

TABLE 8.1

Possible direct and indirect benefits of multiple matings (Pitnick and Hosken, 2010)

Direct Benefits	Indirect Benefits
1. Nuptial gifts	1. Genetically viable offspring
2. Access to male controlled resources	2. Genetically variable offspring
3. Additional Paternal Control	3. Genetically more resistant offspring
4. Sperm replenishment	4. Genetically more attractive and competitive sons
5. Reduced probability of infanticide	

Sperm competition was initially defined as competition between sperm of two or more males in order to fertilize ova (Parker, 1970). A variety of mechanisms are in play in this race for fertilization. Dilution of sperm of previous male via voluminous ejaculate is generally reported in vertebrates where ejaculate size is correlated with degree of polyandry (Gage & Baker, 1991; Birkhead & Møller, 1992). Other tactics involves modification in sperm morphology, via loss of flagellum, multiple flagella, gigantism, polymorphism and conjugation. Physical displacement of sperm of other males using genitalia is also reported (Figure 8.6) (Gack & Peschke, 1994; Simmons, 2001). Many more mechanisms have been discovered and are dealt with in detail later in chapter

However, these postcopulatory male tactics are not alone in determining share, with females now known to modulate male paternity share post copulation, also known as cryptic female choice (Eberhard, 2009).

8.2 Sperm Competition

In most sexually reproducing species, ejaculate of different males compete to fertilize ova of female in a process called sperm competition (Figure 8.7). This is especially true in females with sperm storage organs, as these provide arenas for interaction of sperms of different males, thereby creating

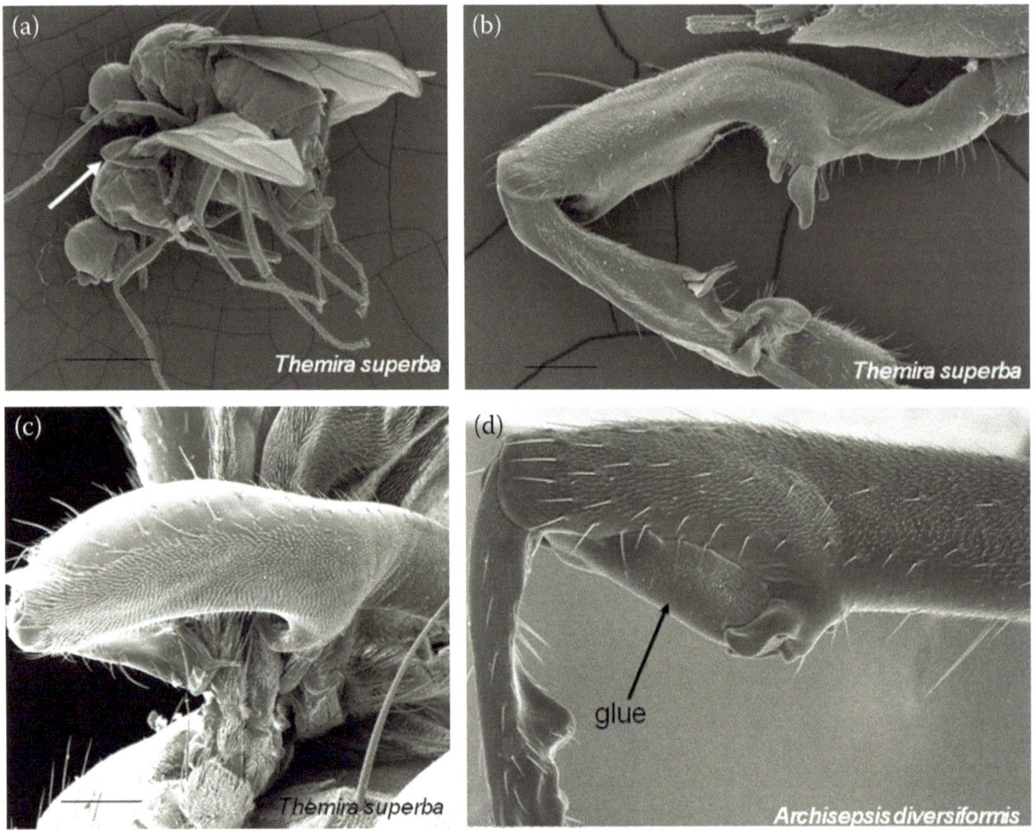

FIGURE 8.6 Species-specific male contact organs which are used to stimulate female. (a) The front leg of sepsid flies which are used to grab female wings before and during copulation (b) in many, such male contact organs, these legs are generally species-specific and sometimes quite complex in form (c) portion of female wing that male grasp is relatively uniform (d) experimental modification of male's femur (arrow) did not reduce his ability to grasp female, but did result in decreased reproductive cooperation from female (Eberhard, 2009 (https://stri-apps.si.edu/docs/publications/pdfs/2002_JIB_A_divers_front_lrg.pdf); Ingram *et al.,* 2008 (https://www.researchgate.net/profile/Rudolf_Meier/publication/229493210_Lack_of_morphological_coevolution_between_male_forelegs_and_female_wings_in_Themira_Sepsidae_Diptera_Insecta/links/5e9d781b92851c2f52b2d9e9/Lack-of-morphological-coevolution-between-male-forelegs-and-female-wings-in-Themira-Sepsidae-Diptera-Insecta.pdf).

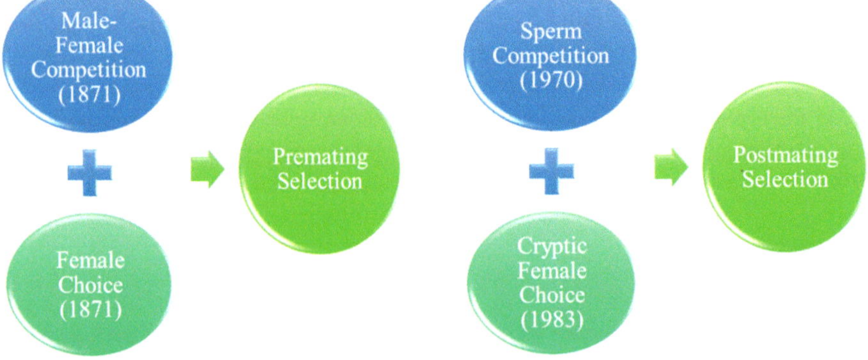

FIGURE 8.7 Mechanism of pre-mating and post-mating sexual selection (Adapted from Firman *et al.*, 2017).

competition and intense selection for male adaptations. Diverse range of taxa show increase in testis size (Harcourt *et al.*, 1995; Hosken, 1997; Hosken & Ward, 2001; Byrne *et al.*, 2003) and sperm numbers (Briskie & Montgomerie, 1992; Dunn *et al.*, 2001) as important sperm competition mechanisms, both of which are not necessarily exclusive (Møller, 1988a, 1988b, 1989).

Female reproductive tract is not necessarily a welcoming environment for sperm, with (1) physical barriers, such as cervix and long ducts, and (2) chemical barriers, such as low pH, viscous mucous. These mechanisms together ensure that a very small number of sperm stand a chance of successfully reaching ova. To understand interactions between ejaculate and female reproductive tract, four non-mutually exclusive hypotheses have been proposed (Birkhead *et al.*, 1993).

Firstly, environment which is selective on sperm may safeguard against pathogens that may enter female reproductive tract, during mating (Birkhead *et al.*, 1993). Secondly, adaptations that discriminate sperm of abnormal morphology (Birkhead *et al.*, 1993) may act as safeguard for future offspring fitness. Thirdly, high sperm mortality may have evolved to prevent polyspermy. Fourth, female tract conditions may be of two types, (i) difficult, to allow fertilization by best quality sperm (Birkhead 1998; Birkhead *et al.*, 1993) or, (ii) supportive, that favour male adaptations that facilitate fertilization.

8.2.1 Morphological Diversity of Sperms

> *The sperm seems never to transgress few rules, which govern production of its fundamental parts, but in arrangement of these parts every sperm (flagellate or non-flagellate) seems to be a law unto itself* (Bowen, 1925).

Tremendous diversity of sperm exists in insects (Jamieson *et al.*, 1999), with loss of flagellum, multiple flagella, gigantism, conjugation, polymorphism and many more, some of which are discussed below (Dallai & Afzelius, 1985; Swallow & Wilkinson, 2002).

8.2.1.1 Within-Male Variations

Within an ejaculate, sperms may be functionally or morphologically abnormal. They may swim erratically, weakly or be non-motile. They may lack head or tail or possess two heads or tails (Cohen, 1967, 1973). While some of these variations are a consequence of immaturity or aging of sperm (Humphreys, 1972), they can also be genetic Cohen, 1973). The error rate in sperm production is directly correlated to numbers of sperms required to ensure fertilization success. Comparative analysis of sperm viability among insect species revealed that in monandrous species hadhigher proportion of dead sperms (Hunter & Birkhead, 2002). However, three aspects of sperm production physiology, limit within ejaculate variation in sperm phenotype, (1) location of testes (Waites, 1970; Carrick & Setchell, 1977), (2) sharing of cytoplasm within spermatids (Braun *et al.*, 1989), and (3) testicular gene expression (Eddy, 2002).

8.2.1.1.1 Among Male Variation

Across taxa variation in sperms of males within a population have been observed (Pitnick *et al.*, 2003); with variations occurring due to both genetic and environmental factors. Kleven *et al.* (2008) have showed that inter-male variation in sperm morphology was lowest in species with sperm competition. Genetic variations of sperm in natural population have been demonstrated by quantitative genetics (Miller *et al.*, 2003; Birkhead *et al.*, 2005). Studies also show factors such as quality or quantity of food affects percentage of abnormal sperm in ejaculates (Hellriegel & Blanckenhorn, 2002; Amitin & Pitnick, 2007).

8.2.1.1.2 Among Population Variation

Studies by Pitnick *et al.* (2003), on eight different populations of fruit fly *Drosophila mojavensis* from entire geographic range of Sonoran Desert found highly significant differences in sperm length.

Snook (2001) also compared sperm length across 11 north American populations of *Drosophila subobscura* and found significant difference in total length of sperm (Snook *et al.*, 1994; Snook & Karr, 1998). However, Hosken *et al.* (2003) found no difference in sperm length in populations of *S. stercoraria.*

8.2.2 Sperm Games

Parker *et al.* (2010) has examined models for evolution of sperm size and number by proposing three mechanisms of sperm competition: (1) raffle with no constraint on space available for competing sperm, (2) direct displacement mainly by seminal fluid, and (3) direct displacement mainly by sperm displacement.

Increase in sperm mass may increase sperm's competitive weight against rival sperm but relationship is not straightforward with possible two trade-offs (Parker, 1993). The first is a direct trade-off, where each male has a fixed mass, M to expend on gametes, and sperm mass and sperm number one trade-offs directly under constraint that $M = ms$ (Parker *et al.*, 1972; Parker, 1982). In indirect trade-offs, each male has a fixed budget for reproductive activity R and energy invested in mate acquisition (mate searching, courtship and fighting for mates) is traded-off against energy spent on ejaculate. Thus, R, n, s and m can vary independently under constraint that $R = n(C + Dms)$; where C is energy cost of gaining a mating and D is a constraint that converts ejaculate expenditure into same units as R and C.

8.2.3 Strategic Ejaculation

Sperm competition is a major selective force acting on male and female behaviour, where reproductive success depends on quantity and quality sperms but also ability to fertilize eggs by successful ejaculation (Parker & Pizzari, 2010). Theory predicts that there will be selection for increased investment in spermatogenesis and hence more risk of sperm competition will positively correlate with spermatogenic investment (Parker, 1970, 1982, 1990a, b). Male ejaculate expenditure is predicted to increase with higher sperm competition risk (Engqvist & Reinhold, 2005; Kelly & Jennions, 2011).

Post adult emergence, time is required to accumulate resources for ejaculate production (Foster, 1967; Oberhauser, 1988) and also post ejaculation prior to re-mating efficiently (Svärd & Wiklund, 1989; Simmons *et al.*, 1993). Intraspecific and interspecific studies revealed that there is trade-off between amount of resources allocated to ejaculate and time to recover from matings (Simmons *et al.*, 1999). The trade-off between mate searching and ejaculate production was observed in *Plodiainter punctella* (Gage, 1995). Similarly, studies of insect parasitism also support trade-off between somatic maintenance and ejaculate production. Table 8.2 shows some examples of various constraints on male ejaculates in insects.

8.2.4 Classification of Sperm Competition

Sperm competition only occurs when there is temporal or spatial overlap of ejaculates from different males. Sperm competition is also directly correlated to value of. However, P_2 can also be high when sperms from previous matings are lost from female's storage organ or in case of insufficient storage, ejection of spermatophores; no temporal overlap and sperm competition.

Tsubaki and Yamagishi (1991) and Yamagishi *et al.* (1992) studied sperm utilization in melon fly, *Dacus cucurbitae,* where they found that sperms stored in storage organ decreased exponentially with time since mating. In double mating trials P_2 increased with interval between first and second matings, reflecting decreased temporal overlapping of sperms, thus decreased sperm competition. It was further noted that competition occurs at inter-ejaculate and intra-ejaculate levels with evolutionary consequences being radically different. Such variation and differing P_2 values lead to classification of sperm competition, as follows:

TABLE 8.2

Some examples of insects with male constraint on ejaculate production

Species	Constraint	Evidence	Reference
Gryllus veletis, G.pennsylvanicus, Gryllodes sigillatus	Gut parasites	Decreased spermatophores production	Zuk, 1987
Diploptera punctata	Male age, duration of larval development	Increased sperm number	Woodhead, 1984
Gerris lateralis	Re-mating interval	Increased sperm number, decreased oviposition stimulation	Arnqvist & Danielsson, 1999
Lucilia cuprina	Multiple matings	Decreased sperm numbers and decreased inhibition of female re-mating	Smith *et al.*, 1990
Drosophila melanogaster	Multiple mating	decreased fertility and lifespan	Prowse & Partridge, 1997
Drosophila nigrospiracula	Ectoparasites	Decreased testes mass	Polak, 1998
Plodia interpunctella	Diet Viral infection, Rearing condition	Decreased sperm numbers, fertility and fecundity	Lum & Flaherty, 1970; Gage & Cook, 1994
Pieris rape	Mating	Decreased spermatophore mass	Bissoondath & Wiklund, 1996; Cook & Wedell, 1996
Pieris napi	Mating	Decreased spermatophores mass	Wiklund & Kaitala, 1995; Bissoondath & Wiklund, 1996
Jalmenus evagoras	Mating	Decreased spermatophores mass	Hughes *et al.*, 2000

8.2.4.1 Sperm Mixing

Mixing of sperms from two or more males takes place within single female, in sperm storage organ, with degree of sperm competition depending on extent of sperm mixing. Intense competition occurs in case of rapid and random mixing of sperms from different males. Slow mixing causes sperm clumping, and random mixing favours sperm with greater competitive ability (Parker, 1990a).

8.2.4.2 Sperm Precedence

This includes non-random utilization of sperm from a particular male, when sperms from two or more males are present. Last male sperm entering female's sperm storage may be first to leave, resulting in high paternity share, also known as last male sperm precedence. However, reverse may happen too. Sperm precedence can be short-term or long-term: former by slow non-random mixing of sperms, while long-term may undertake any of following mechanisms:

8.2.4.2.1 Sperm Stratification

Sometimes sperm can obtain a positional advantage within spermatheca of female, usually last to enter and first to leave. Sperm stratification is consequence of sperm repositioning (Waage, 1984) or mating order (Schlager, 1960).

8.2.4.2.2 Sperm Incapacitation

Sperm incapacitation involves killing or inhibiting sperm of rival males (Silberglied *et al.*, 1984) via physiological adaptations in seminal fluid or morphological changes in sperm ('Kamikaze sperm hypothesis'; Baker & Bellis, 1988).

8.2.4.2.3 Sperm Displacement

This involves spatial displacement of sperm from female sperm storage organ as a result of consecutive mating. This includes basically two mechanisms:

8.2.4.3 Sperm Removal

It includes direct removal of sperm from female storage organ through subsequent ejection facilitated by morphological adaptations (Waage, 1979).

8.2.4.4 Sperm Flushing

Includes indirect removal of sperms, when single copulation is sufficient to fill sperm storage organ. The ejaculate from second male displaces sperm of previous male; due to little overlap there is less competition (Ono *et al.*, 1989; Parker & Simmons, 1991).

8.2.4.4.1 Adaptations to Avoid Sperm Competition

Males have evolved multiple ways (listed below) to reduce sperm competition and increase paternity share.

8.2.4.5 Mate Guarding

It involves guarding of females by males even post sperm transfer usually without genital contact. Competitors that encounter copulating or ovipositing pair are actively countered. In case of continued take-over attempts males re-engage in genital contact (Parker, 1970). Theoretical models predict that sperm competition favours evolution of mate guarding mechanism, when females reman sexually receptive post matings.

Forms of mate guarding differ in different taxa (Table 8.3). In damselfly, *Coenagrion puella*, males remain in direct contact with females carrying them to different locations for oviposition (Banks & Thompson, 1985). In dragonflies, patterning of fore- and hindwings affects mate guarding behaviour (Johansson *et al.*, 2009) (Figure 8.9). In some species, there is non-contact mate guarding (Waage, 1979). Mate guarding can also occur prior to mating, as in *Heliconius* butterflies, where pupae are guarded, since females mate only once and also anti-aphrodisiac pheromones are transferred (Gilbert, 1976; Estrada & Gilbert, 2010).

In *Drosophila* females, post mating, receptivity is reduced due to transference of seminal fluids (Fricke *et al.*, 2009). Experimental evolution studies on *Drosophila* have revealed that productivity of male accessory glands is directly correlated to strength of sperm competition.

8.2.4.6 Prevention of Female Re-mating

In many insects, males transfer substances to females that coagulate within female reproductive tract and act as mating plugs (Simmons, 2001; Mikheyev, 2003; Calabrese *et al.*, 2008). In social insect,

TABLE 8.3

Pre- and post-copulation mate guarding behaviour in different insects

Species	Type of Mate Guarding	Duration	Reference
Locusta migratoria	Pre-copulatory	10 hours	Zhu & Tanaka, 2002
Sphenarium purpurascens	Postcopulatory	17 days	Del Castillo, 2003
Anomala albopilosa sakishimana	Pre-copulatory	Highly variable	Arakaki *et al.*, 2004
Papillia japonica	Postcopulatory	<400 minutes	Saeki *et al.*, 2005
Tenebrio molitor	Postcopulatory	Highly variable	Carazo *et al.*, 2007, 2012
Hypopnera opacior	Pre-copulatory	<800 minutes	Kureck *et al.*, 2011

Bombus terrestris, males block female reproductive tract with mating plug, which is also a nuptial gift (Sauter *et al.*, 2001). However, manufacturing of such mating plugs are costly to males, usually detracting from production of ejaculate components (Simmons, 2001).

8.2.4.7 Traumatic Insemination

This is an extraordinary means of copulation and is majorly observed in bed bugs. Male pierces female abdomen with its lanceolate aedeagus and ejaculates directly into haemocoel (Hinton, 1964; Carayon, 1966); such males increase paternity share. The wounding caused during copulation, compromises a female's ability to maintain water balance and immune system leading to reduced life span (Stutt & Siva-Jothy, 2001; Benoit *et al.*, 2012). This cost is offset by evolution of subcuticular area of tissue called spermalege, into which male ejaculates without much harm to females (Morrow & Arnqvist, 2003; Reinhardt *et al.*, 2003).

8.3 Cryptic Female Choice

In 1996, Eberhard with his idea of CFC, turned sexual selection into a female powered process (Figure 8.8). While evidence for female-driven pre-copulatory sexual selection had been accumulating over years; its occurrence at post copulatory level changed narrative.

Females have ability to reduce or eliminate availability of sperms of inferior males by digesting them or by moving them to such location, where they cannot be accessible to egg. The term cryptic here is used in sense that it is hidden decision made by female after copulation. Female of cassidine plant beetle *Chelymorpha alternans* have a long coiled spermathecal duct, which allows male with long sclerite to successfully sire more offspring (Figure 8.9).

To demonstrate CFC, we first need to identify traits of females or sperm behaviour that affects uptake or utilization of sperm for fertilization. Second, the assessment of female differential response such that sperm of particular males are predictably favoured or disfavoured are based on genotypic or phenotypic factors. This relationship between male traits and pattern of sperm utilization by females can be demonstrated by manipulating male phenotype experimentally or excluding other factors. Any change in sperm utilization or fertilization success is in consistency with CFC (Pilastro *et al.*, 2004). Artificial insemination or in vitro assays of sperm utilization and fertilization can be used to control ejaculate traits and eliminate effect of pre-mating mechanisms (Løvlie *et al.*, 2013; Firman & Simmons, 2015).

Females might also influence paternity by controlling timing and order of insemination. In moth *Ephestia kuehniella*, P2 is influenced by re-mating of females (Xu & Wang, 2010). While ejaculate size is largely under control of males, acceptance and ejection of spermatophores depends on female. In several species, loss of ejaculate depends on several female processes, such as sperm digestion, ejection, and incapacitation, which in turn is dependent on male size (Ala-Honkola *et al.*, 2011), identity of species (Manier *et al.*, 2013) and copulation duration (Peretti & Eberhard, 2010).

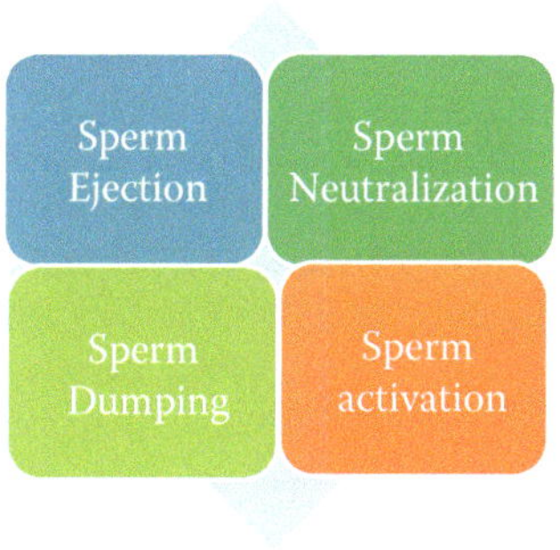

FIGURE 8.8 Sexual Selection by cryptic female choice (adapted from Firman & Simmons, 2015).

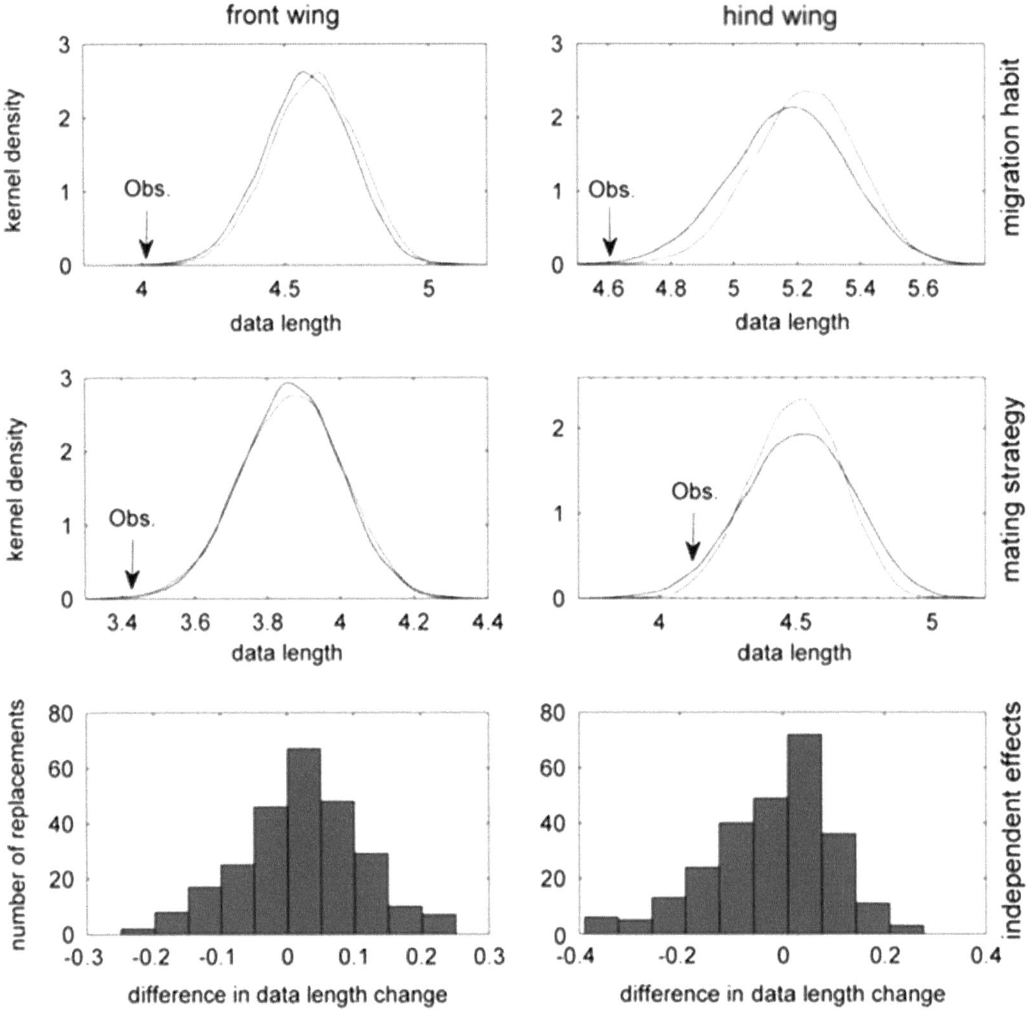

FIGURE 8.9 Distribution of data length and change in data length. Solid lines indicate distribution of data length obtained by randomized wing shape across species. Dotted lines indicate data length distribution by randomizing migration within. (Johansson *et al.*, 2009; https://academic.oup.com/biolinnean/article/97/2/362/2448057).

Differential sperm storage is also likely to influence CFC, with improper storage leading to ejection, digestion, or incapacitation. Eberhard (2009) suggested that complexity of fertilization tubule increases female control of paternity. In dungfly, *Scathophaga stercoraria,* those with four sperm storage have better paternity control than those with three sperm storage organs.

8.3.1 Functional Significance of Cryptic Female Choice

Non-random sperm utilization requires several adaptations which have been proposed 20 years ago. The adaptive significance of CFC falls into two major categories: (i) mainly concerned with females and fitness benefits associated with it are conveyed to female, and (ii) may not be adaptive to females and can be side effects for other female traits, such as sensory bias or male manipulation of sperm utilization. CFC also controls paternity in polyandrous females, where pre-mating choice is difficult. In CFC, females can provide increased offspring fitness, i.e. genetic benefits. These benefits are passed on across generations, i.e. directional selection or female may change in their preferred criteria (non-directional selection). The former criterion supports good genes and Fisherian runaway theory (Lüpold *et al.*, 2016). In post-mating, good genes hypothesis selects ejaculates of male with good genetic quality. The sexy sperm hypothesis (or sexy sons hypothesis) states that males successful in competition can sire sexy sons.

Females of different species show different ways of discrimination. In *Drosophila simulans,* females discriminate against sperm from males by distorting sex ratio (Angelard *et al.*, 2008). The criteria of CFC may differ across male-female combinations. This is often explained by mechanism of genetic compatibility. Heterozygosity is increased, when cryptic female choice favours male genotypes that are less similar to female.

8.3.2 Evolutionary Effects of Cryptic Female Choice

Variations in paternity share provide opportunities for post-mating intersexual selection on males. This selection can be directional or non-directional. Directional CFC works independently and this can provide opportunities for males to evolve alternative mating tactics to attain reproductive success. Though CFC provides benefits to females, there are also unavoidable chances of sexual conflict. Therefore, evolutionary responses of males to CFC comprise strategies that can balance female preference and maximize paternity. These may include strategies to fulfil preference of females or counteract cryptic female choice.

8.4 Sexual Conflict

Sexual conflict occurs when genetic interests of males and females diverge and idea has revolutionized the ways of interpretation of male and female interactions, which leads to their coevolution (Chapman *et al.*, 2003).

There are two main forms of sexual conflict. The one in which trait of particular fitness is expressed differently in males and females is known as intralocus conflict. This differential expression is coupled with genetic correlation between sexes (Parker & Partridge, 1998). This selection for sex-limited expression of genes results in independent evolution in two sexes facilitating them to attain their adaptive peak (Rice, 1984; Rhen, 2000). Studies by Chapman and Bourke (2001) on *D. melanogaster* revealed that there is positive correlation between fitness of males and females at larval stages even when they have same interests, but not so in adults.

Another form of sexual conflict is interlocus conflict where conflict is over outcome of male and female interactions (Holland & Rice, 1997). Such conflict can occur over mating frequency, fertilization, female re-mating behaviour, female reproductive rate, and clutch size. The result is sexually antagonistic between traits in males and females.

8.4.1 Sexual Conflict and Parental Care

Trivers (1972) suggested that sexual conflict is central to parental care; which partner should care, degree of care, etc. (Wachtmeister & Enquist, 2000; Royle *et al.*, 2002). Among biparental species where mating system is variable, females generally monopolize reproductive efforts while paired mates spend considerable time in attracting additional females. In such cases, females exhibit sexually antagonistic adaptations to keep their mates monogamous. Such strategies occur in insects, birds and mammals and this includes: (1) interference in mate-mate attraction by attacking or obstructing male display (Eggert & Sakaluk, 1995), and (2) aggression towards other females particularly during pair formation (Sandell, 1998; Veiga, 1990).

8.4.2 Sexual Conflict and Speciation

Sexual conflict has potential to encourage adaptive divergence (Arnqvist *et al.*, 2000). Recently, two models have been put forward to examine compatibility between males and females, which determine their interactions and promote allopatric and sympatric divergence in reproductive traits (Gavrilets, 2000; Gavrilets & Waxman, 2002).

8.4.3 Sexual Conflict and Evolution of Male Accessory Gland Proteins

Male accessory gland proteins have been studied in much detail in *D. melanogaster*. More than 80 accessory gland proteins (ACPs) are transferred to females. Such proteins modulate postcopulatory female reproductive behaviour and their evolution appears to be influenced by sexual conflict. ACPs include high levels of amino acid polymorphism and interspecific divergence with directional selection for 12 different gene sequences. High frequency of evolutionary changes in loci encoding seminal fluid proteins is common in both between and within *Drosophila* species (Chapman *et al.*, 2001). Genetically, engineering of males has helped study the role of ACPs in conflict (Kalb *et al.*, 1993; Neubaum & Wolfner, 1999; Lung *et al.*, 2002).

8.4.4 Trade-Offs Associated with Postcopulatory Sexual Selection

There are several evidences that primary sexual traits of males, such as testes size, sperm form, and function are subjected to sexual selection (Simmons & Fitzpatrick, 2012). The fundamental assumption is that there is a trade-off between resource allocation and life history traits, such as growth, somatic maintenance, and reproduction (Roff, 2002), with resource allocation bring hierarchical (Figure 8.10). There are two different approaches to measure trade-offs:

8.4.5 Phenotypic Correlations between Traits and Ejaculation

There are studies where phenotypic relationships between secondary sexual traits and ejaculate quality and quantity have been studied (Simmons *et al.*, 2017). Sperm quality can directly get affected by competitive ability of males and is often associated with testes size (Lüpold, 2013; Table 8.4).

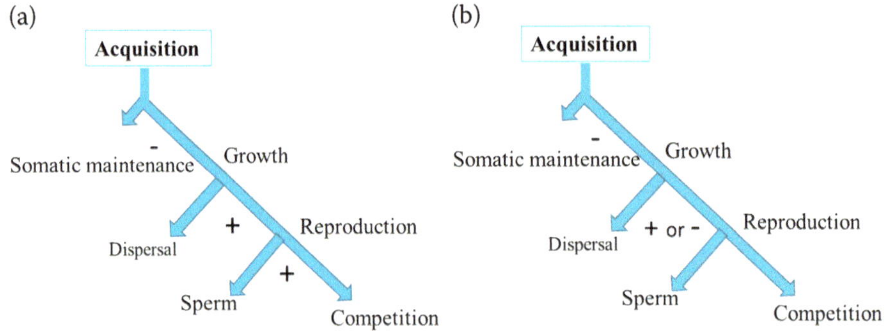

FIGURE 8.10 Resource allocation (adapted from de Jong *et al.* 1993).

TABLE 8.4

Phenotypic correlation between primary (testes size) and secondary sexual traits

Species	Secondary Sexual Traits	Correlation with Testes Size	References
Teleopsis dalmanni (Stalk-eyed flies)	Eye span	Positive	Roge *et al.*, 2008 Cotton *et al.*, 2010
Euoniticellus intermedius (Dung beetle)	Horn length	Positive	Reaney & Knell, 2015
Onthophagus Taurus (Dung beetles)	Horn length	Negative	Simmons *et al.*, 1999
Hemideina crassidens (Tree weta)	Mandible size	Positive	Kelly, 2008
Gnatocerus comutus (Flour beetles)	Mandible size	None	Yamane *et al.*, 2010
Leptoscelis tricolor (Heliconia bug)	Hind Femus	Positive and negative	Somjee *et al.*, 2015
Lasiorhynchus barbicornis (Giraffe weevils)	Rostrum length	None	Painting & Holwell, 2013

Phenotypic correlations are combination of both genetic variances and covariances between traits and environmental influences on those traits. These correlations can be positive or negative. Positive phenotypic correlations between two traits arise when environment influences both traits in same way (Remick, 1992). Negative correlations are evident when organisms are stressed, indicating their environmental dependence.

8.4.6 Genetic Correlations

Genetic correlations provide more robust evidence of trade-off between life-history traits, as they are environmentally independent. In cricket, *Teleogryllus oceanicus*, a negative correlation has been reported between ejaculate quality and energetically expensive sound content of courtship song (Simmons *et al.*, 2010). In sand cricket, *Gryllus firmis,* males can develop hindwings and flight muscles to be macropterous or it can only have vestigial flight morphology (micropterous). The latter males are less attractive to females and also have smaller testes as compared to macropterous ones (Saglam *et al.*, 2008). Despite positive correlation between male song and testes size across morphs, there are negative correlations between flight and male expenditure on calling song duration (Crnokrak & Roff, 1998). It is genetic trade-off between traits, which leads to joint evolution.

8.5 Conclusions

This chapter demonstrated that research on insects has played a central role in advancing understanding of postcopulatory sexual selection. We know that female choice can impose a strong and complex pattern of sexual selection on male sexual traits and their ability to compete with other males. Predictions of evolutionary responses triggered by postcopulatory sexual selection will require quantitative information on: (i) intensity of postcopulatory sexual selection; (ii) mechanisms involved in postcopulatory interactions and selection; (iii) degree of cooperation and conflict between male-driven and female-driven processes, and (iv) efficiency of individual postcopulatory mechanisms.

REFERENCES

Ah-King, M., Barron, A.B. and Herberstein, M.E., 2014. Genital evolution: why are females still understudied?. *PLoS Biology*, *12*(5), 1001851.

Ala-Honkola, O., Manier, M.K., Lüpold, S. and Pitnick, S., 2011. No evidence for postcopulatory inbreeding avoidance in *Drosophila melanogaster*. *Evolution: International Journal of Organic Evolution*, *65*(9), pp. 2699–2705.

Amitin, E.G. and Pitnick, S., 2007. Influence of developmental environment on male-and female-mediated sperm precedence in *Drosophila melanogaster*. *Journal of Evolutionary Biology*, *20*(1), pp. 381–391.

Andersson, M. and Simmons, L.W., 2006. Sexual selection and mate choice. *Trends in Ecology & Evolution*, *21*(6), pp. 296–302.

Angelard, C., Montchamp-Moreau, C. and Joly, D., 2008. Female-driven mechanisms, ejaculate size and quality contribute to lower fertility of sex-ratio distorter males in *Drosophila simulans*. *BMC Evolutionary Biology*, *8*(1), pp. 1–12.

Arakaki, N., Sadoyama, Y., Kishita, M., Nagayama, A., Oyafuso, A., Ishimine, M., Ota, M., Akino, T., Fukaya, M., Hirai, Y. and Yamamura, K., 2004. Mating behavior of scarab beetle *Dasylepida ishigakiensis* (Coleoptera: Scarabaeidae). *Applied Entomology and Zoology*, *39*(4), pp. 669–674.

Arnold, S.J. and Wade, M.J., 1984a. On measurement of natural and sexual selection: theory. *Evolution*, *38*, pp. 709–719.

Arnold, S.J. and Wade, M.J., 1984b. On measurement of natural and sexual selection: applications. *Evolution*, *38*, pp. 720–734.

Arnqvist, G. and Rowe, L., 2005. *Sexual conflict*. Princeton University Press.

Arnqvist, G. (2014). Cryptic female choice. *The evolution of insect mating systems*, pp. 204–220.

Arnqvist, G. and Danielsson, I., 1999. Copulatory behavior, genital morphology, and male fertilization success in water striders. *Evolution*, *53*(1), pp. 147–156.
Arnqvist, G. and Nilsson, T., 2000. The evolution of polyandry: multiple mating and female fitness in insects. *Animal Behaviour*, *60*(2), pp. 145–164.
Arnqvist, G., Edvardsson, M., Friberg, U., & Nilsson, T. 2000. Sexual conflict promotes speciation in insects. 97, pp. 10460–10464.
Baker, R.R. and Bellis, M.A., 1988. 'Kamikaze'sperm in mammals?. *Animal Behaviour*, *36*(3), pp. 936–939.
Banks, M.J. and Thompson, D.J., 1985. Lifetime mating success in damselfly *Coenagrion puella*. *Animal Behaviour*, *33*(4), pp. 1175–1183.
Bateman, A.J., 1948. Intra-sexual selection in *Drosophila*. *Heredity*, *2*(3), pp. 349–368.
Benoit, J.B., Jajack, A.J. and Yoder, J.A., 2012. Multiple traumatic insemination events reduce ability of bed bug females to maintain water balance. *Journal of Comparative Physiology B*, *182*(2), pp. 189–198.
Birkhead, T.R. and Møller, A.P., 1995. Extra-pair copulation and extra-pair paternity in birds. *Animal Behaviour*, *49*(3), pp. 843–848.
Birkhead, T.R. and Møller, A.P., 1992. Numbers and size of sperm storage tubules and duration of sperm storage in birds: a comparative study. *Biological Journal of Linnean Society*, *45*(4), pp. 363–372.
Birkhead, T.R. and Pizzari, T., 2002. Postcopulatory sexual selection. *Nature Reviews Genetics*, *3*(4), pp. 262–273.
Birkhead, T.R., 1998. Cryptic female choice: criteria for establishing female sperm choice. *Evolution*, *52*(4), pp. 1212–1218.
Birkhead, T.R., Møller, A.P. and Sutherland, W.J., 1993. Why do females make it so difficult for males to fertilize their eggs? *Journal of Theoretical Biology*, *161*(1), pp. 51–60.
Birkhead, T.R., Pellatt, E.J., Brekke, P., Yeates, R. and Castillo-Juarez, H., 2005. Genetic effects on sperm design in zebra finch. *Nature*, *434*(7031), pp. 383–387.
Bissoondath, C.J. and Wiklund, C., 1996. Effect of male mating history and body size on ejaculate size and quality in two polyandrous butterflies, *Pieris napi* and *Pieris rapae* (Lepidoptera: Pieridae). *Functional Ecology,* 10, pp. 457–464.
Bowen, R. H.,1925. Further notes on the acrosome of the animal sperm. The homologies of non-flagellate sperms. *The Anatomical Record*, 31, pp. 201–231.
Braun, R.E., Behringer, R.R., Peschon, J.J., Brinster, R.L. and Palmiter, R.D., 1989. Genetically haploid spermatids are phenotypically diploid. *Nature*, *337*(6205), pp. 373–376.
Briskie, J.V. and Montgomerie, R., 1992. Sperm size and sperm competition in birds. *Proceedings of Royal Society of London. Series B: Biological Sciences*, *247*(1319), pp. 89–95.
Byrne, P.G., Simmons, L.W. and Dale Roberts, J., 2003. Sperm competition and evolution of gamete morphology in frogs. *Proceedings of Royal Society of London. Series B: Biological Sciences*, *270*(1528), pp. 2079–2086.
Calabrese, J.M., Ries, L., Matter, S.F., Debinski, D.M., Auckland, J.N., Roland, J. and Fagan, W.F., 2008. Reproductive asynchrony in natural butterfly populations and its consequences for female matelessness. *Journal of Animal Ecology*, *77*, pp. 746–756.
Carayon, J., 1966. Traumatic insemination and paragenital system. *Monograph of Cimicidae (Hemiptera, Heteroptera)*. College Park, MD: Entomological Society of America, pp. 81–166.
Carazo, P., Font, E. and Alfthan, B., 2007. Chemosensory assessment of sperm competition levels and evolution of internal spermatophore guarding. *Proceedings of Royal Society B: Biological Sciences*, *274*(1607), pp. 261–267.
Carrick, F.N. and Setchell, B.P., 1977. The evolution of scrotum. In *Reproduction and evolution* (ed. J.H. Calaby and C.H. Tyndale-Biscoe), pp. 165–170. Canberra City: Australian Academy of Science.
Chapman, J.W., Reynolds, D.R. and Smith, A.D., 2003. Vertical-looking radar: a new tool for monitoring high-altitude insect migration. *Bioscience*, *53*(5), pp. 503–511.
Chapman, T., Herndon, L. A., Heifetz, Y., Partridge, L.andWolfner, M. F.,2001. The Acp26Aa seminal fluid protein is a modulator of early egg hatchability in *Drosophila melanogaster*. *Proceedings of the Royal Society of London. Series B: Biological Sciences*. 268, pp. 1647–1654.
Chapman, R.E. and Bourke, A.F., 2001. The influence of sociality on conservation biology of social insects. *Ecology Letters*, *4*(6), pp. 650–662.

Clutton-Brock, T.H. and Parker, G.A., 1995. Punishment in animal societies. *Nature*, *373*(6511), pp. 209–216.
Cohen, J., 1967. Correlation between sperm "redundancy" and chiasma frequency. *Nature*, *215*(5103), pp. 862–863.
Cohen, J., 1973. Cross-overs, sperm redundancy and their close association. *Heredity*, *31*(3), pp. 408–413.
Cook, P.A. and Wedell, N., 1996. Ejaculate dynamics in butterflies: a strategy for maximizing fertilization success?. *Proceedings of Royal Society of London. Series B: Biological Sciences*, *263*(1373), pp. 1047–1051.
Cotton, S., Small, J., Hashim, R. and Pomiankowski, A., 2010. Eyespan reflects reproductive quality in wild stalk-eyed flies. *Evolutionary Ecology*, *24*(1), pp. 83–95.
Crnokrak, P. and Roff, D.A., 1998. The genetic basis of trade-off between calling and wing morph in males of cricket Gryllus firmus. *Evolution*, *52*(4), pp. 1111–1118.
Dallai, R. and Afzelius, B.A., 1985. Membrane specializations in paired spermatozoa of dytiscid water beetles. *Tissue and Cell*, *17*(4), 561–572.
Danielsson, I., 2001. Antagonistic pre–and post–copulatory sexual selection on male body size in a water strider (Gerris lacustris). *Proceedings of Royal Society of London. Series B: Biological Sciences*, *268*(1462), pp. 77–81.
De Jong, P. W., Verhoog, M. D., and Brakefield, P. M.,1993. Sperm competition and melanic polymorphism in the 2-spot ladybird, Adalia bipunctata (Coleoptera, Coccinellidae). *Heredity*, 70, pp. 172–178.
Del Castillo, R.C., 2003. Body size and multiple copulations in a neotropical grasshopper with an extraordinary mate-guarding duration. *Journal of Insect Behavior*, *16*(4), pp. 503–522.
Dunn, P.O., Whittingham, L.A. and Pitcher, T.E., 2001. Mating systems, sperm competition, and evolution of sexual dimorphism in birds. *Evolution*, *55*(1), pp. 161–175.
Eberhard, W.G. (2009) Postcopulatory sexual selection: Darwin's omission and its consequences. *Proceedings of National Academy of Sciences of USA*, *106*(Supplement 1), pp. 10025–10032.
Eddy, E.M., 2002. Male germ cell gene expression. *Recent Progress in Hormone Research*, *57*(1), pp. 103–128.
Eggert, A.K. and Sakaluk, S.K., 1995. Female-coerced monogamy in burying beetles. *Behavioral Ecology and Sociobiology*, *37*(3), pp. 147–153.
Engqvist, L. and Reinhold, K., 2005. Pitfalls in experiments testing predictions from sperm competition theory. *Journal of Evolutionary Biology*, *18*(1), pp. 116–123.
Estrada, C. and Gilbert, L.E., 2010. Host plants and immatures as mate-searching cues in Heliconius butterflies. *Animal Behaviour*, *80*(2), pp. 231–239.
Firman, R.C. and Simmons, L.W., 2015. Gametic interactions promote inbreeding avoidance in house mice. *Ecology Letters*, *18*(9), pp. 937–943.
Firman, R.C., Gasparini, C., Manier, M.K. and Pizzari, T., 2017. Postmating female control: 20 years of cryptic female choice. *Trends in Ecology & Evolution*, 32, pp. 368–382.
Foster, W., 1967. Hormone-mediated nutritional control of sexual behavior in male dung flies. *Science*, *158*(3808), pp. 1596–1597.
Fricke, C., Wigby, S., Hobbs, R. and Chapman, T., 2009. The benefits of male ejaculate sex peptide transfer in *Drosophila melanogaster*. *Journal of Evolutionary Biology*, *22*(2), pp. 275–286.
Gack, C. and Peschke, K., 1994. Spernathecal morphology, sperm transfer and a novel mechanism of sperm displacement in rove beetle, *Aleochara curtula* (Coleoptera, Staphylinidae). *Zoomorphology*, *114*(4), pp. 227–237.
Gage, M.J. and Baker, R.R., 1991. Ejaculate size varies with socio-sexual situation in an insect. *Ecological Entomology*, *16*(3), pp. 331–337.
Gage, M.J., 1995. Continuous variation in reproductive strategy as an adaptive response to population density in moth *Plodia interpunctella*. *Proceedings of Royal Society of London. Series B: Biological Sciences*, *261*(1360), pp. 25–30.
Gage, M.J.G. and Cook, P.A., 1994. Sperm size or numbers? Effects of nutritional stress upon eupyrene and apyrene sperm production strategies in moth *Plodia interpunctella* (Lepidoptera: Pyralidea). *Functional Ecology*, *8*, pp. 594–599.
Gavrilets, S. and Waxman, D., 2002. Sympatric speciation by sexual conflict. *Proceedings of National Academy of Sciences of USA*, *99*(16), pp. 10533–10538.
Gavrilets, S., 2000. Rapid evolution of reproductive barriers driven by sexual conflict. *Nature*, *403*(6772), pp. 886–889.

Gilbert, L.E., 1976. Postmating female odor in Heliconius butterflies: a male-contributed antiaphrodisiac?. *Science, 193*(4251), pp. 419–420.
Greenway, E.G., Dougherty, L.R. and Shuker, D.M., 2015. Mating failure. *Current Biology, 25*(13), pp. 534–R536.
Gromko, M.H., Newport, M.E.A. and Kortier, M.G., 1984. Sperm dependence of female receptivity to re-mating in *Drosophila melanogaster. Evolution, 38*, pp. 1273–1282.
Harcourt, A.H., Purvis, A. and Liles, L., 1995. Sperm competition: mating system, not breeding season, affects testes size of primates. *Functional Ecology, 9*, pp. 468–476.
Hellriegel, B. and Blanckenhorn, W.U., 2002. Environmental influences on gametic investment of yellow dung fly males. *Evolutionary Ecology, 16*(5), pp. 505–522.
Hinton, H.E., 1964. Sperm transfer in insects and evolution of haemocoelic insemination. *Insect Reproduction*, pp. 95–107.
Holland, B. and Rice, W.R., 1997. Female control: sexual selection by cryptic female choice. *Evolution, 51*(1), pp. 321–325.
Holwell, J.L., Palks, S.I., Blair, A.M., Herdendorf, B.R., Quam, M.S., Roe, M.J. and Blomgren, D.A., Seagate Technology LLC, 2010. Electrical connection for testing head gimbal assemblies. U.S. Patent 7,684,948.
Hosken, D.J. and Ward, P.I., 2001. Experimental evidence for testis size evolution via sperm competition. *Ecology Letters, 4*(1), pp. 10–13.
Hosken, D.J., 1997. Sperm competition in bats. *Proceedings of Royal Society of London. Series B: Biological Sciences, 264*(1380), pp. 385–392.
Hosken, D.J., Garner, T.W.J., Tregenza, T., Wedell, N. and Ward, P.I., 2003. Superior sperm competitors sire higher-quality young. *Proceedings of Royal Society of London. Series B: Biological Sciences, 270*(1527), pp. 1933–1938.
Hosken, D.J., Taylor, M.L., Hoyle, K., Higgins, S. and Wedell, N., 2008. Attractive males have greater success in sperm competition. *Current Biology, 18*(13), pp. 553–R554.
Hughes, L., Chang, B.S.W., Wagner, D. and Pierce, N.E., 2000. Effects of mating history on ejaculate size, fecundity, longevity, and copulation duration in ant-tended lycaenid butterfly, *Jalmenus evagoras. Behavioral Ecology and Sociobiology, 47*(3), pp. 119–128.
Humphreys, L., 1972. *Out of closets: the sociology of homosexual liberation*. Prentice Hall.
Hunter, F.M. and Birkhead, T.R., 2002. Sperm viability and sperm competition in insects. *Current Biology, 12*(2), pp. 121–123
Jamieson, B.G.M., Dallai, R. and Afzelius, B.A., 1999. *Insects: their spermatozoa and phylogeny*. Science Publishers, Inc.
Johansson, F., Söderquist, M. and Bokma, F., 2009. Insect wing shape evolution: independent effects of migratory and mate guarding flight on dragonfly wings. *Biological Journal of Linnean Society, 97*(2), pp. 362–372.
Jones, A.G. and Ratterman, N.L., 2009. Mate choice and sexual selection: what have we learned since Darwin?. *Proceedings of National Academy of Sciences of USA, 106*(Supplement 1), pp. 10001–10008.
Kalb, J.M., DiBenedetto, A.J. and Wolfner, M.F., 1993. Probing function of *Drosophila melanogaster* accessory glands by directed cell ablation. *Proceedings of National Academy of Sciences of USA, 90*(17), pp. 8093–8097.
Kelly, C.D. and Jennions, M.D., 2011. Sexual selection and sperm quantity: meta-analyses of strategic ejaculation. *Biological Reviews, 86*(4), pp. 863–884.
Kelly, C.D., 2008. Sperm investment in relation to weapon size in a male trimorphic insect?. *Behavioral Ecology, 19*(5), pp. 1018–1024.
Kleven, O., Laskemoen, T., Fossøy, F., Robertson, R.J. and Lifjeld, J.T., 2008. Intraspecific variation in sperm length is negatively related to sperm competition in passerine birds. *Evolution: International Journal of Organic Evolution, 62*(2), pp. 494–499.
Kureck, I.M., Neumann, A. and Foitzik, S., 2011. Wingless ant males adjust mate-guarding behaviour to competitive situation in nest. *Animal Behaviour, 82*(2), pp. 339–346.
Løvlie, H., Gillingham, M.A., Worley, K., Pizzari, T. and Richardson, D.S., 2013. Cryptic female choice favours sperm from major histocompatibility complex-dissimilar males. *Proceedings of Royal Society B: Biological Sciences, 280*(1769), p. 20131296.

Lum, P.T.M. and Flaherty, B.R., 1970. Effect of continuous light on potency of *Plodia interpunctella* males (Lepidoptera: Phycitidae). *Annals of Entomological Society of America*, *63*(5), pp. 1470–1471.

Lung, O., Tram, U., Finnerty, C.M., Eipper-Mains, M.A., Kalb, J.M. and Wolfner, M.F., 2002. The *Drosophila melanogaster* seminal fluid protein Acp62F is a protease inhibitor that is toxic upon ectopic expression. *Genetics*, *160*(1), pp. 211–224.

Lüpold, S., 2013. Ejaculate quality and constraints in relation to sperm competition levels among eutherian mammals. *Evolution*, *67*(10), pp. 3052–3060.

Lüpold, S., Manier, M.K., Puniamoorthy, N., Schoff, C., Starmer, W.T., Luepold, S.H.B., Belote, J.M. and Pitnick, S., 2016. How sexual selection can drive evolution of costly sperm ornamentation. *Nature*, *533*(7604), pp. 535–538.

Manier, M.K., Belote, J.M., Berben, K.S., Lüpold, S., Ala-Honkola, O., Collins, W.F. and Pitnick, S., 2013. Rapid diversification of sperm precedence traits and processes among three sibling *Drosophila* species. *Evolution*, *67*(8), pp. 2348–2362.

Manier, M.K., Belote, J.M., Berben, K.S., Novikov, D., Stuart, W.T. and Pitnick, S., 2010. Resolving mechanisms of competitive fertilization success in *Drosophila melanogaster*. *Science*, *328*(5976), pp. 354–357.

Mikheyev, A.S., 2003. Evidence for mating plugs in fire ant *Solenopsis invicta*. *Insectes Sociaux*, *50*(4), pp. 401–402.

Miller, M.A., Ruest, P.J., Kosinski, M., Hanks, S.K. and Greenstein, D., 2003. An Eph receptor sperm-sensing control mechanism for oocyte meiotic maturation in *Caenorhabditis elegans*. *Genes & Development*, *17*(2), pp. 187–200.

Møller, A.P., 1988a. Female choice selects for male sexual tail ornaments in monogamous swallow. *Nature*, *332*(6165), pp. 640–642.

Møller, A.P., 1988b. Badge size in house sparrow *Passer domesticus*: effects of intra-and intersexual selection. *Behavioral Ecology and Sociobiology*, 22, pp. 373–378.

Moller, A.P., 1989. Ejaculate quality, testes size and sperm production in mammals. *Functional Ecology*, *3*, pp. 91–96.

Morrow, E.H. and Arnqvist, G., 2003. Costly traumatic insemination and a female counter-adaptation in bed bugs. *Proceedings of Royal Society of London. Series B: Biological Sciences*, *270*(1531), pp. 2377–2381.

Neubaum, D.M. and Wolfner, M.F., 1999. Mated *Drosophila melanogaster* females require a seminal fluid protein, Acp36DE, to store sperm efficiently. *Genetics*, *153*(2), pp. 845–857.

Oberhauser, K.S., 1988. Male monarch butterfly spermatophore mass and mating strategies. *Animal Behaviour*, *36*(5), pp. 1384–1388.

Oldroyd, B., Smolenski, A., Lawler, S., Estoup, A. and Crozier, R., 1995. Colony aggregations in *Apis mellifera* L. *Apidologie*, *26*(2), pp. 119–130.

Ono, T., Siva-Jothy, M.T. and Kato, A.K.I., 1989. Removal and subsequent ingestion of rivals' semen during copulation in a tree cricket. *Physiological Entomology*, *14*(2), pp. 195–202.

Painting, C.J. and Holwell, G.I., 2013. Exaggerated trait allometry, compensation and trade-offs in New Zealand giraffe weevil (*Lasiorhynchus barbicornis*). *PLoS One*, *8*(11), p. e82467.

Parker, G.A., 1970. Sperm competition and its evolutionary consequences in insects. *Biological Reviews*, *45*(4), pp. 525–567.

Parker, G.A. (1984) Sperm competition and evolution of animal mating strategies. *Sperm competition and evolution of animal mating systems* (ed.R.L. Smith), pp. 1–60.

Parker, G.A. and Partridge, L., 1998. Sexual conflict and speciation. *Philosophical Transactions of Royal Society of London. Series B: Biological Sciences*, *353*(1366), pp. 261–274.

Parker, G.A. and Pizzari, T., 2010. Sperm competition and ejaculate economics. *Biological Reviews*, *85*(4), pp. 897–934.

Parker, G.A. and Simmons, L.W., 1991. A model of constant random sperm displacement during mating: evidence from Scatophaga. *Proceedings of Royal Society of London. Series B: Biological Sciences*, *246*(1316), pp. 107–115.

Parker, G.A., 1970. Sperm competition and its evolutionary consequences in insects. *Biological Reviews*, *45*(4), pp. 525–567.

Parker, G.A., 1982. Why are there so many tiny sperm? Sperm competition and maintenance of two sexes. *Journal of Theoretical Biology*, *96*(2), pp. 281–294.

Parker, G.A., 1990a. Sperm competition games: sneaks and extra-pair copulations. *Proceedings of Royal Society of London. Series B: Biological Sciences*, *242*(1304), pp. 127–133.
Parker, G.A., 1990b. Sperm competition games: raffles and roles. *Proceedings of Royal Society of London. Series B: Biological Sciences*, *242*(1304), pp. 120–126.
Parker, G.A., 1993. Sperm competition games: sperm size and sperm number under adult control. *Proceedings of Royal Society of London. Series B: Biological Sciences*, *253*(1338), pp. 245–254.
Parker, G.A., Baker, R.R. and Smith, V.G.F., 1972. The origin and evolution of gamete dimorphism and male-female phenomenon. *Journal of Theoretical Biology*, *36*(3), pp. 529–553.
Parker, G.A., Immler, S., Pitnick, S. and Birkhead, T.R., 2010. Sperm competition games: sperm size (mass) and number under raffle and displacement, and evolution of P2. *Journal of Theoretical Biology*, *264*(3), pp. 1003–1023.
Peretti, A.V. and Eberhard, W.G., 2010. Cryptic female choice via sperm dumping favours male copulatory courtship in a spider. *Journal of Evolutionary Biology*, *23*(2), pp. 271–281.
Pilastro, A., Simonato, M., Bisazza, A. and Evans, J.P., 2004. Cryptic female preference for colorful males in guppies. *Evolution*, *58*(3), pp. 665–669.
Pischedda, A. and Rice, W.R., 2012. Partitioning sexual selection into its mating success and fertilization success components. *Proceedings of National Academy of Sciences of USA*, *109*(6), pp. 2049–2053.
Pitnick, S. C. O. T. T. & Hosken, D. J., 2010. Postcopulatory sexual selection. *Evolutionary Behavioral Ecology*, pp. 379–399.
Pitnick, S., Miller, G.T., Schneider, K. and Markow, T.A., 2003. Ejaculate-female coevolution in *Drosophila mojavensis*. *Proceedings of Royal Society of London. Series B: Biological Sciences*, *270*(1523), pp. 1507–1512.
Polak, M., 1998. Effects of ectoparasitism on host condition in *Drosophila*–Macrocheles system. *Ecology*, *79*(5), pp. 1807–1817.
Prowse, N. and Partridge, L., 1997. The effects of reproduction on longevity and fertility in male *Drosophila melanogaster*. *Journal of Insect Physiology*, *43*(6), pp. 501–512.
Remick, D. 1992. Measuring costs of reproduction. *Trends in Ecology and Evolution*, *7*(2), pp. 42–45.
Reaney, L.T. and Knell, R.J., 2015. Building a beetle: how larval environment leads to adult performance in a horned beetle. *PLoS one*, *10*(8), p. e0134399.
Reinhardt, K., Naylor, R. and Siva–Jothy, M.T., 2003. Reducing a cost of traumatic insemination: female bedbugs evolve a unique organ. *Proceedings of Royal Society of London. Series B: Biological Sciences*, *270*(1531), pp. 2371–2375.
Rhen, T., 2000. Sex-limited mutations and evolution of sexual dimorphism. *Evolution*, *54*(1), pp. 37–43.
Rice, W.R., 1984. Sex chromosomes and evolution of sexual dimorphism. *Evolution, 38*, pp. 735–742.
Roff, D.A., 2002. Life history evolution (No. 576.54 R6).
Royle, N.J., Hartley, I.R. and Parker, G.A., 2002. Sexual conflict reduces offspring fitness in zebra finches. *Nature*, *416*(6882), pp. 733–736.
Saeki, Y., Kruse, K.C. and Switzer, P.V., 2005. The social environment affects mate guarding behavior in Japanese beetles, *Popillia japonica*. *Journal of Insect Science*, *5*(1), p. 18.
Saglam, I.K., Roff, D.A. and Fairbairn, D.J., 2008. Male sand crickets trade-off flight capability for reproductive potential. *Journal of Evolutionary Biology*, *21*(4), pp. 997–1004.
Sandell, M.I., 1998. Female aggression and maintenance of monogamy: female behaviour predicts male mating status in European starlings. *Proceedings of Royal Society of London. Series B: Biological Sciences*, *265*(1403), pp. 1307–1311.
Sauter, A., Brown, M.J., Baer, B. and Schmid-Hempel, P., 2001. Males of social insects can prevent queens from multiple mating. *Proceedings of Royal Society of London. Series B: Biological Sciences*, *268*(1475), pp. 1449–1454.
Schlager, G., 1960. Sperm precedence in fertilization of eggs in Tribolium castaneum. *Annals of Entomological Society of America*, *53*(5), pp. 557–560.
Silberglied, R.E., Shepherd, J.G. and Dickinson, J.L., 1984. Eunuchs: role of apyrene sperm in Lepidoptera?. *The American Naturalist*, *123*(2), pp. 255–265.
Simmons, L.W., 2001. *Sperm competition and its evolutionary consequences in insects* (*Vol. 68*). Princeton University Press.
Simmons, L.W. and Fitzpatrick, J.L., 2012. Sperm wars and evolution of male fertility. *Reproduction*, *144*(5), p. 519.

Simmons, L.W., Craig, M., Llorens, T., Schinzig, M. and Hosken, D., 1993. Bushcricket spermatophores vary in accord with sperm competition and parental investment theory. *Proceedings of Royal Society of London. Series B: Biological Sciences*, *251*(1332), pp. 183–186.

Simmons, L.W., Lüpold, S. and Fitzpatrick, J.L., 2017. Evolutionary trade-off between secondary sexual traits and ejaculates. *Trends in Ecology & Evolution*, *32*(12), pp. 964–976.

Simmons, L.W., Tinghitella, R.M. and Zuk, M., 2010. Quantitative genetic variation in courtship song and its covariation with immune function and sperm quality in field cricket *Teleogryllus oceanicus*. *Behavioral Ecology*, *21*(6), pp. 1330–1336.

Simmons, L.W., Tomkins, J.L. and Hunt, J., 1999. Sperm competition games played by dimorphic male beetles. *Proceedings of Royal Society of London. Series B: Biological Sciences*, *266*(1415), pp. 145–150.

Smith, P.H., Gillott, C., Browne, L.B. and Gerwen, A.V., 1990. The mating-induced refractoriness of *Lucilia cuprina* females: manipulating male contribution. *Physiological Entomology*, *15*(4), pp. 469–481.

Snook, R.R. and Karr, T.L., 1998. Only long sperm are fertilization-competent in six sperm-heteromorphic *Drosophila* species. *Current Biology*, *8*(5), pp. 291–294.

Snook, R.R., 2001. Absence of latitudinal clines in sperm characters in North American populations of *Drosophila subobscura* (Diptera: Drosophilidae). *The Pan-Pacific Entomologist*, *77*(4), pp. 261–271.

Snook, R.R., Markow, T.A. and Karr, T.L., 1994. Functional nonequivalence of sperm in *Drosophila pseudoobscura*. *Proceedings of National Academy of Sciences of USA*, *91*(23), pp. 11222–11226.

Somjee, U., Allen, P.E. and Miller, C.W., 2015. Different environments lead to a reversal in expression of weapons and testes in heliconia bug *Leptoscelis tricolor* (Hemiptera: Coreidae). *Biological Journal of Linnean Society*, *115*(4), pp. 802–809.

Stutt, A.D. and Siva-Jothy, M.T., 2001. Traumatic insemination and sexual conflict in bed bug *Cimex lectularius*. *Proceedings of National Academy of Sciences of USA*, *98*(10), pp. 5683–5687.

Svärd, L. and Wiklund, C., 1989. Mass and production rate of ejaculates in relation to monandry/polyandry in butterflies. *Behavioral Ecology and Sociobiology*, *24*(6), pp. 395–402.

Swallow, J.G. and Wilkinson, G.S., 2002. The long and short of sperm polymorphisms in insects. *Biological Reviews*, *77*(2), pp. 153–182.

Thornhill, N.W. ed., 1993. *The natural history of inbreeding and outbreeding: theoretical and empirical perspectives*. University of Chicago Press.

Tregenza, T. and Wedell, N., 2002. Polyandrous females avoid costs of inbreeding. *Nature*, *415*(6867), pp. 71–73.

Trivers, R.L., 1972. Parental investment and sexual selection. In *Sexual selection and descent of man* (ed. B. Campbell), *13*(2), pp. 136–179. Chicago: Aldinc.

Tsubaki, Y. and Yamagishi, M., 1991. "Longevity" of sperm within female of melon fly, *Dacus cucurbitae* (Diptera: Tephritidae), and its relevance to sperm competition. *Journal of Insect Behavior*, *4*(2), pp. 243–250.

Veiga, J.P., 1990. Sexual conflict in house sparrow: interference between polygynously mated females versus asymmetric male investment. *Behavioral Ecology and Sociobiology*, *27*(5), pp. 345–350.

Waage, J.K., 1979. Dual function of damselfly penis: sperm removal and transfer. *Science*, *203*(4383), pp. 916–918.

Waage, J.K., 1984. Sperm competition and evolution of odonate mationg systems. *Sperm Competition and Evolution of Animal Mating Systems*, pp. 257–290.

Wachtmeister, C.A. and Enquist, M., 2000. The evolution of courtship rituals in monogamous species. *Behavioral Ecology*, *11*(4), pp. 405–410.

Waites, G.M.H. (1970). Temperature regulation and testis. *The Testis*, *1*, pp. 241–279.

Wedell, N.and Tregenza, T., 1999. Successful fathers sire successful sons. *Evolution*, 53, pp. 620–625.

Wiklund, C. and Kaitala, A., 1995. Sexual selection for large male size in a polyandrous butterfly: effect of body size on male versus female reproductive success in *Pieris napi*. *Behavioral Ecology*, *6*(1), pp. 6–13.

Woodhead, A.P., 1984. Effect of duration of larval development on sexual competence in young adult male *Diploptera punctata*. *Physiological Entomology*, *9*(4), pp. 473–477.

Xu, J. and Wang, Q., 2010. Mechanisms of last male precedence in a moth: sperm displacement at ejaculation and storage sites. *Behavioral Ecology*, *21*(4), pp. 714–721.

Yamagishi, M., Itô, Y. and Tsubaki, Y., 1992. Sperm competition in melon fly, *Bactrocera cucurbitae* (Diptera: Tephritidae): Effects of sperm "longevity" on sperm precedence. *Journal of Insect Behavior*, *5*(5), pp. 599–608.

Yamane, T., Okada, K., Nakayama, S. and Miyatake, T., 2010. Dispersal and ejaculatory strategies associated with exaggeration of weapon in an armed beetle. *Proceedings of Royal Society B: Biological Sciences*, *277*(1688), pp. 1705–1710.

Zhu, D.H. and Tanaka, S., 2002. Prolonged precopulatory mounting increases length of copulation and sperm precedence in *Locusta migratoria* (Orthoptera: Acrididae). *Annals of Entomological Society of America*, *95*(3), pp. 370–373.

Zuk, M., 1987. The effects of gregarine parasites on longevity, weight loss, fecundity and developmental time in field crickets *Gryllus veletis and G. pennsylvanicus*. *Ecological Entomology*, *12*(3), pp. 349–354.

9

Sexual Coercion

Shashwat Singh[1], Priya Singh[2], Geetanjali Mishra[2], and Omkar[2]
[1]*Institute of Plant Protection, Department of Entomology, Agricultural Research Organization, Volcani Center, HaMaccabim Road 68, PO Box 15159, Rishon LeTsiyon, 7505101, Israel*
[2]*Ladybird Research Laboratory, Department of Zoology, University of Lucknow, Lucknow – 226 007, India*

CONTENTS

DOI: 10.1201/9781003043195-9

9.1 Introduction

Reproductive success of any organism depends on multiple factors starting from selection of mates, effective copulation, to successful fertilization of eggs. In many species, males increase their reproductive fitness by obtaining multiple copulations and inseminations. On the contrary, females avoid such matings due to high parental investment on their part (Arnqvist & Nilsson, 2000). Thus, females are likely to prefer limited and selected matings while males are expected to mate multiply and arbitrarily. This disparity in parental investment usually leads to a greater number of reproductively active males than females at a given time (Emlen & Oring, 1977). Thus, females become selective while choosing a mate than males (Trivers, 1972). This lack of alignment in reproductive strategies can give rise to sexual conflict among both sexes.

Sexual conflict among both sexes is a major force for evolution in reproductive performances and can even lead to sexual coercion (Arnqvist *et al.*, 2000; Vahed, 2007; Cummings, 2018). Sexual coercion is use of force or threat of force to a female by male that maximizes chance of mating with coercive male but at some cost to both sexes. Eventually, probability of mating with other males' should theoretically decrease.

Sexual coercion is a third form of sexual selection (Clutton-Brock & Parker, 1995; Brown *et al.*, 1997) along with female choice and inter-male competition (Darwin, 1871). Sexual coercion prevents or constrains free exercise of mate choice by females and entails forced copulation, sexual harassment, and intimidation through punishment of uncooperative females. Sexual coercion is evident among different taxa (Thornhill, 1981; Clutton-Brock, 1989; Sparkes *et al.*, 1996; Poulin & Morand, 2000; Jormalainen *et al.*, 2000).

Insects are some of best-studied organisms for mating systems, mating patterns (Thornhill & Alcock, 1983; Simmons, 2001), sexual conflict, and sexual coercion (Arnqvist & Rowe, 2005). Studies have identified some ultimate fitness consequences of sexual conflict (Arnqvist & Rowe, 2002; Hosken *et al.*, 2003), but proximate mechanisms are still less explored. Morphological characters, such as greater body size and physical strength play a major role in sexual coercion (Clutton-Brock & Parker, 1995). In insects, male intromittent organ may even damage female reproductive tract and sometimes copulation greatly increases chance of female death (Blanckenhorn *et al.*, 2002). Males are especially persuasive towards mating, even with newly emerged (Seeley & Dukas, 2011) delicate and sexually immature females (Seeley & Dukas, 2011).

Some direct costs to females include injury, increased energy expenditure (Watson *et al.*, 1998), increased mortality (Reale *et al.*, 1996), physiological stress (Muller *et al.*, 2007), and decreased reproductive success and life span (den Hollander & Gwynne, 2009; Rossi *et al.*, 2010; Takahashi & Watanabe, 2010). Indirect costs include hindrance in mate choice. The resistive tactics of females towards unwanted matings exert strong selective pressure on males, leading to the evolution of countertactics in males (such as female sensory exploitation, or stimulatory behaviours), particularly in the context of arms race or coercion (Arnqvist & Rowe, 2005). Thus, increased conflict of interests between sexes should lead to an increase in the intensity of the arms race, which could also modify paternal care (Chapman *et al.*, 2003).

Sexual coercion is primarily related to male mating behaviours (forced copulation, harassment, and intimidation) while in a sexual conflict there is interaction between male and female reproductive interest (mating struggle, etc.). Sexual coercion studies generally evaluate female suffering, however, in sexual conflict cost incurred by both sexes matters. Both are so interconnected that it necessitates that we should study these two together. In this chapter, we have summarized sexual conflict, sexual coercion, and their consequences on reproductive fitness including their costs and benefits.

9.2 Sexual Conflict and Coercion

Sexual conflict over mating strategies can be traced to beginning of anisogamy. This difference in gamete size indicates higher physical investment by female and much lower by male. This difference in investment forms primary basis of conflict, thereby evolving biological and behavioural actions of each sex. Females are more inclined to be 'protective' of their gametes and thus are choosy and restricted in their matings. Males on the other hand are more inclined in 'spreading the seed' and more indiscriminate in number of matings. Consequently, possible conflict over whether or not to mate or a number of sex-specific optimal number of mating arises.

In the hypothetical condition in which two sexes have a conflict over number of mating partners, say optimum mating partner in case of male is three but for female is only one. Due to this conflict let us assume somehow female moves away from the optimum number of partner (one) in direction of the male's optimum number of partner (three). Ultimately, male will get some benefit however female will suffer cost reducing her evolutionary fitness. Such a scenario can be considered a classical case of sexual coercion. While going through literature, terms such as premating struggle, harassment, etc. are commonly used in place of or to describe sexual coercion.

9.3 Types of Sexual Conflict

It is unusual that evolutionary interest of two individuals match because of differences in alleles at all loci, which is responsible for competition among them. Thus, conflict in evolutionary interest can be traced to dissimilarity in gene pool. Further anisogamy also results in competition. As a result, there are more chances of sexually antagonistic selection on these traits. Therefore, sexual conflict occurs because of different reproductive interests and tactics of both sexes; that eventually exerts strong selection pressure on many traits (Parker, 1979; Trivers, 1972). It can be broadly classified into major forms:

9.3.1 Intra-locus Sexual Conflict (IASC)

This is when the fitness of particular traits shared by both sexes are different in males and females. In this, antagonist selection in both sexes, in conjugation with genetic correlation between male and female will cause adaptive evolution in one sex by others (Parker & Partridge, 1998). Moreover, independent evolution of traits in both sexes can be possible when there is a selection for sex-limited gene expression, causing acquisition of maximum adaptation in both sexes (Rhen, 2000). IASC was studied in seminal fluid of *D. melanogaster*, where fitness of male and female at larval stage were positively correlated while reverse was found in adults because of different genetic interest of both sexes (Chippindale *et al.*, 2001).

9.3.2 Inter-locus Sexual Conflict (IRSC)

This is most common and occurs due to conflict of reproductive interest between both sexes over mating duration, its frequency, parental care, mating resistance, remating, female reproductive output in terms of oviposition, etc. We can also say that loci in males and females have coevolved in such a way that there is an increase in reproductive success of one sex at cost of other. For instance, males increase their reproductive success by procuring many matings at expense of female, and loci in female increases resistance towards male mating attempts (Trivers, 1972; Parker, 1979). The interlocus sexual conflict may lead to striking 'arms races' with distinct or novel phenotype or traits such as spiny genitalia or toxic ejaculates (Arnqvist & Rowe, 2005). However, mating with these males could also be beneficial to females due to 'sexy sons' effect. The coevolution of novel male traits and female response under the umbrella of sexual conflict usually causes inextricable evolutionary drive.

9.4 Plausible Factors for Sexual Conflict that Lead to Sexual Coercion

In case of extra matings, females bear costs like less foraging time and high energy expenditure (Watson *et al.*, 1998), physical injury (Parker, 1979), high predation risk (Rowe, 1994), longevity, fecundity, and parasite/pathogen infection (Hurst *et al.*, 1995) (Figure 9.1). Thus, females regulate the number of matings in response to phenotype of mates, population density, food availability, predator presence, and operational sex ratio (Heller & Helversen, 1991; Wiklund & Kaitala, 1995). These factors singly or in combination may decide optimal mating rate in females. Thus, while males are likely to force more matings (Clutton-Brock & Parker 1995; Arnqvist, 1997), females are likely to avoid such attempts (Arnqvist & Rowe, 1995). The following are major factors responsible for sexual conflict over mating rate in both sexes that ultimately result in sexual coercion (also see Table 9.1).

9.4.1 Operational Sex Ratio and Density (OSR)

OSR is the ratio of sexually mature males to sexually mature females in a population at a time (Kvarnemo & Ahnesjo 1996), while density is number of individuals in a particular area. A biased sex ratio is likely to create insufficiency of active mates affecting reproductive success of either sex. Studies indicate effects of manipulated OSR on mating strategies (Rowe, 1992; Saxena *et al.*, 2020). OSR represents extent of intra-sexual competition vs. mating opportunities (Clutton-Brock & Parker, 1992). Under male-biased sex ratio, intense male–male competition is obvious for limited female resource and females are likely to experience more sexual coercion (Clutton-Brock & Parker, 1995), which may lead to injury to females (Clutton-Brock *et al.*, 1992) or may hindrance in reproduction (Marchesan, 2002). Higher density induces increased reproductive activity and aggressive behaviour in males (Markow, 2002; Smith, 2007). The strong competition to procure mates may eventually lead to evolution of certain morphological and behavioural characters that increases mating success.

9.4.2 Nuptial Gifts

Material gifts given to females during mating are called as nuptial gifts during mating. These can be plant products, food, prey, oral or glandular secretions, haemolymph, ejaculatory substances, part of body, and even whole body (Dodson *et al.*, 1983; Boggs, 1995; Vahed, 1998). These can work as nutrients, antibiotics, immunostimulatory substances, ions, and minerals (Boggs, 1995; Poiani, 2006). It also speeds up overall fecundity (Lewis & Cratsley, 2008) and oviposition rate (Albo & Costa, 2010).

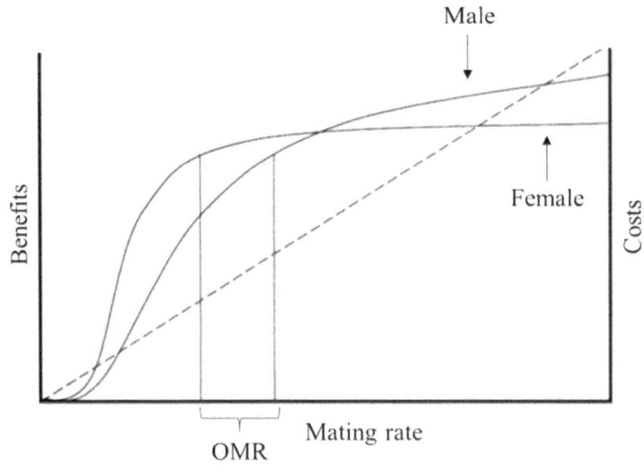

FIGURE 9.1 The dashed line represents costs associated with mating to both sexes, asymmetrical solid lines represents benefits of increased mating rate. Thereby, optimal mating rate (OMR) will vary for both sexes that may lead to sexual conflict over mating rate (Adapted from Arnqvist & Rowe, 2005).

TABLE 9.1

Major coercive adaptations in insects, its benefits to male, associated costs to females, and counter adaptations in females

Male Coercive Adaptations	Benefit to Males	Costs to Females	Counter Adaptations in Females	References
Grasping, genital or intromittent organ spines or claspers (Forced copulation)	Prolonged copulation, success over of rivals in mate procurement	High energy expenditure, Reduce chances of remating, injury, reduced life span	Kicking behaviour, anticlasper kind of morphological adaptations	Arnqvist & Rowe, 1995; Siva-Jothy, 2000; Miller, 2003; Howe, 2008; Rönn & Hotzy, 2012
Aggression /Harassment towards females	Delayed remating chances of females	Reduced foraging activity, low remating chance, Energy loss	Avoidance behaviour, defensive pheromone release	Cordero & Andrés, 2002; Rossi *et al.*, 2010
Seminal fluid receptivity inhibitor and toxin	Delayed or reduced possibility of remating, Kills sperm of previous male	Decreased life span, restrict female mate choice, difficulty in sperm replenishment	Ejaculate rejection behaviour, low sensitivity to seminal substances, degradation of seminal fluid constituents.	Eberhard, 1996; Arnqvist & Nilsson, 2000; Avila *et al.*, 2011
Mate guarding	Prevent other males from mating, delay remating	Energy loss, low foraging opportunities, high predation risk, hindrance in mate choice and oviposition behaviour	Female termination of mating by dislodging males through kicking and simple walkover, antimale morphology, aggression towards males	Parker, 1970; Borgia, 1981; Martens and Rehfelt, 1989; Alcock, 1994; Arnqvist & Rowe, 1995; Zeiss *et al.*, 1999; Chaudhary *et al.*, 2015
Mating plug formation	Prevent ejaculate rejection and remating, works as a doner of seminal signals	Obstruction in oviposition and replenishment of sperm, affect polyandry and mate choice	Removal or dissolution of plug	Eberhard, 1996; Avila *et al.*, 2011

Svensson *et al.* (1990) suggest that both sexes benefit via nuptial gifts; females get nutritive and defensive materials and males enough time for copulation and insemination. Benefits to males enhance chances of coercion by them, while costs to females ensure evolution of resistance tactics (Arnqvist *et al.*, 2000; Arnqvist & Nilsson, 2000).

9.4.3 Phenotype of Mate

Mating experience can also affect future mate selection. Deleterious offspring quality and physical trauma may lead females to adapt strategic mate sampling decisions (Wagner, 1998; Jennions & Petrie, 2000; Dubey *et al.*, 2016). In seaweed fly, *Coelopa frigida*, males usually attempt forced mating and females show resistance towards copulating smaller males causing injury, but not towards larger ones so as to avoid greater injury (Shuker & Day, 2001) and in many cases preference of injured females has evolved towards larger males.

9.4.4 Toxicity of Seminal Fluid

Females benefit from low seminal fluid toxicity, while males' benefit from a high toxicity level as it increases their competitive edge (Johnstone & Keller, 2000). In many insects, seminal fluid proteins reduce female receptivity and remating chances (Eberhard, 1996; Avila *et al.*, 2011), longevity (Lung *et al.*, 2002) or they can also behave as repellents to other males. Such conflicts are likely responsible for evolution of male features that helps in sexual coercion (Sakaluk *et al.*, 1995) and female traits that reduce costs of male harassment (Arnqvist & Rowe, 1995).

9.4.5 Food Intake

In insects, egg production depends on quality and quantity of food consumed (Singh *et al.*, 2016; Singh *et al.*, 2019). Ribeiro and Dickson (2010) demonstrated that mated females of *Drosophila* consumed more protein, while those *Aedes aegypti* required higher blood meals (Houseman & Downe, 1986). This could also be a source of sexual conflict; males that allow mates to feed may be preffered, but higher food consumption reduces female longevity, and exposure to predation an pathogens (Chapman & Partridge, 1996). In water striders and ladybird beetles, food-deprived females show resistance in mating because of conflict with foraging activities (Rowe, 1992; Perry *et al.*, 2009).

9.4.6 Paternal Investment

Asymmetrical investment by sexes turns females into a limited reproductive resource for which males tries to gain sexual access (Trivers, 1972). To anticipate reproductive cost, an evolutionary arms race between both sexes will be favoured; in which males and females would adopt strategies against each other (Parker, 2006). This may sometimes also reduce overall population fitness (Arnqvist & Rowe, 2005).

9.5 Sexual Coercion

Sexual coercion was first defined by Smuts and Smuts (1993) 'force, or threat by males to females to increase chances that she will mate with him at a time when she is likely to be fertile, and also to decrease chances that she will mate with other males, at some cost to female'. The simplest and most accepted form is direct coercion, where chances of mating with female of interest are usually because of involvement of force by a male against resistance by a female. Direct coercion is further categorized into forced copulation, harassment, and intimidation (Clutton-Brock & Parker,1995).

In indirect coercion, male uses force to reduce likelihood of females mating with another male, also coercive mate guarding and is common in promiscuous insects. Coercive behaviour is always associated with costs and benefits to both sexes (Figure 9.2).

9.6 Types of Sexual Coercion

9.6.1 Forced Copulation

This form of sexual coercion is found in a diverse group of animals (Clutton-Brock & Parker, 1995), including insects (Thornhill, 1981; Thornhill & Alcock, 1983; Arnqvist, 1997; Hinn & Niedzlek-Feaver, 2001; Shuker & Day, 2001). Tephritid flies (Smith & Prokopy, 1980), midge, *Culicoides melleus* (Linley & Adams, 1972), *Drosophila* (Manning, 1967), and planthopper *Nilaparvata lugens* (Oh, 1979) are well-studied for sexual coercion. In most insects, forced copulation is assisted by specialized genital structure that helps males to hold resisting females for copulation and insemination. In tettigoniids, during normal copulatory behaviour, cercal spurs of males get inserted into holes present on both sides of sub-genital plate of female (Rentz, 1972). Camel cricket *Pristoceuthophilus marmoratus* males use modified hind legs to hold a female in crook of one hind leg while forcing his genitalia into her genital opening (Haley & Gray, 2012). In alpine bushcricket, *Anonconotus alphinus*, males copulate forcibly by using their anal ceri as pincers to grasp female abdomen (Figures 9.3 and 9.4). In contrast to normal copulatory behaviour, sides of abdomen of females were found to be

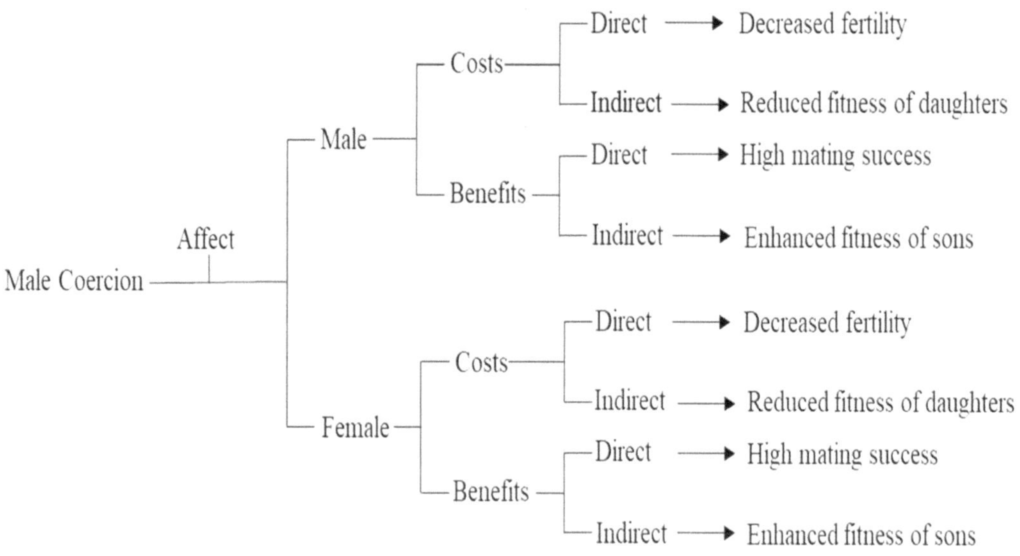

FIGURE 9.2 Major costs and benefits associated with male coercive behaviour (Adapted from Arnqvist & Rowe, 2005).

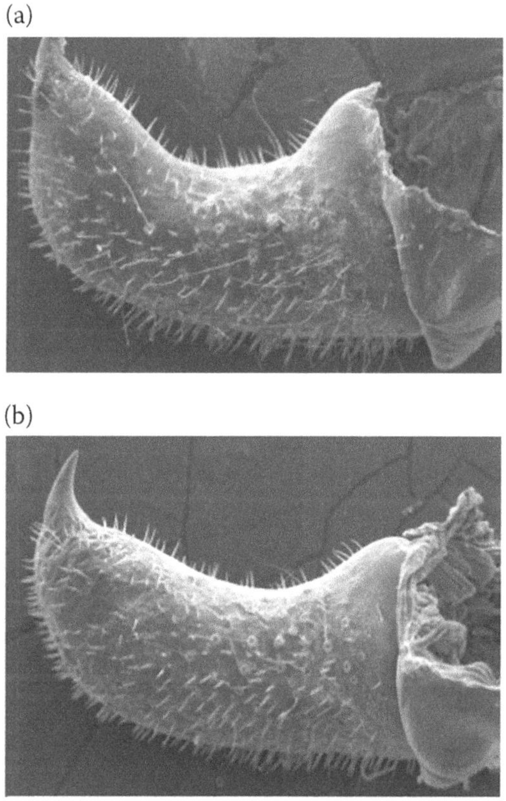

FIGURE 9.3 Scanning electron microscopic image of ercus of male (a) *Anonconotus alpinus ghilianii,* and (b) *Anonconotus alpinus alpinus* (Vahed & Carron, 2008).

bruised by male cercal spurs (Figure 9.5) (Vahed & Carron, 2008). Such insertion of male genitalia into non-genital part of females is called as traumatic insemination and has been observed in bed bugs (Stutt & Siva-Jothy, 2001) (Figure 9.6), bean weevil, *Callosobruchus maculatus* (Crudgington & Siva-Jothy, 2000) (Figure 9.7).

FIGURE 9.4 Abdomen of female *Anonconotus alpinus alpinus* showing an encircled melanization patch post-grasping by male cerci (Vahed & Carron, 2008).

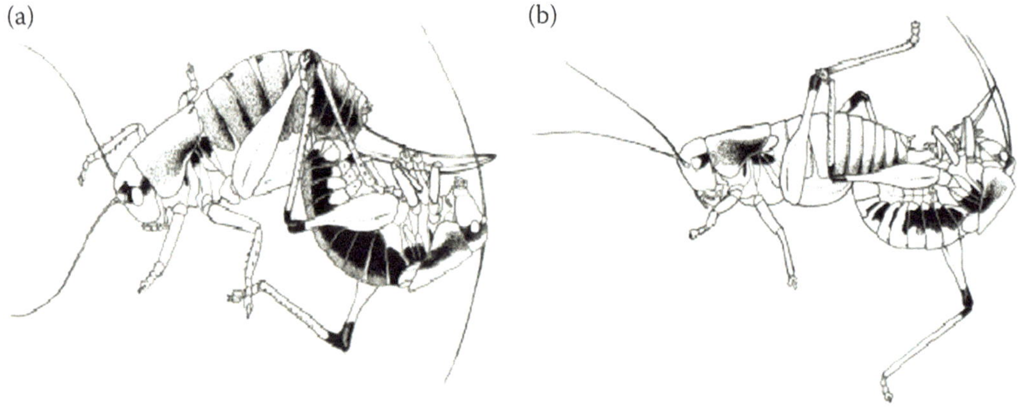

FIGURE 9.5 Normal copulatory behaviour of Alpine bush cricket, (a) *Anonconotus alpinus ghilianii* and, (b) *Anonconotus alpinus alpinus* (Vahed & Carron, 2008).

FIGURE 9.6 Mounting attempt by male bed bug, *Cimex lectularius;* dagger like intromittent organ (shown by arrow) ued for puncturing female ventral side. (Siva-Jothy *et al.,* 2019).

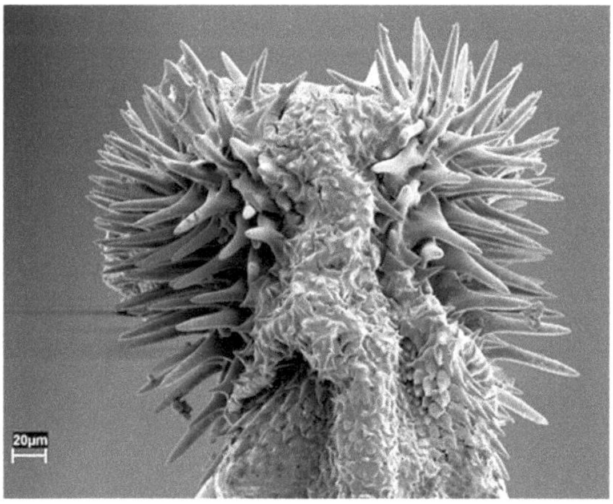

FIGURE 9.7 The sclerotized spine bearing intromittent organ of *Callosobruchus maculatus* (**Picture credit:** Johanna Rönn, Department of Animal Ecology, Uppsala University).

Mating with sexually immature females also comes under forced copulation known as teneral mating. *Drosophila melanogaster* male promptly start mating with females within 1 hour of latters' emergence (Seeley & Dukas, 2011). Similar behaviour has been observed in cockroaches, *Diploptera punctate* and *Oxyhaloa deusta*, (Lizée *et al.*, 2017), butterfly, *Helioconius* (Deinert, 2003), and ladybirds, *Menochilus sexmaculatus* (Fabricius) (Omkar *et al.*, 2006) and *Anegleis cardoni* (Weise) (Omkar *et al.,* 2013). Majerus (1994) also suggested rape hypothesis in ladybird, in which strong male can establish mating even if female does not want to mate. Butterflies also show pupal mating (Deinert, 2003).

Forced copulation is likely to have evolved as a male fertilization strategy. Several predictions can be tested in wild populations: (i) forced copulation attempts should be directed at fertilizable females, (ii) males should have increased effectiveness in finding, recognizing, and copulating with fertilizable females (Mckinney *et al.*, 1983), (iii) older females should be selected if clutch size and reproductive success increase with age, (iv) forced copulation should occur when temporal chances of fertilization are highest (Cheng *et al.*, 1982), (v) forced copulation rates should be positively correlated with annual variations in female reproductive effort and performance, and (vi) males should have tactics that protect their genetic paternity (Mckinney *et al.*, 1983).

9.6.2 Harassment/Aggression

Males performing sexual coercion are believed to be more aggressive, which includes behaviours such as gaining access to mates, defending territory, protecting offspring, etc. with distinct genetic, neurobiological and/or physiological underpinnings (Moyer, 1968). It is more common among higher group of animals (McKinney *et al.*, 1983).

Harassment is usually displayed during or before forced copulation, and may include stalker-like behaviour, in which he follows female at a distance in preparation to attack for mating. It has been observed in numerous insects (Cordero & Andrés, 2002). Aggressively forced copulation may evolve to gain extra mating opportunities (Gowaty & Buschhaus, 1998), but could also be maintained or enahnced because of impact on female remating.

Repeated mating attempts impose several hazardous costs on female, viz. physical injuries (Blanckenhorn *et al.*, 2002), enhanced predation (Mühlhäuser & Blanckenhorn, 2002), low foraging (Stone, 1995), and decreased fecundity (Rossi *et al.*, 2010) performed an aedeagus ablation of bean beetle, *Callosobruchus chinensis* lowered female fitness via harassment by 18.6% (Sakaurai & Kasuya, 2008).

'Sexual Dialectic Theory (SDT)' (Gowaty, 1997) suggests that aggression is more common when (i) male morphological structures can restrain females, (ii) females' movements are restricted, (iii) males defend and broker some resource limiting females' reproduction, and (iv) males are larger than females.

Another theory 'Creation of Dangerous Environment (CODE)', considers forced copulation a non-sexual act of aggression used by a small percentage of reproductively isolated males to modify female behaviour. It predicts that forced copulation: (i) is directed at both fertile and non-fertile females, (ii) does not result in fertilization and is negatively correlated with extra-pair paternity, (iii) is directed at unguarded and unmated females, and (iv) alters mating strategy from polygyny to monogamy.

Immediate Fertilization Enhancement Hypothesis makes following predictions: (i) aggressive copulation attempts are on fertilizable females only, (ii) fertilization success from aggressive copulation is as frequent as in case of females that show lack of resistance, (iii) extra-pair paternity is highest for females most vulnerable to aggressive copulation, and (iv) variance in reproductive success among males is highest for males who socially consort with females most vulnerable to aggressive copulation. Notably, it makes no predictions about modifications of female behaviour for males' advantages. Like forced copulations, harassment can be expected to evolve to a point where it has substantial costs to males as well as to females.

9.6.3 Intimidation

Intimidation is more prominent in higher groups of animals in which male punish females when they decline to mate with them (Clutton-Brock *et al.*, 1982, 1992; Baniel *et al.*, 2017). Generally, male punishes via physical violence (Smuts, 1986) as in case of water strider, *Gerris gracilicornis*, where males intimidate females via rippling water to induce fear of predator; vibrations that will only stop once females expose genitalia for mating (Han & Jablonski, 2009, 2010; Han *et al.*, 2015). Thus, via intimidation future mating probability of males generally increases. Intimidation is more common where males and females form more stable social groups or where chances of encounter of individuals for reproductive event are frequent.

9.7 Morphological and Behavioural Modifications

9.7.1 Grasping and Grappling

Male grasping techniques increase mating duration, overcome female's resistance and restrict females' remating (Arnqvist & Rowe, 2002; Perry & Rowe, 2012), and are similar to mate guarding. Darwin (1871) described males with grasping qualities as having 'organs for prehension'. During forced copulation, water strider, *Rheumatobates rileyi* males grasp females strongly with antennal claspers and modified legs (Arnqvist, 1997; Khila *et al.*, 2012) (Figure 9.8) during and after copulation; energetically expensive to female because she is carrying a strong male and navigating water (Fairbairn, 1993; Arnqvist, 1997). In water strider, *Gerris odontogaster*, males with longer abdominal claspers can endure a lot of somersaults by resistant females and achieve higher mating success (Fairbairn *et al.*, 2003; Rowe & Arnqvist, 2012).

In diving beetles, males rudely grasp females for mating, however, females undergo rapid whirling movements to dislodge males (Cleavall, 2009); males have evolved many grasping apparatuses to avoid dislodgement (Wewalka, 2000; Miller & Bergsten, 2014). In seed beetles, males possess sclerotized spines on their genitalia (Rönn & Hotzy, 2012) (Figure 9.7), which help overcome female resistance, penetration of genitalia, facilitates passage of seminal fluid and as an anchoring device. Sepsis fly males have modifications on their forelegs to help them grasp on to female wing bases (Eberhard, 2002; Figure 9.9). Such modifications are also seen in scorpionflies (Engqvist & Sauer, 2003).

9.7.2 Post-copulatory Guarding

Here, males guard females to exclude other potential rival males from mating (Alcock, 1994; Simmons, 2001) and often involves aggression. The most common mechanisms are to engage females in prolonged copulation (Allen *et al.*, 1994; Wang *et al.*, 2008; Chaudhary *et al.*, 2015; Figures 9.10 and 9.11), mislead rivals away from female (Field & Keller, 1993), and mating plug formation in

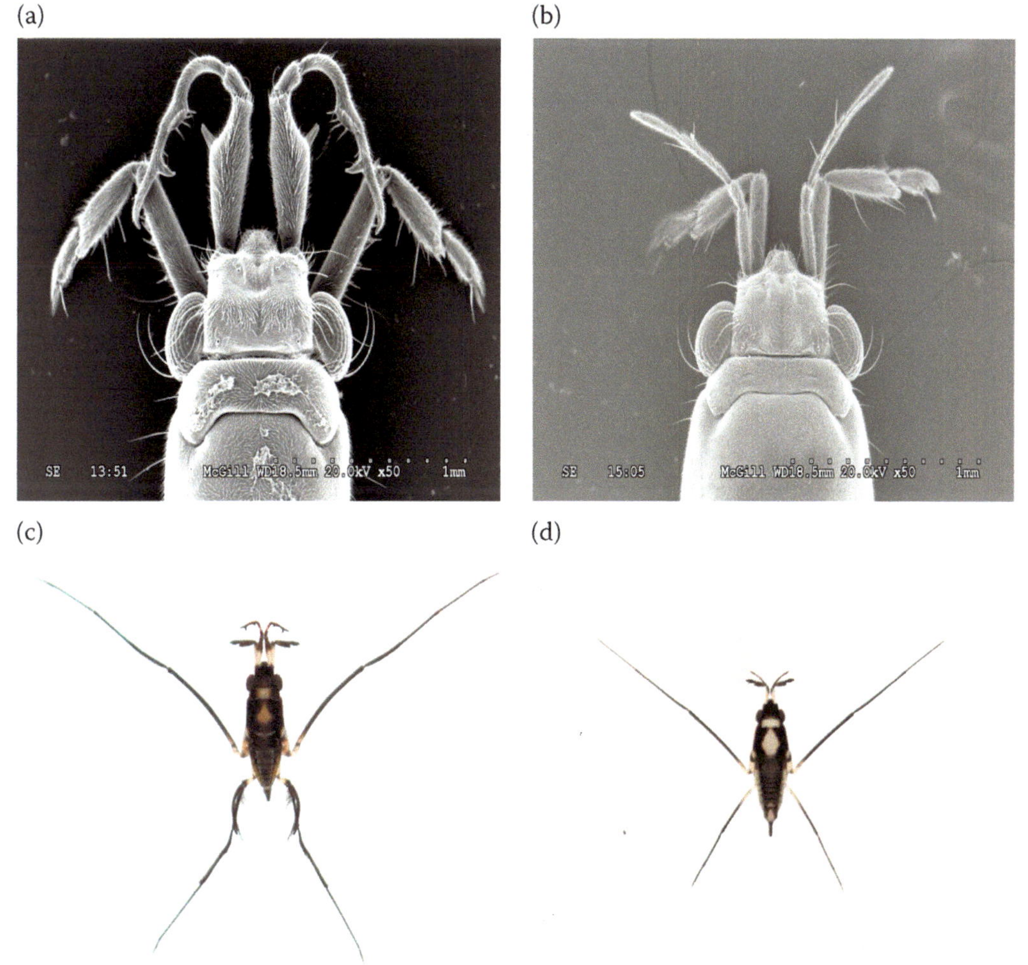

FIGURE 9.8 In water strider *Rheumatobates rileyi*: (a) antenna of male, (b) antenna of female, (c) spiny legs of male, and (d) spiny legs of adult female.

(Picture credit: Abderrahman Khila, Institute of Functional Genomics, Lyon, France).

(Polak *et al.*, 2001). Males of *Lygaeus equestris* procure mating by jumping towards females, after that they use their leg for grasping before actual genital contact (Burdfield-Steel & Shuker, 2014). Guarded females hold spermatophores for a significantly longer period than non-guarded females (Bussiére *et al.,* 2006).

9.8 Costs of Sexual Coercion to Females

If sexual coercion is costly for females, then increased harassment would increase costs of rejecting males and cause females to mate indiscriminately to avoid further harassment (Thornhill & Alcock, 1983). Consequently, opportunity for sexual selection may decrease.

The major direct costs associated with male coercive have been discussed previously. The indirect costs that affect future of females are likely be in form of hindrance in female mate choice thereby affecting offspring fitness. This would lead to increased resistance to coercive males, leading to greater pressure on male to evolve other ornamental features (Jennions *et al.*, 2001), and larger body size in insects (Perry *et al.*, 2009; Dubey *et al.*, 2016).

(a) (b)

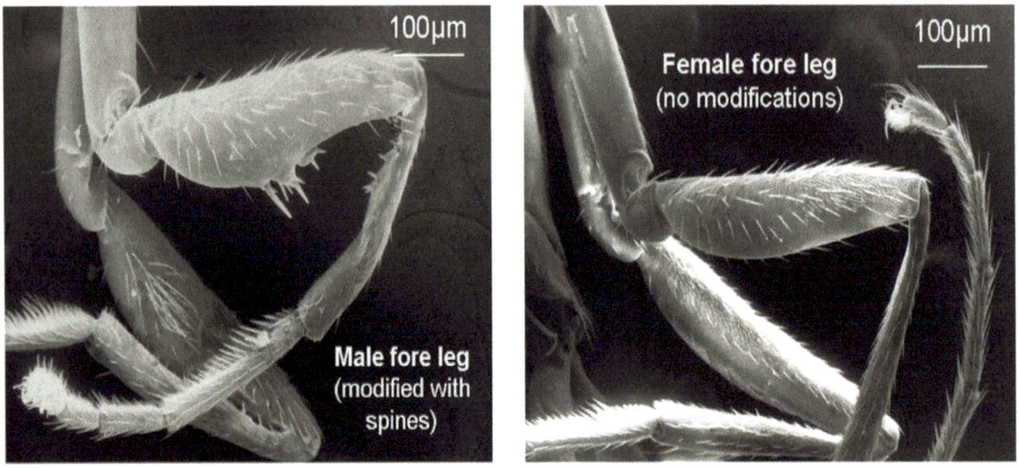

FIGURE 9.9 *Sepsis dissimilis* (a) male foreleg (b) female foreleg (Puniamoorthy *et al.*, 2008).

(a) (b)

FIGURE 9.10 *Lygaeus equestris* (a) commencement of mating by male, (b) pair in copula (**Picture credit:** Liam Dougherty; From Burdfield-Steel & Shuker 2014).

9.9 Female Counter Adaptations to Sexual Coercion

9.9.1 Morphological/Anatomical Protections

Females have evolved morphological adaptations, viz. shield over genitalia (Han & Jablonski, 2009), abdominal spines and curvature (Arnqvist & Rowe, 2002; Ronkainen *et al.*, 2005), anticlasping structure (Arnqvist & Rowe, 1995; Figure 9.12), etc. to counter coercion attempts. Females of *Cimex lectularius* have a specialized organ called 'spemalege' to avoid harmful effect of traumatic insemination and also defends against sexually transmitted pathogens (Reinhardt *et al.*, 2003).

9.9.2 Male Avoidance and Habitat Change

Females of African swallowtail butterfly, *Papilio dardanus*, occur in three distinct morphs: (i) black and white batesian mimic called *hippocoonides*, (ii), black and yellow male-like morph, called *trimeni* and,

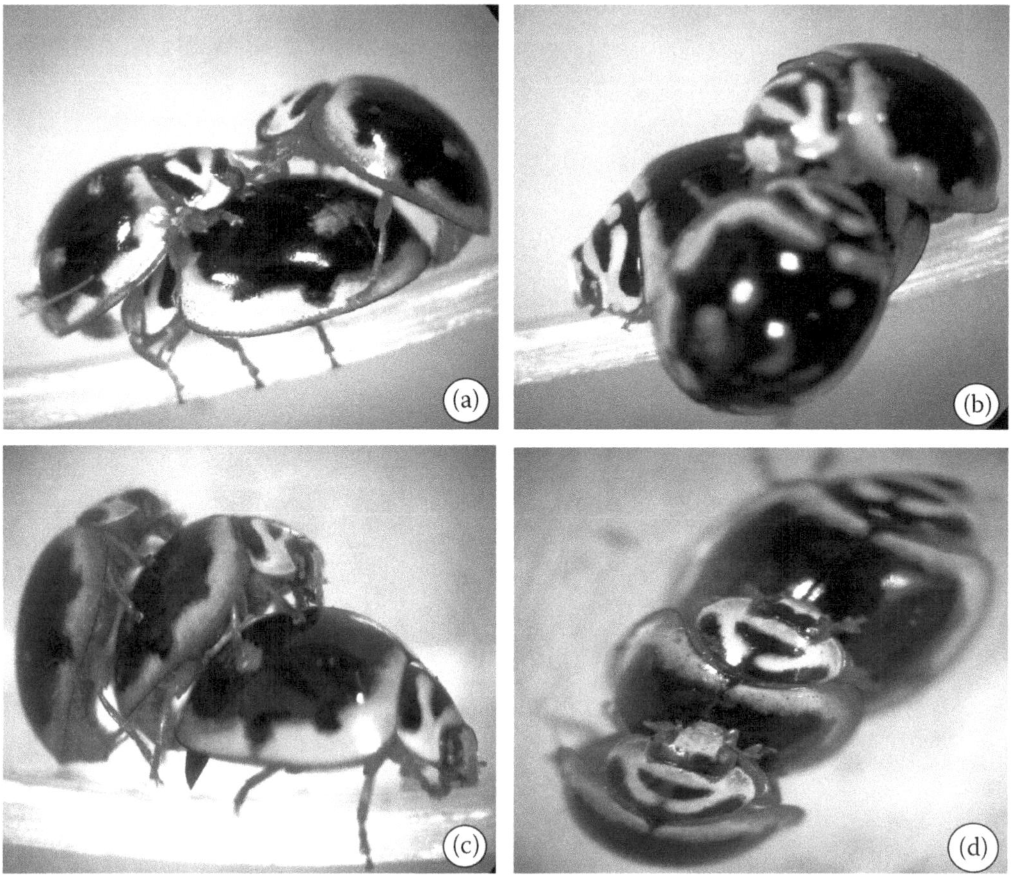

FIGURE 9.11 Mate guarding event (a–d) in zig-zag ladybird beetle, *Menochilus sexmaculatus* (melanic morph) (Chaudhary *et al.*, 2015).

(iii) black and yellow morph with orange-coloured patches on wing, resemblings ageing males, called *lamborni*. Cook *et al.* (1994) demonstrated sexual harassments was lower in *trimeni* morph and least for *lamborni* morph. Similarly, in damselfly *Ischnura elegans*, male-like morph suffered less sexual harassment (Van Gossum *et al.*, 2011).

Female midge *C. melleus* simply walk away and do not raise their abdomen when forced to mate (Linley & Adams, 1972); resistance behaviour is higher in case of smaller males. Habitat change also helps females get away from aggressive males (Smuts & Smuts, 1993).

9.9.3 Resistance/Fighting Back

Mate resistance behaviour is pronounced in insects (Perry *et al.*, 2009; Li *et al.*, 2014; Aldersley & Cator, 2019). This could evolve as a deliberate female strategy where competition is encouraged to choose best male (Westneat & Stewart, 2003; Perry *et al.*, 2009; Dubey *et al.*, 2016,) or via indirect selection; female being inseminated by male capable of overpowering her (Cox & Le Boeuf, 1977; Wiley & Poston, 1996).

However, relative influence of direct and indirect benefits associated with female resistance is poorly understood and it is still controversial as to whether such a scenario can evolve (Kokko *et al.*, 2003). The prevailing view that females benefit from and therefore should encourage extra copulation durations has been questioned (Westneat & Stewart, 2003), raising possibility that females may generally resist extra copulation durations to avoid costs associated with additional copulations.

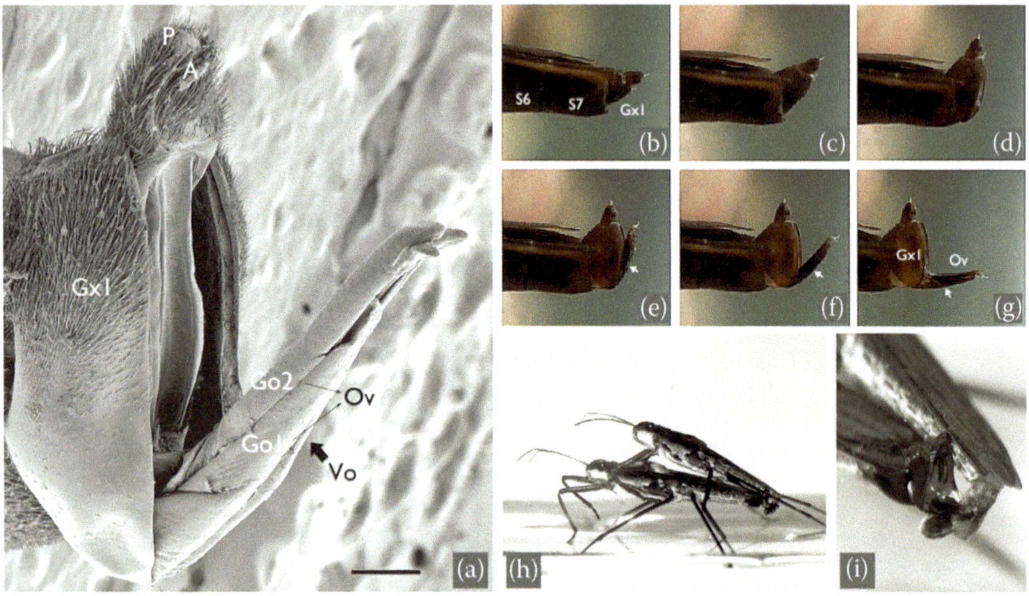

FIGURE 9.12 Anti-grasping traits in female *Gerris gracilicornis*: (a) Scanning electron micrographs of partially inflated genital segment of female *Gerris gracilicornis*, (b) un-inflated gonocoxae 1 in segment 7, (c–f) spreading out of gonocoxae 1 with protruding of ovipositor (gonapophyses), (g) fully inflated female genitalia with protruded ovipositor, (h) copulating pair, and (i) interdigitation of male and female genitalia (where A, anal opening; Gx1, gonocoxa 1; Go1, gonapophysis 1; Go2, gonapophysis 2; Ov, ovipositor; P, proctiger; Vo, vulvar opening, S6, segment 6; S7, segment 7; Ov, ovipositor) (Han & Jablonski, 2009).

FIGURE 9.13 Female kicking behaviour of seed beetle, *Callosobruchus maculatus* (**Picture credit**: Michal Polak, University of Cincinnati).

To counter male coercive traits, resistive behaviour and back fighting are two major strategies employed by many females (Eberhard, 2001; Edvardsson & Canal, 2006; Puniamoorthy *et al.*, 2008; Giesen *et al.*, 2017; Figure 9.13). Several additional female processes, such as ejection of male's sperm, failure to transport them, or flooding of reproductive tract with phagocytes immediately after copulation (Eberhard, 1996) could have same final effects on paternity as prevention of intromission (Gowaty, 1997; Gowaty & Buschhaus, 1998), but they are less likely to have kind of naturally selected payoffs expected from resistance as avoidance.

9.9.4 Acceptance/Submission

When cost incurred by resistance is higher than cost of submitting to mating, females prefer not to struggle (Cordero & Andrés, 2002; Myers *et al.,* 2016). Unfortunately, most studies that deal with submission and acceptance of mate are in higher organisms (Darden *et al.*, 2009).

9.9.5 Protection/Alliances

This is well-documented in a higher group of organisms, where females 'make friendship' with stronger male to get protection from other coercive males (Grueter *et al.*, 2012). Also, females of high rank can create female alliances to fight against aggressive males (Smuts & Smuts, 1993).

9.10 Benefits of Sexual Coercion to Females

A conceivable benefit of sexual coercion can be explained through 'good genes' hypothesis (Kokko, 2001; Edvardsson & Arnqvist, 2006); males able to overcome resistance, must contain good genes (Bergsten & Miller, 2007). By evolving resistance, females may derive benefits such as replenishment of sperm, fertility insurance, seminal feeding, and increased genetic diversity of offspring (Thornhill & Alcock, 1983; Yasui, 1998; Jennions & Petrie, 2000).

9.11 Consequences of Sexual Coercion

When male and females traits were allowed to coevolve under influence of sexual conflict or coercion 'inextricable evolutionary chases' frequently occurred. Parker (1979) suggested that sexually antagonist co-evolution can lead to quick, directionless and random change.

9.11.1 Co-evolutionary Arms Race

Sexual conflict that may lead to sexual coercion by males (Crudgington & Siva-Jothy, 2000) leading to a co-evolutionary arms race via morphological modifications. Arnqvist and Rowe (2002) discuss evolutionary changes in balance of weaponry in insects, like clasping and anticlasping devices, also mating plugs (Sauter *et al.*, 2001), thicker and less delicate genital tracts (Rönn *et al.*, 2007), abdominal spines (Han & Jablonski, 2009, 2010), and behavioural changes (Han & Jablonski, 2009).

9.11.2 Sexual Dimorphism

If optima for a particular traits, for example, mating rate or duration vary between male and female and also there is a pleiotropic genetic association between male and female for it, then there would be strong conflicting selection on alleles at a single locus. Thus, chances of selection for sex limitation and sexual dimorphism would increase (Parker & Partridge, 1998). The female resistance against coercion exerts strong selection pressure on evolution of larger size males (Perry *et al.*, 2009; Dubey *et al.*, 2016). However, there are examples where smaller males achieve higher mating advantages (Gage *et al.*, 2002). Also, in diving beetle, dorsal setose furrows in females and suction cups on male forelegs have evolved due to sexual coercion (Bergsten & Miller, 2007).

9.11.3 Speciation

Sexual conflict can also lead to allopatric (Arnqvist *et al.*, 2000; Gavrilets, 2000) and sometimes sympatric (Gavrilets & Waxman, 2002) divergence. Forced copulation has potential can fuel an arms race would involve strong selection on morphological traits that enhance male's ability to force copulations and female's ability to resist unwanted copulation attempts (Clutton-Brock & Parker, 1995).

The conflict between loci (sex-limited) leads to a change in morphology and/or behaviour which may eventually induce reproductive isolation.

9.12 Conclusions

In most insect species, prime reproductive interest of males is to acquire maximum mating to ensure maximum paternity share, while females are prudent while selecting a mate and show mate preference, creating sexual conflict. Asymmetry in reproductive interest can result in antagonistic coevolution of various reproductive tactics in male and females. Males have evolved with various morphological modifications in form of specialized mating apparatus in males and selective resistance adaptations in females.

When we talk about coercive mating it somehow neglects females' choice for their mate and associated with various costs and risk. Sometimes it becomes very difficult to observe sexual coercion in real for example, in most of cases, it becomes very difficult to distinguish coercive mating from normal one although it has been recorded in many insect species. At all time we cannot see coercive mating as adaptions to sexual coercion it can be related to a specific function in a context other than sexual coercion. For instance, in many species, forceful behaviour of males can be for other sexual rewards like food and shelter. In conclusion, by studying recent advances in sexual conflict, we can think about consequences of sexual coercion that may increase chance of coevolution between male and female and when we should imagine about adaptations that cause least harm to both sexes.

Acknowledgements

S. S. acknowledges an Agricultural Research Organization, The Volcani Center, Rishon Lezion, Israel for providing a postdoctoral fellowship.

REFERENCES

Albo, M.J. & Costa, F.G. (2010). Nuptial gift-giving behaviour and male mating effort in Neotropical spider *Paratrechalea ornata* (Trechaleidae). *Animal Behaviour*, *79*(5), 1031–1036.

Alcock, J. (1994). Postinsemination associations between males and females in insects: mate-guarding hypothesis. *Annual Review of Entomology*, *39*, 1–21.

Aldersley, A. & Cator, L.J. (2019). Female resistance and harmonic convergence influence male mating success in *Aedes aegypti*. *Scientific Reports*, *9*(1), 1–12.

Allen, G.R., Kazmer, D.J. & Luck, R.F. (1994). Post-copulatory male behaviour, sperm precedence and multiple mating in a solitary parasitoid wasp. *Animal Behaviour*, *48*(3), 635–644.

Arnqvist, G. & Nilsson, T. (2000). The evolution of polyandry: multiple mating and female fitness in insects. *Animal behaviour*, *60*(2), 145–164.

Arnqvist, G. & Rowe, L. (1995). Sexual conflict and arms races between sexes: a morphological adaptation for control of mating in a female insect. *Proceedings of Royal Society of London, Series B*, *261*, 123–127.

Arnqvist, G. & Rowe, L. (2002). Antagonistic coevolution between sexes in a group of insects. *Nature*, *415*, 787.

Arnqvist, G. & Rowe, L. (2005). *Sexual conflict*. Princeton: Princeton University Press.

Arnqvist, G. (1997). The evolution of water strider mating systems: causes and consequences of sexual conflicts. In: C. Choe & B.J. Crespi (Eds.), *The Evolution of Mating Systems in Insects and Arachnids*. pp. 146–163. Cambridge: Cambridge University Press.

Arnqvist, G., Edvardsson, M., Friberg, U. & Nilsson, T. (2000). Sexual conflict promotes speciation in insects. *Proceedings of National Academy of Sciences of USA*, *97*(19), 10460–10464.

Avila, F.W., Sirot, L.K., LaFlamme, B.A., Rubinstein, C.D. & Wolfner, M.F. (2011). Insect seminal fluid proteins: identification and function. *Annual Review of Entomology*, *56*, 21– 40.

Baniel, A., Cowlishaw, G. & Huchard, E. (2017). Male violence and sexual intimidation in a wild primate society. *Current Biology*, *27*(14), 2163–2168.
Bergsten, J. & Miller, K.B. (2007). Phylogeny of diving beetles reveals a coevolutionary arms race between sexes. *PLoS One*, *2*, e522
Blanckenhorn, W.U., Hosken, D.J., Martin, O.Y., Reim, C., Teuschl, Y. & Ward, P.I. (2002). The costs of copulating in dung fly *Sepsis cynipsea*. *Behavioral Ecology*, *13*, 353–358.
Boggs, C.L. (1995). Male nuptial gifts: phenotypic consequences and evolutionary implications. In: S.R. Leather & J. Hardie J (Eds.), *Insect Reproduction* (pp. 215–242). Boca Raton: CRC.
Borgia, G. (1981). Mate selection in the fly Scatophaga stercoraria: female choice in a male-controlled system. *Animal Behaviour*, *29*(1), 71–80.
Brown, W.D., Crespi, B.J. & Choe, J.C. (1997). Sexual conflict and evolution of mating systems. In: J.C. Choe & B.J. Crespi (Eds.), *The Evolution of Mating Systems in Insects and Arachnids* (pp. 352–377). Cambridge: Cambridge University Press.
Burdfield-Steel, E.R. & Shuker, D.M. (2014). Mate-guarding in a promiscuous insect: species discrimination influences context-dependent behaviour. *Evolutionary Ecology*, *28*(6), 1031–1042.
Bussière, L.F., Hunt, J., Jennions, M.D. & Brooks, R. (2006). Sexual conflict and cryptic female choice in black field cricket, *Teleogryllus commodus*. *Evolution*, *60*, 792– 800.
Chapman, T. & Partridge, L. (1996). Female fitness in *Drosophila melanogaster*: an interaction between effect of nutrition and of encounter rate with males. *Proceedings of Royal Society of London. Series B: Biological Sciences*, *263*(1371), 755–759.
Chapman, T., Arnqvist, G., Bangham, J. & Rowe, L. (2003). Sexual conflict. *Trends in Ecology & Evolution*, *18*, 41–47.
Chaudhary, D.D., Mishra, G. & Omkar (2015). Prolonged matings in a ladybird, *Menochilus sexmaculatus*: a mate guarding mechanism?. *Journal of Asia-Pacific Entomology*, *18*(3), 453–458.
Cheng, K.M., Burns, J.T. & McKinney, F.R.A.N.K. (1982). Forced copulation in captive mallards (*Anas platyrhynchos*): II. Temporal factors. *Animal Behaviour*, *30*(3), 695–699.
Chippindale, A.K., Gibson, J.R. & Rice, W.R. (2001). Negative genetic correlation for adult fitness between sexes reveals ontogenetic conflict in *Drosophila*. *Proceedings of National Academy of Sciences of USA*, *98*(4), 1671–1675.
Cleavall, L. (2009). *Description of Thermonectus nigrofasciatus and Rhantus binotatus (Coleoptera: Dytiscidae) Mating Behavior*. Albuquerque: University of New Mexico, p 34.
Clutton-Brock, T.H. & Parker, G.A. (1992). Potential reproductive rates and operation of sexual selection. *The Quarterly Review of Biology*, *67*(4), 437–456.
Clutton-Brock, T.H. & Parker, G.A. (1995). Sexual coercion in animal societies. *Animal Behaviour*, *49*(5), 1345–1365.
Clutton-Brock, T.H. (1989). Review lecture: mammalian mating systems. *Proceedings of Royal Society of London. B. Biological Sciences*, *236*(1285), 339–372.
Clutton-Brock, T.H., Guinness, F.E. & Albon, S.D. (1982). *Red Deer: Behavior and Ecology of Two Sexes*. Chicago: University of Chicago Press.
Clutton-Brock, T.H., Price, O.F. & MacColl, A.C. (1992). Mate retention, harassment, and evolution of ungulate leks. *Behavioral Ecology*, *3*(3), 234–242.
Cook, S.E., Vernon, J.G., Bateson, M. & Guilford, T. (1994). Mate choice in polymorphic African swallowtail butterfly, *Papilio dardanus*: male-like females may avoid sexual harassment. *Animal Behaviour*, *47*(2), 389–397.
Cordero, A. & Andrés, J.A. (2002). Male coercion and convenience polyandry in a calopterygid damselfly. *Journal of Insect Science*, *2*(1), 14
Cox, C.R. & Le Boeuf, B.J. (1977). Female incitation of male competition: a mechanism in sexual selection. *The American Naturalist*, *111*(978), 317–335.
Crudgington, H.S. & Siva-Jothy, M.T. (2000). Genital damage, kicking and early death. *Nature*, *407*(6806), 855–856.
Cummings, M.E. (2018). Sexual conflict and sexually dimorphic cognition—reviewing their relationship in poeciliid fishes. *Behavioral Ecology and Sociobiology*, *72*(4), 73
Darden, S.K., James, R., Ramnarine, I.W. & Croft, D.P. (2009). Social implications of battle of sexes: sexual harassment disrupts female sociality and social recognition. *Proceedings of Royal Society B*, *276*(1667), 2651–2656.

Darwin, C. (1871). *The Descent of Man, and Selection in Relation to Sex*. London: Murray, p. 415.

Deinert, E.I. (2003). Mate location and competition for mates in a pupal mating butterfly. In: C.L. Boggs, W.B. Watt & P.R. Ehrlich (Eds.), *Butterflies: Ecology and Evolution Taking Flight* (pp. 91–108). Chicago, IL: University of Chicago Press.

den Hollander, M. & Gwynne, D.T. (2009). Female fitness consequences of male harassment and copulation in seed beetles, *Callosobruchus maculatus*. *Animal Behaviour*, *78*(5), 1061–1070.

Dodson, G.N., Morris, G.K. & Gwynne, D.T. (1983). Mating behavior of primitive orthopteran genus *Cyphoderris* (Haglidae). In: D.T. Gwynne & G.K. Morris (Eds.), *Orthopteran Mating Systems: Sexual Competition in a Diverse Group Of Insects* (pp. 305–318). Boulder: Westview Press.

Dubey, A., Omkar & Mishra, G. (2016). Adult body size drives sexual selection mutually in ladybird, *Menochilus sexmaculatus* (Coleoptera: Coccinellidae). *Acta Entomologica Sinica*, *59*(2), 209–218.

Eberhard W.G. (2001). Species-specific genitalic copulatory courtship in sepsid flies (Diptera, Sepsidae, Microsepsis) and theories of genitalic evolution. *Evolution*, *55*, 93–102.

Eberhard, W.G. (1996). *Female Control: Sexual Selection by Cryptic Female Choice*. Princeton University Press, Princeton, N.J.

Eberhard, W.G. (2002). Physical restraint or stimulation? The function (s) of modified front legs of male *Archisepsis diversiformis* (Diptera, Sepsidae). *Journal of Insect Behavior*, *15*(6), 831–850.

Edvardsson, M. & Arnqvist, G. (2006). No apparent indirect genetic benefits to female red flour beetles preferring males with intense copulatory courtship. *Behavior Genetics*, *36*(5), 775–782.

Edvardsson, M. & Canal, D. (2006). The effects of copulation duration in bruchid beetle *Callosobruchus maculatus*. *Behavioral Ecology*, *17*(3), 430–434.

Emlen, S.T. & Oring, L.W. (1977). Ecology, sexual selection, and evolution of mating systems. *Science*, *197*(4300), 215–223.

Engqvist, L. & Sauer, K.P. (2003). Determinants of sperm transfer in scorpionfly *Panorpa cognata*: male variation, female condition and copulation duration. *Journal of Evolutionary Biology*, *16*(6), 1196–1204.

Fairbairn, D.J. (1993). Costs of loading associated with mate-carrying in waterstrider, *Aquarius remigis*. *Behavioral Ecology*, *4*(3), 224–231.

Fairbairn, D.J., Vermette, R., Kapoor, N.N. & Zahiri, N. (2003). Functional morphology of sexually selected gentalia in water strider *Aquarius remigis*. *Canadian Journal of Zoology*, *81*(3), 400–413.

Field, S.A. & Keller, M.A. (1993). Alternative mating tactics and female mimicry as post-copulatory mate-guarding behaviour in parasitic wasp *Cotesia rubecula*. *Animal Behaviour*, *46*(6), 1183–1189.

Gage, M.J., Parker, G.A., Nylin, S. & Wiklund, C. (2002). Sexual selection and speciation in mammals, butterflies and spiders. *Proceedings of Royal Society of London B*, *269*(1507), 2309–2316.

Gavrilets, S. & Waxman, D. (2002). Sympatric speciation by sexual conflict. *Proceedings of National Academy of Sciences of USA*, *99*(16), 10533–10538.

Gavrilets, S. (2000). Rapid evolution of reproductive barriers driven by sexual conflict. *Nature*, *403*, 886–889.

Giesen, A., Blanckenhorn, W.U. & Schäfer, M.A. (2017). Behavioural mechanisms of reproductive isolation between two hybridizing dung fly species. *Animal Behaviour*, *132*, 155–166.

Gowaty, P.A. & Buschhaus, N. (1998). Ultimate causation of aggressive and forced copulation in birds: female resistance, CODE hypothesis, and social monogamy. *American Zoologist*, *38*(1), 207–225.

Gowaty, P.A. (1997). Principles of females' perspectives in avian behavioral ecology. *Journal of Avian Biology*, *28*, 95–102.

Grueter, C.C., Chapais, B. & Zinner, D. (2012). Evolution of multilevel social systems in nonhuman primates and humans. *International Journal of Primatology*, *33*(5), 1002–1037.

Gwynne, D.T. (2001). *Katydids and Bush Crickets: Reproductive Behavior and Evolution of Tettigoniidae*. Ithaca, NY: Cornell University Press, p. 317.

Haley, E.L. & Gray, D.A. (2012). Mating behavior and dual-purpose armaments in a camel cricket. *Ethology*, *118*(1), 49–56.

Han, C.S. & Jablonski, P.G. (2009). Female genitalia concealment promotes intimate male courtship in a water strider. *PLoS One*, *4*, e5793.

Han, C.S., Jablonski, P.G. & Brooks, R.C. (2015). Intimidating courtship and sex differences in predation risk lead to sex-specific behavioural syndromes. *Animal Behaviour*, *109*, 177–185.

Heller, K.G. & Helversen, D.V. (1991). Operational sex ratio and individual mating frequencies in two bushcricket species (Orthoptera, Tettigonioidea, Poecilimon). *Ethology*, *89*, 211– 228.

Hinn, J.C. & Niedzlek-Feaver, M. (2001). Spermatophore number transferred increases linearly with copulation duration in *Melanoplus differentialis* (Orthoptera: Acrididae). *Journal of Orthoptera Research, 10*(2), 147–152.
Hosken, D.J., Martin, O.Y., Born, J. & Huber, F. (2003). Sexual conflict in Sepsis cynipsea: female reluctance, fertility and mate choice. *Journal of Evolutionary Biology, 16*(3), 485–490.
Houseman, J.G. & Downe, A.E.R. (1986). Methods of measuring blood meal size and proteinase activity for determining effects of mated state on digestive processes of female Aedes aegypti (L.)(Diptera: Culicidae). *The Canadian Entomologist, 118*(3), 241–248.
Hurst, G.D., Sharpe, R.G., Broomfield, A.H., Walker, L.E., Majerus, T.M., Zakharov, I.A. & Majerus, M.E. (1995). Sexually transmitted disease in a promiscuous insect, *Adalia bipunctata. Ecological Entomology, 20*, 230–236.
Howe, W. (2008). Variation in both male and female secondary sexual characters in the diving beetle, *Agabus bipustulatus* (L.) (Coleoptera: Dytiscidae). *The Plymouth Student Scientist, 1*(2), 56–86.
Jennions, M.D., Moller, A.P., & Petrie, M. (2001). Sexually selected traits and adult survival: a meta-analysis. *The Quarterly Review of Biology, 76*(1), 3–36.
Johnstone, R.A. & Keller, L. (2000). How males can gain by harming their mates: sexual conflict, seminal toxins, and cost of mating. *The American Naturalist, 156*(4), 368–377.
Jormalainen, V., Merilaita, S. & Härdling, R. (2000). Dynamics of intersexual conflict over precopulatory mate guarding in two populations of isopod *Idotea baltica. Animal Behaviour, 60*(1), 85–93.
Khila, A., Abouheif, E. & Rowe, L. (2012). Function, developmental genetics, and fitness consequences of a sexually antagonistic trait. *Science, 336*(6081), 585–589
Kokko, H. (2001). Fisherian and "good genes" benefits of mate choice: how (not) to distinguish between them. *Ecology Letters, 4*(4), 322–326.
Kokko, H., Brooks, R., Jennions, M.D. & Morley, J. (2003). The evolution of mate choice and mating biases. *Proceedings of Royal Society of London. Series B: Biological Sciences, 270*(1515), 653–664.
Kvarnemo, C. & Ahnesjo, I. (1996). The dynamics of operational sex ratios and competition for mates. *Trends in Ecology & Evolution, 11*(10), 404–408.
Lewis, S.M. & Cratsley, C.K. (2008). Flash signal evolution, mate choice, and predation in fireflies. *Annual Review of Entomology, 53*, 293–321.
Li, X.W., Jiang, H.X., Zhang, X.C., Shelton, A.M. & Feng, J.N. (2014). Post-mating interactions and their effects on fitness of female and male *Echinothrips americanus* (Thysanoptera: Thripidae), a new insect pest in China. *PLoS One, 9*(1), e87725.
Linley, J.R. & Adams, G.M. (1972). A study of mating behaviour of *Culicoides melleus* (Coquillett) (Diptera: Ceratopogonidae). *Transactions of Royal Entomological Society of London, 124*(2), 81–121.
Lizée, M.H., Barascud, B., Cornec, J.P. & Sreng, L. (2017). Courtship and mating behavior of cockroach *Oxyhaloa deusta* [Thunberg, 1784] (Blaberidae, Oxyhaloinae): Attraction bioassays and morphology of pheromone sources. *Journal of Insect Behavior, 30*(6), 674–694.
Lung, O., Tram, U., Finnerty, C.M., Eipper-Mains, M.A., Kalb, J.M., & Wolfner, M.F. (2002). The Drosophila melanogaster seminal fluid protein Acp62F is a protease inhibitor that is toxic upon ectopic expression. *Genetics, 160*(1), 211–224.
Majerus, M.E.N. (1994). *Ladybirds*. Harper Collins, UK, p. 367.
Manning, A. (1967). The control of sexual receptivity in female *Drosophila. Animal Behaviour, 15*, 239–250.
Marchesan, M. (2002). Operational sex ratio and breeding strategies in Feral Pigeon Columba livia. *Ardea, 90*(2), 249–257.
Martens, A. & Rehfeldt, G. (1989). Female aggregation in Platycypha caligata (Odonata: Chlorocyphidae): a tactic to evade male interference during oviposition. *Animal Behaviour, 38*(3), 369–374.
Markow, T.A. (2002). Perspective: female remating, operational sex ratio, and arena of sexual selection in *Drosophila* species. *Evolution, 56*(9), 1725–1734.
McKinney, F., Derrickson, S.R. & Mineau, P. (1983). Forced copulation in water fowl. *Behaviour, 86*, 250–294.
Miller, K.B. (2003). The phylogeny of diving beetles (Coleoptera: Dytiscidae) and the evolution of sexual conflict.*Biological Journal of the Linnean Society, 79*(3), 359–388.
Miller, K.B. & Bergsten, J. (2014). Predaceous diving beetle sexual systems. In: D.A. Yee (Ed.), *Ecology, Systematics, and Natural History of Predaceous Diving Beetles (Coleoptera: Dytiscidae)* (pp. 199–233). Dordrecht: Springer.

Moyer, K.E. (1968). Kinds of aggression and their physiological basis. *Communications in Behavioral Biology*, *2*(2), 65–87.
Mühlhäuser, C. & Blanckenhorn, W.U. (2002). The costs of avoiding matings in dung fly *Sepsis cynipsea*. *Behavioral Ecology*, *13*(3), 359–365.
Muller, M.N., Kahlenberg, S.M., Emery Thompson, M. & Wrangham, R.W. (2007). Male coercion and costs of promiscuous mating for female chimpanzees. *Proceedings of Royal Society B: Biological Sciences*, *274*(1612), 1009–1014.
Myers, S.S., Buckley, T.R., & Holwell, G.I. (2016). Male genital claspers influence female mate acceptance in the stick insect *Clitarchus hookeri*. *Behavioral Ecology and Sociobiology*, *70*(9), 1547–1556.
Oh, R.J. (1979). Repeated copulation in brown planthopper, *Nilaparvata lugens* Stål (Homoptera; Delphacidae). *Ecological Entomology*, *4*(4), 345–353.
Omkar, Sahu J. & Kumar, G. (2013). Age specific mating incidence and reproductive behavior of ladybird beetle, *Anegleis cardoni* (Weise) (Coleoptera: Coccinellidae). *Journal of Asia-Pacific Entomology*, *16*(3), 263–268.
Omkar, Singh, S.K. & Singh, K. (2006). Effect of age on reproductive attributes of an aphidophagous ladybird, *Cheilomenes sexmaculata*. *Insect Science*, *13*(4), 301–308.
Parker, G.A. (1970). Sperm competition and its evolutionary consequences in the insects. *Biological Reviews*, *45*(4), 525–567.
Parker, G.A. (1979). Sexual selection and sexual conflict. In: M.S. Blum & N.A. Blum (Eds.), *Sexual Selection and Reproductive Competition in Insects* (pp. 123–166). New York: Academic Press.
Parker, G.A. & Partridge, L. (1998). Sexual conflict and speciation. *Philosophical Transactions of Royal Society of London. Series B: Biological Sciences*, *353*(1366), 261–274.
Parker, G.A. (2006). Sexual conflict over mating and fertilization: an overview. *Philosophical Transactions of Royal Society B: Biological Sciences*, *361*(1466), 235–259.
Perry, J.C., Sharpe, D.M. & Rowe, L. (2009). Condition-dependent female remating resistance generates sexual selection on male size in a ladybird beetle. *Animal Behaviour*, *77*(3), 743–748.
Perry, J.C. & Rowe, L. (2012). Sexual conflict and antagonistic coevolution across water strider populations. *Evolution*, *66*, 544–557.
Poiani, A. (2006). Complexity of seminal fluid: a review. *Behavioral Ecology and Sociobiology*, *60*, 289– 310.
Polak, M., Wolf, L.L., Starmer, W.T. & Barker, J.S.F. (2001). Function of mating plug in *Drosophila hibisci* Bock. *Behavioral Ecology and Sociobiology*, *49*, 196–205.
Poulin, R. & Morand, S. (2000). Testes size, body size and male–male competition in acanthocephalan parasites. *Journal of Zoology*, *250*(4), 551–558.
Puniamoorthy, N., Su, K.F.Y. & Meier, R. (2008). Bending for love: losses and gains of sexual dimorphisms are strictly correlated with changes in mounting position of sepsid flies (Sepsidae: Diptera). *BMC Evolutionary Biology*, *8*(1), 155.
Reale, D., Bousses, P. & Chapuis, J.L. (1996). Female-biased mortality induced by male sexual harassment in a feral sheep population. *Canadian Journal of Zoology*, *74*(10), 1812–1818.
Reinhardt, K., Naylor, R. & Siva–Jothy, M.T. (2003). Reducing a cost of traumatic insemination: female bedbugs evolve a unique organ. *Proceedings of Royal Society of London B*, *270*(1531), 2371–2375.
Rentz, D.C. (1972). The lock and key as an isolating mechanism in Katydids: the lock-and-key theory may explain one of means nature uses to assure integrity of species in certain katydids. *American Scientist*, *60*(6), 750–755.
Rhen, T. (2000) Sex-limited mutations and evolution of sexual dimorphism. *Evolution*, *54*, 37–43.
Ribeiro, C. & Dickson, B.J. (2010). Sex peptide receptor and neuronal TOR/S6K signaling modulate nutrient balancing in *Drosophila*. *Current Biology*, *20*(11), 1000–1005.
Ronkainen, K., Kaitala, A. & Huttunen, R. (2005). The effect of abdominal spines on female mating frequency and fecundity in a water strider. *Journal of Insect Behavior*, *18*(5), 619–631.
Rönn, J.L. & Hotzy, C. (2012). Do longer genital spines in male seed beetles function as better anchors during mating?. *Animal Behaviour*, *83*(1), 75–79.
Rönn, J., Katvala, M. & Arnqvist, G. (2007). Coevolution between harmful male genitalia and female resistance in seed beetles. *Proceedings of National Academy of Sciences of USA*, *104*(26), 10921–10925.
Rossi, B.H., Nonacs, P. & Pitts-Singer, T.L. (2010). Sexual harassment by males reduces female fecundity in alfalfa leaf cutting bee, *Megachile rotundata*. *Animal Behaviour*, *79*:165–171.

Rowe, L. & Arnqvist, G. (2012). Sexual selection and evolution of genital shape and complexity in water striders. *Evolution: International Journal of Organic Evolution*, *66*(1), 40–54.
Rowe, L. (1992). Convenience polyandry in a waterstrider: foraging conflicts and female control of copulation frequency and guarding duration. *Animal Behavior*, *44*, 189–202.
Rowe, L. (1994). The costs of mating and mate choice in water striders. *Animal Behaviour*, *48*, 1049–1056.
Sakurai, G., & Kasuya, E. (2008). The costs of harassment in the adzuki bean beetle. *Animal Behaviour*, *75*(4), 1367–1373.
Sakaluk, S.K., Bangert, P.J., Eggert, A.K., Gack, C. & Swanson, L.V. (1995). The gin trap as a device facilitating coercive mating in sagebrush crickets. *Proceedings of Royal Society of London. Series B: Biological Sciences*, *261*(1360), 65–71.
Sauter, A., Brown, M.J., Baer, B. & Schmid-Hempel, P. (2001). Males of social insects can prevent queens from multiple mating. *Proceedings of Royal Society of London B*, *268*(1475), 1449–1454.
Saxena, S., Mishra, G. & Omkar (2020). Operational sex ratio and paternal age sway mating and reproductive performance in *Menochilus sexmaculatus* (Coleoptera: Coccinellidae). *The Canadian Entomologist*, *152*(3), 298–310.
Seeley, C. & Dukas, R. (2011). Teneral matings in fruit flies: male coercion and female response. *Animal Behaviour*, *81*(3), 595–601.
Shuker, D.M. & Day, T.H. (2001). The repeatability of a sexual conflict over mating. *Animal Behaviour*, *61*(4), 755–762.
Simmons, L.W. (2001). *Sperm Competition and its Evolutionary Consequences in Insects*. Princeton: Princeton University Press.
Singh, S., Mishra, G. & Omkar (2016). Perceived prey quantity modulates oviposition in ladybird *Menochilus sexmaculatus*. *Journal of Ethology*, *34*(1), 59–64.
Singh, S., Mishra, G. & Omkar (2019). Oviposition in aphidophagous ladybirds: effect of prey availability and conspecific egg presence. *International Journal of Tropical Insect Science*, *39*(2), 107–114.
Siva-Jothy, M.T., Zhong, W., Naylor, R., Heaton, L., Hentley, W., & Harney, E. (2019). Female bed bugs (*Cimex lectularius* L) anticipate the immunological consequences of traumatic insemination via feeding cues.*Proceedings of the National Academy of Sciences*, *116*(29), 14682–14687.
Smith, C.C. (2007). Independent effects of male and female density on sexual harassment, female fitness, and male competition for mates in western mosquitofish *Gambusia affinis*. *Behavioral Ecology and Sociobiology*, *61*(9), 1349–1358.
Smith, D.C. & Prokopy, R.J. (1980). Mating behavior of *Rhagoletis pomonella* (Diptera: Tephritidae) VI. Site of early-season encounters. *The Canadian Entomologist*, *112*(6), 585–590.
Smuts, B.B. & Smuts, R.W. (1993). Male aggression and sexual coercion of females in nonhuman primates and other mammals: evidence and theoretical implications. *Advances in Study of Behavior*, *22*(22), 1–63.
Smuts, B.B. (1986). Gender, aggression and influence. In: B.B. Smuts, D.L. Cheney, R.M. Seyfarth, W.R. Wrangham & T.T. Struhsaker (Eds.), *Primate Societies* (pp. 400–412). Chicago: University of Chicago Press.
Sparkes, T.C., Keogh, D.P. & Pary, R.A. (1996). Energetic costs of mate guarding behavior in male stream-dwelling isopods. *Oecologia*, *106*(2), 166–171.
Stone, G.N. (1995). Female foraging responses to sexual harassment in solitary bee *Anthophora plumipes*. *Animal Behaviour*, *50*(2), 405–412.
Stutt, A.D. & Siva-Jothy, M.T. (2001). Traumatic insemination and sexual conflict in bed bug *Cimex lectularius*. *Proceedings of National Academy of Sciences of USA*, *98*(10), 5683–5687.
Svensson, B.G., Petersson, E. & Frisk, M. (1990). Nuptial gift size prolongs copulation duration in dance fly *Empis borealis*. *Ecological Entomology*, *15*(2), 225–229.
Takahashi, Y. & Watanabe, M. (2010). Female reproductive success is affected by selective male harassment in damselfly *Ischnura senegalensis*. *Animal Behaviour*, *79*, 211–216.
Thornhill, R. & Alcock, J. (1983). *The Evolution of Insect Mating Systems*. Harvard University Press.
Thornhill, R. (1981). Panorpa (Mecoptera: Panorpidae) scorpionflies: systems for understanding resource-defense polygyny and alternative male reproductive efforts. *Annual Review of Ecology and Systematics*, *12*(1), 355–386.
Trivers, R.L. (1972). Parental investment and sexual selection. In: *Sexual Selection and Descent of Man 1871–1971*. Chicago: Aldine, p. 136–179.

Vahed, K. & Carron, G. (2008). Comparison of forced mating behaviour in four taxa of *Aonconotus*, Alpine bushcricket. *Journal of Zoology*, *276*(3), 313–321.
Vahed, K. (1998). The function of nuptial feeding in insects: a review of empirical studies. *Biological Reviews*, *73*(1), 43–78.
Vahed, K. (2007). All that glisters is not gold: sensory bias, sexual conflict and nuptial feeding in insects and spiders. *Ethology*, *113*(2), 105–127.
Van Gossum, H., Bots, J., Van Heusden, J., Hammers, M., Huyghe, K. & Morehouse, N.I. (2011). Reflectance spectra and mating patterns support intraspecific mimicry in colour polymorphic damselfly *Ischnura elegans*. *Evolutionary Ecology*, *25*(1), 139–154.
Wagner, W.E. (1998). Measuring female mating preferences. *Animal Behaviour*, *55*(4), 1029–1042.
Wang, Q., Yang, L. & Hedderley, D. (2008). Function of prolonged copulation in *Nysius huttoni* White (Heteroptera: Lygaeidae) under male-biased sex ratio and high population density. *Journal of Insect Behavior*, *21*(2), 89–99.
Watson, P.J., Arnqvist, G. & Stallman, R.R. (1998). Sexual conflict and energetic costs of mating and mate choice in water striders. *The American Naturalist*, *151*, 46–58.
Westneat, D.F. & Stewart, I.R. (2003). Extra-pair paternity in birds: causes, correlates, and conflict. *Annual Review of Ecology, Evolution, and Systematics*, *34*(1), 365–396.
Wewalka, G. (2000). Taxonomic revision of *Allopachria* (Coleoptera: Dytiscidae). *Entomological Problems*, *31*(2), 97–128.
Wiklund, C. & Kaitala, A. (1995). Sexual selection for large male size in a polyandrous butterfly: effect of body-size on male versus female reproductive success in *Pieris napi*. *Behavioral Ecology*, *6*, 6–13.
Wiley, R.H. & Poston, J. (1996). Indirect mate choice, competition for mates, and coevolution of sexes. *Evolution*, *50*, 1371–1381.
Yasui, Y. (1998). The "genetic benefits" of female multiple mating reconsidered. *Trends in Ecology & Evolution*, *13*, 246–250.
Zeiss, C., Martens, A. & Rolff, J. (1999). Male mate guarding increases females' predation risk? A case study on tandem oviposition in the damselfly *Coenagrion puella* (Insecta: Odonata). *Canadian Journal of Zoology*, *77*(6), 1013–1016.

10

Sperm Competition

Adolfo Cordero-Rivera
Universidade de Vigo, ECOEVO Lab, Escola de Enxeñaría Forestal, Campus A Xunqueira, 36005 Pontevedra, Spain

CONTENTS

10.1 Introduction

Organisms are continuously interacting using a diverse array of mechanisms, some of them very evident, others subtle, to obtain the resources needed for reproduction. Sexual reproduction is prevalent in Nature, because parthenogenesis needs to surpass several cytological, genomic and even ethological constraints to evolve (Engelstädter 2008), and once sexual reproduction is prevalent, the evolution of parthenogenesis seems geographically restricted to areas without the presence of sexual populations (Cuéllar 1977). Therefore, sexual reproduction and sexual selection are almost ubiquitous in insects.

Historically, sexual selection was defined as the competition between males to mate with selective females (Darwin 1871). Bateman (1948) reasoned that male reproductive success is limited by the number of mates, whereas female reproductive success is limited by resources. Although this appears reasonable, repeated mating seems needed for female insects to realise their full fecundity (Ridley 1988), particularly when males provide nuptial gifts. Since the last third of the 20th century, a new approach to sexual selection extended competition between mates to processes that occur during and after copulation, opening the field of postcopulatory sexual selection (Birkhead & Pizzari 2002). Mirroring the historical development of classic sexual selection, the attention was first focused on male-male competition after copulation, i.e. sperm competition or 'the competition within a single female between the sperm from two or more males for the fertilization of the ova' (Parker 1970a). This phenomenon explains a wide array of adaptations, particularly in behaviour (Alcock 1994). The early reviews (e.g. Smith, 1984) focused in sperm numbers, sperm size, motility, and the patterns of female and male re-mating, & highlighted the need for an integrative approach linking physiology and behavioural ecology (Gomendio & Roldán 1993). Already by the end of 1990, the field was firmly established, and then further included also evidence for sexual conflict, mating costs and benefits (Danielsson 1998).

DOI: 10.1201/9781003043195-10

Simultaneously, the idea that females were modulating postcopulatory competition became established, again, using insects as model organisms (Thornhill 1983; Eberhard 1996). The field of "cryptic female choice" emerged as the equivalent of the Darwinian ideas of females choosing among males during precopulatory displays. Sexual Selection was organised in a dichotomous table, with intrasexual versus intersexual competition and precopulatory versus postcopulatory processes, giving four alternatives: male-male competition, female choice, sperm competition and cryptic female choice (Eberhard 1998).

The experimental evidence and theoretical considerations about sperm competition have been reviewed in detail, both in animals in general (Birkhead & Möller 1998; Cordero-Rivera & Córdoba-Aguilar 2016) and in insects in particular (Danielsson 1998; Simmons and Siva-Jothy 1998; Simmons 2001, 2014a). Furthermore, the interactions between sperm competition and other evolutionary phenomena, particularly in relation to genital evolution (Eberhard 1985; Cordero-Rivera & Córdoba-Aguilar 2010; Leonard & Córdoba-Aguilar 2010; Simmons 2014b; Brennan & Prum 2015), and sexual conflict (Parker 1979; Cordero & Eberhard 2003; Arnqvist & Rowe 2005; Cordero-Rivera & Rivas-Torres 2019) have been also treated in detail in several review papers and books.

In this chapter, I review the literature on sperm competition in insects, with the aim to investigate taxonomic biases. I follow an ecological approach, trying to unify evolutionary and ecological terminologies, to show that sperm competition is a particular case of ecological competition, and shares similar processes and principles.

10.2 A Taxonomic Overview of Sperm Competition in Insects

The particular anatomy of female insects, which store sperm, and the temporal separation of mating and fertilization, were the inspiring facts for the original development of the concept of sperm competition (Parker 1970a). The dung fly (Figure 10.1a), with its fascinating mating behaviour, served to establish the field, thanks to the pioneering work of Geoff Parker (e.g. Parker 1970b). The odonates (Figure 10.1b), were the first animal where sperm removal by male genitalia, from inside the female sperm storage organs, was discovered (Waage 1979). Insects are the most diverse animal class, with over one million described species (Whitfield 1995), and therefore it is logical that sperm competition mechanisms are also varied, and much remains still to be discovered.

I searched the Web of Science for papers containing 'sperm*' in title, and each of insect orders. The number of relevant references is about 1800. However, when also 'competition' is included in the search string, about 300 references remain. Figure 10.2 shows that, in comparison with their species richness, some orders are well studied, particularly Orthoptera, Lepidoptera, and Odonata. For many orders (Diplura, Protura, Archaeognatha, Zygentoma, Ephemeroptera, Plecoptera, Dermaptera, Zoraptera, Embioptera, Grylloblattodea, Mantophasmatodea, Psocodea, Thysanoptera, Neuropterida, Siphonaptera, Trichoptera), I could not find papers on sperm competition, and in other orders there are only 1–3 references (e.g. Collembola, Phasmatodea, Mantodea, Blattodea, Isoptera).

Given this biased taxonomic approach, there are many gaps in our knowledge. Does sperm competition occur in a short-lived group like mayflies, which apparently mate only once? We do not know, but the fact that female mayflies lack a spermatheca, points towards a negative answer. The detailed analyses of mayfly swarms (Figure 10.3) by Harker (1992) indicate that males may mate several times, and that larger males are more successful, but give no light as to whether females can mate multiple times, despite some egg-less females being captured in mating swarms. Similarly, another short-lived group, the stoneflies, is understudied, and no information about females mating more than once is available, but the fact that males show no interest in females after mating points to the absence of sperm competition (Yoshimura 2008). On the other extreme in relation to adult longevity, queens of social insects may live for years, but mate early in their life. In the social Hymenoptera, sperm competition is widespread (Crozier & Fjerdingstad 2001), although many

FIGURE 10.1 Some insects are prominent in sperm competition research, due to their role as model species, like (a) the dung fly (*Scatophaga stercoraria*), which has been intensively used to study competition mechanisms and copulation duration (e.g. Parker 1992), or (b) damselflies of the genus *Calopteryx*, which first served to demonstrate sperm removal by an specialized male intromittent organ (Waage 1979) (here represented by the Iberian *C. xanthostoma*). The Orthoptera (c), a pair of *Gryllus campestris* with the male singing) and the Lepidoptera (d; copulation by the Ecuadorian *Haetera piera*) are two orders prevalent as models in sperm competition research. Pictures by author.

groups mate only once (Strassmann 2001), suggesting that postcopulatory sexual selection is negligible, with the clear exception of honeybees (Franck *et al.*, 2002). On the contrary, male "kings" are present in termites, providing the queens with mating opportunities throughout their life (Jaffe 2008). However, do termite queens mate with several males? Apparently not, but more studies are clearly needed (Hartke & Baer 2011).

Among the understudied orders, praying mantis and their sexually cannibalistic behaviour, are of interest because sexual cannibalism imposes monogamy in males, and therefore first-male advantage might be selected for. However, the paternity analysis of *Pseudomantis albofimbriata* found last-male sperm precedence (Barry 2011).

Although my review of the literature may not detect some papers and may include some false positives, the search indicates that we still have no information about sperm competition mechanisms for many orders, precluding comparative analyses. Questions like how many times has sperm removal by male genitalia evolved, or, how widespread are nuptial feeding or copulatory guarding, need studies dealing with the ethodiversity (Cordero-Rivera, 2017a) of sperm competition mechanisms at level of insect orders.

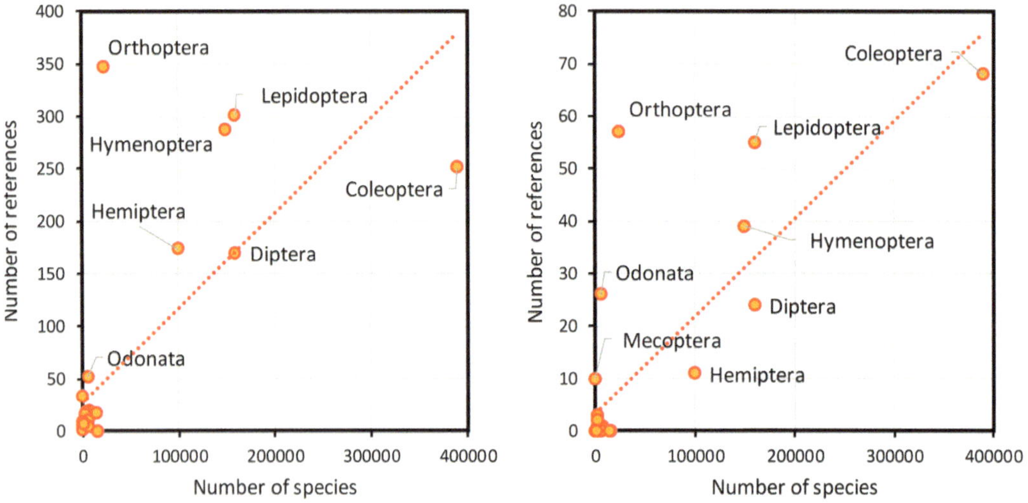

FIGURE 10.2 The relationship between species richness of insect orders and (left) the number of papers in the Web of Sciences that contain "sperm" in their title, and (right) the number of papers from the previous subset including the word "competition". Note that Orthoptera, Lepidoptera and Odonata are more studied than expected given their number of species. Source for species numbers, mainly Gullan & Cranston (1995).

FIGURE 10.3 The mayflies form mating swarms (in this case Leptophyphidae), where males dance up and down (insert) usually in front of trees or other landscape marks. They are functionally a lek, and females enter the swarm to mate. They may be the only insect order where sperm competition does not occur, because females lack spermatheca (Harker 1992). Picture taken at the National Park Armando Bermúdez, Dominican Republic, by author.

10.3 An Ecological View of Sperm Competition in Insects

Wherever resources are limited, individuals will be involved in competition, sometimes very evidently, but also rather cryptically, or even in indirect ways (without contact between the competing individuals). Consider this statement by Klomp (1964) in relation to intraspecific competition: 'The essential requisites of animals for living and maintenance are space, food, cover, and opportunities for reproduction.' This is a classical view, where 'opportunities for reproduction' means resources and sites needed for oviposition. Even modern approaches to ecological competition consider that it 'occurs any time a resource that is essential to growth and reproduction (e.g., food, shelter, nesting sites) occurs in short supply' (Amarasekare, 2009). Surprisingly, this approach does not explicitly mention the most obvious 'resource' for reproduction, which is a mating partner (except in asexual species). The theory of sperm competition started in the field of Ethology, and developed independently of ecological theories, and vice versa. My aim here is to combine both approaches.

Ecological competition can occur by exploitation or interference mechanisms, in scramble or contest situations (Coulson & Godfray, 2007), and may be highly asymmetric, particularly in interspecific interactions. In this section, I address the questions: To what extent are eggs a limiting resource to males? Under which circumstances can sperm production be a limiting factor?

10.3.1 Correlates of Female Fecundity in Insects

Insects are generally short-lived animals as adults. Eggs may mature during the larval period, and the adult female emerges with a fixed number (pro-ovigeny), or alternatively its number increases over time (syn-ovigeny). Multiple mating can be advantageous for syn-ovigenic species, but rarely will be for pro-ovigenic species.

The evidence from many studies indicates that four factors are of crucial relevance for female fecundity: (i) **female body size** is positively correlated with potential fecundity with a slope close to 1 (Honěk, 1993), (ii) **longevity** limits egg production, and may overcome the positiveeffect of size (Leather, 1988), (iii) egg maturation is dependent on **adult nutrition and health** including the effect of parasites and diseases (Leather & Hardie, 1995), and (iv) **female mating frequency** and the resources acquired during mating can also modify female fecundity to the extent that multiple mating in insects with nuptial feeding can increase female fecundity by 35–85% (Arnqvist & Nilsson, 2000). Figure 10.4 shows the relationship between female body size and fecundity in a large sample of insects. It shows that monogamy (or few matings) results in a reduction of 18% in fecundity compared to females that mate several times. However, if there is variation in polyandry propensity within species, the effect of multiple mating would not be so relevant for naturally monogamous females, as a detailed meta-analysis in Lepidoptera has demonstrated (Torres-Vila *et al.*, 2004).

10.3.2 The Rarity of Monandry

The most common mating system in animals is polyandry (Parker & Birkhead, 2013). Monandry can evolve from multiple pathways, including sexual conflict (Hosken *et al.*, 2009). However, even in species where the dominant strategy is monandry, there are reasons for females to re-mate, for instance, if the first male failed to inseminate (García-González, 2004), transferred too few sperm, insufficient for lifetime fertilization, if there is genetic incompatibility, or if sperm cannot be stored for long times. Mating failures are common in insects (Rhainds, 2010), but difficult to estimate in wild populations. If all females of a population were strictly monogamous, then no sperm competition could occur. However, there is increasing evidence for intraspecific variation in the degree of polyandry (Torres-Vila *et al.*, 2004). For instance, in the damselfly *Ischnura elegans*, male-like females mate consistently less often than gynochrome females, over a range of population frequencies (Cordero-Rivera & Sánchez-Guillén, 2007). The housefly typically mates only once, but 2–14% of females re-mate (Andrés & Arnqvist, 2001). Two papers published the same year make contrasting views about the degree of polyandry in social Hymenoptera (Crozier & Fjerdingstad 2001; Strassmann 2001), and there are cases

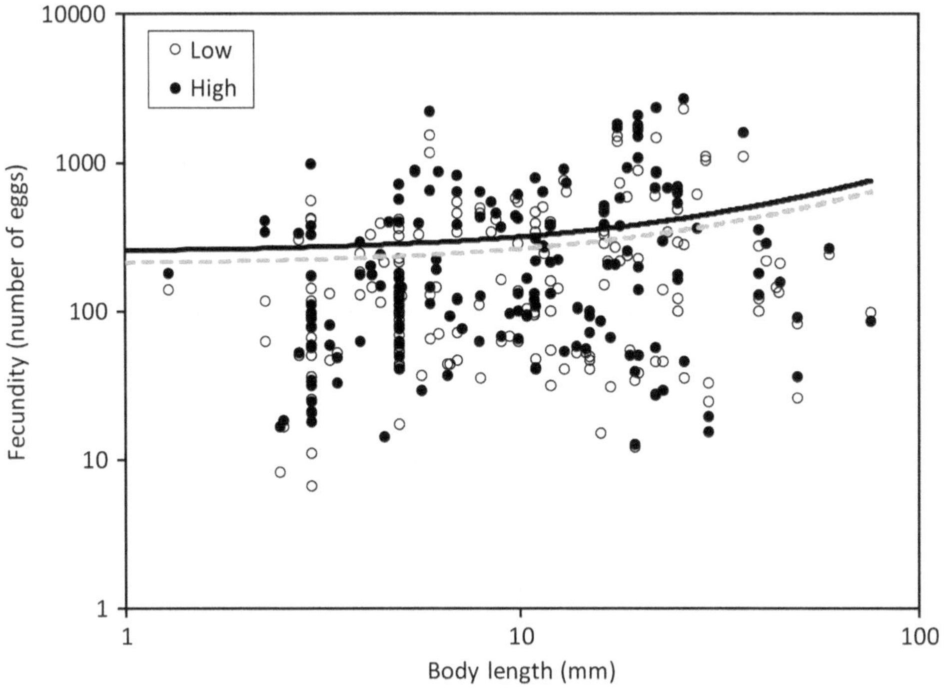

FIGURE 10.4 The relationship between body length and fecundity for female insects mating once (or a few times, Low) compared to mating many times (High). On average, fecundity increases by 18% with multiple mating, indicating that polyandry is beneficial for females. Source of the data: 207 species from 14 orders, including those analysed by Arnqvist and Nilsson (2000) and a review of the literature since 1999. Estimates of body size were obtained from different sources.

where polyandry is clearly detrimental to females (Bybee *et al.*, 2005). Therefore, even in 'monandric' species, there will be occasional cases of multiple mating, thereby producing sperm competition. Ecological conditions, particularly population density and the degree of male monopolization of resources needed for oviposition, will affect the likelihood that female monogamy will evolve.

10.3.3 Mating and Male Fertility

The mating frequency of male insects may be rather high in dense populations, like the rice stink bug who mates 17–44 times (Nilakhe, 1976). In a laboratory study of the seed beetle *Callosobruchus maculatus*, Ofuya (1995) found that males are able to mate up to 73 times in their first 10 days of life, but this has consequences: females mated with older males laid fewer eggs, and the last ones in the sequence were less likely to lay eggs, suggesting that males were losing fertility with age and repeated matings. In the moth *Helicoverpa armigera*, females mated to virgin males had higher reproductive success than those mated to previously mated males (Hou & Sheng 1999). When males transfer a large spermatophore, repeated mating reduces its size (De Morais *et al.*, 2012), and female fecundity and fertility may be reduced to a half (e.g. Callado-Galindo *et al.*, 2013). In the leaf miner *Liriomyza trifolii,* males can copulate with five females in a sequence, inseminating a similar number of sperm, but the last female in the sequence had a higher probability of reproductive failure, suggesting that in the last copulations males were inseminating low quality sperm (Kaspi & Parrella 2008). Very high mating frequency has also been recorded in natural populations, like the case of a male dragonfly (*Perithemis tenera*) mating 127 times in 6 days (Jacobs 1955). All these observations indicate that sperm production is costly, and males cannot fully inseminate many females. This opens the possibility of males cryptically choosing among their mates, and differentially inseminating them (Lüpold *et al.*, 2011). Sperm limitation is therefore a significant factor in sperm competition that needs further study.

10.3.4 Intraspecific Sperm Competition in Insects

Given that female insects store sperm, when a male mates with a previously mated female, or a female that will mate again, it will enter in sperm competition with other males. Two mechanisms operate. In **exploitative competition** (Table 10.1), males will fertilize as many eggs as possible, thereby depriving other males from fertilizing these eggs. This will select for large ejaculates, sometimes associated to long copulation times, repeated mating, nutritional gifts, induction of faster oviposition, etc. These have been termed **defensive mechanisms** in classical sperm competition reviews.

In exploitative competition, the ability to find females as soon as they become available will be selected for. This likely explains the evolution of protandry, like male butterflies who localize female pupae and wait there to be the first to mate (Figure 10.5c) (Gilbert 1976), or male solitary bees that wait at the emergence holes of females (Asís *et al.*, 2006). In scramble competition mating systems, mate-localization traits are expected to evolve. For instance, non-linear selection on mating efficiency has been found for wing length in male non-territorial damselflies, a trait likely related to mate-seeking ability (Cordero-Rivera 2000).

Exploitative sperm competition has been compared to a 'fair raffle', where fertilization is directly proportional to representation of a male's viable sperm in female's sperm storage organ (e.g. García-González & Simmons, 2005). Males are expected to adjust their ejaculate to risk of sperm competition (Table 10.1), so that presence of rivals affects ejaculate quantity and/or quality (Gage & Baker, 1991; Simmons & Kvarnemo, 1997). Another possible adaptation is to copulate frequently or for longer periods, if this allows males to increase amount of sperm transferred. For example, males of the damselfly *Coenagrion scitulum* perform 'cyclic' copulations, repeated 3–7 times, and inseminate the same amount in each cycle (Cordero *et al.*, 1995) (Figure 10.5a). The 'raffle principle' assumes that sperm is cheap to produce, a fact contrary to the evidence derived from multiple studies (García-González, 2004). The number of sperm inseminated in a single copulation is highly variable, with a range between hundreds (e.g. Chevrier & Bressac 2002) to hundreds of millions (Fjerdingstad & Boomsma 1998).

Evidence from odonates, the only insect group with intra-male sperm translocation before copulation, shows that males do not discard unused sperm when a mating is interrupted, and inseminate subsequent females with previously translocated sperm (Rivas-Torres *et al.*, 2019a). This suggests that even in species without nuptial gifs, the ejaculate is costly. Insects that transfer large spermatophores, may show sexual role reversal, and males will become choosy and adjust ejaculates to the quality of females (Gage & Barnard, 1996). In this context, it has been suggested that sperm size polymorphism might be related to the transfer of nutritional substances to the female, sacrificing some sperm cells, increasing in this way male fertilization success (Sasakawa 2009).

Theoretically, faster swimming sperm should be at an advantage in exploitative competition. However, the distance that sperm has to travel inside female insects, from the ejaculation site to the storage organ (spermatheca), is negligible, sometimes even shorter than the sperm itself: it is rarely more than 10 times the length of the sperm (Sivinski, 1980). Therefore, the anatomy of female insects will influence whether swimming speed is selected for in insect spermatozoa. It should be also taken into account that more motile sperm might be shorter lived, and be less competitive (Reinhardt, 2005). In animals with external fertilization, sperm swimming speed is positively related to sperm length (e.g. Fitzpatrick *et al.*, 2010). In the case of insects, studies of sperm swimming speed are scarce, but, in at least one species, males ejaculate bundles of sperm, which swim faster as their size increases (Hayashi 1998a). In territorial odonate species (which have high mating frequencies), males transfer free sperm, whereas in non-territorial species the norm is to transfer sperm bundles (Siva-Jothy 1997), which may contribute to sperm longevity, via nutritional substances (Åbro 2000). The relationship between sperm size and swimming speed needs to be studied in more insect species.

Copulation duration may be affected by sperm competition. If the only function of mating were to transfer the sperm to the female, it should be as short as possible, because mating has costs, in terms of time lost for feeding, predation, infection, etc (Daly, 1978). However, very long copulations are common in insects, lasting for days (Nilakhe, 1976; Santolamazza-Carbone & Cordero-Rivera, 1998). In some cases, long copulations are needed to full sperm transfer (Campbell & Fairbairn, 2001), but in

TABLE 10.1

Mechanisms of exploitative sperm competition in insects, with representative examples

Mechanism	Process Involved	Examples
Long copulation	Increased ejaculation	(Campbell & Fairbairn, 2001)
Repeated copulation	Increased insemination	(Cordero *et al.*, 1995)
Nuptial gift	Increased time for sperm migration inside female	(Svensson *et al.*, 1990; Hayashi, 1998b)
Strategic ejaculation	Large ejaculates at higher competition risk	(Martin & Hosken, 2002)
Oviposition induction	Transfer of accessory seminal products	(Andrés & Arnqvist, 2001)
Sperm function	Sperm size polymorphism	(Sasakawa, 2009)
Sperm bundles	Aggregation of sperm to increase speed swimming and/or longevity	(Hayashi, 1998a; Åbro, 2000)
Mating with newly emerged females	Monopolization of monandric females	(Gilbert, 1976)

FIGURE 10.5 Some examples of exploitative sperm competition. (a) A copulating pair of the South-European damselfly *Coenagrion scitulum*, one of the very few species of odonates with repeated mating, transferring a large ejaculate to the female (Cordero *et al.*, 1995). (b) Copulation by a dipteran from the family Scionidae, from Yanayacu (Ecuador), showing extreme sexual size dimorphism. Males of many dipterans transfer oviposition-inducing peptides to the female. (c) Two males of the butterfly *Heliconius charitonia ramsdeni*, from Cuba, waiting at the pupa of a female, to be the first to mate as soon as the female emerges. Pictures by author.

FIGURE 10.6 Some examples of interference sperm competition. (a) Males of many aquatic Heteroptera (Veliidae) remain over their mate for very long periods, effectively acting as a barrier for further matings by female. (b) Prolonged association between male and female beetles after copulation (*Eucomato ceravittata*, from Yunnan, China). (c) Contact guarding in *Decarloa darlingtoni* from river Yaque del Norte, in Dominican Republic. (d) Oviposition in tandem by damselfly *Neoneura carnatica*, endemic from Cuba, laying eggs in presence of a rival male. Pictures by author.

some cases insemination occurs in the first few minutes (Shuker *et al.*, 2002). The detailed analysis of copulation duration in the dung fly, show that this variable can be modelled to find the conditions that maximize reproductive success (Parker 1992). A similar model of optimal copulation duration in insects, taking into account the influence of sperm competition makes several testable predictions (Nuyts & Michiels 1993). For instance, for odonates it predicted that territoriality should shorten copulation time (Nuyts & Michiels 1994), and this has been corroborated by a review of the literature (Córdoba-Aguilar *et al.*, 2009).

The second mechanism, **interference competition**, occurs when a male has a direct negative effect on rival males, avoiding access to the eggs. Even if females are not limited in number, the fact that some males monopolize the access to them will produce interference. Furthermore, even if two males successfully inseminate the same female, one of them, usually the last in the sequence, can use subtle mechanisms to reduce the chances that the first male fertilizes the eggs. This fact is behind the evolution of mate guarding, sperm removal, sperm repositioning, sperm polymorphism, etc, all of them known as **offensive mechanisms** of sperm competition (Table 10.2). The variety of mechanisms potentially able to delay female re-mating is a proof of the relevance of sperm competition for males' reproductive interests. Many (if not all) of the mechanisms listed in Table 10.2 are also under female control, and this

TABLE 10.2

Mechanisms of interference sperm competition in insects, with representative examples

Mechanism	Process Involved	Examples
Precopulatory guarding	Monopolization of females	(Cordero *et al.*, 1995)
Prolonged copulation	In-copula guarding	(Sillén-Tullberg, 1981; Cordero, 1990)
Sperm removal	Physical elimination of rivals' sperm	(Cordero & Miller, 1992)
Sperm polymorphism (apyrene/eupyrene)	Sperm involved in rivals' ejaculate deactivation and/or delaying female re-mating	(Cook & Wedell, 1999)
Sperm incapacitation	Chemical elimination of rivals' sperm	(Price *et al.*, 1999; but see Snook & Hosken, 2004)
Sperm repositioning	Displacement of rivals' sperm to unfavourable locations inside female sperm storage organs	(Siva-Jothy, 1988; Xu & Wang, 2010)
Giant sperm	Spermathecal clog	(Dybas & Dybas, 1981)
Copulatory wounding	Physical damage to female, diminishing future copulations	(Reinhardt *et al.*, 2014)
Mating plug	Physical avoidance of subsequent copulations	(Aiken, 1992)
Mating delay	Physiological avoidance of subsequent copulations	(Takami *et al.*, 2008; Ortíz-Jiménez & del Castillo, 2015)
Postcopulatory guarding	Physical attachment of male to female	(Arnqvist, 1988)
Postcopulatory genital contact	Male (or detached male genitalia) acting as a mating plug	(Kerr *et al.*, 1962)

fact makes very difficult distinguishing between sperm competition and cryptic female choice, and both generate intense sexual conflict.

Mechanisms that delay further mating by the females are diverse (Table 10.2). The association between males and females after copulation (Figure 10.6), that in many cases continues during oviposition (Figure 10.6d), is probably the most common postcopulatory behaviour in insects (reviewed by Alcock 1994). All these mechanisms have in common that males remain in physical contact or in close association with their mates for a variable time, which is clearly costly in terms of future reproduction. The evolution of specialised grasping apparatus in male genitalia, may serve to immobilise females to achieve copulation but also can avoid take-overs by other males (Cordero-Rivera & Rivas-Torres 2019). Given the costs in future reproduction for males of mate-guarding, males are expected to adjust their behaviour to ecological conditions, particularly population density. For example, males of many *Ischnura* damselflies copulate for several hours, but short copulations are common at very low densities (Cordero 1990). In many cases, the presence of the guarding male is positive for the female as well, because he gives protection in front of other males, or helps detecting predators, or increasing oviposition efficiency. Therefore, the evolution of these mechanisms may be due to sperm competition, cryptic female choice, but also to natural selection pressures.

A second array of mechanisms allow the male to delay female re-mating, without being physically present, like transferring anti-aphrodisiacs, copulatory wounding, mating plugs or detached parts of male genitalia (Table 10.2). In a tropical katydid, the mating gift results in an induction of female monogamy, which lasts for as long as 17 days (Ortíz-Jiménez & del Castillo 2015). In the bumblebee *Bombus terrestris*, a male plug imposes female monogamy, even if experiments with artificially multiple mated queens, demonstrate that sperm diversity increases female fitness (Duvoisin *et al.*, 1999). These mechanisms therefore generate intense sexual conflict because the male 'imposes' monogamy but does not help the female during oviposition. Female counter-adaptations are therefore expected, like anti-aphrodisiac inactivation (Laturney & Billeter 2016), special structures where the copulatory wound is inflicted (Reinhardt *et al.*, 2003), or active removal of the mating plug (of which I could find no evidence in the literature). In some spiders, the male intromittent organ frequently brokens inside the female, and can act as a mating plug (Uhl *et al.*, 2010). This routinely happens also in the honey bee (Koeniger 1990), and occasionally in some earwigs (Kamimura 2003), but do not impede subsequent copulations of the female. Perhaps the only clear example of female monogamy induced by male detached genitalia is that of the bee *Melipona quadrifasciata* (Kerr *et al.*, 1962), but even in this case the evidence is not experimental.

Finally, other mechanisms act at level of ejaculate, and include sperm removal and repositioning, sperm polymorphism, or incapacitation (Table 10.2). The ability for sperm removal and repositioning has been intensively studied in the Odonata (see reviews by Córdoba-Aguilar *et al.*, 2003; Cordero-Rivera & Cordoba-Aguilar 2010). It occurs in the oldest extant damselfly, *Hemiphlebia mirabilis*, suggesting that was acquired early in odonate evolution (Cordero-Rivera 2016). Sperm removal has been also demonstrated in Orthoptera (Ono *et al.*, 1989; Helversen & Helversen 2004), Dermaptera (Kamimura 2000) and Coleoptera (Yokoi 1990) and has been suggested in a millipede (Barnett *et al.*, 1993). The function of apyrene/eupyrene sperm is still unclear but is highly likely an adaptation to sperm competition (Watanabe 2016). The most debatable mechanism is rivals' sperm incapacitation using the ejaculate (Price *et al.*, 1999), which has not been unambiguously demonstrated (Snook & Hosken 2004).

The result of both types of sperm competition, by exploitation and by interference, is an increase of paternity of some males over others. This has been classically studied by double matings, where the variable of interest is the proportion of offspring sired by the last male (P_2). In most cases, P_2 is well over 0.5, with a median of 0.68 in 156 insect species (Simmons, 2014a). The last male is usually the winner, but there is high variability, with some males being consistently more successful even if they are the first mate (e.g. Cordero & Miller, 1992). When the number of males is greater than two, the paternity advantages of the last male usually remain unchanged, but a model of sperm mixing seems to better fit the results in some cases (Simmons, 2001). Given the availability of modern molecular methods to assign paternity, more studies are clearly needed, particularly in orders of insects that are understudied.

The existence of sperm competition predicts that males should be able to detect female mating status and adjust their behaviour. In the damselfly *Ceriagrion tenellum* males perform short copulations with virgin females, and long copulations with mated females, apparently using chemical sensilla in their aedeagus to detect and estimate the amount of sperm stored from previous matings (Andrés & Cordero-Rivera, 2000). This response is shared by at least three Zygopteran families (Uhía & Cordero-Rivera, 2005), and preliminary work suggests that there are neural terminations in the aedeagus of damselflies, which could allow male to detect sperm inside female (pers. obs.). In some butterflies (De Morais *et al.*, 2012), orthopterans (Lugo-Olguín & Cueva Del Castillo, 2007), and beetles (Mbata *et al.*, 1997) the second copulation of a female is much longer than first, but in other insects this is not the case (e.g. Bind, 2007). Further studies are needed to check the generality of this phenomenon.

As indicated above polyandry is common in insects, but very high mating frequencies are usually detrimental to females, due to male harassment. The difference in optimal mating rate between sexes is a source of conflict, and one extreme case is sexual cannibalism. In predatory species, males may incur high risk approaching females, and multiple mating (by males) in these species may be rare. In agreement with theoretical expectations, males have evolved mechanisms to safely inseminate, or even to inseminate after leaving their genitalia attached to the female, like in some spiders (Li *et al.*, 2012). No such extreme cases are known in predatory insects, but the secondary genitalia and indirect sperm transfer of odonates may have evolved as an adaptation to reduce sexual cannibalism (Rivas-Torres *et al.*, 2019b) This, incidentally, may influence sperm competition, by reducing male mating rate.

10.3.5 Evolution of Sperm Size

A detailed analysis of sperm ultrastructure, including morphology, has been discussed by Jamieson *et al.* (1999), and a phylogenetic analysis presented by Gottardo *et al.* (2016). Sperm length is highly variable in insects, ranging from 1–2 micron in Ephemeroptera to 58.3 mm in *Drosophila bifurca*, 20 times longer than the insect itself. Figure 10.7 shows that sperm size evolves independently of male body size. This suggests that sexual selection forces are shaping sperm evolution (Lüpold *et al.*, 2016). A beautiful example is the study of diving beetles by Higginson *et al.* (2012). They found that in two clades of Dytiscidae, the dimensions of the spermatheca and/or the fertilization duct explained a large part of variation in sperm length (even 92% in one case), indicating that female characteristics are crucial forces affecting the evolution of sperm size.

An intriguing phenomenon is the evolution of giant sperm, which is extreme in some *Drosophila* species, with 23.5 mm in *D. hydei* (Pitnick & Markow 1994) and 58.3 mm in *D. bifurca* (Pitnick *et al.*, 1995). Experimental evidence shows that female choice could have fuelled the evolution of giant sperm (Miller & Pitnick 2002). The role of giant sperm in sperm competition is still unresolved. In beetles of the genus *Bambara*, whose sperm is not giant, there is a match between sperm dimensions and the spermathecal duct (Dybas & Dybas 1981). If this allows first-male advantage in paternity, say because giant sperm are more difficult to remove/reposition, it would be a mechanism of interference competition, equivalent to a mating plug.

10.3.6 Interspecific Sperm Competition in Insects

It is normally assumed that males will behave indiscriminately and try to mate with as many females as possible, except when copulation is costly (Simmons, 1995).

When males deliver an ejaculate that is cheap to produce, the optimal strategy is to mate with any female encountered, and this includes females of related species. Interspecific copulations are more common than a naïve approach to this topic would assume. Male insects may or not discriminate females of similar species, and trying to mate even with the 'wrong' female, can provide more benefits than costs. From the perspective of the sperm cell, once it is found in the body of a female, the best strategy is to try to fertilize the egg, no matter the specific identity of the egg. Although it is normally assumed that males courting a heterospecific female are almost always rejected (Charnov, 1984), this is not the case (Figure 10.8). For a male gamete, it pays to produce hybrid progeny rather than no progeny at all. For females, this creates intense sexual conflict.

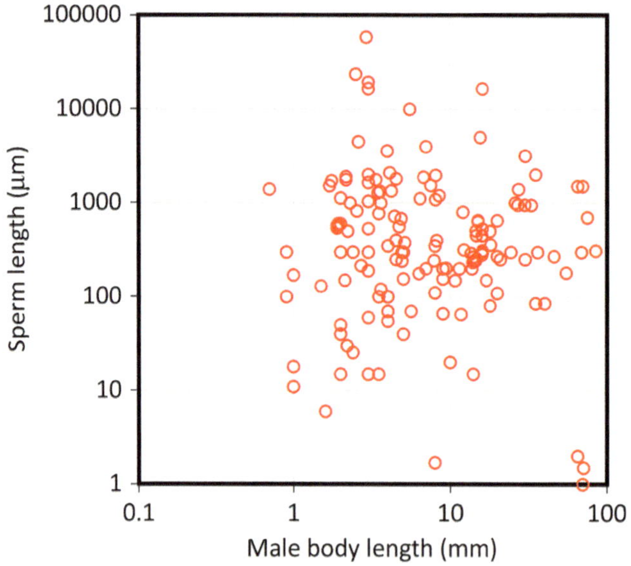

FIGURE 10.7 The relationship between male body length and sperm length in 148 species of insects, including representatives of 28 insect orders. It is evident that sperm size is not correlated with male size ($r = -0.09$, $p = 0.269$). Estimates of sperm length were obtained mainly from Jamieson *et al.* (1999) and Higginson *et al.* (2012), and body size from different sources.

Interspecific sperm competition will be of low relevance when pre-mating isolation barriers are highly effective, but will be a relevant force wherever two related species meet and hybridize, i.e. in hybrid zones (Harrison, 1993). In such cases, the commonest finding is conspecific sperm precedence, the advantage of conspecific over heterospecific sperm (reviewed by Howard, 1999). Well studied examples of postcopulatory barriers to hybridization mediated (at least in part) by conspecific sperm precedence include the crickets of the genus *Allonemobius* (Gregory & Howard, 1994), and *Drosophila* (Price, 1997). In *Allonemobius* heterospecific males sired a maximum of 18% of offspring, but only in 11–42% of trials and none in the rest, indicating a barrier in recognition/movement of heterospecific sperm. Interference competition occurs between the ejaculates of related *Drosophila* species (Price, 1997), to the extent that incapacitation of heterospecific sperm is one of the main mechanisms involved (Price *et al.*, 2000). In the case of the sibling damselfly species *Ischnura graellsii* and *I. elegans*, several barriers, including ethological and mechanical mechanisms, impede most hybrid matings, but not all (Figure 10.8c), producing hybrid specimens that are (partially) fertile. However, when a female is mated by a conspecific and a heterospecific male, the progeny is mainly

FIGURE 10.8 Interspecific copulations occur between related species, opening the scenario for interspecific sperm competition, which usually results in conspecific sperm precedence. These interspecific matings are widely recorded in the Odonata (Utzeri & Belfiore 1990). (a) A male of the European *Platycnemis latipes* in tandem with a female of *Platycnemis acutipennis*, trying to copulate, without success. (b) Intergeneric copulation between a male of *Ischnura elegans* (blue) and a female of *Pyrrhosoma nymphula* (red), photographed in NW Spain. (c) Interspecific copulations between males of *Ischnura elegans* and females of *Ischnura graellsii* are common where both species co-occur in the Iberian Peninsula, not showing conspecific sperm precedence, but last-male sperm precedence (Sánchez-Guillén *et al.*, 2013).

sired by the second male, regardless of whether the male was conspecific or heterospecific, suggesting that there is no conspecific sperm precedence (Sánchez-Guillén *et al.*, 2013). In this last case, males of either species seem able to physically remove rivals' sperm with a similar efficiency. On the contrary, when females of *Calopteryx* were mated to a conspecific and a heterospecific male, sperm removal was consistently more efficient by the heterospecific male (Cordero-Rivera, 2017b). In this example heterospecific males have an advantage, that may explain the presence of hybrids in natural populations (Lorenzo-Carballa *et al.*, 2014).

Given the differences in reproductive physiology of different species, mating with a heterospecific partner is likely to be costly. For example, in the katydids *Orchelimum nigripes* and *O. pulchellum*, oviposition by females mated to heterospecific males is dramatically lower than that of females mated conspecifically, even if hatch rate and offspring viability are not reduced (Shapiro, 2006). In some cases, females mated to heterospecific males may fail to lay eggs (Cordero, 1989). Furthermore, harassment by heterospecific males may be intense and have a detrimental effect on females, even if actual copulation does not occur (McLain & Pratt, 1999). Differences in genitalic structures may also produce damages during interspecific matings (Reinhardt *et al.*, 2014), and this may be a very relevant cost for females in interspecific sperm competition.

To summarize, interspecific sperm competition may follow similar pathways as intraspecific sperm competition, but is certainly understudied, despite its relevance in reproductive isolation and speciation.

10.4 Conclusions

The likelihood and intensity of sperm competition are factors of extreme relevance for male reproductive success. Postcopulatory sexual selection, due to an array of selective forces in the continuum of male-dominated (sperm competition) to female-dominated (cryptic female choice) scenarios, predicts the evolution of behavioural adaptations to increase fertilization success. These include (1) detecting female mating status; (2) evaluating the likelihood of sperm competition; (3) adjusting male investment in relation to the intensity of sperm competition and female re-mating behaviour.

The selective forces based on sperm competition have many evolutionary consequences. In cases where male success is proportional to the number of sperm and/or quality of the ejaculate, an increase in gonad investment is expected. Sperm competition may be also the most relevant force for the evolution of sperm size, a topic that has been studied in detail only in a few groups of insects. The odonates are not studied in this context, and therefore are a good model for future studies.

This review shows that sperm competition is a particular case of a more general phenomenon, ecological competition, and may benefit from a more ecologically oriented agenda, and clearly from a diversity of model species. Despite the progress of last decades, many groups of insects are still completely understudied. I hope this review will stimulate more interest in sperm competition studies of the small insect orders.

Acknowledgements

I would like to thank Rassim Khelifa for his comments on a previous draft of this text. Funding was provided by grants from the Spanish Ministry with competences in Science, including FEDER funds (CGL2014-53140-P, PGC2018-096656-B-I00).

REFERENCES

Åbro A. (2000). Sperm clusters in Zygoptera (Coenagrionidae, Lestidae, Calopterygidae). *Odonatologica*, *29*, 51–56.

Aiken R.B. (1992). The mating behaviour of a boreal water beetle, *Dytiscus alaskanus* (Coleoptera Dytiscidae). *Ethol Ecol Evol*, *4*, 245–254. 10.1080/08927014.1992.9523136

Alcock J. (1994). Postinsemination associations between males and females in insects: mate guarding hypothesis. *Annu Rev Entomol, 39*, 1–21.
Amarasekare P. (2009). Competition and coexistence. In: Levin S.A. (ed.) *The Princeton Guide to Ecology*. Princeton: Princeton University Press, pp. 196–201.
Andersson M. (1994). *Sexual Selection*. Princeton: Princeton University Press, pp. 599.
Andrés J.A., & Arnqvist G. (2001). Genetic divergence of seminal signal-receptor system in houseflies: footprints of sexually antagonistic coevolution? *Proc R Soc B Biol Sci, 268*, 399–405.
Andrés J.A., & Cordero-Rivera A. (2000). Copulation duration and fertilization success in a damselfly: an example of cryptic female choice? *Anim Behav, 59*, 695–703.
Arnqvist G. (1988). Mate guarding and sperm displacement in water strider *Gerris lateralis* Schumm. (Heteroptera: Gerridae). *Freshw Biol, 19*, 269–274.
Arnqvist G., & Rowe L. (2005). *Sexual Conflict*. Princeton, NJ: Princeton University Press.
Arnqvist G., & Nilsson T. (2000). The evolution of polyandry: multiple mating and female fitness in insects. *Anim Behav, 60*, 145–164.
Asís J.D., Tormos J., & Gayubo S.F. (2006). Territorial dynamics and contest behaviour in solitary wasp *Stizus continuus* (Hymenoptera: Apoidea: Crabronidae). *Behaviour, 143*, 83–104.
Barry K.L., Holwell G.I., & Herberstein M.E. (2011). A paternity advantage for speedy males? Sperm precedence patterns and female re-mating frequencies in a sexually cannibalistic praying mantid. *Evol Ecol, 25*, 107–119. 10.1007/s10682-010-9384-3
Bateman A.J. (1948). Intra-sexual selection in *Drosophila*. *Heredity (Edinb), 2*, 349–368.
Begon M., Townsend C.R., & Harper J.L. (2006). *Ecology: From Individuals to Ecosystems*, 4th edn. Oxford: Blackwell Publishing, pp. 738.
Bind R.B. (2007). Reproductive behaviour of a generalist aphidophagous ladybird beetle *Cheilomenes sexmaculata* (Coleoptera: Coccinellidae). *Int J Trop Insect Sci, 27*, 78–84. 10.1017/S1742758407814688
Birkhead T.R., & Möller A.P. (1998). *Sperm Competition and Sexual Selection*. San Diego: Academic Press, pp. 826.
Birkhead T.R., & Pizzari T. (2002). Postcopulatory sexual selection. *Nat Rev Genet, 3*, 262–273.
Brennan P. L.R. & Prum R. O. (2015). Mechanisms and evidence of genital coevolution: the role of natural selection, mate choice, and sexual conflict. In Rice W. R. & Gavrilets S. (eds). *The Genetics and Biology of Sexual Conflict*.Cold Spring Harbor Laboratory Press. pp. 385-413.
Bybee L.F., Millar J.G., Paine T.D., & Hanlon C.C. (2005). Effects of single versus multiple mates: monogamy results in increased fecundity for beetle *Phoracantha semipunctata*. *J Insect Behav, 18*, 513–527. 10.1007/s10905-005-5609-7
Callado-Galindo M.M., Villa-Ayala P., Castrejón-Ayala F., & Jiménez-Pérez A. (2013). Effect of age, body weight and multiple mating on *Copitarsia decolora* (Lepidoptera: Noctuidae) reproductive potential and longevity. *J Insect Behav, 26*, 860–872. 10.1007/s10905-013-9401-9
Campbell V., & Fairbairn D.J. (2001). Prolonged copulation and internal dynamics of sperm transfer in water strider *Aquarius remigis*. *Can J Zool, 79*, 1801–1812. 10.1139/cjz-79-10-1801
Charnov E.L. (1984). Behavioural ecology of plants. In: Krebs J.R., & Davies N.B. (eds.) *Behavioural Ecology. An Evolutionary Approach*. Oxford: Blackwell Scientific Publications. pp. 362– 379.
Cook P.A., & Wedell N. (1999). Non-fertile sperm delay female remating. *Nature, 397*, 486.
Cordero-Rivera A. (2017a). Behavioral diversity (ethodiversity): a neglected level in study of biodiversity. *Front Ecol Evol, 5*, 1–7. 10.3389/fevo.2017.00007
Cordero-Rivera A. (2017b). Sexual conflict and the evolution of genitalia: male damselflies remove more sperm when mating with a heterospecific female. *Sci Rep* 710.1038/s41598-017-08390-3.
Cordero-Rivera A. (2000). An analysis of multivariate selection in a non-territorial damselfly (Odonata: Coenagrionidae). *Etología, 8*, 37–41.
Cordero-Rivera A. (2016). Sperm removal during copulation confirmed in oldest extant damselfly, *Hemiphlebia mirabilis*. *PeerJ, 4*, e2077. 10.7717/peerj.2077
Cordero-Rivera A., & Cordoba-Aguilar A. (2010). Selective forces propelling genitalic evolution in Odonata. In: Leonard J., & Córdoba-Aguilar A. (eds.) *The Evolution of Primary Sexual Characters in Animals*. Oxford: Oxford University Press. pp. 332–352.
Cordero-Rivera A., & Córdoba-Aguilar A. (2016). Selección postcópula: competenciaespermática. In: Carranza J. (ed.) *Etología Adaptativa: el Comportamiento Comoproducto de la Selección Natural*. Córdoba (Spain): UCOpress y Publicaciones de la Universidad de Extremadura, pp. 479–504.

Cordero-Rivera A., & Rivas-Torres A. (2019). Sexual conflict in water striders, dragonflies and diving beetles. In: Del-Claro K., & Guillermo-Ferreira R. (eds) *Aquatic Insects*. Switzerland: Springer Nature, pp. 295–320.

Cordero-Rivera A., & Sánchez-Guillén R.A. (2007). Male-like females of a damselfly are not preferred by males even if they are majority morph. *Anim Behav*, *74*, 247–252.

Cordero A. (1990). The adaptive significance of prolonged copulations of damselfly, *Ischnura graellsii* (Odonata: Coenagrionidae). *Anim Behav*, *40*, 43–48.

Cordero A. (1989). Reproductive behaviour of *Ischnura graellsii* (Rambur) (Zygoptera: Coenagrionidae). *Odonatologica*, *18*, 237–244.

Cordero A., & Miller P.L. (1992). Sperm transfer, displacement and precedence in *Ischnura graellsii* (Odonata: Coenagrionidae). *Behav Ecol Sociobiol*, *30*, 261–267.

Cordero A., Santolamazza-Carbone S., & Utzeri C. (1995). Male disturbance, repeated insemination and sperm competition in damselfly *Coenagrion scitulum* (Zygoptera: Coenagrionidae). *Anim Behav*, *49*, 437–449.

Cordero C., & Eberhard W.G. (2003). Female choice of sexually antagonistic male adaptations: a critical review of some current research. *J Evol Biol*, *16*, 1–6.

Córdoba-Aguilar A., Uhía E., & Cordero-Rivera A. (2003). Sperm competition in Odonata (Insecta): evolution of female sperm storage and rivals' sperm displacement. *J Zool*, *261*, 381–398. 10.1017/S0952836903004357

Coulson T., & Godfray H.C.J. (2007). Single-species dynamics. In: May R.M., McLean A.R. (eds.) *Theoretical Ecology*. Oxford: Oxford University Press, pp. 17–34.

Crozier R.H., & Fjerdingstad E.J. (2001). Polyandry in social Hymenoptera - Disunity in diversity? *Ann Zool Fennici*, *38*, 267–285.

Cuéllar O. (1977). Animal parthenogenesis. *Science (80)*, *197*, 837–843.

Daly M. (1978). The cost of mating. *Am Nat*, *112*, 771–774.

Danielsson I. (1998). Mechanisms of sperm competition in insects. *Ann Zool Fennici*, *35*, 241–257.

Darwin C. (1859). *On Origin of Species by Means of Natural Selection*. London: John Murray, pp. 502.

Darwin C. (1871). *The Descent of Man and Selection in Relation to Sex*. London: John Murray, pp. 523.

De Morais R.M., Redaelli L.R., & Ana J.S. (2012). Age and multiple mating effects on reproductive success of *Grapholita molesta* (Busck) (Lepidoptera, Tortricidae). *Rev Bras Entomol*, 56, 319–324. 10.1590/S0085-56262012005000039

Duvoisin N., Baer B., & Schmid H.P. (1999). Sperm transfer and male competition in a bumblebee. *Anim Behav*, *58*, 743–749.

Dybas L.K., & Dybas H.S. (1981). Coadaptation and saxonomic differentiation of sperm and spermathecae in featherwing beetles. *Evolution (N Y)*, *35*, 168. 10.2307/2407950

Eberhard W.G. (1991). Copulatory courtship and cryptic female choice in insects. *Biol Rev*, *66*, 1–31.

Eberhard W.G. (1996). *Female Control: Sexual Selection by Cryptic Female Choice*. Princeton: Princeton University Press, pp. 501

Eberhard W. G. (1985). *Sexual Selection and Animal Genitalia*. Cambridge, Massachusetts: Harvard University Press.

Eberhard W.G. (1998). Importancia de la elección6femeninacríptica para la Etología. *Etología*, *6*, 1–10.

Engelstädter J. (2008). Constraints on evolution of asexual reproduction. *BioEssays*, *30*, 1138–1150. 10.1002/bies.20833

Franck P., Solignac M., Vautrin D., Cornuet J.M., Koeniger G., & Koeniger N. (2002). Sperm competition and last-male precedence in honeybee. *Anim Behav*, *64*, 503–509. 10.1006/anbe.2002.3078

Gage A.R., & Barnard C.J. (1996). Male crickets increase sperm number in relation to competition and female size. *Behav Ecol Sociobiol*, *38*, 349–353.

Gage M.J.G., & Baker R.R. (1991). Ejaculate size varies with socio-sexual situation in an insect. *Ecol Entomol*, *16*, 331–337. 10.1111/j.1365-2311.1991.tb00224.x

García-González F. (2004). Infertile matings and sperm competition: effect of "nonsperm representation" on intraspecific variation in sperm precedence patterns. *Am Nat*, *164*, 457–472.

García-González F., & Simmons L.W. (2005). Sperm viability matters in insect sperm competition. *Curr Biol*, *15*, 271–275. 10.1016/j/j.cub.2005.01.032

Gilbert L.E. (1976). Postmating female odor in *Heliconius* butterflies: a male-contributed antiaphrodisiac? *Science (80-)*, *193*, 419–420.

Gomendio M., & Roldán E.R.S. (1993). Mechanisms of sperm competition: linking physiology and behavioural ecology. *Trends Ecol Evol*, *8*, 95–100.
Gottardo, M., Dallai, R., Mercati, D., Hörnschemeyer, T., & Beutel, R.G. (2016). The evolution of insect sperm – an unusual character system in a megadiverse group. *J Zool Syst Evol Res*, *54*, 237–256. 10.1111/jzs.12136
Gregory P.G., & Howard D.J. (1994). A postinsemination barrier to fertilization isolates two closely related ground crickets. *Evolution*, *48*, 705–710.
Harker J.E. (1992). Swarm behaviour and mate competition in mayflies (Ephemeroptera). *J Zool*, *228*, 571–587.
Harrison R. (1993). Hybrids and hybrid zones: historical perspectives. In: Harrison R. (ed). *Hybrid Zones and Evolutionary Process*. New York: Oxford University Press, pp. 3–12.
Hartke T.R., & Baer B. (2011). The mating biology of termites: a comparative review. *Anim Behav*, *82*, 927–936. 10.1016/j.anbehav.2011.07.022
Hayashi F. (1998a). Sperm co-operation in Fishfly, *Parachauliodes japonicus*. *Funct Ecol*, *12*, 347–350.
Hayashi F. (1998b). Multiple mating and lifetime reproductive output in female dobsonflies that receive nuptial gifts. *Ecol Res*, *13*, 283–289. 10.1046/j.1440-1703.1998.00272.x
Helversen D Von, & Helversen O Von (1991). Pre-mating sperm removal in bush cricket *Metaplastes orntus* Ramme, 1931 (Orthoptera, Tettigonoidea, Phaneropteridae). *Behav Ecol Sociobiol*, *28*, 391–396.
Higginson D.M., Miller K.B., Segraves K.A., & Pitnick S. (2012). Convergence, recurrence and diversification of complex sperm traits in diving beetles (Dytiscidae). *Evolution (N Y)*, *66*, 1650–1661. 10.1111/j.1558-5646.2011.01532.x
Honěk A. (1993). Intraspecific variation in body size and fecundity in insects: a general relationship. *Oikos*, *66*, 483. 10.2307/3544943
Hosken D.J., Stockley P., Tregenza T., & Wedell N. (2009). Monogamy and battle of sexes. *Annu Rev Entomol*, *54*, 361–378.
Howard D.J. (1999). Conspecific sperm and pollen precedence and speciation. *Annu Rev Ecol Syst*, *30*, 109–132.
Hou M., & Sheng C. (1999). Fecundity and Longevity of *Helicoverpa armigera* (Lepidoptera: Noctuidae): Effects of Multiple Mating. *J Econ Entomol*, *92*, 569–57310.1093/jee/92.3.569.
Jamieson, B.G.M., Dallai, R., & Afzelius B.A. (1999). Insects. *Their Spermatozoa and Phylogeny*. Plymouth: Science Publishers, pp. 555.
Kamimura Y. (2003). Effects of broken male intromittent organs on sperm storage capacity of female earwigs, *Euborellia plebeja*. *J Ethol*, *21*, 29–35. 10.1007/s10164-002-0071-x
Kamimura Y. (2000). Possible removal of rival sperm by elongated genitalia of earwig, *Euborellia plebeja*. *Zoolog Sci*, *17*, 667–672.
Kaspi R., & Parrella M.P. (2008). Polyandry and reproduction in serpentine leaf miner *Liriomyza trifolii* (Diptera: Agromyzidae). *J Insect Behav*, *21*, 323–336. 10.1007/s10905-008-9128-1
Kerr W., Zucchi R., Nakadaira J., & Butolo J. (1962). Reproduction in social bees (Hymenoptera: Apidae). *J New York Entomol Soc*, *70*, 265–276.
Klomp H. (1964). Intraspecific competition and regulation of insect numbers. *Annu Rev Entomol*, *9*, 17–40.
Koeniger G. (1990). The role of mating sign in honey bees, *Apis mellifera* L.: does it hinder or promote multiple mating? *Anim Behav*, *39*, 444–449. 10.1016/S0003-3472(05)80407-5
Laturney M., & Billeter J.C. (2016). *Drosophila melanogaster* females restore their attractiveness after mating by removing male anti-aphrodisiac pheromones. *Nat Commun*, *7*, 12322. 10.1038/ncomms12322
Leather S.R. (1988). Size, reproductive potential and fecundity in insects: things aren't as simple as they seem. *Oikos*, *51*, 386. 10.2307/3565323
Leather S.R., & Hardie J. (1995). *Insect Reproduction*. Boca Raton: CRC Press, pp. 255.
Leonard J.L., & Córdoba-Aguilar A. (2010). *The Evolution of Primary Sexual Characters in Animals*. Oxford: Oxford University Press, pp. 552.
Li D., Oh J., Kralj-Fiser S., & Kuntner M. (2012). Remote copulation: male adaptation to female cannibalism. *Biol Lett*, *8*, 512–515. 10.1098/rsbl.2011.1202
Lorenzo-Carballa M.O., & Cordero-Rivera A. (2009). Thelytokous parthenogenesis in damselfly *Ischnura hastata* (Odonata, Coenagrionidae): genetic mechanisms and lack of bacterial infection. *Heredity (Edinb)*, *103*, 377–384.

Lugo-Olguín S.D., & Cueva Del Castillo R. (2007). Multiple matings, female fecundity, and assessment of sperm competition risk in protandrous Grasshopper *Sphenarium purpurascens* (Orthoptera: Pyrgomorphidae). *Ann Entomol Soc Am, 100*, 591–595. 10.1603/0013-8746(2007)100.

Lüpold S., Manier M.K., Ala-Honkola O., Belote J.M., & Pitnick S. (2011). Male *Drosophila melanogaster* adjust ejaculate size based on female mating status, fecundity, and age. *Behav Ecol, 22*, 185–191. 10.1093/beheco/arq193

Lüpold S., Manier M.K., Puniamoorthy N., Schoff C., Starmer W.T., Luepold S.H.B., Belote J.M., & Pitnick S. (2016). How sexual selection can drive evolution of costly sperm ornamentation. *Nature, 533*, 535–538. 10.1038/nature18005

Martin O.Y., & Hosken D.J. (2002). Strategic ejaculation in common dung fly *Sepsis cynipsea. Anim Behav, 63*, 541–546. 10.1006/anbe.2001.1929

Mbata G.N., Shu S., & Ramaswamy S.B. (1997). Rhythmicity of mating and oviposition in *Callosobruchus subinnotatus* (Pic) (Coleoptera: Bruchidae). *J Insect Behav, 10*, 409–423. 10.1007/BF02765607

McLain D.K., & Pratt A.E. (1999). The cost of sexual coercion and heterospecific sexual harassment on fecundity of a host-specific, seed-eating insect (*Neacoryphus bicrucis*). *Behav Ecol Sociobiol, 46*, 164–170. 10.1007/s002650050606

Miller G.T., & Pitnick S. (2002). Sperm-female coevolution in *Drosophila. Science (80-), 298*, 1230–1233. 10.1126/science.1076968

Nilakhe S.S. (1976). Overwintering, survival, fecundity, and mating behavior of rice stink bug. *Ann Entomol Soc Am, 69*, 717–720. 10.1093/aesa/69.4.717

Nuyts E., & Michiels N.K. (1993). Integration of immediate and long term sperm precedence patterns and mating costs in an optimization model of insect copulation duration. *J Theor Biol, 160*, 271–295.

Nuyts E., & Michiels N.K. (1994). The influence of sperm precedence patterns and mating costs on copulation duration in odonates: predictions and supporting data. *Belgian J Zool, 124*, 11–19.

Ofuya T.I. (1995). Multiple mating and its consequences in males of *Callosobruchus maculatus* (F) (Coleoptera, Bruchidae). *J Stored Prod Res, 31*, 71–75.

Ortíz-Jiménez I., & del Castillo R.C. (2015). Nuptial gifts and female fecundity in neotropical katydid *Conocephalus ictus* (Orthoptera: Tettigonidae). *Insect Sci, 22*, 106–110. 10.1111/1744-7917.12105

Parker G. A. (1979). *Sexual selection and sexual conflict. In: Blum MS, Blum NA (eds) Sexual selection and reproductive competition in insects.*New York: Academic Press, pp. 123–166.

Parker G.A. (1970a). Sperm competition and its evolutionary consequences in insects. *Biol Rev, 45*, 535–567. 10.1111/j.1469-185X.1970.tb01176.x.

Parker G.A. (1992). Marginal value theorem with exploitation time costs - Diet, sperm reserves, and optimal copula duration in dung flies. *Am Nat, 139*, 1237–1256.

Parker G.A. (1970b). Sperm competition and its evolutionary effect on copula duration in fly *Scatophaga stercoraria. J Insect Physiol, 16*, 1301–1328. 10.1016/0022-1910(70)90131-9

Parker G.A., & Birkhead T.R. (2013). Polyandry: history of a revolution. *Philos Trans R Soc London, Ser B BiolSci, 368*, 20120335.

Parker, G.A., Simmons, L.W., Stockley, P., McChristie, D.M., & Charnov, E.L. (1999). Optimal copula duration in yellow dung flies: effects of female size and egg content. *Anim Behav, 57*, 795–805.

Pitnick S., & Markow T.A. (1994). Large-male advantages associated with costs of sperm production in *Drosophila hydei*, a species with giant sperm. *Proc Natl Acad Sci U S A, 91*, 9277–9281. 10.1073/pnas.91.20.9277

Pitnick S., Spicer G.S., & Markow T.A. (1995). How long is a giant sperm? *Nature, 375*, 109.

Price C.S.C. (1997). Conspecific sperm precedence in *Drosophila. Nature, 388*, 663–666. 10.1038/41753

Price C.S.C., Dyer K.A., & Coyne J.A. (1999). Sperm competition between *Drosophila* males involves both displacement and incapacitation. *Nature, 400*, 449–452.

Price C.S.C., Kim C.H., Posluszny J., & Coyne J.A. (2000). Mechanisms of conspecific sperm precedence in *Drosophila. Evolution (N Y), 54*, 2028–2037.

Reinhardt K. (2005). Sperm numbers, sperm storage duration and fertility limitation in Odonata. *Int J Odonatol, 8*, 45–58.

Reinhardt K., Anthes N., & Lange R. (2014). Copulatory wounding and traumatic insemination. In: Rice W.R., & Gavrilets S. (eds.) *The Genetics and Biology of Sexual Conflict.* Cold Spring Harbor: Cold Spring Harbor Laboratory Press, pp. 115–140.

Reinhardt K., Naylor R., & Siva-Jothy M.T. (2003). Reducing a cost of traumatic insemination: female bedbugs evolve a unique organ. *Proc R Soc B Biol Sci*, *270*, 2371–2375. 10.1098/rspb.2003.2515

Rhainds M. (2010). Female mating failures in insects. *Entomol Exp Appl*, *136*, 211–226.

Ridley M. (1988). Mating frequency and fecundity in insects. *Biol Rev*, *63*, 509–549.

Rivas-Torres A., Lorenzo-Caballa M.O., Sánchez-Guillén R.A., & Cordero-Rivera A. (2019a). Variation in intraspecific sperm translocation behaviour in a damselfly and its consequences for sperm viability. *Anim Behav*, *156*, 51–55. 10.1016/j.anbehav.2019.08.001

Rivas-Torres A., Outomuro D., Lorenzo-Carballa M.O., & Cordero-Rivera A. (2019b). The evolution and diversity of intra-male sperm translocation in Odonata: a unique behaviour in animals. *Behav Ecol Sociobiol*, *73*, 54. 10.1007/s00265-019-2660-5

Sánchez-Guillén R.A., Córdoba-Aguilar A., & Cordero-Rivera A. (2013). An examination of competitive gametic isolation mechanisms between damselflies *Ischnura graellsii* and *I. elegans*. *Int J Odonatol*, *16*, 259–267. 10.1080/13887890.2013.821868

Sánchez-Guillén R.A., Wellenreuther M., Cordero-Rivera A., & Hansson B. (2011). Introgression and rapid species turnover in sympatric damselflies. *BMC Evol Biol*, *11*, 210.

Santolamazza-Carbone S., & Cordero-Rivera A. (1998). Sperm competition, cryptic female choice and prolonged mating in Eucalyptus Snout-Beetle, *Gonipterus scutellatus* (Coleoptera, Curculionidae). *Etología*, *6*, 33–40.

Sasakawa K. (2009). Marked sperm dimorphism in ground beetle *Scaritesterricola:* a novel type of insect sperm polymorphism. *Physiol Entomol*, *34*, 387–390.

Shapiro L.H. (2006). Reproductive costs to heterospecific mating between two hybridizing katydids (Orthoptera: Tettigoniidae). *Ann Entomol Soc Am*, *93*, 440–446. 10.1603/0013-8746(2000)093.[0440:rcthmb]2.0.co;2

Shuker D., Bateson N., Breitsprecher H., O'donovan, R., & Taylor, H. (2002). Mating behavior, sexual selection, and copulatory courtship in a promiscuous beetle. *J Insect Behav*, *15*, 617–631.

Sillén-Tullberg B. (1981). Prolonged copulation: a male "postcopulatory" strategy in a promiscuous species, *Lygaeus equestris* (Heteroptera, Lygaeidae). *Behav Ecol Sociobiol*, *9*, 283–289.

Simmons L.W. (2001). *Sperm Competition and its Evolutionary Consequences in Insects*. Princeton: Princeton University Press, pp. 434.

Simmons L.W. (2014a). Sperm competition. In: *The Evolution of Insect Mating Systems*. Oxford: Oxford University Press, pp. 181–203.

Simmons L.W. (2014b). Sexual selection and genital evolution. *Aust J Entomol*, *53*, 1–7.

Simmons L.W. (1995). Male bushcrickets tailor spermatophores in relation to their remating intervals. *Funct Ecol*, *9*, 881–886.

Simmons L.W., & Kvarnemo C. (1997). Ejaculate expenditure by male bushcrickets decreases with sperm competition intensity. *Proc R Soc B Biol Sci*, *264*, 1203–1208.

Simmons L.W., Parker G.A., & Stockley P. (1999). Sperm displacement in Yellow Dung Fly, *Scatophaga stercoraria*: an investigation of male and female processes. *Am Nat*, *153*, 302–314.

Siva-Jothy M.T. (1997). Odonate ejaculate structure and mating systems. *Odonatologica*, *26*, 415–437.

Siva-Jothy M.T. (1988). Sperm "repositioning" in *Crocothemis erythraea*, a libellulid with a brief copulation. *J Insect Behav*, *1*, 235–245.

Simmons L. W. , & Siva-Jothy M. T. (1998). Sperm competition in insects: mechanisms and the potential for selection. In: Birkhead T. R. & Moller AP (eds). *Sperm competition and sexual selection*. San Diego: Academic Press, pp. 341–434.

Sivinski J. (1980). Sexual selection and insect sperm. *Florida Entomol*, *63*, 99–111.

Smith R.L. (1984). *Sperm Competition and Evolution of Animal Mating Systems*. Orlando: Academic Press, pp. 687.

Snook R.R., & Hosken D.J. (2004). Sperm death and dumping in *Drosophila*. *Nature*, *428*, 939–941.

Suomalainen E., Saura A., & Lokki J. (1987). *Cytology and Evolution in Parthenogenesis*. Boca Ratón, Florida: CRC Press, pp. 216.

Svensson B.G., Petersson E., & Frisk M. (1990). Nuptial gift size prolongs copulation duration in dance fly *Empis borealis*. *Ecol Entomol*, *15*, 225–229.

Takami Y., Sasabe M., Nagata N., & Sota T. (2008). Dual function of seminal substances for mate guarding in a ground beetle. *Behav Ecol*, *19*, 1173–1178,

Torres-Vila L.M., Rodríguez-Molina M.C., & Jennions M.D. (2004). Polyandry and fecundity in Lepidoptera: can methodological and conceptual approaches bias outcomes? *Behav Ecol Sociobiol*, *55*, 315–324. 10.1007/s00265-003-0712-2

Thornhill, R. (1983). Cryptic Female Choice and Its Implications in the Scorpionfly *Harpobittacus nigriceps*. *Am Nat 122*, 765–788. 10.1086/284170.

Uhía E., & Cordero-Rivera A. (2005). Male damselflies detect female mating status: importance for post-copulatory sexual selection. *Anim Behav*, *69*, 797–804.

Utzeri C., & Belfiore C. (1990). Tandem anomalifraOdonati. *Fragm Entomol*, *22*, 271–287.

Uhl G., Nessler S. H., & SchneiderJ. M. (2010). Securing paternity in spiders? A review on occurrence and effects of mating plugs and male genital mutilation. *Genetica*, 138, 75–10410.1007/s10709-009-9388-5.

Waage J.K. (1979). Dual function of damselfly penis: sperm removal and transfer. *Science (80)*, *203*, 916–918.

Watanabe M. (2016). *Sperm Competition in Butterflies*. Tokyo: Springer, pp. 174.

Xu J., & Wang Q. (2010). Mechanisms of last male precedence in a moth: sperm displacement at ejaculation and storage sites. *Behav Ecol*, *21*, 714–721. 10.1093/beheco/arq044

Yokoi N. (1990). The sperm removal behavior of yellow spotted longicorn beetle *Psacothea hilaris* (Coleoptera: Cerambycidae). *Appl Entomol Zool*, *25*, 383–388. 10.1303/aez.25.383

Yoshimura M. (2008). Observation of spermatophore transfer in *Stavsolus japonicus* (Plecoptera: Perlodidae) in Japan. *Entomol News*, *119*, 123–129. 10.3157/0013-872x(2008)119 [123:oostis]2.0.co;2

11

Cryptic Female Choice: How It Changes the Narrative of Reproduction?

Payel Biswas, Aradhya Chattopadhyay, and Shampa M. Ghosh
School of Biotechnology, Kalinga Institute of Industrial Technology (KIIT), Bhubaneswar 751024, India

CONTENTS

11.1 Introduction

Reproduction a complex process, and evolutionary forces shaping reproduction can be intriguing. At heart of the theory of natural selection lie two biological phenomena, survival and reproduction. Therefore, reproductive traits are crucial to natural selection. Moreover there exists a special form of selection exclusively responsible for evolution of traits related to reproduction and reproductive success known as sexual selection. The idea of sexual selection was originally proposed by Darwin in his book *The Descent of Man* (1871). Darwin was surprised to find many remarkable traits in organisms that appear extravagant or conspicuous, which are not essential for survival. Such traits

DOI: 10.1201/9781003043195-11

include elaborate ornamentation, conspicuous displays, extravagant weaponry, aggressive behavioural displays, often found in males of a species. Majority of these traits are energetically costly to maintain, yet they not appear essential for survival. Consequently, their evolution cannot be explained by the theory of natural selection. To explain their evolution, Darwin introduced the idea of sexual selection. He proposed that these traits could evolve if they help to increase the bearer's reproductive success, even without any contribution to survival. Examples of such traits include the tail of a peacock or the antler of a deer. These traits can be extravagant and in some cases cumbersome, but they help to attract the opposite sex and increase reproductive success of its bearer. Increased mortality associated with maintenance of these extravagant traits is more than offset by the benefits gained by mate acquisition aided by this trait. In insects, examples of such sexually selected exaggerated traits include very long eyestalks in male Diopsid flies, or enlarged legs and horns in male beetles, etc. (Emlen & Nijhout, 2000; Lavine *et al.*, 2015).

Reproduction is costlier in females than males, because they (a) invest more per gamete production (due to anisogamy), and (b) take care of development of the young and play active role in parental care. Consequently, females are more selective than males in choosing their mates. This gives rise to the phenomenon of female mate choice, an aspect of sexual selection that was overlooked for long. However, it has gradually gained prominence in the past 50 years and the importance of female mate choice in evolution is now well recognized (see Brennan, 2010). A fascinating aspect of female choice that has come to the fore majorly in the past two decades is that females may exercise a choice even for choosing sperm of one mate over that of another to create a fertilization bias (Thornhill, 1983; Eberhard, 1996). Females seem to have evolved complex mechanisms to exercise choice based fertilization deep inside her body, well after mating has ended. Such choice is invisible and hard to detect, and hence is known as cryptic female choice, or CFC (Thornhill, 1983). It can be a powerful component of sexual selection, and can change course of reproductive success of males, however, it has remained largely overlooked till date. As cryptic female choice largely operates during the post-mating stage, mating success of a male may not always lead to fertilization success. CFC thus signifies the role that the female plays in changing fate of reproduction after the completion of mating.

In this chapter, we will discuss the concept of cryptic female choice (CFC), mechanisms of CFC in insects, traits that are potentially under selection driven by CFC, and specific examples. We will also focus on adaptive significance of CFC and future directions for CFC studies. Before we delve into all of these, let us briefly discuss some basic concepts and terms concerning sexual selection.

There are two primary mechanisms of sexual selection: (i) intrasexual selection, *i.e.*, the competition or contest between members of the same sex (usually males) for access to mates, and (ii) intersexual selection, *i.e.*, when members of one sex (usually females) choose members of the opposite sex based on certain criteria (Darwin, 1874). Traits aiding in either of the two processes are acted upon by sexual selection. The idea of evolution of exaggerated traits through intrasexual selection, *i.e.* evolution of traits that help in winning battle against other male competitors to access females was recognized quite early, as soon as Darwin's work was accepted by the biologist fraternity. The idea of mate choice though, did not receive much attention for close to 80 years (Cronin, 1991; Brennan, 2010). However, in the last 50 years our understanding of sexual selection has undergone significant improvement, with huge influx of data from numerous species (see Andersson and Bradbury, 1987; Cronin, 1991). This has led to recognition that intersexual selection signifying mate choice can be a potent mechanism of evolution through sexual selection (see Eberhard, 1996; Andersson & Simmons, 2006; Hosken & House, 2011; Eberhard & Lehmann, 2019).

The choice of mates is far more crucial for the fitness of the female than that of the male. There are two major reasons behind the same. Firstly, given numerous males compete for access to females, females that are choosy and have ability to select genetically superior mates are likely to have greater fitness by producing better quality or number of offspring. Secondly, females invest more per reproductive event, (a) because female gametes are typically larger and less numerous, hence more costly to produce, and (b) secondly, the gestation and care of offspring are commonly female responsibility. Therefore, choosing the wrong mate because of not being discriminatory in mate choice could cost a female lot more than a male.

11.2 Female Mate Choice: Precopulatory *vs.* Postcopulatory Mate Choice

11.2.1 Overview

Mechanisms of mate choice in females are diverse, and they operate at different stages of reproduction, starting from courtship up to fertilization and egg laying. Female mate choice could be precopulatory or postcopulatory in nature. Precopulatory processes include various aspects of mate acquisition. In general, several males compete to mate with females and therefore females can choose whom to copulate with based on which male she perceives to be of good quality, giving rise to precopulatory sexual selection. Female choice exercised before copulation includes circumstances in which females do not respond to certain males or refuse to mate with them, preferring others. Male traits that could be under precopulatory sexual selection driven by female choice thus include the ability to advertise superior quality and attract mates, courtship behaviour, ability to procure good quality nuptial gifts, *etc.* Darwin's idea of sexual selection was centered around this particular aspect of reproductive success *i.e.* individual variation in the ability to obtain mates. Thus it was limited to the idea of precopulatory selection, erroneously assuming mating inevitably leads to reproductive success (Darwin, 1871). Consequently, early studies of sexual selection focused only on processes happening before copulation. However, studies done in last 50 years indicate events during and after copulation also play important roles in determining the outcome of fertilization and reproduction (Birkhead & Pizzari, 2002; Eberhard, 2009). It is now becoming increasingly clear that mating success may not always lead to fertilization success. This has led to the idea of postcopulatory sexual selection, signifying selection operating on events during and after copulation, up to production of progeny.

Of particular relevance to postcopulatory sexual selection is the phenomenon of polyandry in mating systems. Polyandry, *i.e.* a female mating with multiple males is extremely common in nature, and is widespread in many insect species (Arnqvist & Nilsson, 2000; Zeh & Zeh, 2001; Birkhead, 2010). When females mate with multiple males, they can choose (i) males *before* mating, (ii) to complete copulation and sperm transfer only from preferred mates *during* mating, and (iii) preferential fertilization of her eggs using sperm of specific mates *after* mating. Thus, in polyandrous species, sexual selection does not stop at the male's ability to successfully obtain mates (Birkhead & Moller, 1998; Barbosa, 2011). Consequently, sexual selection operates on reproductive events during and after copulation, in addition to precopulatory events. Postcopulatory sexual selection can have profound biological impact in driving evolutionary change (Eberhard, 1996; Birkhead & Pizzari, 2002; Eberhard, 2009). The fraction of sexual selection coming from postcopulatory processes reflects those differences in siring success among males that are independent from their ability to obtain mates, and therefore expected to rely on some aspect of ejaculate quality, dynamics of sperm storage, sperm use within female reproductive tract, and/or biased nurture of embryo.

11.2.2 Postcopulatory Sexual Selection

Postcopulatory sexual selection, like its precopulatory component, has two components: (i) male-male competition, *i.e.* intrasexual selection, and, (ii) female choice, *i.e.* intersexual selection. The concept of female choice during postcopulatory stage has been controversial for long and often overlooked, because it is subtle and far more challenging to document than male-male competition during this stage. However, with careful experimental studies the evidence for both processes have become apparent in several taxa in the recent years, and hence both need to be treated with rigour in order to completely understand the role of sexual selection in evolution. Studies on insects have been particularly useful to unravel the finer aspects of postcopulatory sexual selection.

Postcopulatory male-male competition comprises of sperm competition (Parker 1970). Parker proposed when females mate with multiple males, their ejaculates compete inside the female reproductive tract for access to fertilization. Ever since, detailed studies on sperm competition conducted in many taxa have provided convincing evidence for diverse mechanisms for sperm competition leading to evolution of sperm morphology, number and other ejaculate characteristics (Simmons, 2005; Snook, 2005; Eberhard, 2009; Manier *et al.*, 2010) and CFC (Eberhard, 1996; Arnqvist & Rowe, 2005; Aquiloni & Gherardi, 2008; Eberhard, 2009). Intense sperm competition driving the evolution of superior sperms has even led to some species like *Drosophila bifurca* having giant sperms (Pitnick, 1996; Birkhead & Moller, 1998; Lüpold *et al.*, 2016). However, Parker envisaged the female reproductive tract as a *passive* vessel in which ejaculates of multiple males may compete to access female gametes. This is far from the truth as female also plays active roles in deciding the fate of sperms in her reproductive tract, thus exercising postcopulatory female choice.

11.3 Postcopulatory Female Choice or Cryptic Female Choice (CFC)

11.3.1 The Concept of Cryptic Female Choice (CFC)

When a female mates with multiple males, as in polyandry, female sperm storage organs act as a venue where sperm from different males compete for access to eggs and seminal fluid proteins of different males interact with each other, leading to male-male competition. Female reproductive tract and sperm storage organs also act as an arena where female may exercise a choice to select sperm of one male over other competing males. The latter process is referred to as postmating or postcopulatory female choice, or cryptic female choice (Parker, 1970; Knowlton *et al.*, 1984; Birkhead & Hunter, 1990; Eberhard, 1996; Birkhead & Moller, 1998; Simmons, 2001). Females seem to exert this choice via morphological, chemical and behavioural adaptations.

The idea that females could influence offspring paternity by choosing the fate of sperms transferred in her reproductive tract was first suggested by Randy Thornhill (1983), who coined the term cryptic female choice. The reason behind this nomenclature is two-fold. First, much of the postcopulatory female choice occurs inside the body of a female and cannot be detected from outside (Eberhard, 1996). Moreover, originally female choice was believed to be limited only to precopulatory stage, and choice exercised during or after copulation is 'invisible' or 'cryptic' in a mating system (Eberhard, 1996; Barbosa, 2011).

Another component of CFC is the choice exercised by females of some species *during* mating. It involves instances wherein females terminate mating with certain males, but not others, preventing complete insemination. In some instances the degree of sperm transfer may correlate with the extent of preference the female may have for the mate in question (see Firman *et al.*, 2017).

The idea of CFC has been controversial, largely due to the difficulty of documenting processes operating inside the body of a living female. However, in recent years, several taxa including arthropods, fishes, birds, among other animals have been shown to exhibit processes that strongly suggest occurrence of CFC; albeit, research and discussions about the phenomenon remain inadequate. The majority of data on CFC are available from studies done on diverse insect taxa (*e.g.* Thornhill, 1983; LaMunyon & Eisner, 1993; Qazi *et al.*, 1998; Hellriegel & Ward, 1998; Johnson *et al.*, 1999; Hellriegel & Bernasconi, 2000; Edvardsson & Arnqvist, 2000; Ward, 2000; Tallamy *et al.*, 2002), and in this chapter we will focus on some of the well documented cases of CFC in insects.

11.3.2 Cryptic Female Choice (CFC) and Eberhard's Contribution to The Field

Although Thornhill proposed the idea of CFC in early 1980s, it was not until the mid-1990s when William J. Eberhard published his book "Female Control: Sexual Selection by Cryptic Female Choice" that biologists took notice and began to study CFC (Eberhard, 1996). The field has grown rapidly since then. Eberhard extensively surveyed various mechanisms through which CFC can operate. He was instrumental in demonstrating the roles of 'female biology and choice' in postcopulatory sexual selection, which was traditionally ignored in studies of sexual selection. He reviewed hundreds of examples across taxa, and presented strong evidence in favour of the fact that mechanisms of sexual selection continue to operate well after commencement of copulation. The major takeaways from Eberhard's book as listed by Holland and Rice (1998) are the following:

(1) Female reproductive tract is not a passive vessel where sperm competition takes place; (2) Females determine the "rules of the game" and most of the post-insemination components of sexual selection; (3) CFC drives rapid evolution of many male traits, such as seminal fluid proteins, biology of genitalia, and copulatory courtship, and (4) CFC despite being an important engine of sexual selection is largely ignored (Holland & Rice, 1998).

The idea of CFC has changed our view of sexual selection by providing a counter- point to male-centric notion of sperm competition, and highlighting the female perspective in determining the course of fertilization.

11.3.3 Mechanisms of CFC

Thornhill defined CFC as a form of postcopulatory sexual selection, involving female-mediated morphological, behavioural, or physiological mechanisms that bias fertilization toward the sperm of specific males. Mechanisms of cryptic female choice have been consistently overlooked, due to the following three reasons: (i) traditional focus on sperm competition studies, (ii) difficulty in distinguishing male *vs.* female influence over postcopulatory events, and (iii) logistic difficulties in demonstrating CFC. Another problem is that studies of CFC require knowledge on the mechanisms of insemination, sperm storage in females and fertilization, which are not available for many species.

Despite the challenges, with advent of advanced microscopy, availability of sophisticated tools of molecular biology and genetics, detailed *in vitro* and *in vivo* studies of reproductive biology have become much easier in the past two decades. Such technological advancement has enabled elucidation of proximate mechanisms of fertilization, thereby accelerating studies on postcopulatory sexual selection, including CFC (Firman *et al.*, 2017). Traditional focus on studying reproductive biology in insects coupled with wide diversity of sexual selection mechanisms in them make insects ideal for studying CFC. Incidentally, the vast majority of available data on CFC are available from insects. Eberhard's book suggests there may be over 20 mechanisms of CFC, but experimental evidence is available only for a few. In the subsequent sections of this chapter, we will discuss operational mechanisms of CFC and its wide diversity in specific insects. In this section, we aim to highlight some broad mechanistic aspects of CFC.

Female insects are known to be able to differentiate between sperm from a different species and sperm from their own, through signals present in seminal fluid (Markow, 1997). Females also seem to be able to distinguish and treat sperm from its different mates belonging to the same species. Possible mechanisms include distinguishing the ejaculates depending on order of mating and spatial distribution of the ejaculates in different sperm storage organs. Besides, seminal signals may also show intraspecific variation perceivable by female reproductive tract (see Clark *et al.*, 1999). Selection driven by CFC can occur during two different stages: (i) when females favour use of sperm from preferred mates for fertilizing eggs, *i.e.* prefertilization stage; and, (ii) when they direct nurture towards zygotes fertilized by superior mates; *i.e.* postfertilization stage. CFC before fertilization is exercised by the female by partially reducing or completely eliminating fertilization potential of ejaculates received from "inferior" mates (Barbosa, 2011).

Prefertilization mechanisms of CFC include:

1. transport of sperm to a location away from eggs and sperm use bias
2. sperm dumping or ejection
3. digestion of sperm
4. premature termination of copulation preventing complete insemination
5. early remating without proper usage of sperm from a previous mating

Postfertilization mechanisms of CFC include:

1. selective production of less number of offspring
2. selective abortion
3. biased resource investment for offspring production.

In insects, prefertilization mechanisms of CFC are particularly pronounced. Therefore, we will discuss these mechanisms in detail before moving on to specific case studies.

1. *Transport of sperm to a location away from eggs and sperm use bias*: Reproductive tract of female insects are complex, with different sperm storage organs (SSO) like spermatheca and seminal receptacle (SR), which may vary in number. Some species of female insects seem to store ejaculate of its different mates in different SSO after receiving sperm (Ala-Honkola & Manier, 2016; Firman *et al.*, 2017), This differential spatial distribution of sperm may allow preferential or exclusive use of one mate's sperm over the other's for fertilizing the ova. Sperm from a preferred mate may be stored in vicinity of the eggs and site of fertilization while those from the other mate may remain localized and largely unused in a distant position within the reproductive tract.
2. *Sperm dumping or ejection and sperm displacement*: Shortly after insemination, required amount of sperm is stored in SSO and a large fraction of ejaculate may be ejected out by female insects. In many insects recent sperm precedence has been observed wherein the female flushes out the ejaculate received during the previous mating and fertilizing most eggs with sperm from her most recent mate. The sperm from recent mate may displace sperm of the previous mate from the SSO (Ala-Honkola & Manier, 2016). While sperm competition plays an important role in whole process, some experimental manipulation studies suggest female choice too plays a role in preferential sperm retention and dumping.
3. *Digestion of sperm*: Female reproductive tract could be hostile for sperm, and in some species sperm may get digested or phagocytosed after transfer (Holman & Snook, 2008). Males of different species have evolved counter strategies for safer insemination like transferring spermatophore, i.e. bunch of sperm wrapped in protective sheath, or production of heteromorphic sperm where a bevy of sterile sperm morphs protect functional sperm, *etc.* It is thought females may use this strategy may use to preferentially digest sperm too. Besides, in some species that use spermatophores for sperm transfer, female may dislodge spermatophores from its reproductive tract immediately after the male has released them, and consume them thus not allowing those sperm to take part in fertilization.
4. *Premature termination of copulation preventing complete insemination*: Female insects may terminate copulation early without allowing complete sperm transfer to take place. There are evidences for amount of sperm transfer to correlate with degree of preference for a male (Thornhill, 1983).
5. *Early remating without proper usage of sperm from a previous mating*: Female insects experience a refractory period for remating with a different male followed by copulation, partly due to the effect of seminal fluid proteins. During this period sperm received from the last mating is used for fertilization and may lead to increased ovulation leading to mating success of the last mate. However, it has been observed in some insects, females may remate soon after mating with a male possibly considered "inferior" by the female, thus exercising choice (Pitnick, 1991).

11.3.4 CFC and Male Perspective

The diversity of mechanisms by which males and females impact one another in reproductive success is central to evolution driven by sexual selection. CFC is likely to drive evolution of several male traits, as listed below:

1. *Variation in duration of copulation*: Copulation duration within a species may correlate with amount of sperm transferred to the female and thus, male ability to keep female engaged in copulation until complete sperm transfer can influence paternity.
2. *Seminal fluid composition and quality*: Accessory gland proteins (ACPs) present in seminal fluid are known to influence female response and behaviour in diverse ways post copulation. ACPs may modulate ovulation, oviposition, female attractiveness and willingness to influence remating. Increased oviposition can potentially increase female fitness and thereby cryptic choice for quality

control of seminal fluid proteins might occur (Thornhill, 1983; Chapman & Davies, 2004; Sirot *et al.*, 2009).

3. *Copulatory courtship:* Eberhard's book elaborated many examples of elaborate copulatory courtship that are observed well after sperm transfer is over. Initially it appeared paradoxical since it occurs after insemination has taken place. However, Eberhard explained such behaviour has presumably evolved because it induces greater sperm storage, increased sperm transport toward fertilization site, and increasing the likelihood of fertilization success. Further studies have shown it indeed affects several postcopulatory female processes, like sperm usage, oviposition, sperm ejection and remating (Eberhard, 1991; Edvardsson & Arnqvist, 2000; Tallamy *et al.*, 2002; Barbosa, 2011; Peretti & Eberhard, 2010). Copulatory courtship is widespread in arthropods and provides one of the clear evidences of a behaviour evolved due to CFC.
4. *Ability to present nuptial gifts:* Many species of insects present nuptial gifts during mating, which may comprise of a dead prey or salivary gland or accessory gland secretions of males that female feeds upon. In addition to mate attraction, quality and quantity of nuptial prey influence mate attractiveness, female decision to complete or terminate copulation as well as remating. Therefore the quality of nuptial gifts produced or procured by a male seems to be under selection by CFC (Sakaluk & Eggert, 1996).
5. *Sperm number:* Sperm number would be an obvious trait under selection driven by sperm competition. In addition, the female behaviour of sperm dumping is also likely to contribute to evolution of greater sperm number, driven by CFC (Pitnick *et al.*, 1999).
6. *Sperm size:* Evolution of giant sized sperm in *Drosophila bifurca* results in correlated reduction of sperm number in ejaculates, leading to rapid usage and depletion of sperm for fertilization, and increased female remating rate. Higher remating rate seems to aid females in producing greater brood with higher genetic diversity, possibly leading to higher female fitness (Pitnick *et al.*, 1995; Lüpold *et al.*, 2016). Thus sperm size seems to have evolved in response to sexual selection driven by CFC in this species.
7. *Structure of male genitalia:* Shape and size of male genitalia may be under selection driven by CFC because it may help in different functions, like controlling quality of copulation and copulatory courtship, as well as in some cases depositing sperm deep inside female reproductive tract to maximize fertilization success (Eberhard & Lehmann, 2019).
8. *Body Size:* Body size is a trait of paramount importance in sexual selection. Preference for bigger males is common in premating selection since it is thought to represent superior male quality. Bigger males often have an advantage in sperm competition as well. Besides, male body size has been observed to be a trait under selection driven by CFC too. Females of many insects seem to preferentially store and use sperm of bigger males. Sperm from smaller males may be dumped earlier than usual if the female gets to mate with another, bigger male (Thornhill, 1983). In some cases remating may be delayed if the last male is large in size. The evolutionary significance of a female choice for bigger males would be discussed later in the chapter with specific examples.

It is important to note postcopulatory sexual selection is both complex and difficult to demonstrate or interpret, because of covert nature of mechanisms involved. One of major challenges of studying CFC is that some of mechanisms determining fate of sperm after copulation are jointly controlled by sperm competition and CFC. It might be empirically challenging to tease apart two processes in some cases, and, biologists therefore should exercise caution to interpret cases of CFC.

11.4 Case studies of CFC

CFC is possibly present in many insect taxa, but it has been empirically demonstrated only in a handful owing to technical challenges of documenting the phenomenon. In the past few decades a host of detailed studies have elucidated diverse mechanisms and examples of CFC across insect species, and we will discuss some well documented cases of cryptic choice in females, their mechanisms, and evolutionary significance.

11.4.1 Spermatophores and CFC

Males of many insects use spermatophores to transfer sperm during copulation. A spermatophore is a mass of sperm covered by some Accessory Gland Proteins (ACPs). The mechanism of spermatophore transfer may vary across species the males of many insects, usually deposited by male in female genital tract. Spermatophore may serve different functions. It can act as mating plug preventing further copulation by female ensuring complete sperm transfer to sperm storage organs (SSO) (Firman *et al.*, 2017) Spermatophore can also contain nutritive proteins and can be consumed by the females, thus providing nourishment.

Females in many insects seem to control sperm transfer through spermatophore acceptance or by early termination of copulation. In moth, *Ephestia kuehniella*, females may exercise a cryptic choice to displace spermatophore from SSO and remate sooner (Firman *et al.*, 2017). In field cricket, *Gryllus bimaculatus*, female may bias paternity in favour of certain males by detaching the spermatophore before insemination is complete and may consume the spermatophore early. Thereafter the female may remate soon. Early spermatophore removal occurred more often in case of copulation with smaller males. In contrast, female field crickets may allow spermatophores of some males to stay longer than what is normally required for insemination indicating a bias for that male. In Females of these crickets sometime undergo repeated matings with "preferred" male allowing mutiple ejaculations (Simmons, 1987). Male field crickets also try to guard females post spermatophore transfer to prevent premature rejection, with females presumably assessing, male quality by its mate guarding ability and thereby exercise CFC for males with superior mate guarding ability (Wynn & Vahed, 2004). Tropical house cricket, *Gryllodes sigillatus* females influence offspring paternity through spermatophore removal behaviour; favouring larger males and those providing larger nuptial gifts (Sakaluk & Eggert, 1996).

11.4.2 Nuptial gifts and CFC

In some insects the male provides nuptial gifts to the female before or during copulation. Nuptial gifts may consist of a bolus made of salivary or accessory gland secretion of male or a dead prey and are provided by males to females before or during copulation. They may help in nourishment and production of eggs (Ben-Ari, 2000; Firman *et al.*, 2017).

In scorpion fly, *Harpobittacus nigriceps*, females receive dead insects as nuptial gifts and start feeding on it right away. Copulation is terminated by the female once meal finishes followed by detaching from the male and flying away (Thornhill, 1983); continuous sperm transfer takes place throughout meal. Thus, time to consume gift determines amount of sperm transferred. If the gift comprises of a very tiny insect, female may terminate copulation prematurely without complete insemination. The duration and amount of sperm transfer thus appears to be under female control. ACPs transferred during mating inhibit female from remating with new partners for some hours or even days, while some ACPs enhance oviposition rate often enhancing eggs laid following mating to maximise male fitness. In *H. nigriceps*, both body size of the male and nuptial prey size seem to influence remating interval as well as oviposition rate. Both larger male size and larger nuptial prey size lengthen remating interval and increase oviposition in these scorpionflies, while smaller male or small nuptial gift led to faster remating. Delayed remating following copulation with larger males and enhanced oviposition in females may also be directly mediated by large male ejaculate quality (Thornhill, 1983). Alternatively, it is also possible that females have evolved a mechanism to delay mating and lay more eggs after mating with bigger males because it helps them produce good quality offspring, reflecting a CFC for larger mates. More studies are required to tease apart male *vs.* female control in this matter.

The propensity of female to remate faster after being presented by a smaller nuptial prey size possibly indicates a cryptic female choice for males who are able to procure a bigger and better insect prey for presentation before copulation. This could be a way of judging male genetic quality. In another species of scorpion fly, *Panorpa cognata*, males provide females with saliva secretions as nuptial food gifts (Engqvist, 2006). Female remating interval in this insect correlates with the amount of saliva provided by the male, which results from high paternal investment on part of the male. Not surprisingly, gift size has been shown to correlate with male nutritional condition in these scorpionflies, which may be an indicator of genetic quality of the male. *Panorpa cognata* females possibly benefit from not remating soon following copulations involving large saliva gifts and allowing fertilization of its eggs by sperm of a good quality male.

11.4.2.1 Conspecific sperm precedence and inbreeding avoidance through CFC

Females are able to differentiate between sperm from different species and sperm from their own, through signals present in seminal fluid (Markow, 1997). In hybrid matings of some orthopteran species, evidence for CFC for preferential sperm usage of its own species has been found. In competitive hybrid matings involving two cricket species *Gryllus campestris* and *Gryllus bimaculatus*, postcopulatory conspecific sperm precedence occurs by both preferential storage and sperm use bias (Tyler *et al.*, 2013; Firman *et al.*, 2017). CFC may also aid in inbreeding avoidance. In crickets *Teleogryllus oceanicus* and *G. bimaculatus*, sperm use bias favours fertilization by sperm from unrelated mates, thus avoiding inbreeding (see Firman *et al.*, 2017).

11.4.2.2 CFC in social insects

In honey bees, *Apis mellifera*, spermathecal fluid quality of queen changes with successive matings providing varying environment to sperms received at different times. Virgin female may be less choosy about sperm nourishment in her reproductive tract ensuring fertilization of its eggs. As the female undergoes multiple matings, it may become selective about sperm nourishment and usage indicating CFC (see Firman *et al.*, 2017). In many species of stingless bees, especially *Melipona bicolor*, haplo-diploid system of sex determination makes it possible to select sex of egg during postcopulatory stage (see Firman *et al.*, 2017).

11.4.2.3 Postcopulatory courtship and CFC

In tsetse fly, *Glossina morsitans,* rhythmic stimulation of female by male during copulation triggers longer sperm storage, ovulation and delays in female remating. Species specific genital structures are used by the male to squeeze the abdomen of the female and generate rhythmic stimulation. This leads to longer sperm storage in female, greater ovulation, and delayed remating. Experimental modification of male cerci and sternite so as to prevent rhythmic stimulation led to abolition of sperm storage bias, and reduced remating intervals. This work was successful in inducing a CFC against males who are unable to stimulate the female during copulation (Briceño and Eberhard, 2009). Similar results were obtained by blocking female sensory abilities to detect rhythmic stimulation. The bushcricket *Roeseliana roeselii* also shows similar trends, with male titillator under sexual selection influenced by CFC (Eberhard & Lehmann, 2019). In flour beetle, *Tribolium castaneum,* prevention of tarsal rubbing, reduced paternity success of such males indicating CFC for this behaviour (Edvardsson & Arnqvist, 2000). In soldier fly, *Mewsargus cingulatus,* males perform courtship behaviour throughout copulation, and females oviposit soon after copulation. Courtship involves tapping the female abdomen using hindlegs of the male and waving the legs in air. Males perform courtship behaviour throughout copulation. Copulation without courtship does not trigger immediate oviposition in these flies. This suggests copulatory courtship in *M. cingulatus* is possibly a trait under selection by CFC (see Eberhard, 1996; Barbosa, 2011).

In solitary bee, *Centris pallida*, post copulation and insemination, male strokes female vigorously with his legs, abdomen and antennae and produces a rasping and rattling song. Prevention of this behaviour led to early remating, indicating CFC for postcopulatory courtship (Alcock & Buchmann, 1985).

11.4.2.4 Sperm transfer and CFC

In *T. castaneum*, spermatophores consist of invaginated tube like structures filled with sperms that are transferred to female during act of copulation. These tube everts releasing sperms inside bursa. With experimental manipulation, fed and starved males were generated, and allowed to mate with live females *vs.* those that were freshly killed just before mating. Fed males transferred significantly more sperm than starved males but only if female were alive. The amount of sperm transferred did not differ between fed and starved males while mating with dead females This indicates the female presumably exercises a CFC for amount of sperm transfer depending on male quality (fed vs. starved) (Fedina, 2007).

11.4.3 CFC in Drosophila

The fruit fly, *Drosophila melanogaster*, is an excellent model for studying development, genetics, behaviour and evolution. Reproductive aspects of this insect have been studied for decades, with particular emphasis on its seminal fluid proteins. Postcopulatory sexual selection has been investigated in astonishing details in fruit flies, leading to impressive findings about CFC.

Drosophila females mate with multiple males in succession, whose sperm eventually sire her brood (Ochando *et al.*, 1996; Imhof *et al.*, 1998). *Drosophila* ejaculate contains a bevy of ACPs that are well characterized at the molecular level and have been shown to modulate various aspects of female biology and behaviour (Chapman 2008; Ram and Wolfner 2009). The uterus in *Drosophila* is attached to two different SSOs: two mushroom shaped *spermathecae* and a single seminal receptacle (SR). Sperm is stored in Sperm Storage Organ (SSO) and later used for fertilization. Male *Drosophila* transfers a few thousand sperms during mating along with a host of well characterized ACPs, which modulate females' behaviour, physiology, and immunological responses (Chapman, 2008; Ram & Wolfner, 2009). Male *D. melanogaster* also deposits a gelatinous mating plug at end of copulation, to inhibit further mating of female for few hours, in order to ensure his own paternity. In addition, some ACPs present in the ejaculate manipulate female behaviour and make her less attractive to other males and less receptive to mating for about a week.

However, this behavioural manipulation may not be 100% effective, and recent studies show that remating is also to some extent at discretion of female (Pitnick, 1991). Sperm makes its way to the uterus and collect at the openings of the sperm storage organs, the SR and paired spermathecae. Few hours after copulation, excess sperm and mating plug are ejected out by female, retaining rest in spermathecae and SRs, a process known as "sperm ejection" (see Ala-Honkola & Manier, 2016). Once mating plug is lost, female may engage in mating with a second male, sperm of which displaces that of first mate. This is followed by yet another sperm ejection in which excess sperm of second male and displaced sperms of first male are discarded, bringing an equilibrium in SSO and uterus. In absence of sperm competition and CFC, paternity is expected to correspond to numerical representation of sperms of two males stored at equilibrium. However, investigations indicate frequency of remating, sperm displacement, and sperm ejection are partially under female control, thus potentially creating a paternity bias. Timing of sperm ejection in *D. melanogaster* is at discretion of female, and genetically controlled, with an estimated heritability of 0.36. Timing of sperm ejection following remating can bias fertilization success, and this appears to be one of the major mechanisms to exercise CFC in polyandrous insects (Ala-Honkola & Manier, 2016).

Longer ejection times generally favour second male by allowing his sperm more time to displace rival sperms. In *D. melanogaster*, *D. simulans*, and *D. mauritiana*, proportion of second-male sperm in sperm storage organs (S2) is significantly associated with its paternity success (P2). In competitive hybrid mating involving *D. mauritiana* and *D. simulans*, *D. simulans* females showed much quicker ejection of *D. mauritiana* sperms as a mechanism for conspecific sperm precedence (Manier *et al.*, 2013a). Studies also suggests that female fruit flies may modulate ejection time depending upon relative attractiveness of its second mate (Pitnick *et al.*, 1999; Tram & Wolfner, 1999; Neubaum & Wolfner, 1999). Larger males are generally preferred. Like many other species, female flies are thought to gain greater genetic benefits for their offspring by choosing to mate with larger and healthier males. *D. simulans* female mating first with larger male followed by a smaller one, eject sperm faster to retain sperm of larger males. This suggests a postcopulatory sexual selection reflecting CFC for a larger male (Ala-Honkola & Manier, 2016). However, if order is reverse, no significant change in ejection time is observed as compared to mating with two similar sized males in succession. The absence of delayed ejection in case of a larger second male could be due to the following: a larger male may be able to out compete a previous smaller male merely by its superior sperm number. Therefore, selection pressure to delay sperm ejection by the female following copulation with a larger second male may be lacking. CFC may also involve bias in sperm use for fertilization.

Usage of stored sperm from different mates in direct proportion to their abundance in female reproductive tract represents no sperm use bias. Any significant deviation from this proportion due to disproportionate usage of one male's sperms over others not corresponding to their relative abundance may represent CFC for certain male phenotypes (Manier *et al.*, 2013b, 2013c). Choice of SSO for fertilization seems to be associated with sperm use bias. In *D. melanogaster* sperms from SR are preferred over spermathecae. In *D. simulans*, first male sperm bias is observed in SR while second male sperm is favoured in spermathecae. This mechanism is used for conspecific sperm precedence in *D. simulans*. If *D. simulans* female undergoes heterospecific mating with *D. mauritiana* male first, followed by mating with a conspecific male, sperm is used primarily from ST, which are second-male biased.

Sperm use bias occurs in conspecifics too. *D. simulans* females were allowed to mate with transgenic males producing green fluorescent protein (GFP) tagged sperm *vs.* those producing red fluorescent protein (RFP) tagged sperm. While precopulatory sexual selection was in favour of GFP males, fertilization bias reflecting CFC was found for RFP males (Ala-Honkola & Manier, 2016). This suggests that in principle, it is possible for the females to choose a male phenotype for mating depending on certain traits of the fly, while the sperm choice at the postcopulatory may reflect a different preference, possibly based on some sperm traits. This may look paradoxical at a glance, but it is important to note GFP *vs.* RFP males are not naturally occurring phenotypes, and hence choice at two levels may not reflect an entirely evolved response. The experiment nonetheless indicates existence of sperm use bias in species.

Interesting insights about CFC and sexual selection come from studies of giant sperm in *Drosophila bifurca*. Sperm length and number are two important reproductive traits that are likely targets of sexual selection. Their relevance lies not only in the context of efficient sperm competition, but they have been shown to be important with respect to CFC as well. Anisogamy is norm, with males producing numerous tiny gametes with little resource investment per gamete, and female producing much fewer and larger energy intensive gametes. However, many *Drosophila* species produce sperm much longer than their body, e.g. *Drosophila bifurca*, with sperm 20 times longer than its body length, the longest sperms known in any species so far (Pitnick *et al.*, 1995). Normally sperm length is expected to trade off with sperm number and reduced sperm number is likely to impair sperm competition particularly at post-copulatory stage within female reproductive tract (Pitnick, 1996; Immler *et al.*, 2011). It has been found *D. bifurca* owing to its giant sperm transfer only about 6 sperms in its ejaculate during copulation. Previously evolution of such exaggerated sperm length coupled with greatly reduced sperm number in *D. bifurca* was considered a paradox as it reduces sperm number but is also energetically costly. Scott Pitnick and fellow researchers thus envisaged evolution of such exaggerated length in sperms could be favoured if there is some hidden fitness benefit in terms of sexual selection. Subsequent research by Pitnick's group has revealed some interesting connections between sperm length evolution and CFC. The length of SR is correlated with sperm length across taxa. Long SR favours long sperms, both because longer sperms are superior at displacing, as well as more resistant to be displaced by shorter sperms within SR. Longer SRs thus can drive sperm-length evolution by enhancing this competitive advantage. An experiment involving isogenic lines of *D. melanogaster* showed SR length is also negatively correlated with remating interval in female, and positively correlated with timing of sperm ejection post remating (Lüpold *et al.*, 2013, 2016). Species with long SRs house long sperms but they are also fewer in numbers and are exhausted for fertilization faster than species with shorter and numerous sperms. Therefore, it creates a selection pressure to remate faster to replenish sperms in female reproductive tract. This explains negative correlation between SR length and remating interval. Rapid remating leads to repeated opportunities for pre- and postmating sexual selection. Delayed ejection, on other hand, creates ideal environment for longer sperms to exert their superiority and more time to win at sperm competition. This is underlying reason of observed positive correlation between SR length, sperm length, and time of sperm ejection. Shorter interval of female remating in species producing from bigger and fewer sperms increases mating frequency of female, endowing greater opportunity for exercising mate choice thus, increasing strength of precopulatory sexual selection. This also exerts increased pressure on males to replenish their sperm reserves repeatedly, a task that only good quality males can achieve. Thus, by exercising a cryptic choice for bigger sperms, females are likely to select bigger and better male partners and gain genetic benefits for their progeny.

Finally, sperm might face a hostile environment inside the female reproductive tract of some species, where it may be attacked by immune responses, phagocytosed, digested, or inactivated by spermicidal actions. A prominent example of hostile environment inside female reproductive tract is observed in *D. pseudoobscura,* in which significant sperm death occurs after insemination. Males of this species has evolved a unique adaptation to overcome spermicide by producing two different types of sperm, namely fertile long eusperm, and short sterile parasperm that seem to protect eusperm from spermicidal activity of reproductive tract (Holman & Snook, 2008; Firman *et al.*, 2017). Such instance of sexually antagonistic coevolution might have evolved due to CFC for males, who can afford to invest in producing not only functional sperm but also strong protective acumen, like parasperm that eventually increase fertilization success of male.

It is important to note CFC results from close interactions between male and female aspects of reproduction. The sperm and seminal fluid proteins manipulate female response while female behaviour, and physico-chemical environment and biology of female reproductive tract influence fate of sperms. Disentangling male and female effects in mechanisms of CFC, and who influences whom to evolve during cascade of events from insemination to fertilization are difficult to interpret. Therefore, one needs to be cautious while teasing apart male and female perspectives of sexual selection in general (Firman *et al.*, 2017).

11.5 Adaptive Role of Female Choice in General and CFC

Female choice is central to theories addressing evolution of exaggerated traits in male organisms (Darwin, 1871; Andersson 1994; Kokko *et al.*, 2002). By being choosy, females can directly increase their reproductive success by mating with superior quality males and gain *direct benefits* (Andersson, 1994; Moller & Alatalo, 1999; Byers & Waits, 2006). Right choice of mates can provide better access to food, shelter, protection from harassing males, or help in raising offspring Mating with healthy mates also causes avoidance of parasites or other diseases. Females may also gain *indirect benefits* by choosing good quality males from whom her offspring can inherit genes responsible for higher fitness (Petrie, 1994; see Barbosa, 2011). Male assessment occurs via traits and displays advertising their physical and genetic benefits. Two major models concerning evolution of female mate choice have been proposed: (i) good genes model, and (ii) Fisherian runaway process.

Under 'good genes' model, phenotypic variation in male traits can be honest indicators of underlying variation in genetic quality (Manning, 1985; Kokko & Lindstrom, 1996; Kokko, 1998). By choosing right males, females can pass on good quality genes to her progeny such that offspring have better viability and reproductive success. Preference may include genes to avoid infection after copulation, or compatible genes. These advertisements may get so exaggerated that they may become a burden to maintain. However, if such a cost is offset by benefits gained in mate acquisition, both trait and female choice for it would be maintained by sexual selection.

The Fisherian Runaway model is named after Ronald A. Fisher (Fisher, 1930). Under this model, females may evolve preference for an arbitrary trait that may not have any direct fitness benefits under natural selection. However, due to evolutionary history of species female choice might have evolved, and males bearing that trait are favoured by females. This leads to assortative mating between males bearing trait and females with preference for it. Male offspring from such matings would be attractive to females and favoured by sexual selection. This would lead to a genetic correlation for trait and preference. Thus females with choice for the said arbitrary trait would produce 'sexy sons', as commonly known in the sexual selection literature. This may lead to further exaggeration of trait, which in turn would reinforce preference, resulting in a runaway feedback loop (see Brennan, 2010).

Though good genes and Fisherian runaway model both offer convincing arguments for evolution of female choice, evidences in favour of models are still few, as it is often extremely difficult to empirically quantify fitness benefits of mate choice. A third model called sensory exploitation model has been proposed to explain adaptive role of mate choice. According to this model, female preference for a trait might have evolved as a pleiotropic effect of some pre-existing sensory bias (West-Eberhard, 1984;

Endler, 1992; Cummings, 2007). Males who play to sensory bias of female are preferred and this reinforces trait, however clear evidence is lacking (Fuller, 2009). Theoretical modelling as well as careful empirical studies addressing mate choice and its evolutionary benefits can fill the gap in our understanding of adaptive significance of mate choice.

Females showing choice for whom to mate with may have direct or indirect fitness benefits as proposed by models discussed. An additional level of control for choosing which mate's sperm to use for fertilization is likely to be advantageous, where precopulatory mate choice is constrained, and males are able to coerce females into unwanted copulations, as found in species where individuals mate multiple times within a reproductive cycle. Investigations of such mating systems have been promising and suggest that, in such instances, CFC can lead to evolution of exaggerated male traits, and influence course of coevolutionary dynamics, particularly in polyandrous species.

11.6 Conclusions

CFC has important evolutionary consequences for both males and females. It plays a significant role in shaping behaviour, morphology and physiology of both sexes. With examples discussed in this chapter it is obvious that CFC is common in insects. In fact what we know about CFC could just be tip of iceberg and there might be many more taxa in which this phenomenon is prevalent. Careful observational studies of mating and reproduction across several species coupled with experimental manipulation may potentially enhance our understanding of CFC.

A particular aspect for which there is hardly any data available is neural basis of CFC. Only a handful of studies suggest neurons innervating female reproductive tract control postcopulatory behaviour including mating. With availability of modern biological tools, neural basis of cryptic choice in females can be hopefully addressed in finer detail in future. Transgenic fruit flies can be potentially helpful in this regard. Another unexplored aspect is assessment of cost of choice. Being choosy has its own risk, and CFC can also cause fitness cost in females, in principle. CFC makes female choosy about sperm use for fertilizing its gametes such that it can opt out from producing offspring with suboptimal sires. However, that can happen at expense of reduced offspring production due to rejecting sperms. CFC thus may constrain fertilization, in principle, imposing a battle between maintaining quality *vs.* quantity of progeny, and females need to strike a fine balance to optimize its fitness. In addition, immunological reaction against ejaculates of one's own species can predispose females to risks of autoimmunity (Firman *et al.*, 2017). Therefore, like any feature of adaptive evolution, prerequisite for CFC to evolve is that benefits of it must offset costs incurred by risks mentioned. To conclude, mechanisms of CFC, its neural basis, fitness benefits and costs associated with phenomenon, all of these remain largely unexplored. However, with increasing attention to biology of mate choice and postcopulatory sexual selection in recent years, we can hope research focusing on female perspective in reproductive success would gain further attention in future.

Acknowledgments

Preparation of this manuscript was supported by a DST Women Scientist A grant SR/WOS-A/LS-1179/2015(G) and a DST-SERB grant SRG/2020/001573 to SG.

REFERENCES

Ala-Honkola, O., and Manier, M.K. (2016). Multiple mechanisms of cryptic female choice act on intraspecific male variation in *Drosophila simulans*. *Behavioral Ecology and Sociobiology*, *70*(4), 519–532. doi: 10.1007/s00265-016-2069-3.

Andersson, M.B., & Bradbury, J.W. (1987). Sexual Selection: Testing the Alternatives: Report of the Dahlem Workshop on Sexual Selection: Testing the Alternatives Berlin, 1986 August 31–September 5, J. Wiley.

Andersson, M., and Simmons, L.W. (2006). Sexual selection and mate choice. *Trends in Ecology & Evolution*, *21*(6), 296–302.

Alcock, J., and Buchmann, S.L. (1985). The Significance of Post-insemination Display by Male *Centris pallida* (Hymenoptera: Anthophoridae). *Zeitschrift Für Tierpsychologie*, *68*(3), 231–243. doi: 10.1111/j.1439-0310.1985.tb00126.x.

Andersson, M. (1994). *Sexual Selection*. Princeton University Press. Princeton, NJ.

Aquiloni, L., and Gherardi, F. (2008). Evidence of female cryptic choice in crayfish. *Biology Letters*, *4*(2), 163–165. doi: 10.1098/rsbl.2007.0590.

Arnqvist, G., and Rowe, L. (2005). *Sexual Conflict*. Princeton University Press, Princeton, NJ.

Arnqvist, G., and Nilsson, T. (2000). The evolution of polyandry: multiple mating and female fitness in insects. *Animal Behaviour*, *60*(2), 145–164. doi: 10.1006/anbe.2000.1446.

Barbosa, F. (2011). *Postcopulatory sexual selection in soldier fly Merosargus cingulatus*. PhD diss., University of Missouri, Columbia.

Ben-Ari, E.T. (2000). Choosy Females. *BioScience*, *50*(1), 8. doi: 10.1641/0006-3568(2000)050[0008:cf]2.3.co;2.

Birkhead, T.R. (2010). How stupid not to have thought of that: post-copulatory sexual selection. *Journal of Zoology*, *281*(2), 78–93. doi: 10.1111/j.1469-7998.2010.00701.x.

Birkhead, T.R., and Hunter, F.M. (1990). Mechanisms of sperm competition. *Trends in Ecology & Evolution*, 5, 48–52.

Birkhead, T.R., and Moller, A.P. (1995). Extra-pair copulation and extra-pair paternity in birds. *Animal Behaviour*, 49, 843–848.

Birkhead, T.R., and Moller, A.P. (eds.) (1998). *Sperm competition and sexual selection*. eds. San Diego: Academic Press.

Birkhead, T.R., and Pizzari, T. (2002). Postcopulatory sexual selection. *Nature Reviews Genetics*, *3*(4), 262–273. doi: 10.1038/nrg774.

Brennan, P. (2010). Sexual Selection. *Nature Education Knowledge*, *1*(8), 24.

Briceno, R.D., and Eberhard, W.G. (2009). Experimental demonstration of possible cryptic female choice on male tsetse fly genitalia. *Journal of Insect Physiology*, *55*(11), 989–996. doi: 10.1016/j.jinsphys.2009.07.001.

Byers, J.A., and Waits, L. (2006). Good genes sexual selection in nature. *Proceedings of National Academy of Sciences*, *103*(44), 16343–16345. doi: 10.1073/pnas.0608184103.

Chapman, T. (2008). The Soup in My Fly: Evolution, Form and Function of Seminal Fluid Proteins. *PLoS Biology*, *6*(7), e179. doi: 10.1371/journal.pbio.0060179.

Chapman, T., and Davies, S.J. (2004). Functions and analysis of seminal fluid proteins of male *Drosophila melanogaster* fruit flies. *Peptides*, *25*(9), 1477–1490. doi:10.1016/j.peptides.2003.10.023.

Clark, A.G., Begun, D.J., and Prout, T. (1999). Female × Male Interactions in *Drosophila* Sperm Competition. *Science*, *283*(5399), 217–220. doi: 10.1126/science.283.5399.217.

Cronin, H. (1991). *The ant and peacock: altruism and sexual selection from Darwin to today*. Cambridge University Press. United Kingdom.

Cummings, M.E. (2007). Sensory trade-offs predict signal divergence in surfperch. *Evolution*, *61*(3), 530–545. doi: 10.1111/j.1558-5646.2007.00047.x.

Darwin, C. (1871). *The descent of man and selection in relation to sex*. London: John Murray

Darwin, C. (1874). *The descent of man and selection in relation to sex* (2nd ed.). London: John Murray.

Eberhard, W.G. (1991). Copulatory courtship and cryptic female choice in insects. *Biological Reviews*, 66, 1–31.

Eberhard, W.G. (1996). *Female control: sexual selection by cryptic female choice*. Princeton University Press.Princeton, NJ.

Eberhard, W.G. (2009). Postcopulatory sexual selection: Darwin's omission and its consequences. *Proceedings of National Academy of Sciences*, *106*(Supplement_1), 10025–10032. doi:10.1073/pnas.0901217106.

Eberhard, W.G., and Lehmann, G.U.C. (2019). Demonstrating sexual selection by cryptic female choice on male genitalia: what is enough? *Evolution*. doi: 10.1111/evo.13863.

Edvardsson, M., and Arnqvist, G. (2000). Copulatory Courtship and Cryptic Female Choice in Red Flour Beetles *Tribolium castaneum*. *Proceedings of Royal Society B: Biological Sciences, Vol., 267*, No., 1443, 559–563. doi: 10.2307/2665975.

Emlen, D.J., and Nijhout, H.F. (2000). The Development and Evolution of Exaggerated Morphologies in Insects. *Annual Review of Entomology*, *45*(1), 661–708. doi: 10.1146/annurev.ento.45.1.661.
Endler, J.A. (1992). Signals, Signal Conditions, and Direction of Evolution. *The American Naturalist*, *139*, S125–S153. doi: 10.1086/285308.
Engqvist, L. (2006). Nuptial gift consumption influences female remating in a scorpionfly: male or female control of mating rate? *Evolutionary Ecology*, *21*(1), 49–61. doi:10.1007/s10682-006-9123-y.
Fedina, T.Y. (2007). Cryptic female choice during spermatophore transfer in *Tribolium castaneum* (Coleoptera: Tenebrionidae). *Journal of Insect Physiology*, *53*(1), 93–98. doi: 10.1016/j.jinsphys.2006.10.011.
Firman, R.C., Gasparini, C., Manier, M.K., and Pizzari, T. (2017). Postmating Female Control:, 20 Years of Cryptic Female Choice. *Trends in Ecology & Evolution*, *32*(5), 368–382. doi: 10.1016/j.tree.2017.02.010.
Fisher, R.A. (1930). *The genetical theory of natural selection*. Clarendon Press. 10.5962/bhl.title.27468.
Fuller, R.C. (2009). A test of critical assumption of sensory bias model for evolution of female mating preference using neural networks. *Evolution*, *63*:1697–1711. doi: 10.2307/40306248.
Hellriegel, B., and Bernasconi, G. (2000). Female-mediated differential sperm storage in a fly with complex spermathecae. *Animal Behaviour*, 59, 311–317.
Hellriegel, B., and Ward, P.I. (1998). Complex Female Reproductive Tract Morphology: Its Possible Use in Postcopulatory Female Choice. *Journal of Theoretical Biology*, *190*(2), 179–186. doi: 10.1006/jtbi.1997.0546.
Holland, B., and Rice, W.R. (1998). Perspective: Chase-Away Sexual Selection: Antagonistic Seduction Versus Resistance. *Evolution*, *52*(1), 1. doi: 10.2307/2410914.
Holman, L., and Snook, R.R. (2008). A Sterile Sperm Caste Protects Brother Fertile Sperm from Female-Mediated Death in *Drosophila pseudoobscura*. *Current Biology*, *18*(4), 292–296. doi:10.1016/j.cub.2008.01.048.
Hosken, D.J., and House, C.M. (2011). Sexual selection. *Current Biology*, *21*(2), R62–R65. doi: 10.1016/j.cub.2010.11.053
Imhof, M., Harr, B., Brem, G., and Schlotterer, C. (1998). Multiple mating in wild *Drosophila melanogaster* revisited by microsatellite analysis. *Molecular Ecology*, *7*(7), 915–917. doi: 10.1046/j.1365-294x.1998.00382.x.
Immler, S., Pitnick, S., Parker, G.A., Durrant, K.L., Lüpold, S., Calhim, S., and Birkhead, T.R. (2011). Resolving variation in reproductive tradeoff between sperm size and number. *Proceedings of National Academy of Sciences*, *108*(13), 5325–5330. doi: 10.1073/pnas.1009059108.
Jennions, M.D., and Petrie, M. (2000). Why do females mate multiply? A review of the genetic benefits. *Biological Reviews of the Cambridge Philosophical Society*, 75, 21–64.
Johnson, J.C., Tracie, M., and Sakaluk, S.K. (1999). Female remating propensity contingent on sexual cannibalism in sagebrush crickets, *Cyphoderris strepitans*: a mechanism of cryptic female choice. *Behavioral Ecology*, 10, 227–233.
Knowlton, N., Greenwell, S.R., and Mating (1984). Male sperm competition avoidance mechanisms: the influence of female interests'. In *Sperm Competition and the Evolution of Animal systems*. Academic Press, Florida.
Kokko, H. (1998). Good genes, old age and life-history trade-offs. *Evolutionary Ecology*, *12*(6), 739–750. doi: 10.1023/a:1006541701002.
Kokko, H., and Lindstrom, J. (1996). Evolution of Female Preference for Old Mates. *Proceedings of Royal Society B: Biological Sciences*, *263*(1376), 1533–1538. doi: 10.1098/rspb.1996.0224.
Kokko, H., Ranta, E., Ruxton, G., and Lundberg, P. (2002). Sexually Transmitted Disease and evolution of mating systems. *Evolution*, *56*(6), 1091–1100. doi: 10.1111/j.0014-3820.2002.tb01423.x.
LaMunyon, C.W., and Eisner, T. (1993). Postcopulatory sexual selection in an arctiid moth (*Utetheisa ornatrix*). *Proceedings of National Academy of Sciences*, *90*(10), 4689–4692. doi: 10.1073/pnas.90.10.4689.
Lavine, L., Gotoh, H., Brent, C.S., Dworkin, I., and Emlen, D.J. (2015). Exaggerated Trait Growth in Insects. *Annual Review of Entomology*, *60*(1), 453–472. doi: 10.1146/annurev-ento-010814-021045.
Lüpold, S., Manier, M.K., Puniamoorthy, N., Schoff, C., Starmer, W.T., Luepold, S.H.B., Belote, J.M., and Pitnick, S. (2016). How sexual selection can drive evolution of costly sperm ornamentation. *Nature*, *533*(7604), 535–538. doi: 10.1038/nature18005.

Lüpold, S., Pitnick, S., Berben, K.S., Blengini, C.S., Belote, J.M., and Manier, M.K. (2013). Female mediation of competitive fertilization success in *Drosophila melanogaster. Proceedings of National Academy of Sciences of United States of America, 110*(26), 10693–10698. doi: 10.1073/pnas.1300954110.

Manier, M.K., Belote, J.M., Berben, K.S., Novikov, D., Stuart, W.T., and Pitnick, S. (2010). Resolving Mechanisms of Competitive Fertilization Success in *Drosophila melanogaster. Science, 328*(5976), 354–357. doi: 10.1126/science.1187096.

Manier, M.K., Belote, J.M., Berben, K.S., Lüpold, S., Ala-Honkola, O., Collins, W.F., and Pitnick, S. (2013a). Rapid diversification of sperm precedence traits and processes among three sibling *Drosophila* species. *Evolution, 67*(8), 2348–2362. doi: 10.1111/evo.12117.

Manier, M.K., Lüpold, S., Belote, J.M., Starmer, W.T., Berben, K.S., Ala-Honkola, O., William, F.C., and Pitnick, S. (2013b). Postcopulatory Sexual Selection Generates Speciation Phenotypes in *Drosophila. Current Biology, 23*(19), 1853–1862. doi: 10.1016/j.cub.2013.07.086.

Manier, M.K., Lüpold, S., Pitnick, S., and Starmer, W.T. (2013c). An Analytical Framework for Estimating Fertilization Bias and Fertilization Set from Multiple Sperm-Storage Organs. *The American Naturalist, 182*(4), 552–561. doi: 10.1086/671782.

Manning, J.T. (1985). Choosy females and correlates of male age. *Journal of Theoretical Biology, 116*(3), 349–354. doi: 10.1016/s0022-5193(85)80273-3.

Markow, T.A. (1997). Assortative fertilization in *Drosophila. Proceedings of National Academy of Sciences, 94*(15), 7756–7760. doi: 10.1073/pnas.94.15.7756.

Moller, A.P., and Alatalo, R.V. (1999). Good-genes effects in sexual selection. *266*(1414), 85–91.

Neubaum, D.M., and Wolfner, M.F. (1999). Mated *Drosophila melanogaster* females require a seminal fluid protein, Acp 36DE to store sperm efficiently. *Genetics, 153*, 845–857.

Ochando, M.D., Reyes, A., and Ayala, F.J. (1996). Multiple paternity in two natural populations (orchard and vineyard) of *Drosophila. Proceedings of National Academy of Sciences, 93*(21), 11769–11773. doi: 10.1073/pnas.93.21.11769.

Parker, G.A. (1970). Sperm competition and its evolutionary consequences in insects. *Biological Reviews, 45*(4), 525–567. doi: 10.1111/j.1469-185x.1970.tb01176.x.

Peretti, A.V., and Eberhard, W.G. (2010). Cryptic female choice via sperm dumping favours male copulatory courtship in a spider. *Journal of Evolutionary Biology, 23*(2), 271–281. doi: 10.1111/j.1420-9101.2009.01900.x.

Petrie, M. (1994). Improved growth and survival of offspring of peacocks with more elaborate trains. *Nature, 371*(6498), 598–599. doi: 10.1038/371598a0.

Pitnick, S. (1991). Male size influences mate fecundity and remating interval in *Drosophila melanogaster. Animal Behaviour, 41*(5), 735–745. doi: 10.1016/s0003-3472(05)80340-9.

Pitnick, S. (1996). Investment in Testes and the Cost of Making Long Sperm in *Drosophila. The American Naturalist*, 148, 57–80.

Pitnick, S., Markow, T.A., and Spicer, G.S. (1995). Delayed male maturity is a cost of producing large sperm in *Drosophila. Proceedings of National Academy of Sciences, 92*, 10614–10618.

Pitnick, S., Markow, T., and Spicer, G.S. (1999). Evolution of multiple kinds of female sperm-storage organs in *Drosophila. Evolution, 53*(6), 1804–1822. doi: 10.1111/j.1558-5646.1999.tb04564.x.

Qazi, M.C.B., Aprille, J.R., and Lewis, S.M. (1998). Female role in sperm storage in the red flour beetle, *Tribolium castaneum. Comparative Biochemistry and Physiology Part A: Molecular & Integrative Physiology*, 120, 641–647.

Ram, K.R., and Wolfner, M.F. (2009). A network of interactions among seminal proteins underlies long-term postmating response in *Drosophila. Proceedings of National Academy of Sciences, 106*(36), 15384–15389. doi:10.1073/pnas.0902923106.

Sakaluk, S.K., and Eggert, A.K. (1996). Female Control of Sperm Transfer and Intraspecific Variation in Sperm Precedence: Antecedents to Evolution of a Courtship Food Gift. *Evolution, 50*(2), 694. doi: 10.2307/2410842.

Simmons, L. (2001). *Sperm Competition and Its Evolutionary Consequences in Insects*. Princeton; Oxford: Princeton University Press. doi: 10.2307/j.ctvs32sr2.

Simmons, L.W. (1987). Sperm competition as a mechanism of female choice in field cricket, *Gryllus bimaculatus. Behavioral Ecology and Sociobiology, 21*(3), 197–202. doi:10.1007/bf00303211.

Simmons, L.W. (2005). The Evolution of Polyandry: Sperm Competition, Sperm Selection, and Offspring Viability. *Annual Review of Ecology, Evolution, and Systematics*, *36*(1), 125–146. doi: 10.1146/annurev.ecolsys.36.102403.11250.

Sirot, L.K., LaFlamme, B.A., Sitnik, J.L., Rubinstein, C.D., Avila, F.W., Chow, C.Y., and Wolfner, M.F. (2009). Molecular Social Interactions. *Advances in Genetics*, 23–56. doi: 10.1016/s0065-2660(09)68002-0.

Snook, R. (2005). Sperm in competition: not playing by the numbers. *Trends in Ecology & Evolution*, 20, 46–53.

Tallamy, D.W., Powell, B.E., and McClafferty, J.A. (2002). Male traits under cryptic female choice in spotted cucumber beetle (Coleoptera: Chrysomelidae). *Behavioral Ecology*, *13*(4), 511–518. doi: 10.1093/beheco/13.4.511.

Thornhill, R. (1983). Cryptic Female Choice and Its Implications in Scorpionfly *Harpobittacus nigriceps*. *The American Naturalist*, *122*(6), 765–788. doi: 10.1086/284170.

Tram, U., and Wolfner, M.F. (1999). Male seminal fluid proteins are essential for sperm storage in *Drosophila melanogaster*. *Genetics*, *153*, 837–844.

Tyler, F., Harrison, X.A., Bretman, A., Veen, T., Rodríguez-Muñoz, R., and Tregenza, T. (2013). Multiple post-mating barriers to hybridization in field crickets. *Molecular Ecology*, *22*(6), 1640–1649. doi: 10.1111/mec.12187.

Ward, P.I. (2000). Cryptic female choice in yellow dung fly *Scathophaga stercoraria* (L). *Evolution*, *54*(5), 1680–1686. doi: 10.1111/j.0014-3820.2000.tb00712.x.

West-Eberhard, M.J. (1984). Sexual selection, competitive communication and species specific signals in insects. *Insect Communication*, 283–324.

Wynn, H., and Vahed, K. (2004). Male *Gryllus bimaculatus* Guard Females to Delay Them from Mating with Rival Males and to Obtain Repeated Copulations. *Journal of Insect Behavior*, *17*(1), 53–66. doi: 10.1023/b:joir.0000025132.02196.da.

Zeh, J.A., and Zeh, D.W. (2001). Reproductive mode and genetic benefits of polyandry. *Animal Behaviour*, *61*(6), 1051–1063. doi: 10.1006/anbe.2000.1705.

12

Semiochemicals

Suresh Nebapure[1], P. S. Soumia[2], Yogesh Yele[3], G. Guru-Pirasanna-Pandi[4], and N. R. Prasannakumar[5]
[1]*ICAR-Indian Agricultural Research Institute, New Delhi, India*
[2]*ICAR-Directorate of Onion and Garlic Research, Pune, India*
[3]*ICAR-National Institute of Biotic Stress Management, Raipur, India*
[4]*ICAR-National Rice Research Institute, Cuttack, India*
[5]*ICAR-Indian Institute of Horticulture Research, Bengaluru, India*

CONTENTS

12.1 Introduction

Chemical communication, both within and between species, is the most ubiquitous form of information exchange studied extensively in insects. Such volatile organic compounds are called semiochemicals, which enable insects in food seeking and preference, orientation, defence, predator recognition, and mate finding. The term 'Semiochemical' is derived from the Greek word '*semeon*', which means signal or sign. These semiochemicals are classified based on function or role played by them and type of

DOI: 10.1201/9781003043195-12

interaction facilitated. Semiochemicals-mediating intraspecific interactions are called 'pheromones', whereas those mediating interspecific interactions are called 'allelochemicals'. Allelochemicals are further divided into allomones (emitter benefitted not receiver), kairomones (receiver benefitted not emitter), and synomones (both receiver and emitter benefitted). Apneumones are also useful in chemical communication and are emitted by non-living material and are beneficial to receiver but harmful to second inhabitant of living material.

12.2 Pheromones

Jean-Henri Fabre during the 19th century noticed that male great peacock moths were attracted to a seductive scent released by female peacock moth. In 1959, the term 'pheromone' (Greek; *Pherin* (to transfer) and *hormon* (to excite)) was proposed (Karlson & Luscher, 1959). In the same year, first insect pheromone was chemically unravelled by Adolph Butenandt as bombykol [(Z, E)-10,12-hexadecadienol]; a major component in *Bombyx mori* sex pheromone emitted by females to attract males (Butenandt *et al.,* 1959). Since then, hundreds of pheromone blends have been identified in many organisms with myriad functions (Yew & Chung, 2015).

12.2.1 Major Types of Pheromones

Pheromones may have long-term physiological changes (primer pheromones) or short-term or immediate behavioural response (releaser pheromones). Releaser pheromones are further classified into different groups based on interaction they mediate (Wyatt, 2014) and are discussed below.

a. **Primer Pheromones:** These are detected through gustatory sensilla, which, in turn, triggers a cascade of physiological changes in recipient without any immediate change in behaviour, like caste determination and reproduction in social insects, like ants, bees, wasps, and termites.
b. **Releaser Pheromones:** These are detected through olfactory sensillae, which directly act on central nervous system of recipient and produce immediate behavioural change. Releaser pheromones may be further subdivided into further categories:

1. **Sex Pheromones** – These are involved in species recognition, gender identification, and partner evaluation. They are usually produced by females to stimulate behavioural response in opposite sex usually over long distances, e.g. bombykol (Butenandt *et al.*, 1961). However, some male-specific sex pheromones are produced by cotton boll weevil, *Anthonomas grandis*, cabbage looper, *Trichoplusia ni*, and Mediterranean fruitfly, *Ceratitis capitata*. In *D. melanogaster*, only volatile pheromone, cis-vaccenyl acetate (cVA), is male-specific, and mediates courtship inhibition and aggression (Kurtovic *et al.*, 2007Azmir 2013, Kurtovic 2007; Benton *et al.*, 2007), whereas non-volatile cuticular hydrocarbon pheromones are both male-specific [7,11-heptacosadiene (7,11-HD) as aphrodisiac] and female-specific (Blomquist & Vogt 2003). There is vast diversity in chemical nature of insect sex pheromones.

Pheromones can consist of saturated or unsaturated hydrocarbon chains carrying various functional groups, such as aldehyde, alcohol or acetate, polyenic hydrocarbons, methyl branched hydrocarbons, epoxides, ketones, terpenes and terpenoids, iridoids and alkaloids. Primitive lepidopteran species have mainly short-chain alcohols and ketones (Zhu *et al.*, 1995), while higher moths have generally aldehydes, alcohols, and acetates. Female sex pheromones of polyphagous pest, *Helicoverpa armigera,* are aldehyde compounds with double bond at different positions (Kehat and Dunkelblum, 1990; Zhang *et al.*, 2012) among which (Z)-11-Hexadecenal and (Z)-9-Hexadecenal are major ones. Pheromone of diamond back moth, *Plutella xyllostella,* is a binary mixture of (Z)-11-hexadecenal and (Z)-11-hexadecenyl acetates (Tamaki *et al.*, 1977).

The sex pheromone of brinjal shoot and fruit borer, *Leucinodes orbonalis,* is a blend of an acetate, (E)-11-Hexadecenyl acetate (E11-16:Ac) and minor component, an alcohol, (E)-11-hexadecen-1-ol

(E11-16:OH) (Zhu *et al.*, 1987; Srinivasan & Babu, 2000). Decadienal and dodecanal are components of sex pheromone of *Zootermopsis nevadensis* excreted from sterna glands of female and male (Bordereau *et al.*, 2006), dodecatrienol from female tergal glands of *Prorhinotermes simplex* (Hanus *et al.*, 2009), and *Cornitermes bequaerti* (Bordereau *et al.*, 2006), sternal glands of *Reticulotermes santonensis* (Laduguie *et al.*, 1994), *Pseudacanthotermes spiniger*, and *P. militares* (Bordereau *et al.*, 1993).

2. **Alarm Pheromone** – The second most commonly produced pheromones are alarm pheromones (Barbier, 1982). Alarm pheromones are chemical compounds that are emitted specifically in defense against predators or released upon injury (Wyatt, 2003). Chemically they are low molecular weight organic compounds (terpenoids or aliphatic ketones and esters) that are highly volatile and dissipate rapidly upon emission (Payne, 1974).

Tremendous diversity of alarm signaling systems occur in presocial (aphids) and social (ants, termites, honeybees) insects. In aphids, (E)-β-Farnesene, a sesquiterpene secreted from cornicles evokes fright reactions, whereas in various bee species, isopentyl acetate and over 20 other substances stimulate recruitment and aggression behaviour. In worker honeybees, alarm pheromone (2-heptanone) is released from both mandibular glands and Koschevnikov glands associated with sting apparatus.

Ant alarm pheromones are usually short chain aliphatic carbons, like ketones, alcohols, esters, aldehydes, alkylpyrazines, terpenes, and formic acid, secreted from mandibular, pygidial, metapleural, anal, and Dufour's glands. Alarm pheromones produced by cephalic glands of termite soldiers contain limonene, terpinolene, and α- and β-pinenes, which evoke aggressive excitement behaviour in other soldier and non-soldier nest members (Zablotny, 2009).

3. **Trail-marking Pheromone** – Trail pheromones help colony members navigate to distant locations to exploit food sources. Trail communication is commonly based on a multicomponent system, which is secreted from different glands or a blend of pheromones produced by same gland (Jackson *et al.*, 2006). Methyl-4-methylpyrrole-2-carboxylate was first trail pheromone identified from Texas leafcutter ant, *Atta texana* (Tumlinson 1972). A terpenoid compound, faranal, has been identified as trail pheromone of ant, *Monomorium pharaonis* (Ritter *et al.*, 1977) from Dufor's glands. Other trail pheromones from termite species are decadienol (Sillam-Dussès *et al.*, 2007), C18 aldehyde (Bordereau & Pasteels, 2010), dodecenol (Sillam-Dussès *et al.*, 2009b), and dodecatrienol (Sillam-Dussès *et al.*, 2006).
4. **Aggregation Pheromones** – These are produced by individuals or groups of either or both sexes to attract both sexes to a particular spot for mate location, host plant location, or overcome host defences. Aggregation pheromones have been reported from members of Coleoptera (especially Curculionidae family), Dictyoptera, Hemiptera, Homoptera, and Orthoptera. *Anthonomus grandis* produces grandisol and grandisal as aggregation pheromone, *Carpophilus hemipterus,* produces (2E, 4E, 6E, 8E)-3, 5, 7- trimethyl-2, 4, 6, 8-decatetraene, *Rhynchophorus ferrugineus* produces ferrugineol (4-methyl-5-nonanol), and *Oryctes rhinoceros* produces ethyl-4-methyloctanoate. A mixture of three compounds: hexadecanoic acid, pentadecanoic acid, and pentaethylene glycol induced aggregation in *Periplaneta americana* (Imen *et al.*, 2015).
5. **Oviposition-Deterring or Epideictic Pheromones (ODP)/Oviposition-stimulating Pheromone** – These pheromones either act as attractant or repellant, thereby influence spacing pattern through marking territories resulting in lowering intraspecific competition (Stelinski *et al.*, 2007). Pepper weevil, *Anthonomus eugenii* females deposit an oviposition plug that deters egg-laying by its conspecifics (Addesso *et al.*, 2007) as is also seen in *Callosobruchus chinensis* (Tanaka, 2000). Conversely, oviposition stimulating pheromones either attract or stimulate females for egg depositing on site/ host that are already being oviposited by conspecifics. *Schistocerca gregaria, Ephestia kuehniella* and *Plodia interpunctella* produce oviposition stimulating pheromones (Saini *et al.*, 1995; Anderson & Löfqvist, 1996).

12.2.2 Pheromone-producing Glands

Pheromones are ectohormones produced by ectodermal glands that are situated in different parts of insect's body, especially on abdomen, mandibles of some hymenopterans, and wings of lepidopteran insects; and discharged via external ducts. Principal pheromone-producing glands are Dufour's, poison, pygidial, frontal, mandibular, salivary or labial, sternal, and tergal glands. Different types of glands and types of pheromones released are summarized in Table 12.1.

12.2.3 Morphology of Pheromone Glands

The structure or morphology of pheromone producing glands of some insects is discussed below.

TABLE 12.1

Pheromones production sites across different insect orders

Insect Order/Insects	Gland	Type of Pheromone Secreted	Behavioural Responses
Lepidoptera			
Female Lepidoptera	Gland located at tip of abdomen	Sex pheromone	Aggregation and mating
Males of Monarch Butterfly	Modified scales on wings	Aphrodisiac pheromones	Mating function
Ephestia kuehniella and *Plodia interpunctella*	Mandibular glands	Oviposition-stimulating pheromone	Flight and oviposition behaviour
Hymenoptera			
Honeybee queen	Mandibular glands	Queen mandibular pheromone (QMP)	Maintains reproductive division of labour inside colony
	Dufour's gland	Similar to QMP	Reproduction and egg-laying in queens
Honeybee workers	Koschevnikov gland	Alarm pheromone	Aggression
	Nasonov glands	Nasonov pheromone	Assemblage of worker bees
	Dufour's gland	Similar to QMP	Produces sting lubricant, neutralizes acid secretion remains in sting, and inhibits ovarian development in workers
Honeybee drones	Mandibular glands	Drone pheromones	Attract other flying drones to congregation areas
	Tarsal glands		Mating function
Triagona recursa workers	Labial glands	Trail pheromones	Recruit workers to locate food source
Ants: Formicid genera	Mandibular, pygidial, metapleural and Dufour's glands	Alarm pheromones	Aggression
	Mandibular, Dufour's glands and Poison sac (in fire ants)	Queen pheromone (cuticular hydrocarbons)	Regulates reproductive division of labour
Ants Genus *Crematogaster*	Tibial glands are located in hind legs	Trail pheromone	Foraging trails and aggression
Leaf cutting ants, genus *Atta*	Mandibular glands	Alarm pheromone	Aggression
Pheidole embolopyx worker ants	Pygidial glands	Trail and alarm pheromones	Fright reactions or recruitment and aggression
Arboreal east Asian ant: *Dolichoderus thoracicus*	Pavan glands	Trail pheromones	Foraging trails

TABLE 12.1 (Continued)

Pheromones production sites across different insect orders

Insect Order/Insects	Gland	Type of Pheromone Secreted	Behavioural Responses
Isoptera			
Termites	Sternal glands	Trail pheromone	Raiding, nest migration and foraging behaviour
	Frontal gland	Alarm pheromone	Alarm and defense (involve altruism)
	Tergal and posterior sternal glands	Sex pheromones (produced by alate females)	Nuptial promenade and pairing of reproductives
Subterranean termites, *Coptotermes gestroi*	Salivary glands	Aggregation and phagostimulating pheromones	Aggregation and feeding stimulants
Dictyoptera			
German Cockroach: *Blattella germanica*	Rectal glands	Aggregation pheromone	Aggregation behaviour
Orthoptera			
Desert locust: *Schistocerca gregaria*	Epidermal gland cells on forewings and legs	Phenylacetonitrile pheromone	Courtship inhibition, anti- predatory activity, induces male sexual maturity and intense sexual competition (in gregarious phase)
	Comstock-Kellog Glands	Sex pheromone	Stimulates premating behaviour
Hemiptera			
Green Peach Aphid, *Myzus persicae*	Secreted from cornicles	Alarm pheromone	Fright reactions
Coleoptera			
Boll weevil, *Anthonomus grandis*	Dufour's gland	Aggregation pheromone (Male-produced)	Aggregation behaviour
Trogoderma spp.	Sternal glands	Sex pheromone	Mating function
Red flour beetle, *Tribolium castaneum*	Modified epidermal cells on prothoracic femur	Aggregation pheromone (Male-produced)	Aggregation behaviour
Coconut rhinoceros beetle *Oryctes rhinoceros*	Dufour's gland	Aggregation pheromone (Male-produced)	Aggregation behaviour
Red palm weevil, *Rhynchophorus ferrugineus*	Thoracic gland	Aggregation pheromone (Male-produced)	Aggregation behaviour
White grub, *Holotrichia consanguinea*	Ball-shaped abdominal sac	Aggregation pheromone	Aggregation behaviour

12.2.3.1 Lepidoptera

The diversity of pheromone glands in females belonging to family Arctiidae has been reported (Bendib & Minet, 1998). Basically, these glands are dorsal, externally opening invaginated tubular sacs with internal spines present in A8–A9 intersegmental membrane. There exists distinct difference between glands possessed by moths belonging to subfamily Lithosiinae and non-lithosiine arctiids. Lithosiinae moths have unpaired dorsal glands with short lobes, whereas non-lithosiine possesses long lobes. Similar structural features have been reported in pheromone glands of noctuiid moths (Vogel *et al.*, 2010; Dos Santos *et al.*, 2017). Exception to gland position at eighth and ninth abdominal segments are seen in moths from familiy Psychidae (thoracic and first abdominal segment), Zygaenidae (intertergal membrane of dorsal abdomen), Geometridae (pair of glands on ninth abdominal segment (Hallberg & Poppy, 2003).

12.2.3.2 Coleoptera

Epithelial cells present at the inner surface of anal plate and two apical sternites have been reported to act as pheromone producing organs in females of beetle, *Anomala albopilosa albopilosa*, *A. cuprea*, *Popillia japonica,* and *Heptophylla picea* (Scarabaeidae: Coleoptera) (Tada & Leal, 1997).

Other scarabaeid beetles possesses a ball-shaped abdominal sac consisting of cuticular epithelium of columnar cells (Leal *et al.*, 1996; Ward *et al.*, 2002). Weevils produce aggregation pheromone via pair of symmetrical glands with alveolar structure without a reservoir and cells arranged in a typical acinus pattern (Sánchez *et al.*, 1998).

12.2.3.3 Hymenoptera

In worker and queen honeybees, Koschevnikov gland is situated at sting shaft. In workers, it secretes alarm pheromone, whereas in queen it facilitates formation of clusters of court bees. Other glands, viz. Nasonov glands, Dufor's glands, mandibular glands, etc. are present in honeybee and also other social insects. Trail pheromone- releasing glands of ants of different species have been found to present at different sites on body, such as Pavan's glands or glandular ventral organ (Wilson & Pavan, 1959). Sac like part of these glands represents reservoir, whereas secretory cells are located in epithelium of seventh abdominal sternite (Billen, 1986).

12.2.3.4 Isoptera

The principle semiochemical producing glands in termites (Isoptera) are frontal, mandibular, sternal, tergal, and salivary or labial glands (Costa-Leonardo & Haifig, 2010). Unpaired frontal glands are mainly found in soldiers of family Termitidae, Serritermitidae, and Rhinotermitidae and in some imagos of family Termitidae and Rhinotermitidae. Mandibular or cephalic glands are paired glands present in head of all castes and species opening to body exterior between mandible and maxilla (Costa-Leonardo & Haifig, 2010). Salivary glands, present in thorax, are also paired glands consisting of two reservoirs, which open into common duct having exterior opening at base of labium in head (Costa-Leonardo, 1987; Noirot, 1969). Sternal gland is single, unpaired present mainly under fifth sternite (families Kalotermitidae, Rhinotermitidae, Serritermitidae, and Termitidae) or fourth sternite (Hodotermitidae and Termopsidae) whereas in *Mastotermes darwiniensis* (Mastotermitidae), it is present under third, fourth, and fifth sternite (Noirot, 1969; Ampion & Quennedey, 1981). Tergal glands are composed of class 1 and class 2 cells and are an epidermal thickening, present in females or in both sexes and in variable numbers (Costa-Leonardo & Haifig, 2010).

12.2.3.5 Dictyoptera

In cockroaches, rectal glands are identified as source of aggregation pheromones. Rectal pads with single layer of cells showing glandular nature are found in German cockroach, *Blattella germanica* (Ishii & Kuwahara, 1967). Sreng (1984) has studied and reported morphology of sternal (three to six) and tergal glands (four to seven) from male cockroaches of seven species.

12.2.4 Pheromone Perception and Reception

Pheromone communication involves: (a) pheromone production site (exocrine glands), (b) medium (air or water) through which pheromones travel/emit, and (c) pheromone receptors (olfactory and gustatory sense organs (Deisig *et al.*, 2014; Renou, 2014). Olfactory sensilla present mainly on antennae detect volatile compounds (gaseous phase) from a long range, while gustatory receptors located on proboscis and legs receive contact chemoreception from non-volatile compounds (Joseph & Carlson, 2015). All sensilla have a cuticular and a cellular component. Olfactory sensilla are multiporous while gustatory sensilla have one or two pores at tip. These pores lead into lumen of sensillum filled with a fluid called sensillum liquor that baths sensory neurons (dendrites). Pheromone molecules are transported via liquor by odourant binding proteins (OBPs) to receptor site located on dendrite, where different degrading enzymes initiate transduction (transformation of chemical stimuli into electrical/nervous impulses). The nervous impulse is transducted to central nervous system and analyzed to generate specific response (Fleischer & Krieger, 2018). Insects receive olfactory signals through three main families of chemosensory receptor proteins: odourant receptors (ORs), gustatory receptors (GRs), and ionotropic receptors (IRs) (Wicher, 2015; Fleischer & Krieger, 2018), with former most identified.

12.3 Analytical Methods/Tools Used for Semiochemical Detection

Extracting semiochemicals and identifying particular compounds responsible for behaviour manipulation is a difficult task with several techniques (discussed below) having been used.

12.3.1 Semiochemical Extraction Techniques

Mainly there are two methods by which insect pheromones are extracted: solvent assisted extraction (SAE) and air entrapment method. Dichloromethane (DCM), hexane, and diethyl ether are preferred solvents for pheromone extraction because of their high volatility at low temperatures which facilitates easy concentration of extracts (Jacobson, 1972). 95% ethanol and pentane are also efficient for pheromone extraction (Mutis *et al.*, 2009). While solvent extraction is able to identify pheromonal compounds present in glands or reservoirs, it is not able to shed light on compounds actually released during courtship (Visser *et al.*, 2018). Dynamic headspace collection using an appropriate adsorbent can help in trapping and assessing released pheromones (Visser *et al.*, 2018). This method is also an efficient way to study volatiles used by insect herbivores to locate host plant or plant volatiles released by plants upon herbivory, often referred as herbivore induce plant volatiles (HIPVs), which plays role in either attraction of other individuals of species or to repel conspecifics or to attract natural enemies of herbivore (Azmir *et al.*, 2013; Figure 12.1). Being of non-destructive nature, this sampling method gives a real time picture and is more ecologically relevant. In this technique usually plant or plant part is enclosed in a chamber of proper size.

Commonly used adsorbents are Porapak (ethylvinylbenzene-divinylbenzene) Q (80 to 100 mesh size), Tenax (2,6-diphenyl-p-phenylene oxide) GC(60 to 80 mesh), and activated charcoal (Carbotrap or Carbopak B). Sensitivity of static headspace collection and detection can be enhanced by using solid phage micro extraction (SPME) method. It is better than liquid (solvent) extraction method as it does not involve use of highly pure solvents and detection of volatiles can be achieved at ppbv (parts per billion by volume) level. SPME assembly is similar to syringe with needle (Figure 12.2). The needle of syringe contains an inert fiber coated with different adsorbents or absorbents, which can be pushed out or

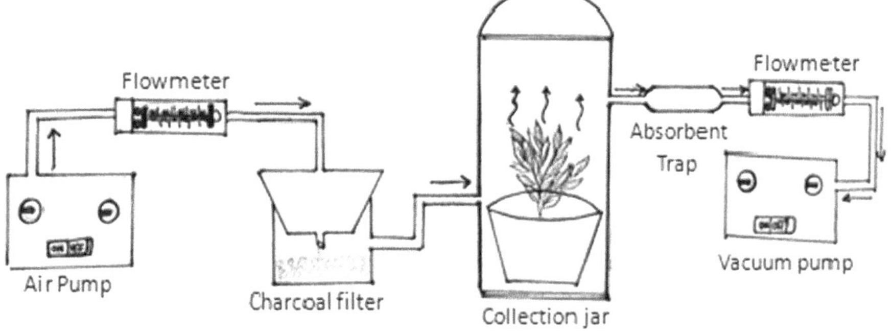

FIGURE 12.1 Setup of dynamic headspace collection system.

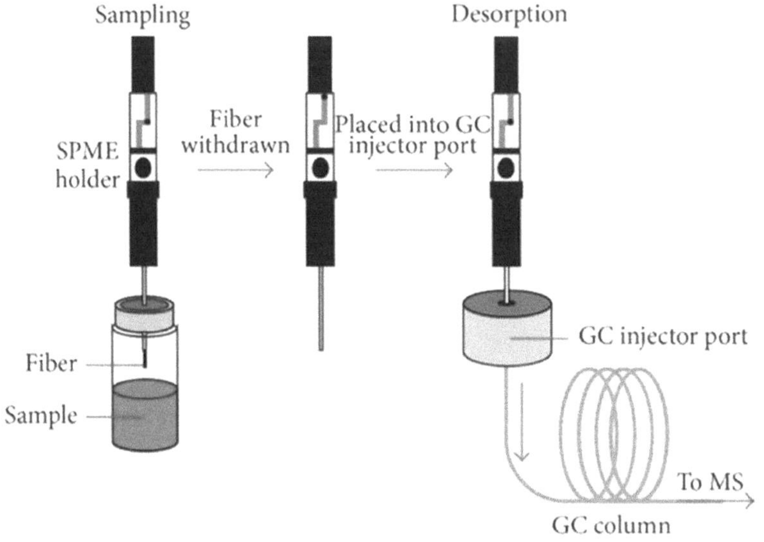

FIGURE 12.2 Solid-phase microextraction.

retracted with help of plunger and adsorbent, can be chosen based on polarity of targeted volatiles. While sampling plant volatiles, needle is inserted through a septum into airtight container housing plant, leading to adsorption of volatiles, which are then transferred to gas chromatography or gas chromatography-mass spectrometry for thermal desorption.

Solvent-assisted extraction, an alternative method, involves bringing plant part in contact with solvents. The metabolites extracted in solvent are separated and concentrated to yield crude extract. Extraction efficacy can be enhanced by intermittent stirring or manipulating temperature. One such example is Soxhlet extraction (Figure 12.3) where solvent in contact with plant material is re-distilled several times to dissolve volatiles in solvent as much as possible by repeated cycling of same solvent and heating (De Castro & Priego-Capote, 2010).

Microwave-assisted solvent extraction (MASE) involves reheating with microwave radiations, where solvent can absorb radiation; efficient for extraction of aromatic hydrocarbons (Tomaniova *et al.*, 1998). Supercritical fluid extraction is another method where CO_2 is employed for volatile extraction.

12.3.2 Identification of Behaviour-eliciting Compounds

To identify compound/s responsible for insect behaviour, it is necessary to first identify compounds which are eliciting response from insect antennae as most of time olfactory sensilla are located on antennae. Before going for identification of behaviour-eliciting compounds profiling or characterization of extracts need to be done and it is done with help of gas chromatography (GC) and gas chromatography-mass spectroscopy (GC-MS).

12.3.2.1 Gas Chromatography (GC)

GC is commonly used for separating and analyzing compounds in extracts. A predetermined amount of filtered extract (usually below 2 µl) is injected into its injection port where it is vapourized and then sent to column with help of inert gas like helium. Thermolabile compounds are likely to degrade and give error. Selection of column with particular polarity depends on sample type and polarity of compounds in sample. Compounds are separated based on their polarity and those eluted singly are detected. Detectors usually used in semiochemical studies on insects are mass spectrometry detector (MSD) for molecular identification of compounds, flame ionization detector (FID) for quantification, and electro-antennography detector (EAD) for in-line analysis of biologically active molecules.

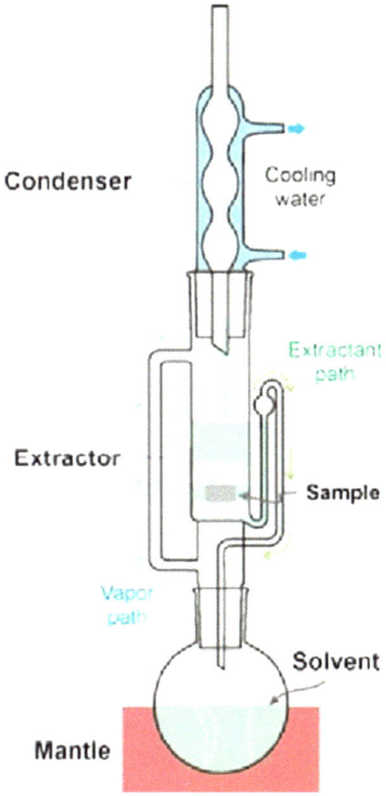

FIGURE 12.3 Soxhlet apparatus.

12.3.2.2 Electroantennography (EAG)

Schneider (1957) reported on perception of volatile pheromone, there is small fluctuation in voltage between tip and base of antennae due to depolarization of olfactory neurons. This finding led to electroantennography, which consists of electrodes (one recording and other reference), amplifier, IDAC signal collection box, stimulus controller, stimulus mixing tube, and source tube and software for recording electrograph in computer (Van der Pers, (1998); Figure 12.4). Insect responses can be

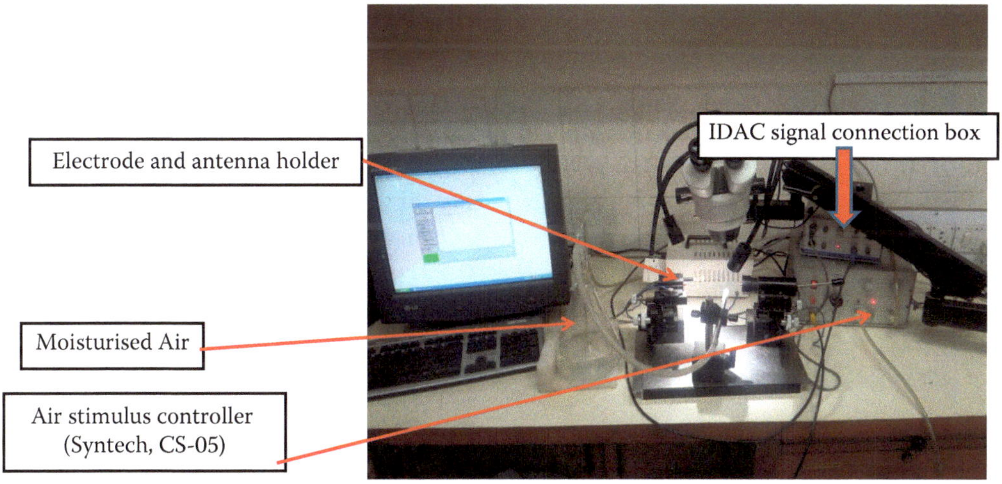

FIGURE 12.4 Electroantennogram (EAG) system.

recorded from cut antenna or head with antenna or from whole body preparation (Syntech, 2015). Electroantennograms are observed on exposure to various volatile compounds. The amplitude of electrogram or peak depends on the type of volatile compound, concentration of stimulus, insect species, sex of species, physiological maturity, and mating status of individual insect, time of recording, etc.

12.3.2.3 Gas Chromatography Coupled with Electroantennography Detector (GC-EAD)

GC-EAD is used to distinguish or identify only those VOCs which are eliciting response from insect antennae (Figure 12.5). This instrument is a coupling of GC and EAG through an effluent conditioning assembly or heat transfer line Syntech (2015). Eluent of GC column is split into two parts by splitting column with a Y-connector. One branch leads eluted compound to GC detector and other part is sent to EAG antennal preparation through heated transfer line followed by filtered and humidified air. As long as GC runs, the antenna is continuously exposed to eluting compounds and compound peaks from GC detector and response of insect antenna to compound are simultaneously recorded. As both peaks or signals are synchronized, it is possible to distinguish response-eliciting GC peaks. However, chemical identification of compounds has to be done through gas chromatography-mass spectroscopy (GC-MS) or nuclear magnetic resonance (NMR).

12.3.2.4 Behavioural Bioassays

Post identification of response-eliciting or biologically active volatile compounds, behavioural assays are required to confirm role played by single compound or mixture of compounds.

12.3.2.5 Y-Tube Olfactometer

Y-tube olfactometer or dual choice olfactometer is a glass tube with a 'Y'-shaped bifurcation where an insect walks and decides between two choices, usually clean air versus source odour (Figure 12.6). A suction pump, attached at the base of Y-tube regulates flow of odour and clean air through source and control arm, respectively of 'Y'. Pre-starved insects are released and their choice for arm observed. Success or failure of experiment depends upon physiological state (age, sex, nutritional state, etc.) of insects to be used in experiment. Y-tube olfactometer experiments can provide a limited and simplified picture of actual decision making during short distance orientation by insects.

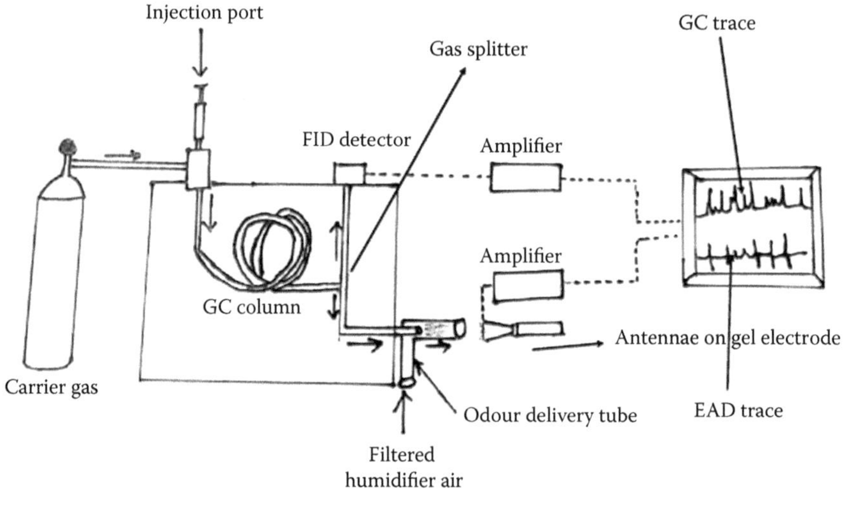

FIGURE 12.5 Setup of gas chromatography coupled with electroantennography detector (GC-EAD).

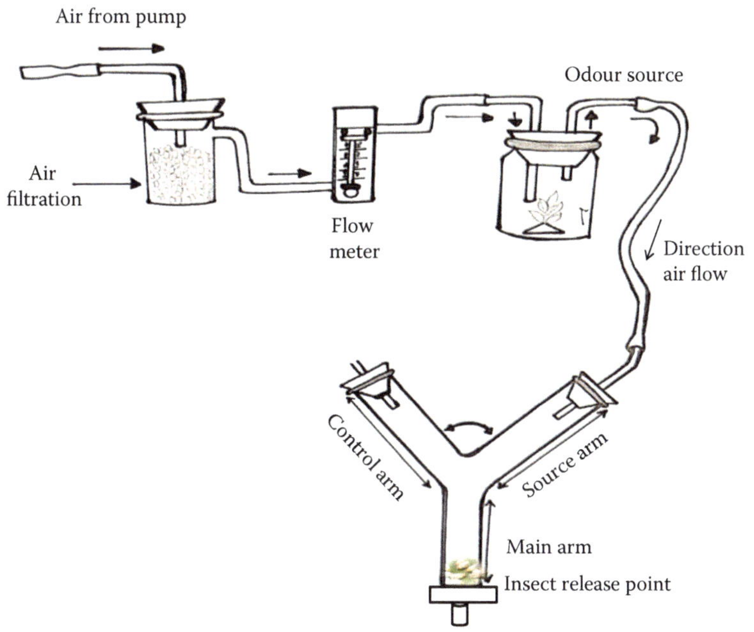

FIGURE 12.6 Y-tube olfactometer and accessories.

12.3.2.6 Wind Tunnel

Wind tunnel (Figure 12.7) is an olfactometer used to study orientation response of flying insects primarily lepidopterans. It is a physical model of environment, allowing experimental manipulation of one variable at a time. It allows flight bioassays indoors using 'sustained-flight' tunnels, and has advantages over outdoor flight bioassays. Temperature, humidity, wind velocity, and chemical plume conditions can be controlled and reproduced over a long period of time. Sustained flight over a distance of hundreds of metres can be mimicked in wind tunnel by rotating a visual floor pattern.

12.4 Role of Pheromone in Reproductive Isolation and Speciation

In insects, reproductive isolation that leads to speciation is often achieved by shifts of sex-pheromone chemistries.

Species within a genus that co-occur in same habitat provide useful indicators of relative importance of pheromones as reproductive isolating mechanisms. Mutation at a desaturase locus in *D. melanogaster* leads to formation of two populations, which differ in their pheromone mixtures and have developed premating isolation. Closely related species, *D. melanogaster* and *D. simulans*, differ in their female pheromonal cuticular hydrocarbons. *D. melanogaster* is sexually dimorphic for hydrocarbons, with males having high levels of 23-carbon compound cis-7-tricosene (7-T). *D. melanogaster* females have very little 7-T but high levels of two other hydrocarbons: 27-carbon compound cis, cis-7,11- heptacosadiene (7,11-HD), and 29-carbon compound cis,- cis-7,11-nonacosadiene (7,11-ND) (Jallon & David, 1987; Cobb *et al.*, 1989; Pechine *et al.*, 1995). *D. simulans* on other hand, is sexually monomorphic, with both sexes having 7-T as predominant hydrocarbon (Jallon & David, 1987; Coyne *et al.*, 1994). 7,11-HD in *D. melanogaster* and 7-T in *D. simulans* are known to induce male courtship (Jallon, 1984). Role of hydrocarbons in sexual isolation of interspecific differences was suggested by Cobb and Jallon (1990) after comparing interspecific mating behaviour with interspecific differences in pheromones among eight species of *D. melanogaster* subgroup. These workers reported that sexually monomorphic males court females only from other monomorphic species, while males of sexually dimorphic species court females of all eight species. Cobb and Jallon (1990) suggested that a male will

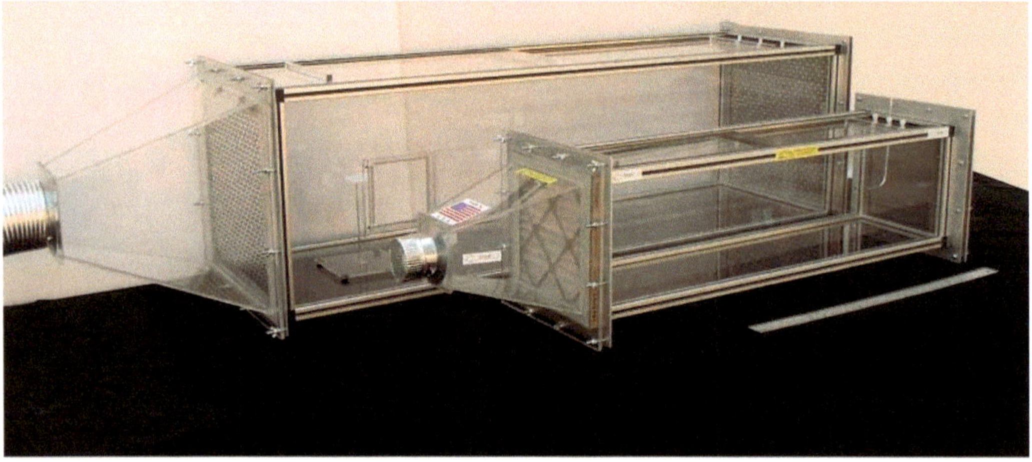

FIGURE 12.7 Wind tunnels.

court a heterospecific female only if she shares either his own predominant hydrocarbon or that of his conspecific females, which has been experimentally supported. This pheromonal difference is due to two species- and female-specific genes, *desatF* and *eloF*. Activity of *desatF* could account for an effective barrier between these species (Wicker-Thomas, 2011).

In northeastern North American tuffed apple budmoth, *Platynota idaeusafis* (Walker) and *P. flavedana* (Clemens), overlap broadly in host plants and seasonal cycles often emerging on same host trees (Chapman & Lienk, 1971). *P. idaeusalis* pheromone is approximately a 1:1 blend of (E)-ll-tetradecenyl acetate and (E)-ll-tetradecenyl-l-ol (Hill *et al.*, 1974), whereas *P. flavendana* employs a 85: 15 mix of (E)- and (Z)-ll-tetradecen-l-ol (Hill *et al.*, 1977).

12.5 Conclusions

The decades of research in insect semiochemicals has generated plethora of information and have given in-depth knowledge of insect's interaction with environment. Species specificity of these semiochemicals, especially pheromones, has played great role in insect evolution and speciation. Several technologies based on this knowledge are successfully deployed for ecofriendly pest management. Still there are certain insects for which pheromones and other semiochemicals are either not identified or identified but not successfully utilized in fields. With advancement of techniques and awareness among farmers we can hope that semiochemicals will play a significant role in achieving ecofriendly sustainable pest management in the near future.

REFERENCES

Addesso, K.M., McAuslane, H.J., Stansly, P.A. & Schuster, D.J. (2007). Host-marking by female pepper weevils, *Anthonomus eugenii*. *Entomologia Experimentalis et Applicata*, *125*: 269–276.

Ampion, M., & Quennedey, A. (1981). The abdominal epidermal glands of termites and their phylogenetic significance. In P.E. Howse and J.-L. Clément (Eds.), *Biosystematics of Social Insects*, 249–261, Academic Press, London, United Kingdom.

Anderson, P., & Löfqvist, J. (1996). Asymmetric oviposition behaviour and influence of larval competition in two pyralid moths *Ephestia kuehniella* and *Plodia interpunctella*. *Oikos*, 47–56.

Azmir, J., Zaidul, I.S.M., Rahman, M.M., Sharif, K.M., Mohamed, A., Sahena, F., Jahurulm, M.H.A., Ghafoor, K., Norulaini, N.A.N., & Omar, A.K.M. (2013). Techniques for extraction of bioactive compounds from plant materials: A review. *Journal of Food Engineering*, *117*: 426–436.

Barbier, M. (1982). *Les phe´romones. Aspects biochimiqueset biologiques*. Masson, France.

Bendib, A., & Minet, J. (1998). Female pheromone glands in Arctiidae (Lepidoptera). Evolution and phylogenetic significance. *Comptes Rendus de l'Académie des Sciences-Series III-Sciences de la Vie, 321*(12): 1007–1014.

Benton, R., Vannice, K.S., & Vosshall, L.B. (2007). An essential role fora CD36-related receptor in pheromone detection in *Drosophila. Nature, 450*(7167), 289–293.

Billen, J. (1986). Morphology and ultrastructure of abdominal glands in Dolichoderine ants (Hymenoptera, Formicidae). *Insectes Sociaux, 33*(3): 278–295.

Blomquist, G.J., & Vogt, R.G. (2003). Insect Pheromone Biochemistry and Molecular Biology. Elsevier Academic Press, London, UK.

Bordereau, C., & Pasteels, J.M. (2010). Pheromones and chemical ecology of dispersal and foraging in termites. In D.E. Bignell, Y. Roisin, and N. Lo, (Eds.), *Biology of Termites: A Modern Synthesis.* Heidelberg: Springer. in press.

Bordereau, C., Lacey, M., Ghostin, J.C., Braekman, J.C., Sillam-Dussès, D., Robert, A.,Shellman, J.S., & Sèmon, E. (2006). Sex pheromones and trail-following pheromones in *Zootermopsis nevadensis* and *Z. angusticollis* (Isoptera, Termopsidae). Proc. XV Congr. IUSSI, Washington, DC, USA.

Bordereau, C., Robert, A., Laduguie, N., Bonnard, O., Le Quère, J.L., & Yamaoka, R. (1993). Dètection du (Z,Z,E)-3,6,8-dodecatrien-1-ol par les ouvriers et les essaimants dedeux espe`ces de termites champignonnistes: Pseudacanthotermes spiniger et P. militaris (Termitidae, Macrotermitinae). *Actes. Coll. Ins. Soc. 8*, 145–149.

Butenandt, V.A., Beckmann, R., Stamm, D. & Hecker E. (1959). Über den Sexual-Lockstoff des Seidenspinners *Bombyx mori* – Reindarstellung und Konstitution. *Z. Naturforsch, 14*:283–284.

Butenandt, A., Beckmann, R., & Stamm, D. (1961). Uber den Sexual lockstoff des Seidenspinners. II. Konstitution und Konfiguration des Bombykols. *Hoppe Seylers Z Physiol Ch, 324*, 84–87.

Chapman, P.J. & Lienk, S.E. (1971). Tortricid fauna of apple in New York; including an account of apple's occurrence in state, especially as a naturalized plant. *NY State Agric. Exp. Stn.(Geneva, NY) Spec Publ, 142.*

Cobb, M., & Jallon, J.-M. (1990). Pheromones, mate recognition and courtship stimulation in *Drosophila melanogaster* species sub-group. *Animal Behaviour, 39*(6): 1058–1067.

Cobb, M., Burnet, B., Blizard, R. *et al.* (1989). Courtship in *Drosophila sechellia*: its structure, functional aspects, and relationship to those of other members of the *Drosophila melanogaster* species subgroup. *Journal of Insect Behaviour, 2*: 63–89.

Costa-Leonardo, A.M. (1987). Morfologia das gla^ndulas salivares de Cornitermes cumulanse Armitermes euamignathus (Isoptera, Termitidae, Nasutitermitinae). *Naturalia, 11*(12): 71–76.

Costa-Leonardo, A.M., & Haifig, I. (2010). Pheromones and exocrine glands in Isoptera. In *Vitamins & Hormones* (*Vol. 83*, pp. 521–549). Academic Press.

Coyne, J.A., Crittenden, A.P. & Mah, K., (1994). Genetics of a pheromonal difference contributing to reproductive isolation in *Drosophila. Science, 265*: 1461–1464.

De Castro, M.D.L., & Priego-Capote, F. (2010). Soxhlet extraction: Past and present panacea. *Journal of Chromatography A, 1217* (16): 2383–2389.

Deisig, N., Dupuy, F., Anton, S., & Renou, M. (2014). Responses to pheromones in a complex odor world: sensory processing and behavior. *Insects, 5*(2): 399–422.

Dos Santos, J.S., Coracini, M.D.A., Kunz, R., Roggia, S. & Brancalhão, R. (2017). Morphological description of sex pheromone gland of Helicoverpa armigera (Hübner) (Lepidoptera: noctuidae). *Embrapa Soja-Artigo em periódico indexado (ALICE).*

Fleischer, J., & Krieger, J. (2018). Insect pheromone receptors–key elements in sensing intraspecific chemical signals. *Frontiers in Cellular Neuroscience, 12*: 425.

Hallberg, E., & Poppy, G. (2003). Exocrine glands: chemical communication and chemical defense. In N.P. Kristensen, (Eds.), *Lepidoptera, Moths and Butterflies Vol 2: Morphology, Physiology, and Development.*

Hanus, R., Luxová, A., Šobotník, J., Kalinová, B., Jiroš, P., Křeček, J., Bourguignon, T., & Bordereau, C. (2009). Sexual communication in *Prorhinotermes simplex* (Isoptera,Rhinotermitidae) mediated by a pheromone from female tergal glands. *Insectes Sociaux, 56*, 111–118.

Hill, A., Carde, R., Bode, W., & Roelof, W. (1977). Sex pheromone components of varigated leafroller moth, *Platynota flavedana. Journal of Cherical Ecology, 3*: 369.

Hill, A., Carde, R., Comeau, A., Bode, W. & Roelofs, W. (1974). Sex pheromones of tufted apple bud moth (*Platynota ideausalis*). *Environmental Entomology, 3*: 249.
Imen, S., Christian, M., Virginie, D., & Colette, R. (2015). Intraspecific signals inducing aggregation in Periplaneta americana (Insecta: Dictyoptera). *Environmental Entomology, 44*(3), 713–723.
Ishii, S., & Kuwahara, Y. (1967). An aggregation pheromone of German cockroach *Blattella germanica* L. (Orthoptera: Blattellidae): I. Site of pheromone production. *Applied Entomology and Zoology, 2*(4): 203–217.
Jacobson, M. (1972). Insect Sex Attractants, Academic Press, NewYork.
Jackson, D.E., Martin, S.J., Holcombe, M., & Ratnieks, F.L. (2006). Longevity and detection of persistent foraging trails in Pharaoh's ants, *Monomorium pharaonis* (L.). *Animal Behaviour, 71*(2): 351–359.
Jallon, J.M. & David, J.R. (1987). Variation in cuticular hydrocarbons among eight species of *Drosophila melanogaster* subgroup. *Evolution, 41*: 294–302.
Jallon, J.M. (1984). A few chemical words exchanged by *Drosophila* during courtship and copulation. *Behaviour Genetics, 14*: 441–478.
Joseph, R.M., & Carlson, J.R. (2015). Drosophila chemoreceptors: a molecular interface between chemical world and brain. *Trends in Genetics, 31*(12): 683–695.
Karlson, P. & Luscher, M. (1959). Pheromones: a new term for a class of biologically active substances. *Nature 183*: 55–56.
Kehat, M., & Dunkelblum, E. (1990). Behavioral responses of male *Heliothis armigera* (Lepidoptera: Noctuidae) moths in a flight tunnel to combinations of components identified from female sex pheromone glands. *Journal of Insect Behavior, 3*(1): 75–83.
Kurtovic, A., Widmer, A., & Dickson, B.J. (2007). A single class of olfactory neurons mediates behavioural responses to a Drosophila sex pheromone. *Nature, 446*, 542–546.
Laduguie, N., Robert, A., Bonnard, O., Vieau, F., Le Quéré, J.L., Sémon, E., & Bordereau, C. (1994). Isolation and identification of (3Z,6Z,8E)-3,6,8-dodecatrien-1-ol in *Reticulitermes santonensis* Feytaud (Isoptera, Rhinotermitidae): roles in worker trail following and in alate sex-attraction behavior. *Journal of Insect Physiology, 40*: 781–787.
Leal, W.S., Yadava, C.P., & Vijayvergia, J.N. (1996). Aggregation of scarab beetle *Holotrichia consanguinea* in response to female-released pheromone suggests secondary function hypothesis for semiochemical. *Journal of Chemical Ecology, 22*(8): 1557–1566.
Mutis, A., Parra, L., & Palma, R. (2009). Evidence of contact pheromone use in mating behavior of the raspberry weevil (Coleoptera: Curculionidae). *Environmental Entomology, 38* : 192–197.
Noirot, C. (1969). Glands and secretions. In K. Krishna and F.M. Weesner, (Eds.), *Biology of Termites* (Vol. *1*, pp. 89–123). New York: Academic Press.
Payne, T.L. (1974). Pheromone perception. In M.C. Birch, (Ed.), *Pheromones*. North Holland Publishing Company.
Pechine J.M., Perez, F., Antony C. & Jallon J.M. (1995). A further characterization of Drosophila cuticular monoenes using a mass spectrometry method to localize double bonds in complex mixtures. *Analytical Biochemistry, 145*: 177–182.
Pers, J. N. C.V.D. (1998). Electroantennography: A Practical Introduction. Syntech Inc. Hilversum, The Netherlands.
Renou, M. (2014). Pheromones and general odor perception in insects. *Neurobiology of Chemical Communication*, 23–55.
Ritter, F.J., Bruogemann-Rotgans, I.E.M., Verwiel, P.E.J., Persoons, C.J., & Talman, E. (1977). Trail pheromone of pharaoh's ant *Monomorium pharaonis*; isolation and identification of faranal. A terpenoid related to juvenile hormone 1I. *Tetrahedron Letters, 30*: 2617–2618.
Saini, R.K., Rai, M.M., Hassanali, A., Wawiye, J., & Odongo, H. (1995). Semiochemicals from froth of egg pods attract ovipositing female *Schistocerca gregaria. Journal of Insect Physiology, 41*(8): 711–716.
Sánchez, P.A., Sánchez, F., Jaffé, K., & Caetano, F.H. (1998). Ultrastructure of gland responsible for synthesis of an aggregation pheromone in *Rhynchophorus palmarum* (L.) (Coleoptera: Curculionidae). *Coleopterists' Bulletin*, 297–305.
Schneider, D. (1957). Elektrophysiologische Untersuchungen von Chemound Mechanorezetoren der Antenne des Seidenspinners. *Bombyx mori* L. *Zeitschrift Fur Vergleichende Physiologieze, 40*: 8–41.
Sillam-Dusse`s, D., Robert, A., Se´mon, E., Lacey, M., & Bordereau, C. (2006). Trail following pheromones and phylogeny in termites. Proc. 15th Congr IUSSI, Washington, DC.

Sillam-Dusse`s, D., Se´mon, E., Lacey, M.J., Robert, A., Lenz, M., & Bordereau, C. (2007). Trail-following pheromones in basal termites, with special reference to *Mastotermes darwiniensis*. *Journal of Chemical Ecology*, *33*: 1960–1977.
Sillam-Dussès, D., Sèmon, E., Robert, A., & Bordereau, C. (2009b). (Z)-Dodec-3-en-1-ol, a common major component of trail-following pheromone in termites Kalotermitidae. *Chemoecology*, *19*: 103–108.
Sreng, L. (1984). Morphology of sternal and tergal glands producing sexual pheromones and aphrodisiacs among cockroaches of subfamily Oxyhaloinae. *Journal of Morphology*, *182*(3): 279–294.
Srinivasan, G., & Babu, P.C.S. (2000). Sex pheromones for brinjal shoot and fruit borer, *Leucinodes orbonalis*. *Indian Journal of Entomology*, *62*: 225–226.
Stelinski, L.L., Oakleaf, R., & Rodriguez-Saona, C. (2007). Oviposition-deterring pheromone deposited on blueberry fruit by parasitic wasp, *Diachasma alloeum*. *Behaviour*, *144*(4): 429–446.
Tada, S., & Leal, W.S. (1997). Localization and morphology of sex pheromone glands in scarab beetles. *Journal of Chemical Ecology*, *23*(4): 903–915.
Tamaki, Y., Kawasaki, K., Yamada, H., Koshihara, T., Osaki, N., Ando, T., Yoshida, S., & Kakinohana, H. (1977). Z)-11-hexadecenal and (Z)-11-hexadecenyl acetates: sex-pheromone components of the diamondback moth (Lepidoptera: Plutellidae). *Applied Entomology and Zoology*, 12: 208–210.
Tanaka, Y. (2000). Realized heritability of behavioral responsiveness to an oviposition deterring pheromone in azuki bean weevil *Callosobruchus chinensis*. *Entomologia Experimentalis et Applicata*, *96*(3): 239–243.
Tomaniova, M., Hajslova, J., PavelkaJr., J., Kocourek, V., Holadova, K., & Klimova, I. (1998). Microwave-assisted solvent extraction - a new method for isolation of polynuclear aromatic hydrocarbons from plants. *Journal of Chromatography A*, *827* : 21–29.
Tumlinson, J.H., Moser, J.C., Silverstein, R.M., Brownlee, R.G., & Ruth, J.M. (1972). A volatile trail pheromone of leaf-cutting ant, *Atta texana*.1. *Insect Physiology*, *18*: 809–814.
Van der Pers, J. N. C. (1998). Electroant ennography: A practical introduction. Syntech Inc, Hilversum, The Netherlands.
Visser, B., Ian, A.N.D., Stphanie, H., Florent, L., Paul, M.B.B., Georges, L., & Caroline, M.N. (2018). Common Practice Solvent Extraction Does not Reflect Actual Emission of a Sex Pheromone During Butterfly Courtship. *Frontiers in Ecology and Evolution*, *6:* 154.
Vogel, H., Heidel, A.J., Heckel, D.G., & Groot, A.T. (2010). Transcriptome analysis of sex pheromone gland of noctuid moth *Heliothis virescens*. *BMC Genomics*, *11*(1): 29.
Ward, A., Moore, C., Anitha, V., Wightman, J., & Rogers, D.J. (2002). Identification of sex pheromone of *Holotrichia reynaudi*. *Journal of Chemical Ecology*, *28*(3): 515–522.
Wicher, D. (2015). Olfactory signaling in insects. In *Progress in Molecular Biology and Translational Science* (*Vol. 130*, pp. 37–54). Academic Press.
Wicker-Thomas, C. (2011). Evolution of insect pheromones and their role in reproductive isolation and speciation. In *Annales de la Société entomologique de France* (*Vol. 47*, No. 1-2, pp. 55–62). Taylor & Francis Group.
Wilson, E.O., & Pavan, M. (1959). Glandular sources and specificity of some chemical releasers of social behavior in dolichoderine ants. *Psyche*, *66*(4): 70–76.
Wyatt, T.D. (2003). *Pheromones and Animal Behaviour: Communication by Smell and Taste*. Cambridge, U.K.: Cambridge University. Press.
Wyatt, T.D. (2014). *Pheromones and Animal Behavior: Chemical Signals and Signatures*. Cambridge University Press, United Kingdom.
Yew, J.Y., & Chung, H. (2015). Insect pheromones: an overview of function, form, and discovery. *Progress in Lipid Research*, *59*: 88–105.
Zablotny, J.E. (2009). Sociality. In *Encyclopedia of Insects* (pp. 928–935). Academic Press.
Zhang, J.P., Salcedo, C., Fang, Y.L., Zhang, R.J., & Zhang, Z.N. (2012). An overlooked component:(Z)-9-tetradecenal as a sex pheromone in *Helicoverpa armigera*. *Journal of Insect Physiology*, *58*(9): 1209–1216.
Zhu, J., Kozlov, M.V., Philipp, P., Francke, W., & Löfstedt, C. (1995). Identification of a novel moth sex pheromone in *Eriocrania cicatricella* (Zett.)(Lepidoptera: Eriocraniidae) and its phylogenetic implications. *Journal of Chemical Ecology*, *21*(1): 29–43.
Zhu, P.C., Kong, F.L., Yu, S.D., Yu, Y.Q., Jin, S.P., Hu, X.H., & J.W. Xu. (1987). Identification of sex pheromone of eggplant borer *Leucinodes orbonalis* Guenée (Lepidoptera: Pyralidae). *Zeitschrift für Naturforschung C*, *42*: 1347–1348.

13

Reproductive Strategies in Aphids

Rajendra Singh[1] and Garima Singh[2]
[1]*Department of Zoology, Deendayal Upadhyay Gorakhpur University, Gorakhpur, India*
[2]*Department of Zoology, University of Rajasthan, Jaipur, India*

CONTENTS

13.1 Introduction

Aphids are soft-bodied plant sap-sucking insects from family Aphididae (Order Hemiptera, Suborder Sternorrhyncha, Infraorder Aphidomorpha, Superfamily Aphidoidea). They are small (<7 mm) sap sucking insects. They possess proboscis, which originates between and behind forelegs, their antennae

DOI: 10.1201/9781003043195-13

has two thick basal segments and a flagellum composed of up to four segments, last of which is divided into a proximal part and a thinner distal part called a 'process terminalis'. They bear two compound eyes and two ocular tubercles made up of three lenses each, which are situated behind and above compound eyes. They have two tarsal segments. The wings when present have only one prominent longitudinal vein. The fifth abdominal segment bears a pair of upward and backwardly pointing tubes on dorsal surface called siphunculi or cornicles. A cauda is usually present on last abdominal segment. They are cosmopolitan but most abundant in temperate climates.

Aphids are also known as greenfly, blackfly, plant-lice, ant-cow, etc. Globally, more than 250 species of aphids are notorious pests of agricultural and horticultural crops (Singh & Singh, 2016, 2019). The damage in some crops (cereals, potatoes, and rapeseed mustard) is extensive with nothing remaining to harvest. Aphids directly damage crops via (i) sucking their nutrients causing reduced vigour, and (ii) saliva toxic to plants (Behura & Das, 1976). Some aphids make different kinds of leaf and stem galls (Chakrabarti, 2007). In addition, aphids also damage crops indirectly by (i) secreting high amount of honeydew that blocks stomata hampers normal plant physiological processes, (ii) aiding growth of black sooty mould (*Capnodium* spp.) which reduces photosynthesis, and (iii) transmitting viral diseases. Honeydew also reduces effectiveness of fungicides by obstructing their absorption (Dika & Van Pelt, 1992). However, honeydew attracts bees, wasps, and ants that may protect them from their natural enemies, such as parasitoids and predators. Viral diseases caused by viruses cause most economic loss (Eastop, 1977). Aphid-transmitted viruses account for approximately 50% of 600 known viruses with an invertebrate vector (Hooks & Fereres, 2006), with majority from subfamily Aphidinae (Blackman & Eastop, 2000). The polyphagous nature of aphids, parthenogentic reproduction, and stylet to pierce cell walls are together responsible for success of aphids as vectors of plant viruses (Pettersson *et al.*, 2017). A list of virus pathogens transmitted by various aphid species, regarded as vector of one or more plant viruses has been listed by Ghosh *et al.* (2017).

13.2 Reproductive Strategy in Aphids

Being a good source of food for predators and parasitoids, success of aphids lie in their highly evolved reproductive mechanisms that alternates between asexual and sexual reproduction. In optimal environmental conditions and without any ecological resistance, a single female can reproduce 600 billion offspring in 1 season (Marshall, 2017).

It has already been demonstrated that ancestors of aphids reproduce sexually like other hemipterans via laying eggs (Ogawa & Miura, 2014) and during evolutionary history while retaining oviparity, they evolved to alternate sexual reproduction with one or more generations of asexual reproduction. Unlike other hemipterans, usually aphids lay eggs (sexual reproduction) usually in autumn, and reproduce parthenogenetically (viviparous) for rest of the year. Most aphids are holocyclic species with some anholocyclic (3% of all species).

It is interesting that aphids using same set of genes are able to reproduce by both sexual and asexual modes of reproduction. The strictly asexual species (or clones) tend to be distributed in low-latitude regions, suggesting that sexual reproduction in aphids is an adaptation to harsh winters, and that parthenogenesis occurs in areas where nymphal or adult aphids can overwinter (Dixon, 1998). Aphids have probably evolved cyclical parthenogenesis to facilitate asexual viviparous reproduction in optimum climatic conditions and overwintering by diapausing eggs (Dixon, 1998) (Figure 13.1).

13.2.1 Anholocyclic Reproduction in Aphids

According to aphid evolution theory, anholocycly originated from holocycly during fourth glacial epoch (Moran, 1992) and parthenogenesis is the first key of aphid success. With a few exceptions, aphids in spring and summer are all females. The first apterous morph (fundatrices) that hatch from eggs in early spring (from eggs laid late prior year to overwinter), are viviparous and give birth to young ones (diploid parthenogenesis) during spring and summer (Figure 13.2). Their offspring also quickly mature and reproduce within a few days and soon after that, second, third, fourth, fifth, and so on, generations appear

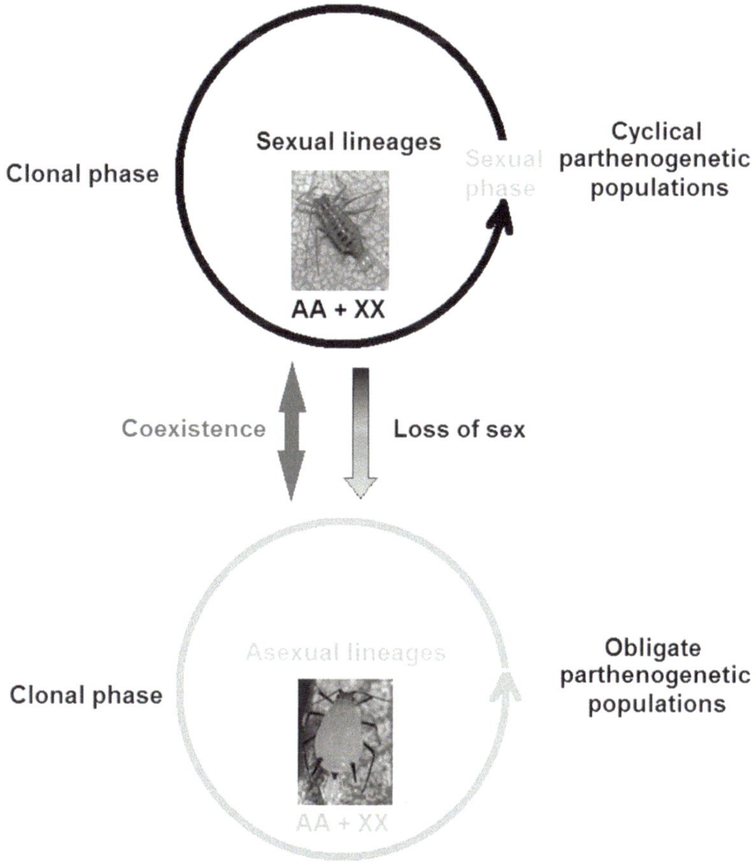

FIGURE 13.1 Representation of sexual (cyclical parthenogenetic) and asexual (obligate parthenogentic) mode of reproduction in aphids.

FIGURE 13.2 A parthenogenetic aphid colony consisting of individuals of different morphs (alate and apterous) and viriginoparae giving birth along with attending ants.

that cause exponential population expansion by a single female. In winter females and males both develop. The life cycle is spent feeding on a host plant with need not to go further, so walking is sufficient enough and hence they do not produce wings. The development of wings is protein-intensive task and aphids very wisely conserve protein for reproduction. Only on overcrowding of host plant or depletion of food resources, they produce wings to migrate to other food resources (Dixon, 1998; Hardie, 2017).

13.2.2 Holocyclic Reproduction in Aphids

Aphid life cycles are complex and may be either monoecious or dioecious. In a simple and generalized monocious holocyclic aphid life cycle (Figure 13.3a), aphids feed on a single host-plant species throughout year. The sexual morphs are produced in following autumn, in response to decreasing photoperiod. Then mating takes place between males and female (oviparae) producing eggs that overwinter on host plant. In spring, fundatrix emerges from overwintering eggs, mature parthenogenetically, and gives birth to nymphs that mature to viviparae and continue reproducing parthenogenetically (Sorensen, 2009). The viviparae may be apterae (wingless) or alatae (winged), but in

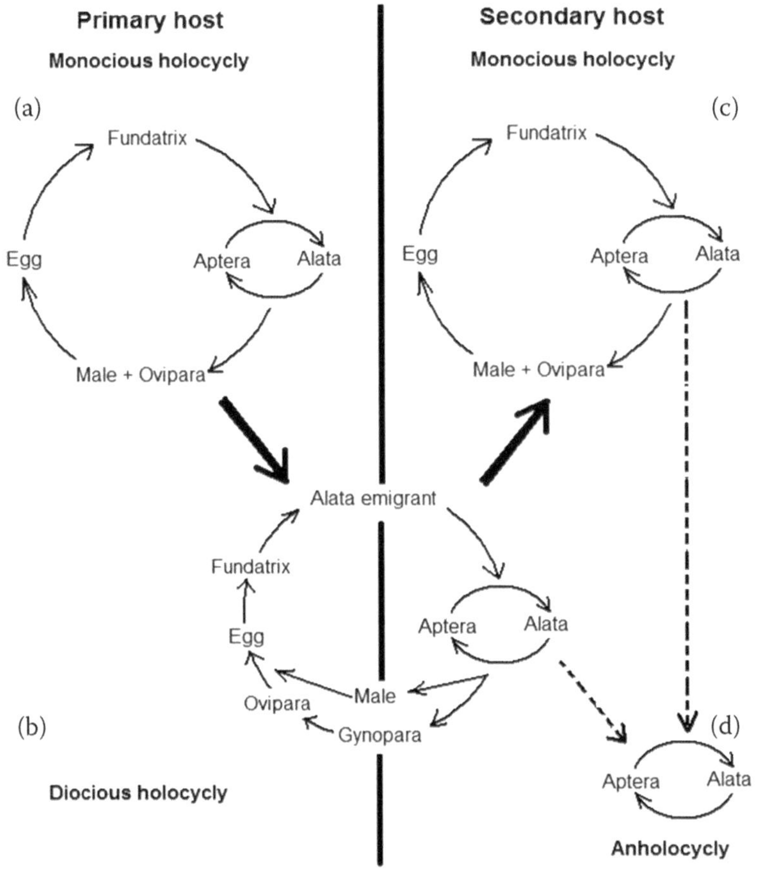

FIGURE 13.3 Evolutionary development of generalized aphid life cycles. Initially, aphids developed monoecious holocycly on an ancestral woody primary host, where aestivation occurred because sap amino acids were unavailable during summer growth cessation (a). Next, multiple subfamilies independently evolved dioecious holocycly, where viviparae moved to summer-growing herbaceous secondary hosts but returned to their ancestral host in autumn (b). In some aphids, secondarily monoecious holocycly developed on secondary host when primary host was lost (d). Often in warm areas, where selection for an overwintering egg is not imposed, some populations of dioecious and secondarily monoecious holocyclic aphids may lapse into facultative anholocycly on their secondary hosts; this condition may become obligate anholocycly if ability to produce sexuals is lost (d) (Sorensen, 2009).

some all viviparae are alate. Apterae produce more offspring per female than do alatae. Crowding or lack of nutrient levels turns apterae to alatae, which migrate and reproduce on new plants. The production of viviparae continues until autumn when conditions stimulate them to produce sexuals.

A second, more complicated dioecious life cycle (Figure 13.3b) has independently evolved probably in response to seasonally inadequate supply of nutrients, especially amino acids and shows seasonal alternation between differing hosts. Aphid groups feeding on woody plants that transport amino acids only in spring suffer nutrient deficit in summer. This may lead to development of an aestivating nymph that halts growth until autumn, while others (e.g. Aphidinae) escape to new host plants. Spring alatae migrate to herbaceous fast growing secondary hosts as emigrants during summer, and return to woody plants in autumn by producing migrating males and gynoparae, which mate to lay overwintering eggs. Secondary host plants may be specific or of several plant species, but primary hosts are often specific to a plant genus. In warmer climates,where overwintering eggs are not required, holocyclic dioecious or monoecious populations may drop facultatively into anholocycly on their secondary hosts. If such populations remain anholocyclic for a longer period, they evolve into obligate anholocycly by losing capacity to produce sexual morphs (Figure 13.3c & d; Dixon, 1998).

A generalized life-cycle pattern of heteroceous life cycle of bird-cherry aphid, also known as oat aphid, *Rhopalosiphum padi* is illustrated in Figure 13.4; whose several populations reproduce in temperate regions with cold winters by combining holocycly and, alternating between primary winter trees (bird cherry) and secondary summer host plants (Poaceae). However, same aphid in areas with mild winters show anholocyclic life cycle on poaceous plants.

Discovery of apterous oviparous females of mustard aphid *Lipaphis erysimi* in arid and semi-arid region of Jaipur (India) (Ghosh & Rajendran, 1988) suggests that factors other than low temperature, short day-length, and physical condition of host plants are also regulatory (Figure 13.5). Medda *et al.* (1997) studied different modes of life cycles of some aphid species infesting *Salix* in India (Figure 13.6).

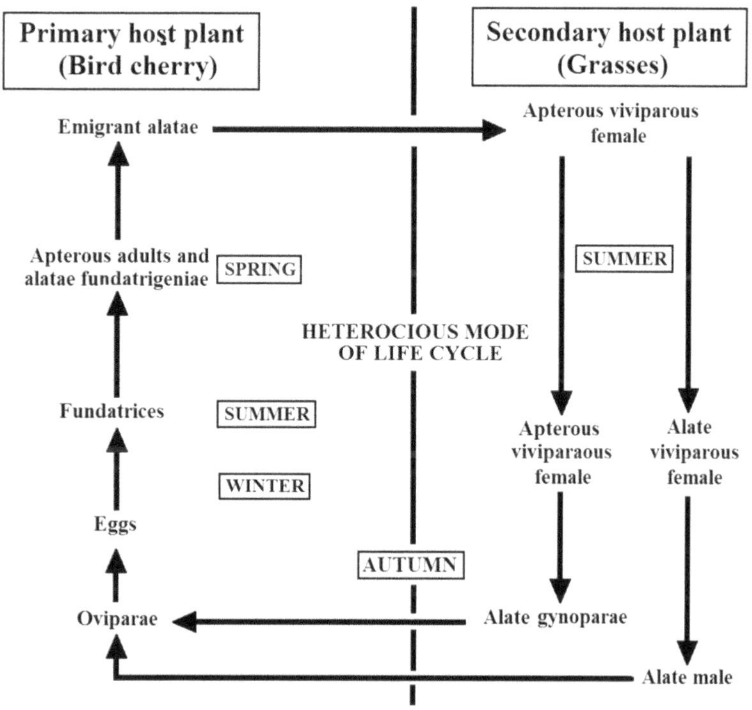

FIGURE 13.4 A generalized heteroceous life cycle of oat aphid, *Rhopalosiphum padi.*

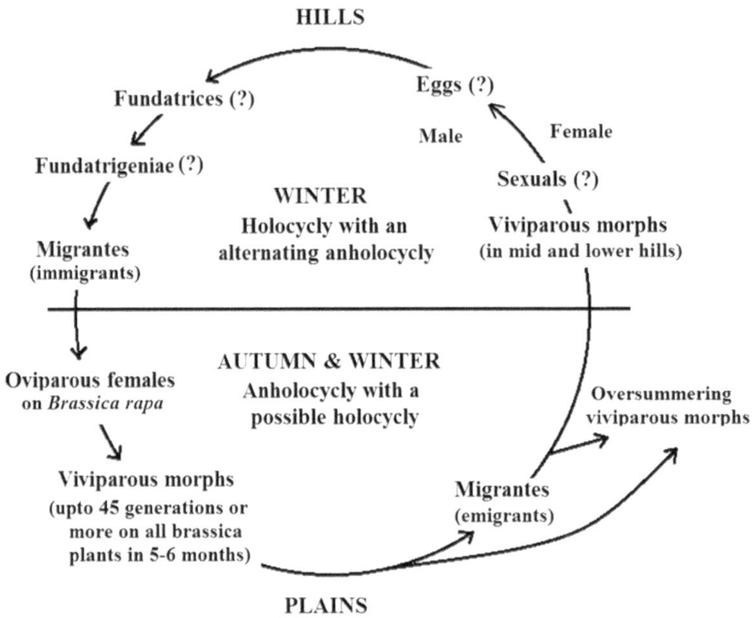

FIGURE 13.5 Possible life cycle of mustard aphid, *Lipaphis erysimi* in India (Singh & Singh, 2016).

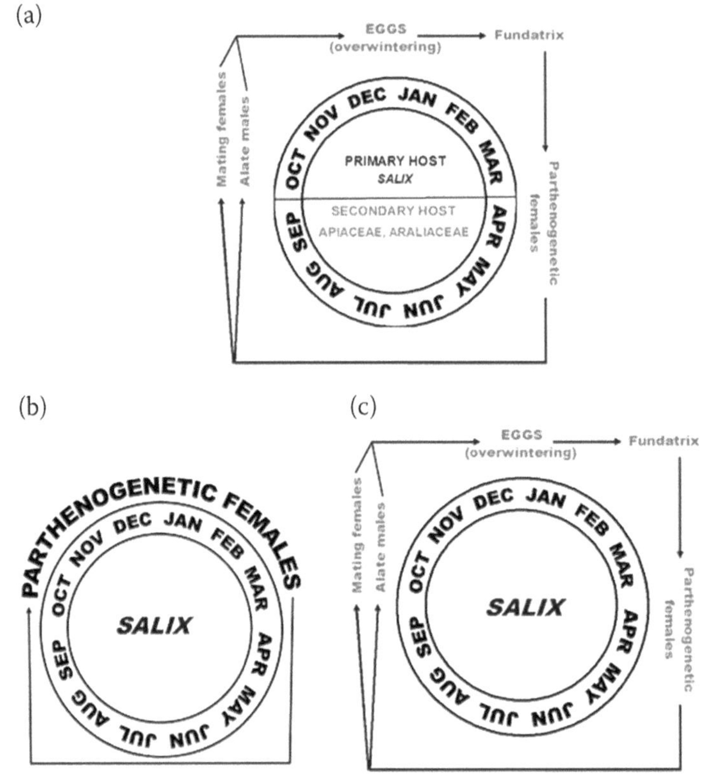

FIGURE 13.6 Life cycle pattern of *Salix* infesting aphids in India: (a) Host alternating and cyclical parthenogenetic, (b) non-host alternating and cyclical parthenogenetic, and (c) non-host alternating and permanently parthenogenetic (Medda *et al.*, 1997).

The life cycle patterns and phenotypes of aphids can be exemplified and understood by following four examples:

a. **Nipplewort aphid, *Hyperomyzus lampsanae***: Found throughout year on underside of lower leaves of nipplewort (*Lapsana communis*) has asexual forms reproducing parthenogenetically via viviparous females. Its primary host for sexual reproduction has been lost with nymphs developing into alatae or apterae adults, which can produce both wingless (Figure 13.7a) or winged (Figure 13.7b) offspring (Blackman & Eastop, 2006).

b. **Sycamore aphid, *Drepanosiphum platanoidis***: It is a single-host species, which reproduces by cyclical parthenogenesis. For most of year, all adults are green- winged viviparous (Figure 13.8a) but in late spring nymphs mature into pale-winged, viviparae adults (Figure 13.8b) that diapause during summer, then produce normal winged viviparae. In autumn, viviparae produce nymphs that mature to sexual morphs; either winged sexual males (Figure 13.8c) or wingless egg-laying sexual females called as oviparae (Figure 13.8d).

c. **Bird cherry-oat aphid, *Rhopalosiphum padi***: The bird cherry-oat aphid alternates host seasonally: primary host being trees of *Prunus* spp. and secondary hosts are major cereals. Oviparous female oviposits 10–15 eggs in autumn in narrow gap between axillary buds and stem of primary host (Figure 13.9a). Newly emerged nymphs move to host plant leaves, where they develop into very large pale-green with rust red tint aroud cornicles fundatrices (Figure 13.9b); giving rise to wax-covered second generation (Leather & Dixon, 1981; Figure 13.9c). Their feeding induces rolled leaf galls, which turn multicoloured from initial green. From wax-covered nymphs develop large, highly fecund, apterous adults that give birth to alate emigrants that migrate from primary to secondary hosts before end of summer (Figure 13.9d). Dixon (1971) observed that apterous viviparous females prefer oats, while sexuales prefer bird cherry leaves and that primary host were repellent to emigrants (Glinwood & Pettersson, 2000). When emigrants land on secondary host, they feed and give rise to parthenogenetic generations either of wingless or winged females, which return to primary host, where their nymphs mature to wingless oviparae female (Figure 13.9g) and sometimes winged males in case of short days (Figure 13.9f).

d. ***Paracletus cimiciformis***: This aphid is a host-alternating species (primary host: turpentine tree; secondary hosts: grasses or cereals) with a complex life cycle of 2 years. For most year, apterae viriginoparae reproduced on secondary host plants, but in late summer, alate males and females are produced, which fly to primary hosts, where they mate, oviposit overwintering eggs. In spring, eggs hatch and wingless females (fundatrices) develop, produce leaf-galls, and wingless viviparae until autumn. In late summer, they produce winged female emigrants that fly off and colonize roots of secondary hosts. On secondary hosts nymphs mature to ones with wingless, round green

FIGURE 13.7 (a) Parthenogenetic wingless and (b) winged morph of nipplewort aphid, *Hyperomyzus lampsanae* on nipplewort (*Lapsana communis*).(Courtesy InfluentialPoints.com)

FIGURE 13.8 *Drepanosiphum platanoidis*. (a) Green parthenogenetic winged viviparous female, (b) pale-winged viviparous female, (c) winged sexual male, and (d) wingless sexual oviparous female on sycamore, *Acer pseudoplatanus*. (Courtesy InfluentialPoints.com)

body or ones with flattened white or brown body. The former feed host plant sap, and produce honeydew, while other produce pheromones mimicking ant larvae. As a result latter nymphs are carried into ant nest, where they suck on haemolymph of ant larvae and overwinter. In spring they are carried back to surface where they produce round-bodied aphids which feed on grass roots and reproduce asexually. In late summer, winged form appears that fly on its primary host, and produces wingless sexual-forms (males and oviparae). The whole cycle takes 2 years.

e. ***Nipponaphis monzeni***: This aphid is also a gall making anholocyclic and alternates between its primary host, *Distylium racemosum* and secondary host, evergreen oaks. It induces galls on primary host with 600–800 alates. Late in autumn, these alates fly off in search of secondary hosts, wherethey give birth to first-instar nymphs, which become scale-like apterous adults (Sorin, 1958). Offspring of these adults develop into alate sexuparae in spring and fly back to primary host (Takahashi, 1962), where males and sexual females are produced, which lay eggs that hatch into first-instar fundatrices, which induce galls. Galls may develop over 3, 4, or 5 years.

13.3 Overwintering Strategy of Aphids

Aphids that live in temperate or boreal regions have evolved two ways to overwinter: (i) pass winter as active adult and/or nymph (*Brevicoryne brassicae*), or (ii) as eggs (*Rhopaloishum padi*). Despite their soft bodies and fragile appearance, aphids have quite low supercooling point values, such as −26°C (Griffiths & Wratten, 2011), due to presence of cryoprotectant contents (glucose, mannitol, myo-inositol, trehalose) (Saeidi *et al.*, 2012). Some overwinter in roots (Durak, 2014), while others turn into a special form called a 'hiemalis' with poorly developed ova and excess fat bodies rich in triglycerides (Phillips *et al.*, 2000).

FIGURE 13.9 *Rhopalosiphum padi*. (a) Eggs laid by oviparaous females, (b) fundatrix with young nymphs inside gall of bird-cherry, (c) wax-coated apterae viriginoparae on bird-cherry in colony in leaf gall, (d) emigrant alate viviparous female, (e) apterae viriginoparae females on oat, (f) male sexuparae (preserved), and (g) alate sexuparae female. (Courtesy InfluentialPoints.com)

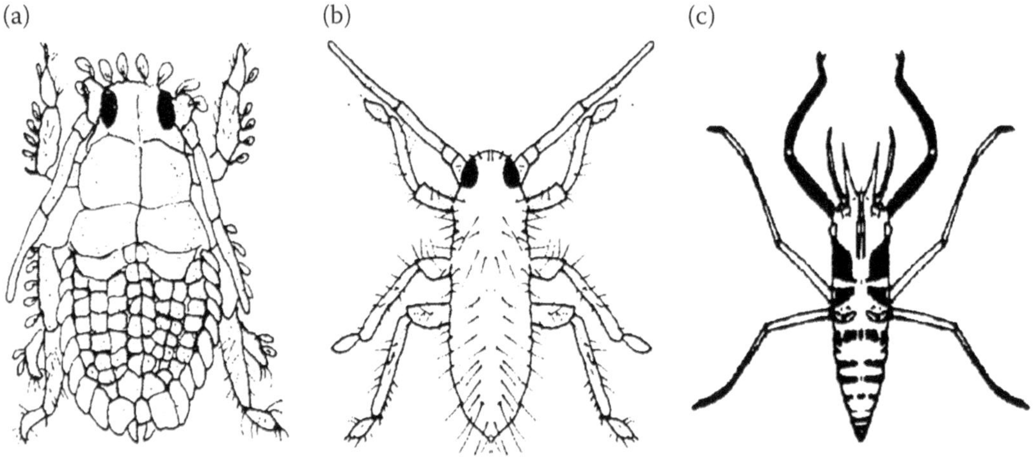

FIGURE 13.10 First instar nymph aestivating in summer (a), first instar normal nymph produced in autumn (b) of *Periphyllus* sp. infesting acer plants in Himalayas; sterile first instar soldier phenotype of a bamboo-feeding aphid. *Ceratovacuna* sp. showing frontal horns in head (Agarwala, 2007).

In some holocyclic aphids, apterae viviparae produce very small and stout nymphs, 'midget' that hide themselves in bark and overwinter without moulting until next spring (Aoki, 1980). Similarly, several aphid species, aestivate during summer by producing dormant first instar nymphs, which are flattened and covered with plates, and remain glued to leaves until autumn (Miyazaki, 1987; Figure 13.10a & b).

Some species of aphid that make galls in divergent taxa have evolved sterile morphs, soldiers, or defenders, which remain forever at larval stages (Figure 13.10c). Soldiers are heavily sclerotized to defend aphid colonies from attacks from their enemies and ants (Hattori *et al.*, 2013), and do not contribute to reproduction (Itô, 1989), dispersal, etc., but increase survival value of parental colonies.

13.4 Development of Aphids: Gametogenesis and Embryogenesis

Le Trionnaire *et al.* (2008) nicely reviewed and discussed how aphids alternate their reproductive mode from viviparous parthenogenesis in spring and summer (short nights, shorter scotophase) to oviparous sexual reproduction in autumn (long nights, longer scotophase), and thus enable themselves for survival in extreme climatic changes in different seasons (Le Trionnaire *et al.*, 2008, 2012) (Figure 13.3–13.5).

13.4.1 Reproductive System of Sexual and Asexual Females

The reproductive system of aphids consists of paired ovaries with several telotrophic ovarioles connected by lateral oviducts to common or median oviduct, which opens posteriorly into a genital chamber, to which is attached a spermathecae with a pair of accessory glands in sexuales. There is no difference in organization of reproductive system of sexual and asexual females, but there are significant differences in their functioning: (a) all ovarioles develop in asexual viviparous females, whereas some of them degenerate in sexual oviparous females, (b) germaria in sexual females are larger than in asexual females (Tsitsipis & Mittler, 1976), (c) number of germ cells in each ovariole is usually greater in sexual females than in asexual females, (d) nuclei of nurse cells in germarium drastically enlarge by endomitosis only in sexual females, (e) development of oocytes stops after previtellogenesis in vivipare but continues to choriogenesis in ovipare, (f) future oocytes at basal part of germaria enter meiosis in oviparae but not in viviparae, and (g) lipid droplets and yolk granules accumulate in oocyte cytoplasm of ovipare, but only lipid droplets in viviparae (Le Trionnaire *et al.*, 2008; Michalik *et al.*, 2013). Asexual females maintain diploidy via a modified meiosis that skips chromosomal reduction (Blackman, 1987).

13.4.2 Oogenesis and Embryogenesis in Sexual Females

In future sexual females, future oocytes remain blocked in metaphase-I during formation of germarium (Blackman, 1976) and are passed down in a follicle chamber, where they enter a growth phase with accumulation of yolk in their cytoplasm. Second maturation division is completed after entry of male nucleus (Blackman, 1987). Diploid eggs less than 1 mm in length and filled with yolk are laid. After 2–3 days, germ band falls into yolk, and by the end of day 15, embryo completes anatrepsis (first inversion of anterior–posterior and dorsal–ventral axes of embryo) and is fully segmented (Miura *et al.*, 2003). At this stage, embryo enters diapause in response to photoperiodic signals. The period of temperature-independent development ends with katatrepsis, i.e. reversal of anatrepsis, after about 50 days of oviposition.

13.4.3 Oogenesis and Embryogenesis in Asexual Females

In parthenogenetic females, embryos develop within ovariole from a diploid oocyte after a single modified meiosis-II division in absence of chromosome recombination. The future oocytes, located in posterior part of germarium, rapidly differentiate one by one (Le Trionnaire *et al.*, 2008). A number of morphological changes take place in these cells. According to Blackman (1987), cells are arrested at an early prophase stage, within germarium, but future oocytes selected to differentiate increase in size and enter prophase of meiosis-II-like division and are eleased from germarium and form a new follicular chamber, where chromosomes rapidly condense (Bermingham & Wilkinson, 2009). A single maturation division occurs, giving a polar body and a diploid oocyte, which undergoes a series of mitotic divisions to initiate embryo development. Immediately after release of one oocyte from germarium, another oocyte starts to differentiate. Thus, within an ovariole, there is a sequential chain of a germarium, differentiating oocyte-like cells and developing embryos, each included in a different follicular chamber (Le Trionnaire *et al.*, 2008). Oocytes of parthenogenetic aphids self-organize microtubule-based asters, necessary for first mitotic spindle in absence of male centrosomes (Riparbelli *et al.*, 2005). Detail blastulation and gastrulation in aphids was reviewed by Le Trionnaire *et al.* (2008). Even embryos inside parthenogenetic females may contain emboys, i.e. a mother can have in its ovarioles developing embryos, which in turn also contain embryos, future granddaughters. Thus, there is a telescopic generation due to parthenogenesis and viviparity in aphids (Minks & Harrewijn, 1987) (Figure 13.11). This type of viviparity in aphids is referred to as pseudoplacental viviparity in which nutritional input is provided by maternal hemolymph through cells of ovariole sheath (Bermingham & Wilkinson, 2009).

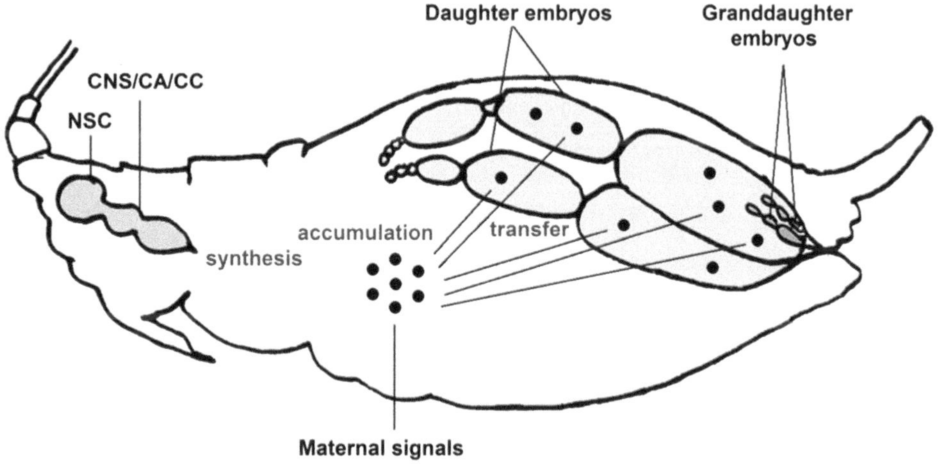

FIGURE 13.11 Transfer of maternal signals in viviparous aphids. The multi-step events leading to production of different aphid polyphenisms are hypothesized in this reproductive unit: (1) mother perceives environmental information and integrates information, probably via neurosecretory cells (NSC) or central nervous system (CNS); (2) mother converts stimuli into a transmissible signal, probably in corpora allata (CA) or corpora cardiaca (CC); (3) this signal is transmitted to daughter embryos developing in ovarioles of mother; (4) daughter embryos respond to signal (Ogawa & Miura, 2014).

13.4.4 Oogenesis and Embryogenesis: Difference between Sexual and Asexual Females

The fertilized oocytes of sexual females begin embryogenesis after being deposited on leaves of secondary host plant while oocytes of asexual females complete embryogenesis within mother. Compared with oviparous development, viviparous development involves a smaller transient oocyte surrounded by fewer somatic epithelial cells and a smaller early embryo that comprises fewer cells (Bickel *et al.*, 2013; Table 13.1).

13.4.5 Regulation of Reproduction

Climatic factors, such as temperature and photoperiods, and biotic factors, such as host plant quality, its species, growth stage, and health, directly affect of reproduction in aphids from asexual to sexual and vice versa (Cortés *et al.*, 2008).

13.4.5.1 Photoperiod

Marcovitch (1923, 1924) was first to observe regulation of sexual reproduction by night length in *Aphis forbesi,* with long night-inducing sexual morphs. Thereafter, a number of workers have shown regulation of reproductive modes by photoperiodic changes in aphids (Lamb & Pointing, 1972; Hardie, 1990; Margaritopoulos & Tsitsipis, 2002; Campbell & Tregidga, 2005). The sexual reproduction is triggered by scotophases longer than 9–10 hour corresponding to early autumn (Chen *et al.*, 2019). The critical photoperiod for male production is usually longer than for females (Lamb & Pointing, 1972). Perception of photoperiod is maternal even when she herself is an embryo (Lees, 1966). Photoperiod is perceived by photoreceptors in anterior ventral neuropile region of protocerebrum with is no role of medial neurosecretory cells (Gao *et al.*, 1999). A biological clock to measure scotophase length rather than photophase length, and a counter mechanism that integrates information from clock may exist to induce polyphenism, with different expression levels of circadian clock genes between scotophase and photophase (Shuo *et al.*, 2020). The outputs from photoperiodic clock affect endocrine systems that release humoral factors that control developmental programmes (Saunders, 2002, 2009) (Figure 13.11). The photoperiodic clock has been considered to consist of a two-step process: one step being

TABLE 13.1

Differences in oogenesis and embryogenesis between sexual and asexual female aphids (from Miura *et al.*, 2003; Le Trionnaire *et al.*, 2008; Bickel *et al.*, 2013)

Particulars	Sexual Reproduction	Asexual Reproduction
Oocytes	Oocytes blocked in metaphase of meiosis I in germarium and vitellarium	Pre-oocytes blocked in prophase, within germarium
	Meiosis resumes after fertilization (expulsion of two polar bodies)	First mitotic division (expulsion of one polar body) occurs within first follicle chamber
Eggs/embryos	One or two eggs per ovariole	Six to ten embryos per ovariole
	Egg = 1 mm long	Embryo = 60 µm long
	Large amount of yolk	Little if any yolk. No serosal cuticle, vitellogenesis or chorion
	Duration of syncytium stage longer	
Development	Rate of development: 100 days	Rate of development: 10 days
	Discontinuous: a rapid early step and a late longer step	Continuous in duration
Primary symbiotic bacteria	ration by posterior pole through multi-opening in follicular epithelium. Bacteria are then integrated	Integrated after cellularization

time-measurement system, which measures day or night portion of a day and other is photoperiodic counter. In the latter step, photoperiodic information is successively received during sensitive period and it then accumulates to an internal threshold. Once this threshold is passed, endocrine events controlling development are triggered (Shiga, 2013). Juvenile hormone (JH) is a likely candidate for regulating transition from asexual to sexual reproduction after measuring photoperiod (Mittler *et al.*, 1976; Corbitt & Hardie, 1985; Ishikawa *et al.*, 2013). Ishikawa *et al.* (2012) reported that expression of JH esterase, responsible for JH degradation, was significantly higher in aphids reared under shortday conditions leading to lower JH titres but this is not a universal finding (Schwartzberg *et al.*, 2008)

13.4.5.2 Temperature

Temperature has been found to modulate photoperiodic response (Lees, 1959); high temperature may make long scotophase ineffective in some species (Lamb & Pointing, 1972; Poupoulidou *et al.*, 2006). TMale production is higher at 17°C than at 20°C in Utsunomiya clone of *Aphis glycines* (Oka *et al.*, 2018). Aphids with coexisting sexual and asexual reproduction are more common at lower latitudes (Dedryver *et al.*, 2001; Delmotte *et al.*, 2002; Kanbe & Akimoto, 2009).

13.4.5.3 Internal Timer or Intrinsic Factors

In some aphid species, sexual reproduction occurs in response to termination of host plant shoot growth (Dixon, 1998). Root-feeding aphids that live under constant darkness at relatively constant temperature also produce sexual morphs when host plants become dormant (Dixon, 1998). Dixon (1972) observed that young clones founded by fundatrices and reared at short-photophase and low-temperature could not produce sexual individuals as expected. This type of control mechanism, called 'interval timer', is considered to be an adaptation to avoid unexpected production of a sexual generation under shortday low-temperature condition in early spring (Dixon, 1972, 1998). Alate females and progeny descended from it hardly produce alate progeny (Mackay & Wellington, 1977). However, apterae progeny derived from apterae mothers respond strongly to wing-inducing stimuli (Mackay & Wellington, 1977; MacKay & Lamb, 1979). These findings suggest that, in aphids, environmental information perceived by ancestors can be passed down through generations and also that developmental fate must be determined following integration of such successive information (Ogawa & Miura, 2014).

The perception of photoperiodic signal to production of clonal or sexual forms remain largely unknown (Simon *et al.*, 2010), with some issues regarding switch of reproductive modes in aphids, requiring clarification (Shuo *et al.*, 2020): (a) unknown photoreceptors, (b) regulation of photoperiodism by circadian clock, (c) role of JH in photoperiodic transduction, and (d) genes and proteins involved in switch of reproductive modes (Mandrioli *et al.*, 2019).

13.5 Sex-Determination System in Aphids

All aphids make use of XO sex-determination system (Wilson *et al.*, 1997), with sexual females genetically identical to their asexual mothers with a full complement of homozygous XX chromosomes and males are hemizygous for X chromosome(XO). Sperm produced by XO male without X sex chromosome degenerate very rapidly and never contribute to an embryo (Blackman, 1985). All offspring of a sexual mating must have XX as their sex chromosomes, because females always contribute an X chromosome and, therefore, all aphids resulting from sexual matings are females (fundatrices) (Figure 13.12).

13.6 Phenotypic Plasticity among Aphids

Aphids show great phenotypic polymorphism depending upon climatic conditions, seasonality of their host plants, food quality, climate, natural enemy associations (Singh & Singh, 2016), and even

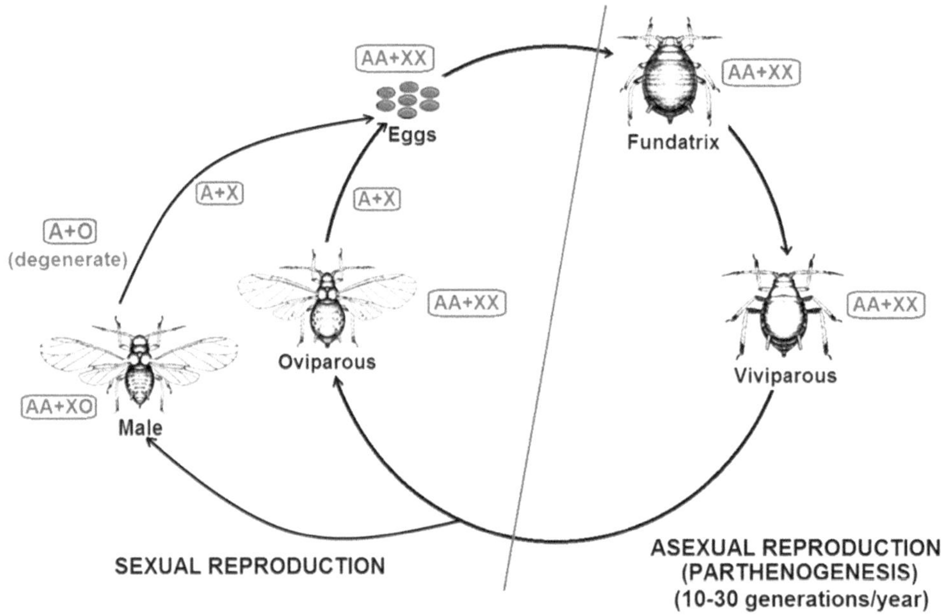

FIGURE 13.12 Sex determination in aphids. Aphids employs XO sex-determination system. Fundatrices have full complement of chromosomes including homozygous for X chromosome, i.e. AA+XX and males are hemizygous for X chromosome, i.e. AA+XO (only X sex chromosome). It implies that males are produced parthenogenetically with random loss of one X chromosome during maturation division, which is modified meiosis without chromosomal reduction. Oviparous females produce haploid oocytes (A+X) and males produce haploid sperms (A+X and A+O) by reductive meiosis. Sperm having an A+X set of chromosome are only viable and sperms lacking X chromosome (A+O) degenerate). Fertilization results diploid eggs (AA+XX), hence, next generation, which will hatch as fundatrices in spring, is entirely female (AA+XX).

symbionts (Leonardo & Mondor, 2006) that vary in space and time, and also differ greatly in morphology, behaviour, physiology, and biochemistry to cope with situations. This indicates exceptional flexibility of aphid genome at both structural and expressional levels (Casacuberta, 2004).

Aphids can perceive changes in food quality through gustatory mechanisms and can respond via morphogenetic changes (Dubey & Singh, 2008; Singh & Singh, 2015; Tiwari & Singh, 2016), though change is not directional. Crowding also induces production of alate aphids (Martínez & Costamagna, 2018), which leads to dispersal and colonization to newer optimal environments (Purandare *et al.*, 2014), but again this response is not universal (Mehrparvar *et al.*, 2013). However, life cycle stage at which crowding has most influence differs with species (Kawada, 1965; Shaw, 1970). Alterations in endocrine balance during embryonic and post-embryonic development are likely responsible for morphogenetic changes, as are entomopathogenic fungi (Hatano *et al.*, 2012).

It has been observed that magnitude and direction of transgenerational phenotypic responses to predators vary among individuals and/or populations of same species (Weisser *et al.*, 1999), with some producing winged offspring (Balog *et al.*, 2013; Sentis *et al.*, 2019), while others do not (Mehrparvar *et al.*, 2013). Weisser *et al.* (1999) reported that kairomones emitted by predators cause plasticity in morphology, life cycle, and behavior of their prey.

Females may have up to eight genetically identical distinct phenotypes that are described below

13.6.1 Fundatrices – Stem Mother or Foundress

These are usually apterous, viviparous, viriginoparae, or parthenogenetic females, which emerge in spring from overwintered eggs (Figure 13.9b) on primary host. This morph is characteristic of egg-laying holocyclic aphids. The legs, antennae, and maybe siphunculi are not so well-developed in succeeding apterous parthenogenetic generations. The reduction of parts is apparently correlated with conservation of nutrients to increase fecundity, longevity, and weapons for intra- and interspecific competition (Roff & Fairbairn, 1991). Fundatrices in Drepanosiphinae and Phyllaphidinae are however alate (Dixon, 1998).

13.6.2 Fundatrigeniae or Viriginoparae – Apterous Viviparous Female

These are apterous, parthenogenetic, viviparous progeny of fundatrices that live on primary host and give rise to further wingless forms or to migrantes (Figure 13.9c). In heteroecious species, where sexual and asexual generations are on plants of unrelated taxa, this morph is distinguishable into fundatrigeniae proper and alienicolae; latter being produced by former on secondary hosts (Dixon, 1998).

13.6.3 Migrantes – Alate Viviparous Female

The migrantes develop in second, third, or later generations of fundatrigeniae and consist of alate parthenogenetic viviparous females (Figure 13.9d). Wings maybe incompletely developed (brachypterae) though some are macropterous (Moritsu 1983). They have more extensive sclerotization of head and thorax, more developed sensory structures, such as compound eyes, ocelli, longer antennae equipped with more rhinaria/sensoria, and occasionally larger siphunculi and cauda (Ishikawa & Miura, 2007; Ogawa *et al.*, 2012). They develop on primary host in beginning of spring, called spring migrants or emigrants and subsequently fly to secondary host. The return migrantes to primary hosts are sexuparae or sexuales (Dixon, 1998). Alate aphids have longer nymphal development period, pre-reproductive adult period, reproductive period, longevity and lower fecundity (Campbell & Mackauer, 1977; Ogawa *et al.*, 2012), and are more resistant to starvation (Hazell *et al.*, 2005).

13.6.4 Alienicolae

Alienicolae are parthenogenetic, viviparous females (also exule) developing largely on secondary host (Figure 13.9e) as winged or wingless morphs (Braendle *et al.*, 2006; Brisson, 2010). In tribe Macrosiphini, wing polymorphisms have been attributed to both genetic and environmental factors, even within a single species (Brisson *et al.*, 2007, 2010). They often differ markedly from fundatrices and migrantes; 10–30 generations may be produced comprising both wingless and winged forms. Temperature, photoperiod, and host plant; all induce alary polymorphism in aphids (Schaefers & Judge, 1971).

13.6.5 Sexuparae

These are specialized parthenogenetic viviparous females developing on secondary hosts that produce sexual phenotypes in next generation in holocyclic species; terminating alenicolae by giving rise to sexuales (Dixon, 1998).

13.6.6 Sexuales

The sexuales appear only once in life cycle. In host-alternating species, they produce gynoparae first, and then androparae (Figure 13.9f). The gynoparae (Figure 13.9g) are an intermediate stage producing oviparae on primary host (Dixon, 1998), and are apterous, and have thickened hindleg tibiae and longer body. The males are either alate or apterous. Intermediates between alate and apterous (brachypterous) forms also occur. The sexuales exhibit various types of specialization among different genera (Lubiarz, 1980; Gautam & Verma, 1983).

In non-migratory species the terms migrantes and alienicolae are not applicable. Here, winged and wingless viviparous females are referred to as fundatrigeniae alatae or apterae as case may be, and either one or other may give rise to sexuparae.

13.7 Colour Polymorph in Aphids

Among insects, aphids display remarkable colour polymorphism (Caillaud & Losey, 2010), with variations between genotypes as well as clones, members of a colony, or associated with a particular sex or generation. The extent of colour variation may be limited (Figure 13.13) (Caillaud & Losey, 2010), or extensive matching coloration of different parts of host plants (Behura, 1996; Forsman *et al.*, 2008). Colour development is regulated by environmental (photoperiod and temperature), nutritional (plant and type of diet in general), bacterial symbionts, infection by pathogens, genetic factors, or a combination of one or more of these (Hille Ris Lambers, 1966). Water soluble pigment protoaphin, a glucoside derived from plant sapdetermines colouration of aphid (Bowie *et al.*, 1966; Behura, 1996). Aphids are also frequently enveloped entirely or partly with white or gray wax as protective cover (Figure 13.14). Colour polymorphism in aphids is believed to be an adaptive response to environmental factors, in particular natural enemies and host plant quality.

Ahsaei *et al.* (2013) observed a higher percentage content of water-soluble carbohydrates and lipids in red morphs and higher percentage content of protein in green morphs of pea aphid, with latter being more fecund and former producing more winged progenies when crowded or surrounded by natural enemies and are also more active (Farhoudi, 2012) (Braendle & Weisser, 2001). Protein is used for for development, maintenance, morphogenesis, and reproduction (Slanski & Scriber, 1985; Karowe & Martin, 1989), with three generations within one viviparous female

Colour also changes susceptibility to predators, with ladybird beetles eating more pink morphs on green plants and parasitoid wasps preferentially attacking green morph (Losey *et al.*, 1997). Differntial ability has been seen in colour morphs to survive and exploit on varying hostplants, with some studies showing no difference (Losey & Eubanks, 2000) and others showing significant difference (Simon *et al.*, 2003; Farhoudi, 2012). It seems that colour polymorphism in pea aphid may be under a complex set of selective pressures from both higher and lower trophic levels, and appears to be associated with behavioural differences as well as host plant preferences (Caillaud & Losey, 2010).

FIGURE 13.13 Red and green morph of pea aphid, *Acyrthosiphon pisum* (Courtesy of Charles Hedgcock, R.B.P.).

FIGURE 13.14 Colony of sugarcane wooly aphid (*Ceratovacuna lanigera*) (a), and apple wooly aphid (*Eriosoma lanigerum*) (b) with wax-covered body for protection against natural enemies and harsh environmental conditions.

13.7.1 Genetic Regulation of Colour Morph

Müller (1979, 1987) concluded that genetically based variation in body colour in aphids was result of a single two-allele locus, one allele being dominant over other but with very little supporting genetic data. Caillaud and Losey (2010) revealed that colour polymorphism in pea aphids is determined by a single biallelic autosomal locus, *colorama*, with alleles *P* and *p*, pink being dominant to green (the putative genotypes are *Pp* or *PP* for pink morphs, and *pp* for green morphs), and also found no evidence of influence of direction of cross on colour inheritance demonstrating that cytoplasmic effects and/or maternally inherited symbionts play no role in inheritance of colour polymorphism in pea aphids.

13.7.2 Endosymbiont-induced Colour Morph

Moran and Jarvik (2010) observed that aphid genome contains genes of fungal origin for carotenoid synthesis that seem to have been acquired during evolution via ancient lateral transfer. One gene is involved in synthesis of red colour pigments, with its absence leading to green colouration. Bacterium, *Rickettsiella viridis,* an endosymbiont of pea aphid modifies their body colour of pea aphids influencing prey-predator/host-parasitoid and endosymbiont interactions (Tsuchida *et al.*, 2010, 2011; Bilodeau *et al.*, 2013). Protection against parasitoids is also conferred by other bacterial symbionts, viz. *Hamiltonella defensa, Serratia symbiotica*, and *Regiella insecticola* (Vorburger *et al.*, 2010) and also against fungal pathogens (Łukasik *et al.*, 2013). Although maternally inherited bacterial symbionts and colour genes could be associated with such changes (Jenkins *et al.,* 1999).

13.7.3 Environment-induced Colour Morph

Colour polymorphism is known to occur in response to changes in diet (Watt & Hales, 1996) and temperature (Fröhlich, 1962; Caillaud & Losey, 2010).

13.8 Effects of Inbreeding and Outbreeding on Reproduction of Aphids

An outcome of cyclic parthenogenesis in aphids is inbreeding (Helden & Dixon, 1997), which may have influenced local adaptations, evolution of sex ratios, and mating systems in holocyclic populations. In mixed aphid populations in temperate regions, consisting of sexual, asexual, and intermediate clones, effects of inbreeding may be restricted. Inbreeding is expected to occur frequently in species with low migratory ability (Thornhill, 1993) as well as in those without host alternation (Komatsu & Akimoto, 1995). In Eriosomatinae and Hormaphidinae, inbreeding may arise because sexuals are stout, remain on host trunk after birth, and mate without feeding (Dixon, 1998). Akimoto (2006) observed that when males and females of same clone of *Prociphilus oriens* are confined in a small cage, they mate readily and produce inbred eggs which hatch less successfully outbred eggs. However, impact of inbreeding depression varies among aphid species. Huang and Caillaud (2012) indicated severe inbreeding depression in *Acyrthosiphon pisum* and opined that inbreeding avoidance may involve cryptic female choice.

Inbreeding may also influence aphid sex-ratios and mating systems with female-biased sex ratio (Yamaguchi, 1985; Foster, 2002) explained by local mate competition. Akimoto (2006) observed that *Prociphilus oriens* sexuparae consist of two types: one that produced both sexes simultaneously in abdomen ('M+F' type), while other produced only females ('F' type). It is possible that 'F' type sexuparae have advantage of avoiding inbreeding.

13.9 Conclusions

The aphids as a group of sucking insects are very fascinating with regard to their life cycle patterns, host selection behaviour, host-plant interaction, etc. Out of about 5110 species described, about 250 species are considered as notorious agricultural and horticultural pests. They may be monophagous, oligophagous, or highly polyphagous. Aphids excrete considerable amount of sugary liquid, honeydew, on which sooty mould usually develop hampering photosynthetic and respiratory ability of plants. The aphids are unique on account of their peculiar mode of reproduction, development, and polymorphism. They may reproduce either by parthenogenesis (asexual reproduction) or zygogenesis (sexual reproduction). They may either be oviparous or viviparous both may be winged or wingless. Parthenogenetic reproduction allows rapid increase in numbers and results in populations consisting of clones. Some species reproduce both parthenogenetically and sexually (holocyclic species), whereas only a few reproduce parthenogenetically (anholocyclic species) only. The aphids are polymorphic and both winged (alate) and wingless (aptera) morphs may be found in same colony. Several factors, both abiotic and biotic, have effects on formation of different phenotypes and each phenotype performs different ecological roles in their life history. Aphids are frequently engaged in mutualistic associations with bacterial endosymbionts that not only provide essential amino acids to them, but also provide them protection from natural enemies, protection from extreme temperatures, and development of resistance to a fungal pathogen and ability to use a greater diversity of resources.

REFERENCES

Agarwala, B.K. (2007). Phenotypic plasticity in aphids (Homoptera: Insecta): components of variation and causative factors. *Current Science*, *93*: 308–313.

Ahsaei, S.M., Tabadkani, S.M., Hosseininaveh, V., Allahyari, H. & Bigham, M. (2013). Differential accumulation of energy by colour morphs of pea aphid *Acyrthosiphon pisum* (Hemiptera: Aphididae) mirrors their ecological adaptations. *European Journal of Entomology*, *110*(2): 241–245.

Akimoto, S. (2006). Inbreeding depression, increased phenotypic variance, and a trade-off between gonads and appendages in selfed progeny of aphid *Prociphilus oriens*. *Evolution*, *60*: 77–86.

Aoki, S. (1980. Life cycles of two *Colophina* aphids (Homoptera: Pernphigidae) producing soldiers. *Kontyû* *48*, 464–476.

Balog, A., Mehrparvar, M. & Weisser, W.W. (2013). Polyphagous predatory rove beetles (Coleoptera: Staphylinidae) induce winged morphs in pea aphid *Acyrthosiphon pisum* (Hemiptera: Aphididae). *European Journal of Entomology*, *110*: 153–157.

Behura, B.K. & Das, M.M. (1976). Aphid and their role in agriculture. *Proceedings of National Academy of Sciences, India*, *46*: 261–265.

Behura, B.K. (1996). The colour of aphids (Homoptera: Aphididae): a mini-review and bibliography. *Journal of Aphidology, 10*: 61–66.
Bermingham, J. & Wilkinson, T.L. (2009). Embryo nutrition in parthenogenetic viviparous aphids. *Physiological Entomology, 34*: 103–109.
Bickel, R.D., Cleveland, H.C., Barkas, J., Jeschke, C.C., Raz, A.A., Stern, D.L. & Davis, G.K. (2013). The pea aphid uses a version of terminal system during oviparous, but not viviparous, development. *EvoDevo, 4*: 10.
Bilodeau, E., Guay, J.-F., Turgeon, J. & Cloutier, C. (2013). Survival to parasitoids in an insect hosting defensive symbionts: a multivariate approach to polymorphic traits affecting host use by its natural enemy. *PLoS One, 8*: e60708.
Blackman, R.L. (1976). Cytogenetics of two species of *Euceraphis* (Homoptera, Aphididae). *Chromosoma, 56*: 393–408.
Blackman, R.L. (1985). Spermatogenesis in aphid *Amphorophora tuberculata* (Homoptera, Aphididae). *Chromosoma, 92*: 357–362.
Blackman, R.L. (1987) Reproduction, cytogenetics and development. In: *Aphids: their Biology, Natural Enemies and Control* (Eds. Minks, A.K. & Harrewijn, A.P.), *Vol. 2A*, Amsterdam: Elsevier, pp. 163–195.
Blackman, R.L. & Eastop, V.F. (2006). *Aphids on World's Herbaceous Plants and Shrubs. Vols 1 & 2.* Chichester, U.K: John Wiley & Sons, p. 1460.
Blackman, R.L. & Eastop, V.F. (2000). *Aphids on World's Crops: An Identification and Information Guide*, 2nd edn., London, UK: John Wiley & Sons, p. 476.
Bonnet, C. (1745). *Traité d'insectologie ou observations sur les pucerons.* Paris: Chez Durand, Libraire, p. 236.
Bowie, J.H., Cameron, D.W., Findlay, J.A. & Quartey, J.A.K. (1966). Haemolymph pigments of aphids. *Nature (London), 210*(5034): 395–397.
Braendle, C. & Weisser, W.W. (2001). Variation in escape behavior of red and green clones of pea aphid. *Journal of Insect Behavior, 14*: 497–509.
Braendle, C., Davis, G.K., Brisson, J.A. & Stern, D.L. (2006). Wing dimorphism in aphids. *Heredity, 97*: 192–199.
Brisson, J.A. (2010). Aphid wing dimorphisms: linking environmental and genetic control of trait variation. *Philosophical Transactions of Royal Society B: Biological Sciences, 365*: 605–616.
Brisson, J.A., Davis, G.K. & Stern, D.L. (2007). Common genome-wide patterns of transcript accumulation underlying wing polyphenism and polymorphism in pea aphid. *Evolution & Development, 9*: 338–346.
Brisson, J.A., Ishikawa, A. & Miura, T. (2010). Wing development genes of pea aphid and differential gene expression between winged and unwinged morphs. *Insect Molecular Biology, 19*: 63–73.
Caillaud, M.C. & Losey, J.E. (2010). Genetics of color polymorphism in pea aphid, *Acyrthosiphon pisum. Journal of Insect Science, 10*: 95.
Campbell, A. & Mackauer, M. (1977). Reproduction and population growth of pea aphid (Homoptera: Aphididae) under laboratory and field conditions. *Canadian Entomologists, 109*: 277–284.
Campbell, C.A.M. & Tregidga, E. (2005). Photoperiodic determination of gynoparae and males of damson-hop aphid *Phorodon humuli. Physiological Entomology, 30*: 189–194.
Casacuberta, J.M. (2004). *El Genoma Fluido.* Barcelona, Spain: La Voz de Galicia
Chakrabarti, S. (2007). Diversity and biosystematics of gall-inducing aphids (Hemiptera: Aphididae) and their galls in Himalaya. *Oriental Insects, 41*: 35–54.
Chen, Y., Verheggen, F.J., Sun, D., Wang, Z., Francis, F. & He, K. (2019). Differential wing polyphenism adaptation across life stages under extreme high temperatures in corn leaf aphid. *Scientific Reports, 9*: 8744.
Corbitt, T.S. & Hardie, J. (1985). Juvenile-hormone effects on polymorphism in pea aphid, *Acyrthosiphon pisum. Entomologia Experimentalis et Applicata, 38*: 131–135.
Cortés, T., Tagu, D., Simon, J.-C., Moya, A. & Martínez-Torres, D. (2008). Sex versus parthenogenesis: a transcriptomic approach of photoperiod response in model aphid *Acyrthosiphon pisum* (Hemiptera: Aphididae). *Gene, 408*: 146–156.
Dedryver, C.A., Hullé, M., Le Gallic, J.F., Caillaud, M.C. & Simon, J.C. (2001). Coexistence in space and time of sexual and asexual populations of cereal aphid *Sitobion avenae. Oecologia, 128*: 379–388.

Delmotte, F., Leterme, N., Gauthier, J.P., Rispe, C. & Simon, J.C. (2002). Genetic architecture of sexual and asexual populations of aphid *Rhopalosiphum padi* based on allozyme and microsatellite markers. *Molecular Ecology*, *11*: 711–723.

Dika, J. & Van Pelt, J.A. (1992). Interaction between phyllosphere yeasts, aphid honeydew and fungicide effectiveness in wheat under field conditions. *Plant Pathology*, *41*(6): 661–675.

Dixon, A.F.G. (1971). The life cycle and host preferences of bird cherry-oat aphid, *Rhopalosiphum padi* L. & their bearing on theories of host alternation in aphids. *Annals of Applied Biology*, *68*(2): 135–147.

Dixon, A.F.G. (1972). The interval timer, photoperiod and temperature in seasonal development of parthenogenetic and sexual morphs in lime aphid, *Eucallipterus tiliae* L. *Oecologia*, *9*: 301–310.

Dixon, A.F.G. (1998). *Aphid Ecology: An Optimization Approach*. London: Chapman and Hall, pp. 300.

Dubey, S. & Singh, R. (2008). Host-plant induced variations in life-table statistics of *Aphis spiraecola* Patch (Homoptera: Aphididae). *Journal of Aphidology*, *23*: 53–60.

Durak, R. (2014). The overwintering strategy of anholocyclic aphid *Cinara tujafilina*. *Physiological Entomology*, *39*(4): 313–321.

Eastop, V.F. (1977). World wide importance of aphids as virus vectors. In: *Aphids as Virus Vectors* (Eds. Harris, F. & Marmorosch, K.). London: Academic Press, pp. 4–62.

Farhoudi, F. (2012). A comparison on developmental reproductive parameters of red and green morphs of pea aphid *Acyrthosiphon pisum* (Harris) (Hemiptera: Aphididae) and its defensive behavior against ladybird, *Hippodamia variegata*. MSc thesis, University of Tehran, pp. 73.

Forsman, A., Ahnesjö, J., Caesar, S. & Karlsson, M. (2008). A model of ecological and evolutionary consequences of color polymorphism. *Ecology*, *89*: 34–40.

Foster, W.A. (2002). Aphid sex ratios. In: *Sex Ratios: Concepts and Research Methods* (Ed. Hardy, I.C.W.). Cambridge: Cambridge University Press, pp. 254–265.

Fröhlich, G. (1962). Das Verhalten der grünen Erbsenlaus *Acyrthosiphon pisum* (Harris). *Zeitschift für Angewandte Entomologie*, *51*: 55–68.

Gao, N., von Schantz, M., Foster, R.G. & Hardie, J. (1999). The putative brain photoperiodic photoreceptors in vetch aphid, *Megoura viciae*. *Journal of Insect Physiology*, *45*: 1011–1019.

Gautam, D.C. & Verma, L.R. (1983). Life history of sexuparae and sexual morphs of woolly apple aphid (*Eriosoma lanigerum* Hausmann). *Proceedings of Indian Academy of Sciences (Animal Science)*, *92*(3): 247–251.

Ghosh, A., Chakrabarti, S., Mandal, B. & Krishna Kumar, N.K. (2017). Aphids as vectors of plant viruses in India. In: *A Century of Plant Virology in India* (Eds. Mandal, B., Rao, G., Baranwal, V. & Jain, R.). Singapore: Springer, pp. 515–536.

Ghosh, L.K. & Rajendran, T.P. (1988). First record of an aphid sexual (Homoptera: Aphididae) from Rajasthan, India. *Journal of Aphidology*, *2*: 51–58.

Glinwood, R. and Pettersson, J. (2000). Host choice and host leaving in *Rhopalosiphum padi* (Hemiptera: Aphididae) emigrants and repellency of aphid colonies on winter host. *Bulletin of Entomological Research*, *90*: 57–61.

Griffiths, E. & Wratten, S.D. (2011). Intra- and inter-specific differences in cereal aphid low-temperature tolerance. *Entomologia Experimentalis et Applicata*, *26*(2): 161–167.

Hardie, J. (1990). The photoperiodic counter, quantitative day-length effects and scotophase timing in vetch aphid *Megoura viciae*. *Journal of Insect Physiology*, *36*: 939–949.

Hardie, J. (2017). Life cycle and polyphenism. In: *Aphids as Crop Pests* (van Emden, H.F. & Harrington, R.), 2nd edn., Uttarakhand: CABI, pp. 81–97.

Hatano, E., Baverstock, J., Kunert, G., Pell, J.K. & Weisser, W.W. (2012). Entomopathogenic fungi stimulate transgenerational wing induction in pea aphids, *Acyrthosiphon pisum* (Hemiptera: Aphididae). *Ecological Entomology*, *37*: 75–82.

Hattori, M., Kishida, O. & Itino, T. (2013). Soldiers with large weapons in predator-abundant midsummer: reproductive plasticity in a eusocial aphid. *Evolutionary Ecology*, *27*: 847–862.

Hazell, S.P., Gwynn, D.M., Ceccarelli, S. & Fellowes, M.D.E. (2005). Competition and dispersal in pea aphid: clonal variation and correlations across traits. *Ecological Entomology*, *30*: 293–298.

Helden, A.J. & Dixon, A.F.G. (1997). Inbreeding and egg hatching success in *Sitobion avenae*. *Ecological Entomology*, *22*(1): 124–126.

Hille Ris Lambers, D. (1966). Polymorphism in Aphididae. *Annual Review of Entomology*, *11*: 47–78.

Hooks, C.R.R. & Fereres, A. (2006). Protecting crops from non-persistently aphid-transmitted viruses: a review on use of barrier plants as a management tool. *Virus Research*, *120*: 1–16.
Huang, M.H. & Caillaud, M.C. (2012). Inbreeding avoidance by recognition of close kin in pea aphid, *Acyrthosiphon pisum*. *Journal of Insect Science*, *12*: 39.
Ishikawa, A. & Miura, T. (2007). Morphological differences between wing morphs of two Macrosiphini aphid species, *Acyrthosiphon pisum* and *Megoura crassicauda* (Hemiptera, Aphididae). *Sociobiology*, *50*; 881–893.
Ishikawa, A., Gotoh, H., Abe, T. & Miura, T. (2013). Juvenile hormone titer and wing-morph differentiation in vetch aphid *Megoura crassicauda*. *Journal of Insect Physiology*, *59*: 444–449.
Ishikawa, A., Ogawa, K., Gotoh, H., Walsh, T.K., Tagu, D., Brisson, J.A., Rispe, C., Jaubert-Possamai, S., Kanbe, T., Tsubota, T., Shiotsuki, T. & Miura, T. (2012). Juvenile hormone titre and related gene expression during change of reproductive modes in pea aphid. *Insect Molecular Biology*, *21*(1): 49–60.
Itô, Y. (1989). The evolutionary biology of sterile soldiers in aphids. *Trends in Ecology & Evolution*, *4*: 69–73.
Jenkins, R.L., Loxdale, H.D., Brooks, C.P. & Dixon, A.F.G. (1999). The major carotenoid pigments of grain aphid, *Sitobion avenae* (F.) (Hemiptera: Aphididae). *Physiological Entomology*, *24*: 171–178.
Kanbe, T. & Akimoto, S.I. (2009). Allelic and genotypic diversity in long-term asexual populations of pea aphid, *Acyrthosiphon pisum* in comparison with sexual populations. *Molecular Ecology*, *18*: 801–816.
Karowe, D.N. & Martin, M.M. (1989). The effects of quantity and quality of nitrogen diet on growth, efficiency of food utilization, nitrogen budget, and metabolic rate of fifth-instar Spodoptera eridania larvae (Lepidoptera: Noctuidae). *Journal of Insect Physiology*, *35*: 699–708.
Kawada, K. (1965). The development of winged forms in cabbage aphid *Brevicoryne brassicue* Linnaeus II. The period of determination of wing development. *Berichte de Ohara Instituts*, *13*(1): 1–5.
Komatsu, T. & Akimoto, S. (1995). Genetic differentiation as a result of adaptation to phenologies of individual host trees in galling aphid *Kaltenbachiella japonica*. *Ecological Entomology*, *20*: 33–42.
Lamb, R.J. & Pointing, P.J. (1972). Sexual morph determination in aphid, *Acyrthosiphon pisum*. *Journal of Insect Physiology*, *18*: 2029–2042.
Le Trionnaire, G., Hardie, J., Jaubert-Possamai, S., Simon, J.C. & Tagu, D. (2008). Shifting from clonal to sexual reproduction in aphids: physiological and developmental aspects. *Biology of Cell*, *100*: 441–451.
Le Trionnaire, G., Jaubert-Possamai, S., Bonhomme, J., Gauthier, J.P., Guernec, G., Le Cam, A., Legeai, F., Monfort, J. & Tagu, D. (2012). Transcriptomic profiling of reproductive mode switch in pea aphid in response to natural autumnal photoperiod. *Journal of Insect Physiology*, *58*(12):1517–1524.
Leather, S.R. & Dixon, A.F.G. (1981). Growth, survival and reproduction of bird-cherry aphid, *Rhopalosiphum padi*, on its primary host. *Annals of Applied Biology*, *99*: 115–118.
Lees, A.D. (1959). The role of photoperiod and temperature in determination of parthenogenetic and sexual forms in aphid *Megoura viciae* Buckton. I. The influence of these factors on apterous virginoparae and their progeny. *Journal of Insect Physiology*, *3*: 92–117.
Lees, A.D. (1966). The control of polymorphism in aphids. *Advances in Insect Physiology*, *3*: 207–277.
Leonardo, T.E. & Mondor, E.B. (2006). Symbiont modifies host life-history traits that affect gene flow. *Proceedings of Royal Society B: Biological Sciences*, *273*: 1079–1084.
Losey, J.E. & Eubanks, M.D. (2000). Implications of pea aphid host-plant specialization for potential colonization of vegetables following post-harvest emigration from forage crops. *Environmental Entomology*, *29*: 1283–1288.
Losey, J.E., Harmon, J., Ballantyne, F. & Brown, C. (1997). A polymorphism maintained by opposite patterns of parasitism and predation. *Nature*, *388*: 269–272.
Lubiarz, M. (1980). Dynamics of numbers and bionomy of *Lachnus roboris* (L.) on pedunculate oak (*Quercus robur* L.) in natural and degraded landscape. *Aphids and other Hemipterous Insects*, *13*: 129–135.
Łukasik, P., van Asch, M., Guo, H., Ferrari, J. & Godfray, H.C.J. (2013). Unrelated facultative endosymbionts protect aphids against a fungal pathogen. *Ecology Letters*, *16*: 214–218.
Mackay, P.A. & Wellington, W.G. (1977). Maternal age as a source of variation in ability of an aphid to produce dispersing forms. *Researches on Population Ecology*, *18*: 195–209.
MacKay, P.A. & Lamb, R.J. (1979). Migratory tendency in aging populations of pea aphid, *Acyrthosiphon pisum*. *Oecologia*, *39*: 301–308.

Mandrioli, M., Zanetti, E., Nardelli, A. & Manicardi, G.C. (2019). Potential role of heat shock protein 90 (hsp90) in buffering mutations to favour cyclical parthenogenesis in peach potato aphid *Myzus persicae* (Aphididae, Hemiptera). *Bulletin of Entomological Research*, *109*(4): 426–434.

Marcovitch, S. (1923). Plant lice and light exposure. *Science*, *58*: 537–538.

Marcovitch, S. (1924). The migration of Aphididae and appearance of sexual forms as affected by relative length of daily light exposure. *Journal of Agricultural Research*, *27*: 513–522.

Margaritopoulos, J.T. & Tsitsipis, J.A. (2002). Production of sexual morphs by apterous virginoparae of *Myzus persicae* (Hemiptera: Aphididae) in relation to preand postnatal exposure to short day conditions. *Bulletin of Entomological Research*, *92*: 321–330.

Marshall, S.A. (2017). *Insects: Their Natural History and Diversity*. Firefly Books Ltd; 2nd revised edn., pp. 736.

Martínez, A.F.R. & Costamagna, A.C. (2018). Effects of crowding and host-plant quality on morph determination in soybean aphid, *Aphis glycines*. *Entomologia Experimentalis et Applicata*, *166*: 53–62.

Medda, P.K., Sarkar, S. & Chakrabarti, S. (1997). Willow infesting aphids (Homoptera: Aphididae) of India and adjacent countries, *Journal of Aphidology*, *11*(1): 83–97.

Mehrparvar, M., Zytynska, S.E. & Weisser, W.W. (2013). Multiple cues for winged morph production in an aphid metacommunity. *PLoS One*, *8*: e58323.

Michalik, A., Szklarzewicz, T., Wegirek, P. & Wieczorek, K. (2013). The ovaries of aphids (Hemiptera, Sternorrhyncha, Aphidoidea): morphology and phylogenetic implications. *Invertebrate Biology*, *132*(3): 226–240.

Minks, A.K. & Harrewijn, P. (1987). *Aphids: Their Biology, Natural Enemies and Control, Vol. A*. Amsterdam: Elsevier Science Publisher B.V., pp. 450.

Mittler, T.E. (1973). Aphid polymorphism as affected by diet. In: *Perspectives in Aphid Biology* (Ed. Lowe, A.D.), Bull. No. 2, New Zealand: Entomological Society, pp. 65–75.

Mittler, T.E., Nassar, S.G. & Staal, G.B. (1976). Wing development and parthenogenesis induced in progenies of kinoprene-treated gynoparae of *Aphis fabae* and *Myzus persicae*. *Journal of Insect Physiology*, *22*: 1717–1725.

Miura, T., Braendle, C., Shingleton, A., Sisk, G., Kambhampati, S. & Stern, D.L. (2003). A comparison of parthenogenetic and sexual embryogenesis of pea aphid *Acyrthosiphon pisum* (Hemiptera: Aphidoidea). *Journal of Experimental Zoology Part B: Molecular and Developmental Evolution*, *295*: 59–81.

Miyazaki, M. (1987). Forms and morphs of aphids. In: *Aphids: Their Biology, Natural Enemies and Control* (Eds. Minks, A.K. & Harrewijn, P.), *Vol. A.*, Amsterdam: Elsevier Science Publisher B.V., pp. 27–50.

Moran, N.A. & Jarvik, T. (2010). Lateral transfer of genes from fungi underlies carotenoid production in aphids. *Science*, *328*: 624–627.

Moran, N.A. (1992). The evolution of aphid life-cycles. *Annual Review of Entomology*, *37*, 321–348.

Moritsu, M. (1983). *Nihon Genshoku Aburamushi Zukan (in Japanese)*. Tokyo: Zenkoku Noson Kyoiku Kyokai.

Müller, F.P. (1979). Eine gelbe mutante der schwarzen Blattlaus *Aphis fabae cirsiiacanthoidis* Scopoli und Bastardierungsversuche. *Biologie Zentralbibliotek*, *104*: 225–237.

Müller, F.P. (1987). Faunistich-okologische untersuchungen uber aphiden im westlichen Erzgebirge und Vogtland. *Faunistische Ablandlungen Staatliches Museum für Tierkunde Dresden*, *14*:105–129.

Ogawa, K. & Miura, T. (2014). Aphid polyphenisms: trans-generational developmental regulation through viviparity. *Frontiers in Physiology*, *5*(1): 1–11.

Ogawa, K., Ishikawa, A., Kanbe, T., Akimoto, S. & Miura, T. (2012). Male-specific flight apparatus development in *Acyrthosiphon pisum* (Aphididae, Hemiptera, Insecta): comparison with female wing polyphenism. *Zoomorphology*, *131*: 197–207.

Oka, Y., Kagami-Yashima, C., Kagawa, K., Sonoda, S. & Murai, T. (2018). Clonal variation of sexual morph production in response to temperature and photoperiod in soybean aphid, *Aphis glycines* (Hemiptera: Aphididae). *Applied Entomology and Zoology*, *53*: 509–517.

Pettersson, J., Tjallingii, W.F. & Hardie, J. (2017). In: *Aphids as Crop Pests* (Eds. van Emden, H.F. & Harrington, R.), 2nd edn., U.K.: CABI, pp. 173–195.

Phillips, S.W., Bale, J.S. & Tatchell, M. (2000). Overwintering adaptations in lettuce root aphid *Pemphigus bursarius* (L.). *Journal of Insect Physiology*, *46*(3): 353–363.

Poupoulidou, D., Margaritopoulos, J.T., Kephalogianni, T.E., Zarpas, K.D. & Tsitsipis, J.A. (2006). Effect of temperature and photoperiod on life cycle in lineages of *Myzus persicae nicotianae* and *Myzus persicae s.str.* (Hemiptera: Aphididae). *European Journal of Entomology*, *103*: 337–346.
Purandare, S.R., Bickel, R.D., Jaquiery, J., Rispe, C. & Brisson, J.A. (2014). Accelerated evolution of morph-biased genes in pea aphids. *Molecular Biology & Evolution*, *31*: 2073–2083.
Riparbelli, M.G., Tagu, D., Bonhomme, J. & Callaini, G. (2005). Aster self-organization at meiosis: a conserved mechanism in insect parthenogenesis? *Developmental Biology*, *278*: 220–230.
Roff, D.A. & Fairbairn, D.J. (1991). Wing dimorphisms and evolution of migratory polymorphisms among Insecta. *American Zoologist*, *31*: 243–251.
Saeidi, F., Moharramipour, S. & Barzegar, M. (2012). Seasonal patterns of cold hardiness and cryoprotectant profiles in *Brevicoryne brassicae* (Hemiptera: Aphididae). *Environmental Entomology*, *41*(6): 1638–1643.
Saunders, D.S. (2002). *Insect Clocks*, 3rd edn. The Netherlands: Elsevier, pp. 576.
Saunders, D.S. (2009). Circadian rhythms and evolution of photoperiodic timing in insects. *Physiological Entomology*, *34*: 301–308.
Schaefers, G.A. & Judge, F.D. (1971). Effects of temperature, photoperiod, and host plant on alary polymorphism in aphid, *Chaetosiphon fragaefolii*. *Journal of Insect Physiology*, *17*: 365–379.
Schwartzberg, E.G., Kunert, G., Westerlund, S.A., Hoffmann, K.H. & Weisser, W.W. (2008). Juvenile hormone titres and winged offspring production do not correlate in pea aphid, *Acyrthosiphon pisum*. *Journal of Insect Physiology*, *54*: 1332–1336.
Sentis, A., Bertram, R., Dardenne, N., Ramon-Portugal, F., Louit, I., Le Trionnaire, G., Simon, J.C., Magro, A., Pujol, B., Hemptinne, J.L. & Danchin, E. (2019). Different phenotypic plastic responses to predators observed among aphid lineages specialized on different host-plants. *Scientific Reports*, *9*: 9017.
Shaw, M.J.P. (1970). Effect of population density on alienocolae of *Aphis fabae* Scop.I. The effect of crowding on production of alatae in laboratory. *Annals of Applied Biology*, *65*: 191–196.
Shiga, S. (2013). Photoperiodic plasticity in circadian clock neurons in insects. *Frontiers in Physiology*, *4*: 69.
Shuo, Y., Wan-Xing, W. & Jie, S. (2020). Reproductive polyphenism and its advantages in aphids: Switching between sexual and asexual reproduction. *Journal of Integrative Agriculture*, *19*(6): 1447–1457.
Simon, J.-C., Carre, S., Boutin, M., Prunier-Leterme, N., Sabater-Munoz, B., Latorre, A. & Bournoville, R. (2003). Host-based divergence in populations of pea aphid: insights from nuclear markers and prevalence of facultative symbionts. *Proceedings of Royal Society of London B Biological Science*, *270*:1703–1712.
Simon, J.-C., Stoeckel, S. & Tagu, D. (2010). Evolutionary and functional insights into reproductive strategies of aphids. *Comptes Rendus Biologies*, *333*: 488–496.
Singh, K. & Singh, R. (2015). Effect of temperature on life history traits of *Aphis gossypii* Glover (Homoptera: Aphididae) on bottle gourd, *Laginaria siceraria* (Molina) Standl. (Cucurbitaceae). *International Journal of Life Sciences Biotechnology & Pharma Research*, *4*(4): 179–183.
Singh, R. & Singh, G. (2016). Aphids and their biocontrol. In: *Ecofriendly Pest Management for Food Security* (Ed. Omkar). Academic Press, pp. 63–108.
Singh, R. & Singh, G. (2019). Species diversity of Indian aphids (Hemiptera: Aphididae. *International Journal of Biological Innovations*, *1*(1): 27–33.
Slanski, F. Jr. & Scriber, M.R. (1985): Food consumption and utilization. In: *Comprehensive Insect Physiology, Biochemistry and Pharmacology* (Eds. Kerkut, S.A. & Gilbert, J.F.). Oxford: Pergamon Press, pp. 87–151.
Sorensen, J.T. (2009). Aphids. In: *Encyclopedia of Insects* (Eds. Resh, V.H. & Carde, R.T.). U.K.: Academic Press, pp. 27–31.
Sorin, M. (1958). Life cycles of 2 aphids causing galls on *Distylium racemosum*. *Akitu*, *7*: 89–92.
Takahashi, R. (1962). Aphidscausing galls on Distylium racemosum in Japan, with descriptions of two new related species (Aphididae, Homoptera). Bulletin of the University of Osaka Prefecture, Series B, 13, 1–11.
Thornhill, N.W. (1993). *The Natural History of Inbreeding and Outbreeding*. Chicago: University of Chicago Press, pp. 575.
Tiwari, A.K. & Singh, R. (2016). Effect of host-plants on morphology of green peach aphid, *Myzus persicae* (Sulzer) (Homoptera: Aphididae). *International Journal of Zoological Investigations*, *2*(1): 133–146.

Tsitsipis, J.A. & Mittler, T.E. (1976). Embryogenesis in parthenogenetic and sexual females of *Aphis fabae*. *Entomologia Experimentalis et Applicata*, *19*: 263–270.

Tsuchida, T., Koga, R., Horikawa, M., Tsunoda, T., Maoka, T., Matsumoto, S., Simon, J. & Fukatsu, T. (2010). Symbiotic bacterium modifies aphid body color. *Science*, *330* (6007): 1102–1104.

Tsuchida, T., Koga, R., Matsumoto, S. & Fukatsu, T. (2011). Interspecific symbiont transfection confers a novel ecological trait to recipient insect. *The Biological Letters*, *7*: 245–248.

Vorburger, C., Gehrer, L. & Rodriguez, P. (2010). A strain of bacterial symbiont *Regiella insecticola* protects aphids against parasitoids. *The Biological Letters*, *6*(1): 109–111.

Watt, M. & Hales, D.F. (1996). Dwarf phenotype of cotton aphid, *Aphis gossypii* Glover (Hemiptera: Aphididae. *Australian Journal of Entomology*, *35*: 153–159.

Weisser, W., Braendle, C. & Minoretti, N. (1999. Predator-induced morphological shift in pea aphid. *Proceedings of Royal Society B: Biological Sciences*, *266*: 1175–1181.

Wilson, A.C.C., Sunnucks, P., & Hales, D.F. (1997). Random loss of X chromosome at male determination in an aphid, *Sitobion* near *fragariae*, detected using an X-linked polymorphic microsatellite marker. *Genetics Research*, *69*, 233–236.

Yamaguchi, Y. (1985). Sex ratios of an aphid subject to local mate competition with variable maternal condition. *Nature*, *318*: 460.

14

Reproductive Strategies in Parasitoids

Richa Varshney, Omprakash Navik, and Sushil K. Jalali
ICAR-National Bureau of Agricultural Insect Resources, Post Bag No. 2491, H. A. Farm Post, Hebbal, Bangalore 560024, Karnataka, India

CONTENTS

14.1 Introduction

Parasitoid is defined as 'an organism, which develops on or in another single (host) organism, extracts nourishment from it, and kills it as a direct or indirect result of that development' (Eggleton & Gaston, 1990). Réaumur (1738) was perhaps the first biologist to state enormous diversity of fecundities seen among insects. However, we are no closer to understanding dynamics of insect reproduction despite increasing data (Price, 1975). Knowledge of selective factors acting on egg production and results of likely selection is obligatory for a full understanding of co-evolution of host-parasitoid community ecology.

DOI: 10.1201/9781003043195-14

More than a hundred thousand species of insects are parasitoids and are found in Hymenoptera, Diptera, Coleoptera, Lepidoptera, and Neuroptera; hymenopteran parasitoids form 78% of known parasitoids (Silveira *et al.*, 2019). Some are ectoparasitoids, others endoparasitoids, some solitary, some gregarious.

Gregariousness of parasitoids has evolved by several female parasitoids attacking a single host or more than one egg laid by female or by polyembryony (Segoli *et al.*, 2010). In polyembryony, a single egg is placed by female within host body, but this single egg produced consists of many identical embryos that finally develop into adult parasitoids (Grbic *et al.*, 1998).

Parasitoids are often classified based on juvenile stages of holometabolous insects that attacked from egg, larval, pupal parasitoids, and also a few parasitoids attack adults. Many parasitoids attack nymphs and adults, wherein no pupal stage is found, such as hemimetabolous insects (Godfray, 1994). Some parasitoids develop as hyperparasitoids that parasitize other parasitoids in nature to survive and reproduce. Mostly Hymenopteran families develop as hyperparasitoids either externally or internally on/in host insects (Gordh *et al.*, 1999).

Diverse group of parasitoids develop different parasitism strategies and have co-evolved with their hosts for survival (Lopez-Vaamonde *et al.*, 2005). Parasitoids adapt different reproductive strategies consisting of a series of decision-making (Maynard Smith, 1978; Stearns, 1992). These reproductive strategies help making decisions for host selection, to parasitize or not, allocation of egg load per host, and sex of progeny (Charnov & Skinner, 1984; Godfray, 1994). Parasitoids adapt to different reproductive modes, like arrhenotoky or thelytoky, to manipulate sex of future offspring mainly in Hymenoptera. Several symbiotic bacteria, such as *Wolbachia*, etc., influence reproduction either producing higher female sex roles or male killing in parasitoids (Sazama *et al.*, 2019).

14.2 Evolution and Reproductive Strategies

Newly emerged female parasitoids have to overcome a series of problems, *viz.* (1) where to search; (2) which host to attack; (3) which host to eat; (4) clutch size; (5) egg size; (6) allocation of sex ratio; and (7) host should be active or should be paralyzed before oviposition (Charnov & Skinner, 1984).

Several factors determine fitness of a female parasitoid, time it takes to mate; rate at which it finds and parasitizes suitable host, quality of a host, number of eggs it lays, sex ratio, and ability of its offspring to survive and complete their development in parasitized hosts. The production of offspring in parasitoid wasps involves at least two general components: host habitat and host selection and determination of clutch size. Size and nutritional quality of host affect growth, development, and survival of parasitoid and its progeny (Godfray, 1994; Mackauer *et al.*, 1997).

Many studies have shown correlation between host size and adult fitness attributes, like mating capacity, dispersal efficiency, longevity, and lifetime reproductive success (Godfray, 1994; Visser, 1994). However, very less information is available on effects of development time on parasitoid fitness or conditions that might select for rapid development at expense of reduced adult size (Strand, 2000; Harvey & Strand, 2002).

14.2.1 Host Finding and Acceptance

Relations involving insect parasitoids and their hosts result in premature death of hosts, and are mandatory for parasitoid development. This perceptibly places strong stress on hosts to avoid recognition by parasitoids, and on parasitoids themselves to improve encounter rates with suitable hosts. To deal with this, parasitoids have developed a mixture of searching strategies that depend on a huge range of environmental signal (Turlings *et al.*, 1993).

Parasitoid often uses chemical cues, i.e. kairomones, while some employ sight, sound, light, heat, and vibration as cues to locate presence of host. Godfray (1994) suggested three parasitoid strategies to locate host: stimuli received by host microhabitat or food-plant, cues from activity of host, and stimuli directly emitted by host itself. Most hymenopteran parasitoids use first two strategies indicated above, whereas dipteran parasitoids communicate directly with their hosts (Feener & Brown, 1997). Parasitoids prefer to oviposit in best quality of host and reject host where probability of immature progeny survival is less (Godfray, 1994).

14.2.2 Clutch Size

How many eggs should be laid on host is a key question for parasitoids. Solitary parasitoids get rid of rivals, including sibs, for its own survival, while in gregarious species, parasitic larvae accept each other and several individuals can arise from a single insect host. In switch from solitary to gregarious form of development, conspecific superparasitism is a possible intermediate step, and will be preferred because of surviving larvae being that of one female. Under same conditions, theory predicts that gregarious species should reduce their clutch size (Pexton & Mayhew, 2005). Thus, the amount of resource allocated to each developing parasitoid decides ultimate fitness (Hardy, 1991).

The fitness of surviving brood represents a dynamic trade-off between number of offspring produced from a clutch and size of offspring (Skinner, 1985; Godfray, 1994). The exploitive competition between siblings in greater clutch size leads to smaller offspring adults (Vet *et al.*, 1994).

In gregarious parasitoids, female adjust clutch size based on host size (Dijkstra, 1986; Godfray, 1987); host quality, competition between parent females, mortality schedules of parent females, egg maturation time, and host encounter rate (Charnov & Skinner, 1984; Wilson & Lessells, 1994). At lower clutch size, survival and size of individual offspring increases more relative to greater clutch size (Hardy, 1991; Godfray, 1995) (Figure 14.1).

Solitary parasitoid larvae are more aggressive and stronger and usually kill their rival sibling, so that only single individual can successfully develop on host (Mock & Parker, 1997). In solitary species, where risk of superparasitism is high, female adapts to lay multiple-egg clutches in a single host (Visser & Rosenheim, 1998). In competition for a host, only single offspring will develop as an adult from allocated host. Further, solitary parasitoids lay more than one egg to increase survival probability of their offspring (Pexton & Mayhew, 2005).

14.2.3 Sex Ratio

Parasitoids can control fertilization of eggs and, thus, can control sex of offspring (van Dijken & Waage, 1987); permitting evolution of sex ratios adapted to environmental variables (environment, host quality, and local mate offsprings) (Nunney, 1985; Nunney & Luck, 1988).

Usually, large host produces more female offspring due to maternal control of offspring sex ratio (Cook *et al.*, 1994). The host-size model suggested by Charnov *et al.* (1981) explains evolution of host size-dependent sex allocation in parasitoid wasps. Offspring with less developmental time in larger host are mostly female (King, 1987). Parasitoid female allocate more female eggs in larger hosts and more male eggs in smaller hosts (Ueno & Tanaka, 1994, 1997).

(a) (b)

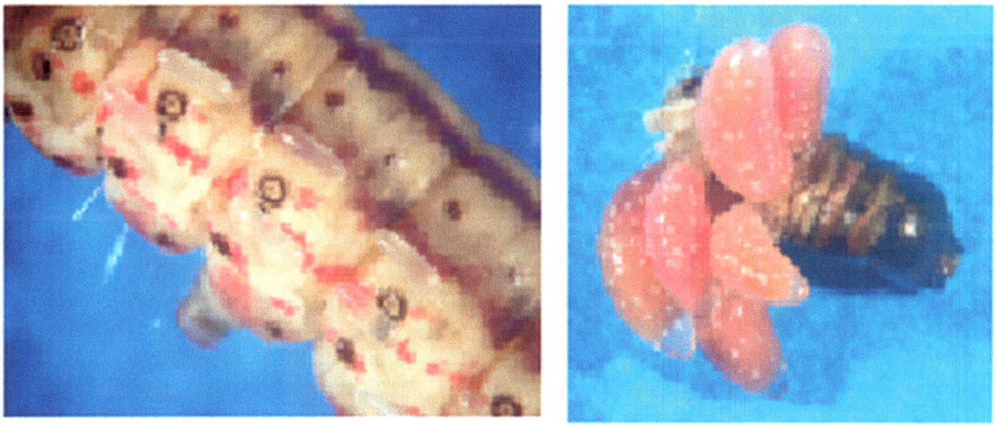

FIGURE 14.1 *Goniozus nephantidis* (Muesebeck) eggs (a) and larvae (b) on *Opisina arenosella* Walker larva.

Photo source: ICAR-NBAIR database.

Ueno (1998) reported that males are less efficient than females in consuming larger hosts; a probable evolutionary consequence to selection favouring female larvae as efficient feeders. The fecundity of female *Heterospilis* increased 20 times when developing on bigger host compared to small host (Charnov *et al.*, 1981). *In Laelius pedatus* (Hymenoptera: Bethylidae), female parasitoid lays single male egg on the smallest host, on slightly larger hosts lays single female egg and larger hosts gregarious clutches are laid which constitute single and more than one female's eggs along with single male egg (Mayhew and Godfray, 1997). Waage (1982) hypothesized that size-dependent sex ratios are likely to evolve in idiobiont parasitoids but not in koinobiont parasitoids.

Lopez-Vaamonde *et al.* (2005) observed that three types of brood production occur in *Achrysocharoides*: (1) solitary offspring, (2) mixed sex broods, and (3) separate (split) sex broods, and suggested that split sex broods have evolved twice and provide evidence for a transition from solitary to mixed sex broods, via split sex broods.

14.2.4 Mating

Mating in gregarious and quasi-gregarious parasitoids often involve protandrous mating at emergence site (Boulton *et al.*, 2014). However, mating is required only for producing female progeny (Godfray, 1994). Therefore, decision between oviposition and mating affects wasp fitness (Kant *et al.*, 2012). Storage of sperm and their selective use allow females to produce females (Thornhill & Alcock, 1983) and control sex ratio (Leach *et al.*, 2009).

The reproductive success of female depends on number of viable eggs produced while that of males depends on number of female offspring, and number of females inseminated by them. Females are expected to increase their fitness by adjusting sex ratio of their offspring in response to environmental conditions (Steiner & Ruther, 2009). However, their net reproductive gain from mating may depend largely on mating history of males they mate with (Marcotte *et al.*, 2007). According to LMC (Local mate competition) (Shuker *et al.*, 2005, 2006), a female can control sex ratio of its offspring by reducing its male progeny competition. This is an evolutionarily stable strategy where males are not dispersing sex and mating takes place at natal patch (Hamilton, 1967).

The foundress female shifts progeny sex ratio strongly towards females to reduce competition among its male progeny, when patch population consists of offspring of a single female. Further, competition between unrelated males increases when number of foundress increases in a patch and thus selection favours females to produce more male offspring. (Hamilton, 1967).

14.3 Climate Change and Reproductive Behaviour of Parasitoids

Environmental stimuli, especially global warming, may act directly on parasitoid, indirectly through its host or on both (Gilman *et al.*, 2010; Tougeron *et al.*, 2020).

14.3.1 Adaptation to Extreme Temperatures

Adaptations of parasitoids to extreme temperatures are similar to those of most ectotherms, with acclimatization constrained by responses of hosts as well as endosymbiont populations. Differences between host and parasitoid lead to a disruption of temporal or geographical synchronization, and may lead to insect pest outbreaks (Hance *et al.*, 2007).

Diapause is a strategy by which insects face adverse conditions and they terminate it sometime after the return of favourable conditions. The insects manage changing environmental conditions by many of plastic changes like diapause, life history, physiological and biochemical changes (Sgrò *et al.*, 2016). Several traits of parasitoids are affected by temperature change and are given hereunder.

14.3.1.1 Longevity

Low temperature exposure reduces longevity (Rundle *et al.*, 2004; Pandey & Johnson, 2005). Adults of *Trissolcus basalis* (Wollaston) and *Telenomus podisi* Ashmead, exposed to low temperature showed decrease longevity on return to normal temperature (Foerster & Doetzer, 2006).

14.3.1.2 Fecundity

When immature stages are exposed to low temperature, reduction in adult fecundity occurs (Levie *et al.*, 2005) and this effect depends on both, temperature, and duration of exposure (Bayram *et al.*, 2005). However, if immature stages undergo diapause, effects are negated (Hance *et al.*, 2007).

14.3.1.3 Sex Allocation

Temperature stress influences sex allocation, especially in arrhenotokous parasitoids. Low temperature usually results in male-biased sex ratio (Denlinger & Lee, 1998; Pandey & Johnson, 2005; Foerster & Doetzer, 2006) with decreased mobility in either sex, as well as sperm incapacitation reducing their fitness (Denlinger & Lee, 1998). Change in sex ratio can also be because of change in endosymbiont population due to extreme temperature.

14.3.2 Effect of Temperature on Host Parasitoid Interaction

Climate change can lead to phenological asynchrony between parasitoids and their hosts (Jeffs & Lewis, 2013; Furlong & Zalucki, 2017). Wetherington *et al.* (2017) observed that extreme heat waves accompanying climate change could disrupt host utilization (emerald ash borer) and synchronization behaviour of parasitoid (*Oobius agrili*). Furlong and Zalucki (2017) in their study on diamondback moth and parasitoid, *Diadegma semiclausum* Hellén found that higher temperatures will have greater negative impact on distribution of parasitoid than on host. Increased temperature enhances host immunity, thereby decreasing probability of completion of development by immature stages of parasitoids (Hance *et al.*, 2007).

14.4 Host Resource Utilization and Reproductive Strategies

Parasitoids are adapted to fluctuating host resources for their survival and reproduction. The selection of host species depends on many host factors, *viz.* species, developmental stages, sizes, nutrition, immunoreactions, escape responses, competition from other natural enemies, associated symbionts or mutualists, host plants, learning behaviour of parasitoids, and symbiotic microorganisms of parasitoids, etc. It also depends on ability of parasitoid to regulate host biology via measures like injecting immune-depressing venom inside host (Poirié *et al.*, 2014) or polydnavirus (Bredlau *et al.*, 2019) for ensuring offspring fitness.

Idiobionts are acclimatized to exploit fixed or rapidly diminishing host resources (Mackauer *et al.*, 1997). More female progeny and large clutch size are expected in response to adaptive responses of idiobiont parasitoids to variable host resources (young and/or large hosts) (Waage & Ng, 1984; Urano & Hijii, 1995).

In koinobiont parasitoids, a flexible growth rate is one of key adaptive features which enables parasitoids to exploit host growth's potential. (Mackauer & Sequeira, 1993; Harvey *et al.*, 1994). Harvey *et al.* (1994) observed that developmental duration in *Venturia canescens* decreased progressively in older hosts, with accelerated growth in L5. Low-quality host also delays growth (Mackauer & Sequeira, 1993). Dimorphism can also result from sex-specific trade-offs between development time and size at maturity, or from one sex exploiting host resources more efficiently than other (Mackauer *et al.*, 1997).

The close relationship with insect hosts has a potential for divergent selection of parasitoid populations using different hosts and could be a factor involved in the adaptive evolution (Abrahamson & Blair, 2008; Feder & Forbes, 2010). Sequential radiation hypothesis (Abrahamson & Blair, 2008) proposes association of parasitoids with host race (ecological race through the adaptation to new host plants) promotes adaptation and diversification of parasitoid populations (Forbes *et al.*, 2009). However, in absence of a preferred host, host adaptation could also involve a cost associated with reducing phenotypic plasticity in parasitoid population (Hufbauer, 2002).

14.5 Reproductive Modes in Parasitoids

Hymenopteran parasitoids are extensively studied and display haplo-diploid reproductive strategy (Zuparko, 2008), with some showing thelytokous parthenogenesis (Slobodchikoff & Daly, 1971).

Thelytokous parthenogenesis is frequently induced by endosymbionts (Hagimori *et al.*, 2006; Giorgini *et al.*, 2009). Deuterotokous parthenogenesis is least common and occurs in a few species, wherein unfertilized eggs can produce both males and females. In amphitokous parthenogenesis or ampherotokous parthenogenesis (Flanders, 1945; Doutt, 1959), unfertilized eggs may produce males or females (White, 1973) and males are not necessary for reproduction even though they may be reproductively viable (Rossler & DeBach 1973; Walter, 1983).

14.5.1 Arrhenotoky

In Hymenoptera, most sexually reproducing species are arrhenotokous (haplo-diploid), with males developing from unfertilized haploid eggs, and females developing from fertilized diploid eggs (Table 14.1). Arrhenotoky allows production of offspring with a diversity of genotypes, production of males, which cannot themselves produce offspring and an inability of females to reproduce without males (Normark & Kirkendall, 2009). Males from haploid unfertilized eggs are 100% related to their mother, while fertilized eggs develop into diploid females and share a haploid complement of both parents (Leach *et al.,* 2009). Details of this mode of reproduction are available in chapter 3.

14.5.2 Thelytoky

Thelytoky involves production of diploid females from unfertilized eggs and male offspring are absent (Normark & Kirkendall, 2009) and could be potentially useful in augmentative biological control programs with higher proportion of females significantly lowering costs of mass rearing (Ramirez-Romero *et al.*, 2011). Thelytoky is an evolved form found in several group of insects, unlike arrhenotoky, which is considered to be an ancestral mode of reproduction (Leach *et al.*, 2009). Two classes of thelytoky can be distinguished based on cause; it can be genetic, which is controlled by nuclear gene and other is via cytoplasmic genes including microorganisms (Leach *et al.*, 2009), details of which are discussed in Chapter 3.

14.5.3 Symbionts in Reproduction

14.5.3.1 Wolbachia

Wolbachia is an alpha proteobacterium, inherited maternally through egg cytoplasm, and found in 40% terrestrial arthropod and 66% of insect species (Werren *et al.*, 2008; Zug & Hammerstein, 2012). This endosymbiont enhances its spread by promoting reproductive success of infected

TABLE 14.1

Reproduction modes of hymenopteran parasitoids

Family	Species	Mode of Reproduction	References
Trichogrammatidae	*Trichogramma semifumatum*	Thelytoky	Bowen and Stern (1966)
	T. acacioi	Arrhenotoky	Zucchi and Monteiro (1997)
			Pratissoli and Fornazier (1999)
	T. atopovirilia	Arrhenotoky & Thelytoky	Zucchi (1985)
			Pinto (1998)
	T. bourarache	Arrhenotoky	Gonçalves *et al.* (2006)
	T. brassicae	Arrhenotoky	Almeida (2004)
	T. breuicapillum	Arrhenotoky & Thelytoky	Stouthamer *et al.* (1990a)
			Werren *et al.* (1995)
	T. bruni	Arrhenotoky	Zucchi and Monteiro (1997)
			Silva (2002)
	T. cacoeciae	Thelytoky	Almeida and Stouthamer (2003)
			Roopa *et al.* (2009)
	T. cordubensis	Thelytoky	Cabello and Vargas (1985)
			Stouthamer *et al.* (1990b); (1993)
	T. chilonis	Arrhenotoky & Thelytoky	Stouthamer *et al.* (1990a)
			Roopa *et al.* (2009)
			Chen *et al.* (1992)
	T. deion	Arrhenotoky & Thelytoky	Stouthamer *et al.* (1990, 1993)
			Schilthuizen and Stouthamer (1997)
			Tagami *et al.* (2001)
	T. dendrolimi	Arrhenotoky & Thelytoky	Whu and Valdivieso (1999)
			Liu *et al.* (2019)
	T embryophagum	Arrhenotoky	Nagaraja (1987)
	T. esalqueanum	Arrhenotoky	Querino and Zucchi (2003)
	T. evanescence	Arrhenotoky & Thelytoky	Roopa *et al.* (2009)
			Tagami *et al.* (2001)
	T. exiguum	Arrhenotoky	Zucchi and Monteiro (1997), Basso *et al.* (1999)
			Whu and Valdivieso (1999)
	T. fuentesi	Arrhenotoky	Whu and Valdivieso (1999)
			Silva (2002)
	T. galloi	Arrhenotoky	Zucchi (1988)
			Zucchi and Monteiro (1997)
	T. iracildae	Arrhenotoky	Querino and Zucchi (2003)
	T. japonicum	Arrhenotoky	Roopa *et al.* (2009)
	T. kaykai	Arrhenotoky & Thelytoky	Tagami *et al.* (2001)
	T. lasallei	Arrhenotoky	Foerster *et al.* (2000)
	T. lopezandinesis	Arrhenotoky	Sarmiento (1993)
	T. minutum	Arrhenotoky & Thelytoky	Wang and Smith (1996)
	T. nerudai	Arrhenotoky	Pintureau *et al.* (1999)
	T. nubilale	Thelytoky	Schilthuizen and Stouthamer (1997)
			van Meer *et al.* (1999)
	T. oleae	Thelytoky	Louis *et al.* (1993)
	T. pintoi	Arrhenotoky	Whu and Valdivieso (1999)
			Silva (2002)

(Continued)

TABLE 14.1 (Continued)

Reproduction modes of hymenopteran parasitoids

Family	Species	Mode of Reproduction	References
	T. platneri	Thelytoky	Stouthamer *et al.* (1990a)
	T. pretiosum	Arrhenotoky & Thelytoky	Zucchi *et al.* (1991)
			Prabhulinga *et al.* (2016)
	T. rojasi	Arrhenotoky	Nagaraja and Nagarkatti (1973)
	T. sibericum	Thelytoky	Schilthuizen and Stouthamer (1997)
	T. telengai	Thelytoky	Sorakina (1987)
	T. turkestanica	Arrhenotoky & Thelytoky	Gonçalves *et al.* (2006)
	Trichogrammatoidea armigera	Arrhenotoky	Roopa *et al.* (2009)
	Tr. lutea	Arrhenotoky	Kfir (1982)
Mymaridae	*Anagrus sophiae*	Arrhenotoky & Thelytoky	Segoli *et al.* (2013)
Ichneumonidae	*Venturia canescens*	Arrhenotoky & Thelytoky	Speicher (1937)
			Foray *et al.* (2013)
Braconidae	*Perilitus coccinellae*	Thelytoky	Balduf (1926)
	Bracon hebetor	Thelytoky	Speicher (1934)
Signiphoridae	*Thysanus* spp.	Thelytoky	DeBach (1969)
	Signiphora borinquensis	Thelytoky	Quezada *et al.* (1973)
	Signiphora spp.	Thelytoky	DeBach (1969)
Eulophidae	*Prospaltella perniciosi*	Thelytoky	Flanders (1953)
	Thripoctenus javae	Thelytoky	Monti *et al.* (2016)
	Galeopsomyia fausta	Thelytoky	Argov *et al.* (2000)
	Baryscapus servadeii	Thelytoky	Simonato *et al.* (2019)
	Neochrysocharis formosa	Thelytoky	Hagimori *et al.* (2006)
	Quadrastichus mendeli	Thelytoky	Gualtieri *et al.* (2017)
	Tamarixia radiate	Arrhenotoky	Chen and Stansly (2014)
Aphelinidae	*Aphytis chilensis*	Thelytoky	Gottlieb *et al.* (1998)
	A. chrysomphali	Thelytoky	Gottlieb *et al.* (1998)
	A. diaspidis	Thelytoky	Zchori-Fein *et al.* (1995)
	A. lignanensis	Thelytoky	Zchori-Fein *et al.* (1995); (1994)
	A. mytilaspidis	Arrhenotoky & Thelytoky	Rossler and DeBach (1973)
	A. yanonensis	Thelytoky	Werren *et al.* (1995)
	Encarsia formosa	Thelytoky	Zchori-Fein *et al.* (1992)
			Werren *et al.* (1995)
	Eretmocerus stauffen	Thelytoky	Van Meer *et al.* (1999)
	Eretmocerus mundus	Arrhenotoky & Thelytoky	De Barro *et al.* (2000)
			Ardeh *et al.* (2005)
	Eretmocerus eremicus	Thelytoky	Ardeh *et al.* (2005)
	Aphelinus asynchus	Thelytoky	Schlinger and Hall (1959)
Encyrtidae	*Aponanagyms diverskornis*	Thelytoky	Pijls *et al.* (1996)
	Coccidoxenoides peregrinus	Thelytoky	Flanders (1953)
	Ooencyrtus submetalluum	Thelytoky	Wilson and Woolcock (1960)
	O. fecundus	Thelytoky	Laraichi (1978)
	Tropidophryne melvillei	Thelytoky	Doutt and Smith (1950)
	Pauridia peregrina	Thelytoky	Flanders (1953)
	Habrolepis rouxi	Thelytoky	Flanders (1946)
	Encyrtus fulginosus	Thelytoky	Flanders (1943)
	Trechnites psyllae	Thelytoky	Slobodchikoff and Daly (1971)

TABLE 14.1 (Continued)

Reproduction modes of hymenopteran parasitoids

Family	Species	Mode of Reproduction	References
	Plagiomerus diaspidis	Thelytoky	Gordh and Lacey (1976)
Platygasteridae	*Amitus fuscipennis*	Thelytoky	Van Meer *et al.* (1999)
	Telonomus nawai	Arrhenotoky	Arakaki *et al.* (2000)
Bethylidae	*Scleroderma immigrans*	Thelytoky	Keeler (1929)
Pteromalidae	*Muscidifurax uniraptor*	Thelytoky	Legner (1985a, b)
			Stouthamer *et al.* (1993); (1994)
	Spalangia fuscipes	Thelytoky	Werren *et al.* (1995)
			van Meer *et al.* (1999)
Figitidae	*Leptopilina australis*	Thelytoky	Werren *et al.* (1995)
			van Meer *et al.* (1999)
	L. clavipes	Thelytoky	Werren *et al.* (1995)
	Gronotoma micromorpha	Thelytoky	Arakaki *et al.* (2001)

females, via parthenogenesis, male-killing, feminization, or induction of cytoplasmic incompatibility (Engelstädter & Hurst, 2009). The involvement of *Wolbachia* in thelytokous parthenogenesis is reported in 70 haplodiploid species, out of which 18 are egg parasitoids (Huigens & Stouthamer, 2003), details are provided in Chapter 3.

Thelytoky has also been induced in two completely infected and 17 partly infected *Trichogramma* species with *Wolbachia* (Stouthamer *et al.*, 1993). However, effect of *Wolbachia* is not highly diversified in *Trichogramma*. In *Trichogramma bourarachae* (Pintureau & Babault) infection has not induced thelytoky, but strongly increased fecundity of female parasitoids (Vavre *et al.*, 1999). *Wolbachia* often also has a negative effect or sometimes no effect on host fitness at all. The partially infected *Trichogramma* species has negative effects or no effect on parasitoid fitness, while completely infected species showed a positive effect involving an increase in fecundity, longevity in absence of a host, emergence rate, dispersal capacity, and parasitism rate under greenhouse.

14.5.3.2 *Cardinium*

Cardinium is an intracellular bacterial symbiont in Bacteroidetes, infecting many species of arthropods (Zchori-Fein *et al.*, 2004). *Cardinium* causes (1) feminization (Weeks *et al.*, 2001), (2) cytoplasmic incompatibility, and (3) thelytokous parthenogenesis (Matalon *et al.*, 2007) (Table 14.2). *Cardinium* infection is common in genus *Encarsia*, which are primary endoparasitoids of whiteflies or armoured scale insects (Perlman *et al.*, 2006).

14.5.3.3 *Rickettsia*

Rickettsia is obligate intracellular symbiont that resides in cytoplasm of eukaryotic cells (Weinert, 2015). In insects, *Rickettsia* symbionts influence host fitness to manipulate reproduction of parasitoids (Table 14.2) causing thelytokous parthenogenesis. *Rickettsia* (Hagimori *et al.*, 2006; Giorgini *et al.*, 2010) may be present alone or in conjunction with *Wolbachia* (Semiatizki *et al.*, 2020).

14.5.3.4 *Arsenophonus*

The bacterial symbionts of genus *Arsenophonus* are male-killing gamma-proteobacterial symbiont known to infect approximately 5% of arthropods (Duron *et al.*, 2008) (Table 14.2). It potentially infects a wide range of chalcid parasitoids (Duron *et al.*, 2010; Taylor *et al.*, 2011) and is male-killing

TABLE 14.2

Influence of symbionts on reproduction of parasitoids

Family	Species	Symbionts	Cause	References
Trichogrammatidae	*Trichogramma atopovirilia*	W	P	Ciociola *et al.* (2001)
	T. brassicae	W	P	Farrokhi *et al.* (2010)
	T. brevicapillum	W	P	Stouthamer *et al.* (1990a)
	T. chilonis	W	P	Schilthuizen and Stouthamer (1997)
				Chen *et al.* (1992)
	T. cordubensis	W	P	Stolk and Stouthamer (1996)
	T. dendrolimi	W	P	Dong *et al.* (2017)
	T. deion	W	P	Stouthamer *et al.* (1993)
	T. embryophagum	W	P	Stouthamer *et al.* (1990a)
	T. evanescens	W	P	Stouthamer *et al.* (1990a)
	T. kaykai	W	P	Stouthamer and Kazmer (1994)
	T. nubilale	W	?	van Meer *et al.* (1999)
	T. oleae	W	?	Rousset *et al.* (1992)
	T. ostriniae	W	?	Wu *et al.* (2016)
	T. platneri	W	P	Schilthuizen and Stouthamer (1997)
	T. pretiosum	W	P	Stouthamer *et al.* (1990a)
	T. sibericum	W	?	van Meer *et al.* (1999)
	T. semblidis	W	?	Pintureau *et al.* (2000)
	T. turkestanica	W	P	Gonçalves *et al.* (2006)
Platygasteridae	*Telonomus nawai*	W	P	Arakaki *et al.* (2000)
	Amitus fuscipennis	W, R	P	Manzano *et al.* (2000)
Mymaridae	*Anagrus sophiae*	W	P	Segoli *et al.* (2013)
Encyrtidae	*Apoanagyrus diversicornis*	W	P	Pijls *et al.* (1996)
	Diaphorencyrtus aligarhensis	W	P	Meyer and Hoy (2007)
	Coxxidoxenoides peregrinus	W	P	van Meer (1999)
	Plagiomerus diaspidis	C	P	Matalon *et al.* (2007)
	Tachinaephagus zealandicus	W	?	Kyei-Poku *et al.* (2006)
	Ixodiphagus hookeri	A	?	Bohacsova *et al.* (2016)
Aphelinidae	*Aphytis chilensis*	W	?	Cook and Butcher (1999)
	A. chrysomphali,	W	?	Cook and Butcher (1999)
	A. diaspidis,	W	?	Cook and Butcher (1999)
	A. lingnanensis	W	?	Cook and Butcher (1999)
	A. melinus	W	CI	Vasquez *et al.* (2011)
	A. mytilaspidis	W	P	Rossler and DeBach (1973)
	A. yanonensis	W	?	Zchori-Fein *et al.* (1998)
	Encarsia berlesei	C	P	Zchori-Fein *et al.* (2004)
	E. citrine	C	P	Zchori-Fein *et al.* (2004)
	E. hispida	C	P	Giorgini *et al.* (2009)
	E. pergandiella	C	P	Zchori-Fein *et al.* (2004)
	E. perniciosi,	C	P	Zchori-Fein *et al.* (2004)
	E. protransvena	C	P	Zchori-Fein *et al.* (2004)
	E. formosa	W	P	De Barro and Hart (2001)
	E. lutea	W	?	Karut *et al.* (2020)
	E. inaron	W, C	CI	White *et al.* (2011)
	Eretmocerus mundus	W	P?	De Barro and Hart (2001)
		R		Karut *et al.* (2020)

TABLE 14.2 (Continued)

Influence of symbionts on reproduction of parasitoids

Family	Species	Symbionts	Cause	References
	Eretmocerus staufferi	W	?	van Meer (1999)
	Marietta sp.	C	?	Zchori-Fein *et al.* (2004)
Eulophidae	*Neochrysocharis formosa*	R	P	Hagimori *et al.* (2006)
	Pnigalio soemius	R	P	Giorgini *et al.* (2010)
	Melittobia acasta	A	MK	Duron *et al.* (2010)
	Tetrastichus coeruleu	W	P	Reumer *et al.* (2013)
	Baryscapus servadeii	R	P	Simonato *et al.* (2019)
	Quadrastichus mendeli	R	P	Gualtieri *et al.* (2017)
Pteromalidae	*Muscidifurax uniraptor*	W	P	Gottlieb and Zchori-Fein (2001)
	Muscidifurax raptor	W	?	Betelman *et al.* (2017)
		A	MK	Duron *et al.* (2010)
	Muscidifurax raptorellus	A	MK	Duron *et al.* (2010)
	Nasonia longicornis	W	CI	Breeuwer *et al.* (1992)
		A	MK	
	Nasonia vitripenni	W	CI	Stolk and Stouthamer (1996)
		A	MK	Duron *et al.* (2010)
	N. giraulti	A	MK	Duron *et al.* (2010)
	Pteromalus venustus	A	MK	Goerzen and Erlandson (2018)
	Spalangia fuscipes	W	?	Werren *et al.* (1995)
	S. cameroni	W	CI	Betelman *et al.* (2017)
		R	CI	Duron *et al.* (2010
		A	MK	Taylor *et al.* (2011)
	S. endius	W	CI	Betelman *et al.* (2017)
		R	CI	Semiatizki *et al.* (2020)
		A	MK	Taylor *et al.* (2011)
	S. nigra	W	?	Kyei-Poku *et al.* (2006)
	S. nigroaenea	W	?	Kyei-Poku *et al.* (2006)
	Trichomalopsis sarcophagae	W	?	Kyei-Poku *et al.* (2006)
	Urolepis rufipes	W	CI	Kyei-poku *et al.* (2003)
Chalcididae	*Brachymeria podagrica*	W	?	Kyei-Poku *et al.* (2006)
	Dirhinus himalayanus	W	?	Kyei-Poku *et al.* (2006)
Eupelmidae	*Eupelmus vesicularis*	A	MK	Taylor *et al.* (2011)
Ichneumonidae	*Hyposoter horticola*	W	?	Duplouy *et al.* (2015)
	Diplazon annulatus	W	?	Klopfstein *et al.* (2016)
	D. bradleyi	W	?	Klopfstein *et al.* (2016)
	D. deletus	W	?	Klopfstein *et al.* (2016)
	D. flixi	W	?	Klopfstein *et al.* (2016)
	D. parvus	W	?	Klopfstein *et al.* (2016)
	D. tetragonus	W	?	Klopfstein *et al.* (2016)
	D. varicoxa	W	?	Klopfstein *et al.* (2016)
Braconidae	*Aphaereta pallipes*	W	?	Kyei-Poku *et al.* (2006)
	Asobara tabida	W	CI	Dedeine *et al.* (2005)
	A. japonica	W	P	Furihata *et al.* (2015)
	Cotesia sesamiae	W	CI	Mochiah *et al.* (2002)
	C. vestalis	W	P	Murthy *et al.* (2015)
	Psyttalia rhagoleticola	W	?	Schuler *et al.* (2016)

(Continued)

TABLE 14.2 (Continued)

Influence of symbionts on reproduction of parasitoids

Family	Species	Symbionts	Cause	References
	Habrobracon hebetor	W	CI	Bagheria *et al.* (2019)
Bethylidae	*Prorops nasuta*	W	?	Navarro *et al.* (2019)
Figitidae	*Gronotoma micromorpha*	W	P	Arakaki *et al.* (2001)
	Leptopilina heterotoma	W	CI	Fytrou *et al.* (2006)
	L. clavipes	W	P	Pannebakker *et al.* (2004)
Diapriidae	*Diapria conica*	W	?	Kyei-Poku *et al.* (2006)
	Trichopria nigra	W	?	Kyei-Poku *et al.* (2006)

Parasitoid symbionts: W – *Wolbachia*; C – *Cardinium*; R – *Rickettsia*; A – *Arsenophonus*; Symbionts cause in parasitoids: P – Parthenogenesis; CI – Cytoplasmic incompatibility; MK – Male killing?; – Role unknown.

reproductive parasite (Bohacsova *et al.*, 2016). It transmits both vertically from mother to offspring and horizontally to unrelated parasitoids wasps developing in same host (Taylor *et al.*, 2011).

14.5.3.5 Microbes in Fitness Attributes of Parasitoids

Several bacterial species assume diverse symbiotic roles in insects, like providing nutrition (Douglas, 1998), modulating defense mechanisms (Nyabuga *et al.*, 2010), reproduction, and development (Taylor *et al.*, 2011), and in enhancing prospectives for biological control (Tagami *et al.*, 2006).

Nagarkatti and Nagaraja (1978) observed that field-collected populations of *Trichogramma* were not only robust, but had a higher biological fitness, which these wasps tend to lose after prolonged laboratory rearing, which was later attributed to better symbiont population in field collected individuals (Srinatha *et al.*, 2015).

Ye *et al.* (2018) found support for species-specific facultative endosymbiont–parasitoid effects; parasitoids with higher fecundity and lower longevity are more likely to oviposit unselectively, whereas species with a lower fecundity and higher longevity are more likely to overcome endosymbiont protection or to oviposit selectively.

14.6 Implication of Reproductive Behaviour in Biological Control

Parasitoids are employed in biological control programs throughout world to suppress populations of insect pests and are environmentally safe alternatives to chemical pesticides.

For success of biological control, large-scale production of parasitoids with high fitness and female biased population is prerequisite. Thus, thelytokous parasitoids are better suited in biological control programs (Ramirez-Romero *et al.*, 2011). Thelytokous females develop faster and produce higher progeny than counterpart arrhenotokous female (Prabhulinga *et al.*, 2016).

Wolbachia received an interest in biological control because it induces thelytoky in parasitoids (Stouthamer & Kazmer, 1994; Silva *et al.*, 2000; Figure 14.2).

The production potential and female biased parasitoid for biological control needs high host quality maintenance (Lampson *et al.*, 1996). The sex ratio of many Hymenoptera parasitoids used in biological control have reported male-biased sex ratios during mass rearing in laboratory that is largely influenced by host quality and rearing conditions (Heimpel & Lundgren, 2000). In commercial insectaries, higher female population by reducing males may allow producing higher quality of parasitoids at equal cost. Although cost of production of parasitoid may not reduce but biocontrol users may realize this effect in controlling pest by using same degree or lesser number of parasitoid releases. This would minimize overall production cost with effective biocontrol of pests. By manipulating the host size to small and larger alternatively increased the female progeny and saved 50% of the production cost of female and showed a level of pest control without compromising the quality of female parasitoids (Chow & Heinz, 2005). Thus, 'host size distribution manipulation technique' has a great

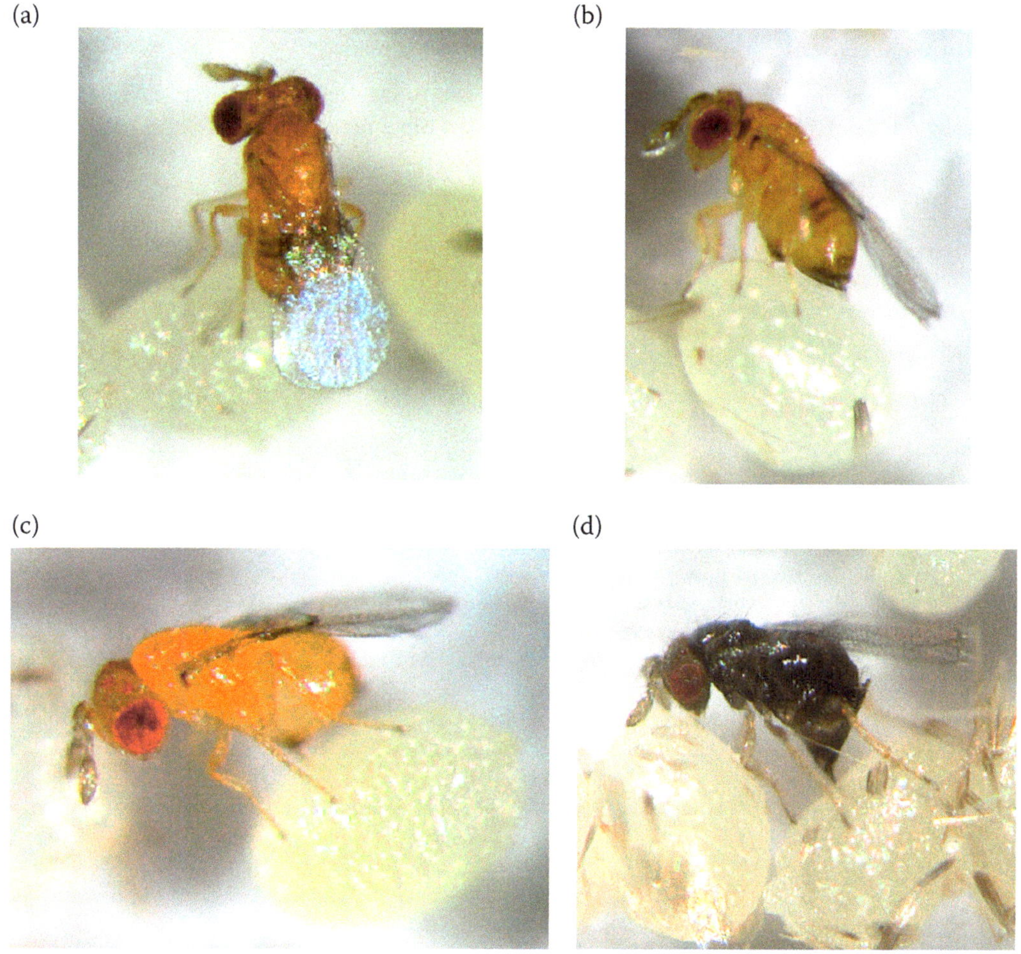

FIGURE 14.2 Different species of *Trichogramma* parastizing *Corcyra cephalonica* (Stainton) eggs, (a) *Trichogramma pretiosum,* Riley, (b) *Trichogramma chilonis,* Ishii, (c) *Trichogramma embryophagum,* Hartig, (d) *Trichogramma japonicum.* Ashmead.

Photo source: ICAR-NBAIR database.

role in production of female biased and potential parasitoids used in biological control programs that involve mass-rearing of solitary idiobiont parasitoids (Ode & Hardy, 2008).

14.7 Conclusions

The reproductive strategies followed by different groups of parasitoids are complex and have co-evolved with their hosts under strong natural selection. Reproductive strategies, like clutch size, sex ratio, and precise allocation of offspring sexes and their numbers on host based on their size are main strategies to maximize reproductive success of parasitoids under fluctuating population of hosts in surrounding environment. The reproductive modes adapted by parasitoids, especially thelytoky, which produces higher female-biased individuals has importance in biological control. The manipulation of sex ratio either through endosymbionts or by using host size-based sex manipulation techniques have been employed to produce higher female parasitoids in commercial insectaries for management of crop pests through augmentative biological control program. Thus, production of higher female parasitoids reduces cost of production and provides efficient control of pest by using lesser numbers of parasitoid releases.

Acknowledgement

The authors are thankful to Director, ICAR-National Bureau of Agricultural Insect Resources, Bengaluru, India, for constant support for research and writing review papers.

REFERENCES

Abrahamson, W.G. & Blair, C.P. (2008). Sequential radiation through host-race formation: Herbivore diversity leads to diversity in natural enemies. In K.J. Tilmon (Ed.), *Specialization, speciation and radiation: The evolutionary biology of herbivorous insects*. Berkeley, California: University of California Press, pp. 188–202.

Almeida, R.P. (2004). *Trichogramma and its relationship with Wolbachia: Identification of Trichogramma species, phylogeny, transfer and cost of Wolbachia symbionts*. PhD. Thesis. The Netherlands: Wagenignen University, p. 142.

de Almeida, R.P. & Stouthamer, R. (2003). Molecular identification of *Trichogramma cacoecieae* Marchal (Hymenoptera: Trichogrammatidae): A new record for Peru. *Neotropical Entomology*, *32*, 269–272.

Arakaki, N., Oishi, T. & Noda, H. (2001). Parthenogenesis induced by *Wolbachia* in *Gronotoma micromorpha* (Hymenoptera: Eucoilidae). *Entomological Science*, *4*: 9–15

Arakaki, N., Noda, H. & Yamagishi, K. (2000). Wolbachia-induced parthenogenesis in egg parasitoid Telenomus nawai. *Entomologia Experimentalis et Applicata*, *96*(2): 177–184.

Ardeh, M.J., de Jong, P.W. & van Lenteren, J.C. (2005). Selection of *Bemisia* nymphal stages for oviposition or feeding, and host-handling times of arrhenotokous and thelytokous *Eretmocerus mundus* and arrhenotokous *E. eremicus*. *BioControl*, *50*(3), 449–463.

Argov, Y., Gottlieb, Y., Amin-Spector, S. & Zchori-Fein, E. (2000). Possible symbiont-induced thelytoky in *Galeopsomyia fausta*, a parasitoid of citrus leafminer *Phyllocnistis citrella*. *Phytoparasitica*, *28*(3): 212–218.

Bagheria, Z., Talebia, A.A., Asgarib, S. & Mehrabad, M. (2019). Wolbachia induce cytoplasmic incompatibility and affect mate preference in *Habrobracon hebetor* to increase chance of its transmission to next generation. *Journal of Invertebrate Pathology*, *163*: 1–7

Balduf, W. V. (1926). The bionomics of *Dinocampus coccinellae* Schrank. *Annals of the Entomological Society of America*, *19*: 465–498.

Basso, C., Pintureau, B. & Grille, G. (1999). Taxonomic study of two *Trichogramma* species from Uruguay (Hymenoptera: Trichogrammatidae). *Boletín de Sanidad Vegetal Plagas*, *25*: 372–382.

Bayram, A., Ozcan, H. & Kornosor, S. (2005). Effect of cold storage on performance of *Telenomus busseolae* Gahan (Hymenoptera: Scelionidae), an egg parasitoid of *Sesamia nonagrioides* (Lefebvre) (Lepidoptera: Noctuidae). *Biological Control*, *35*: 68–77.

Betelman, K., Caspi-Fluger, A., Shamir, M. & Chiel, E. (2017). Identification and characterization of bacterial symbionts in three species of filth fly parasitoids. *FEMS Microbiology Ecology*, *93*(9). https://doi.org/10.1093/femsec/fix107

Bohacsova, M., Mediannikov, O., Kazimirova, M., Raoult, D. & Sekeyova, Z. (2016). *Arsenophonus nasoniae* and *Rickettsiae* infection of *Ixodesricinus* due to parasitic wasp *Ixodiphagus hookeri*. *PLoS ONE*, *11*: e0149950.

Boulton, R.A., Collins, L.A. & Shuker, D.M. (2014). Beyond sex allocation: Role of mating systems in sexual selection in parasitoid wasps. *Biological Reviews of Cambridge Philosophical Society*, *90*: 599–627.

Bowen, W.R. & Stern, V.M. (1966). Effect of temperature on production of males and sexual mosaics in a uniparental race of *Trichogramma semifumatum* (Hymenoptera: Trichogrammatidae). *Annals of Entomological Society of America*, *59*: 823–834.

Bredlau, J.P., Kuhar, D., Gundersen-Rindal, D.E. & Kester, K.M. (2019). The parasitic wasp, *Cotesia congregata* (Say), consists of two incipient species isolated by asymmetric reproductive incompatibility and hybrid inability to overcome host defenses. *Frontiers in Ecology and Evolution*, *7*: 187. https://doi.org/10.3389/fevo.2019.00187.

Breeuwer, J.A.J., Stouthamer, R., Barns, S.M., Pelletier, D.A., Weisburg, W.G. & Werren, J.H. (1992). Phylogeny of cytoplasmic incompatibility microorganisms in parasitoid wasp genus *Nasonia* (Hymenoptera: Pteromalidae) based on 16S ribosomal DNA sequences. *Insect Molecular Biology*, *1*(1): 25–36.

Cabello, T. & Vargas, P. (1985). Temperature as a factor influencing form of reproduction of *Trichogramma cordubensis*. *Zeitschrift fur angewandte Entomologie*, *100*, 434–441.
Charnov E.L., los-den Hartogh, R.L., Jones W.T. & van den Assem J. (1981). Sex ratio evolution in a variable environment. *Nature*, *289*: 27–33.
Charnov, E.L. & Skinner, S.W. (1984). Evolution of host selection and clutch size in parasitoid wasps. *Florida Entomologist*, *67*: 5–21.
Chen, B., Kfir, R. & Chen, C.N. (1992). The thelytokous *Trichogramma chilonis* in Taiwan. *Entomologia Experimentalis et Applicata*, *65*(2): 187–194.
Chen, X. & Stansly, P.A. (2014). Biology of *Tamarixia radiate* (Hymenoptera: Eulophidae), parasitoid of citrus greening disease vector *Diaphorina citri* (Hemiptera: Psylloidea): A mini review. *Florida Entomologist*, *97*(4): 1404–1413.
Chow, A. & Heinz, K.M. (2005). Using hosts of mixed sizes to reduce male-biased sex ratio in the parasitoid wasp, *Diglyphus isaea*. *Entomologia Experimentalis et Applicata*, *117*: 193–199.
Ciociola Jr., A.I., De Almeida, R.P., Zucchi, R.A. & Stouthamer, R. (2001). Detacção de *Wolbachia* em uma população Telítoca de *Trichogramma atopovirrilia* Oatman and Platner (Hymenoptera: Trichogrammatidae) via PCR com o primer específico wsp. *Neotropical Entomology*, *30*: 489–491.
Cook, J.M. & Butcher, R. (1999). The transmission and effects of *Wolbachia* bacteria in parasitoids. *Researches on Population Ecology*, *41*: 15–28.
Cook, J.M., Rivero Lynch, A.P. & Godfray, H.C.J. (1994). Sex ratio and foundress number in parasitoid wasp *Bracon hebetor*. *Animal Behaviour*, *47*: 687–696.
De Barro, P.J. & Hart, P.J. (2001). Antibiotic curing of parthenogenesis in *Eretmocerus mundus* (Australian parthenogenic form). *Entomologia Experimentalis et Applicata*, *99*(2): 225–230.
De Barro, P.J., Hart, P.J. & Morton, R. (2000). The biology of two *Eretmocerus* spp. (Haldeman) and three *Encarsia* spp. (Hymenoptera: Aphelinidae) Forster and their potential as biological control agents of *Bemisia tabaci* biotype B (Homoptera: Aleyrodidae) in Australia. *Entomologia Experimentalis et Applicata*, *94*: 93–102.
DeBach, P. (1969). Uniparental, sibling and semi-species in relation to taxonomy and biological control. *Israel Journal of Entomology*, *4*(1): 1–2.
Dedeine, F., Bouletreau, M. & Vavre, F. (2005). *Wolbachia* requirement for oogenesis: Occurrence within genus *Asobara* (Hymenoptera, Braconidae) and evidence for intraspecific variation in *A. tabida*. *Heredity*, *95*(5): 394–400.
Denlinger, D.L. & Lee, R.E. (1998) Physiology of cold sensitivity. In G.J. Hallman & D.L. Denlinger (Eds.), *Temperature sensitivityin insects and application in integrated pest management*. Boulder, CO: Westwiew, pp. 55–96.
Dijkstra, L.J. (1986). Optimal selection and exploitation of hosts in parasitic wasp *Colpoclypeus florus* (Hymenoptera: Eulophidae). *Netherlands Journal of Zoology*, *36*: 177–301.
Dong, H., Liu, Q., Xie, L., Cong, B. & Wang, H. (2017). Functional response of *Wolbachia* -infected and uninfected *Trichogramma dendrolimi* Matsumura (Hymenoptera: Trichogrammatidae) to Asian corn borer, *Ostrinia furnacalis* Guenée (Lepidoptera: Pyralidae) eggs. *Journal of Asia-Pacific Entomology*, *20*(3): 787–793.
Douglas, A.E. (1998). Nutritional interactions in insect-microbial symbioses: Aphids and their symbiotic bacteria *Buchnera*. *Annual Review of Entomology*, *43*: 17–37.
Doutt, R.L. (1959). The biology of parasitic Hymenoptera. *Annual Review of Entomology*, *4*: 161–182.
Doutt, R.L. & Smith, R.A. (1950). Males and intersexes in a normally thelytokous insect, *Tropidophryne melvillei* Compere (Hymenoptera: Encyrtidae). *Canadian Entomologist*, *82*: 165–170.
Duplouy, A., Couchoux, C., Hanski, I. & van Nouhuys, S. (2015). *Wolbachia* Infection in a Natural Parasitoid Wasp Population. *PLoS One*, *10*, e0134843.
Duron, O., Bouchon, D., Boutin, S., Bellamy, L., Zhou, L.Q., Jan, E. & Hurst, G.D. (2008). The diversity of reproductive parasites among arthropods: *Wolbachia* do not walk alone. *BMC Biology*, *6*: 12.
Duron, O., Wilkes, T.E. & Hurst, G.D.D. (2010). Interspecific transmission of a male killing bacterium on an ecological timescale. *Ecology Letter*, *13*: 1139–1148.
Eggleton, P. & Gaston, K. J. (1990). Parasitoid species and assemblages: Convenient definitions or misleading compromises? *Oikos*, *59*(3): 417–421.
Engelstädter, J. & Hurst, G.D.D. (2009). The ecology and evolution of microbes that manipulate Host reproduction. *Annual Review of Ecology, Evolution, and Systematics*, *40*(1): 127–149.

Farrokhi, S., Ashouri, A., Shirazi, J., Allahvari, H. & Huigens, M.E. (2010). A comparative study on functional response of *Wolbachia*-infected and uninfected forms of parasitoid wasp *Trichogramma brassicae*. *Journal of Insect Science*, *10*: 167.
Feder, J.L. & Forbes, A.A. (2010). Sequential speciation and diversity of parasitic insects. *Ecological Entomology*, *35*: 67–76.
Feener, D.H. & Brown, B. V. (1997). Diptera as parasitoids. *Annual Review of Entomology*, *42*(1): 73–97.
Flanders, S.E. (1945). The bisexuality of uniparental Hymenoptera, a function of environment. *The American Naturalist*, *79*: 122–141.
Flanders, S.E. (1943). The role of mating in reproduction of parasitic Hymenoptera. *Journal of Economic Entomology*, *36*, 802–803.
Flanders, S.E. (1946). Control of sex and sex-limited polymorphism in Hymenoptera. *The Quarterly Review of Biology*, *21*: 135–143.
Flanders, S.E. (1953). Aphelinid biologies with implications for taxonomy. *Annals of Entomological Society of America*, *46*: 84–94.
Foerster, L.A. & Doetzer, A.K. (2006). Cold storage of egg parasitoids *Trissolcus basalis* (Wollaston) and *Telenomus podisi* Ashmead (Hymenoptera: Scelionidae). *Biological Control*, *36*: 232–237.
Foerster, L.A., Avanci, M.R.F., Martin, J.M.R. & Cañete, C.L. (2000). Egg parasitoids of *Anticarsia gemmatalis* (Lep.: Noctuidae) on soybean in Brazil. In: Proceedings of XXI International Congress of Entomology, Foz do Iguaçu, p. 385.
Foray, V., Helene, H., Martinez, S., Gibert, P. & Desouhant, E. (2013). Occurrence of arrhenotoky and thelytoky in a parasitic wasp *Venturia canescens* (Hymenoptera: Ichneumonidae): Effect of endosymbionts or existence of two distinct reproductive modes? *European Journal of Entomology*, *110*(1): 103–107.
Forbes, A.A., Powell, T.H.Q., Stelinski, L.L., Smith, J.J. & Feder, J.L. (2009). Sequential sympatric speciation across trophic levels. *Science*, *323*: 776–778.
Furihata, S., Hirata, M., Matsumoto, H. & Hayakawa, Y. (2015). Bacteria endosymbiont, *Wolbachia*, promotes parasitism of parasitoid wasp *Asobara japonica*. *PLoS One*, *10*(10): e0140914.
Furlong, M.J. & Zalucki, M.P. (2017). Climate change and biologicalcontrol: Consequences of increasing temperatures onhost–parasitoid interactions. *Current Opinion in Insect Science*, *20*: 39–44.
Fytrou, A., Schofield, P.G., Kraaijeveld, A.R. & Hubbard, S.F. (2006). *Wolbachia* infection suppresses both host defence and parasitoid counter-defence. *Proceedings of Royal Society B*: *Biological Sciences*, *273*(1588): 791–796.
Gilman, S.E., Urban, M.C., Tewksbury, J., Gilchrist, G.W. & Holt, R.D. (2010). A framework for community interactions under climate change. *Trends in Ecology & Evolution*, *25*: 325–331.
Giorgini, M., Bernardo, U., Monti, M.M., Nappo, A.G. & Gebiola, M. (2010). Rickettsia symbionts cause parthenogenetic reproduction in parasitoid wasp *Pnigalio soemius* (Hymenoptera: Eulophidae). *Applied and Environmental Microbiology*, *76*(8): 2589–2599.
Giorgini, M., Monti, M.M., Caprio, E., Stouthamer, R. & Hunter, M.S. (2009). Feminization and collapse of haplodiploidy in an asexual parasitoid wasp harboring bacterial symbiont *Cardinium*. *Heredity*, *102*: 365–371.
Godfray, H.C.J. (1987). The evolution of clutch size in parasitic wasps. *American Naturalist*, *129*: 221–233.
Godfray, H.C. J. (1994). *Parasitoids*: *Behavioral and evolutionary ecology*. Princeton University Press, p. 473.
Godfray, H.C.J. (1995). Evolutionary theory of parent off-spring conflict. *Nature*, *376*: 133–138.
Goerzen, D.W. & Erlandson, M.A. (2018). Infection of chalcid parasitoid *Pteromalus venustus* Walker (Hymenoptera: Pteromalidae) with male-killing symbiont *Arsenophonus nasoniae* (Gamma-Proteobacteria: Enterobacteriaceae). *Journal of Invertebrate Pathology*, *154*: 24–28.
Gonçalves, C.I., Huigens, M.E., Verbaarschot, P., Duarte, S., Mexia, A. & Tavares, J. (2006). Natural occurrence of *Wolbachia*-infected and uninfected *Trichogramma* species in tomato fields in Portugal. *Biological Control*, *37*(3): 375–381.
Gordh, G. & Lacey, L. (1976). Biological studies of *Plagiomerus diaspidis,* a primary internal parasite of diaspidid scale insects. *Proceedings of Entomological Society of Washington*, *78*: 132–144.
Gordh, G., Legner, E.F. & Caltagirone, L.E. (1999). Biology of parasitic Hymenoptera. In *Handbook of Biological Control: Principles and Applications of Biological Control* by Fisher, T.W., Bellows, T.S., Caltagirone, L.E., Dahlsten, D.L., Huffaker, C.B. & Gordh, G., Elsevier, 355–381.

Gottlieb, Y. & Zchori-Fein, E. (2001). Irreversible thelytokous reproduction in *Muscidifurax uniraptor*. *Entomologia Experimentalis et Applicata*, *100*: 271–278.
Gottlieb, Y., Zchori-Fein, O. & Rosen, D. (1998). Phylogenetic analysis of parthenogenetic-inducing *Wolbachia* in genus *Aphytis* (Hymenoptera: Aphelinidae). *Insect Molecular Biology*, *7*: 393–396.
Grbic, M., Nagy, L. & Strand, M. (1998). Development of polyembryonic insects: a major departure from typical insect embryogenesis. *Development Genes and Evolution*, *208*, 69–81.
Gualtieri, L., Nugnes, F., Nappo, A., Gebiola, M. & Bernardo, U. (2017). Life inside a gall: Closeness does not favour horizontal transmission of *Rickettsia* between a gall wasp and its parasitoid. *FEMS Microbiology Ecology*, *93*(7): 1–11.
Hagimori, T., Abe, Y., Date, S. & Miura, K. (2006). The first finding of a *Rickettsia* bacterium associated with parthenogenesis induction among insects. *Current Microbiology*, *52*(2): 97–101.
Hamilton, W.D. (1967). Extraordinary sex ratios. *Science*, *156*: 477–488.
Hance, T., van Baaren, J., Vernon, P. & Boivin, G. (2007). Impact of extreme temperatures on parasitoids in a climate change perspective. *Annual Review of Entomology*, *52*: 107–126.
Hardy, I.C.W. (1991). *Reproductive strategies in parasitic wasps*. PhD. Thesis. London: University of London, p. 129.
Harvey, J.A. & Strand, M.R. (2002) The developmental strategies of endoparasitoid wasps vary with host feeding ecology. *Ecology*, *83*: 2439–2451.
Harvey, J.A., Harvey, I.F. & Thompson, D.J. (1994). Flexible larval growth allows use of a range of host sizes by a parasitoid wasp. *Ecology*, *75*: 1420–1428.
Heimpel, G.E. & Lundgren, J.G. (2000). Sex ratios of commercially reared biological control agents. *Biological Control*, *19*: 77–93.
Hufbauer, R.A. (2002). Evidence for nonadaptive evolution in parasitoid virulence following a biological control introduction. *Ecological Applications*, *12*: 66–78.
Huigens, M.E. & Stouthamer, R. (2003). Parthenogenesis associated with *Wolbachia*. In K. Bourtzis & T.A. Miller (Eds.), *Insect symbiosis*. Boca Raton, FL, USA: CRC Press, pp. 247–266.
Jeffs, C.T. & Lewis, O.T. (2013). Effects of climate warming on host-parasitoid interactions: Effects of climate warming. *Ecological Entomology*, *38*: 209–218.
Kant, R., Minor, M.A. & Trewick, S.A. (2012). Mating or ovipositing? A crucial decision in life history of cabbage aphid parasitoid *Diaeretiella rapae* (McIntosh). *Ecological Entomology*, *37*: 169–174.
Karut, K., Castle, S.J., Karut, Ş.T. & Karaca, M.M. (2020). Secondary endosymbiont diversity of *Bemisia tabaci* and its parasitoids. *Infection, Genetics and Evolution*, *78*: 104104.
Keeler, C.E. (1929). Thelytoky in *Scleroderma immigrans*. *Psyche*, *36*: 41–44.
Kfir, R. (1982). Reproduction characteristics of *Trichogramma brasiliensis* and *T. lutea*, parasitizing eggs of *Heliothis armigera*. *Entomologia Experimentalis et Applicata 32*: 249–255.
King, B.H. (1987). Offspring sex ratios in parasitoid wasps. *The Quarterly Review of Biology*, *62*: 367–396.
Klopfstein, S., Kropf, C. & Baur, H. (2016). *Wolbachia* endosymbionts distort DNA barcoding in parasitoid wasp genus *Diplazon* (Hymenoptera: Ichneumonidae). *Zoological Journal of Linnean Society*, *177*(3): 541–557.
Kyei-Poku, G.K., Floate, K.D., Benkel, B. & Goettel, M.S. (2003). Elimination of *Wolbachia* from *Urolepis rufipes* (Hymenoptera: Pteromalidae) with heat and antibiotic treatments: Implications for host reproduction. *Biocontrol Science and Technology*, *13*(3): 341–354.
Kyei-Poku, G.K., Giladi, M., Coghlin, P., Mokady, O., Zchori-Fein, E. & Floate, K.D. (2006). *Wolbachia* in wasps parasitic on filth Flies with emphasis on *Spalangia cameroni*. *Entomologia Experimentalis et Applicata*, *121*: 123–135.
Lampson, L.J., Morse, J.G. & Luck, R.F. (1996). Host selection, sex allocation, and host feeding by *Metaphycus helvolus* (Hymenoptera: Encyrtidae) on *Saissetia oleae* (Homoptera: Coccidae) and its effect on parasitoid size, sex, and quality. *Environmental Entomology*, *25*: 283–294.
Laraichi, M. (1978). L'effect de hautes temperatures sur le taux sexuel de *ooencyrtus fecundus* Hymenoptera: encyrtidae. *Entomologia Experimentalis et Applicata*, *23*(3): 237–242.
Leach, I.M., Pannebakker, B.A., Schneider, M.V., Driessen, G., van de Zande, L. & Beukeboom, L.W. (2009). Thelytoky in hymenoptera with *Venturia canescens* and *Leptopilina clavipes* as case studies. In I. Schön *et al.* (Eds.), *Lost sex*, Springer, Dordrecht, pp. 347–375.
Legner, E.F. (1985a) Effects of scheduled high temperature on male production in thelytokous *Muscidifurax uniraptor*. *Canadian Entomologist*, *117*: 383–389.

Legner, E.F. (1985b). Natural and induced sex ratio changes in populations of thelytokous *Muscidifurax uniraptor*. *Annals of Entomological Society of America*, *78*: 398–402.
Levie, A., Vernon, P. & Hance, T. (2005). Consequences of acclimation on survival and reproductive capacities of cold-stored mummies of *Aphidius rhopalosiphi* (Hymenoptera: Aphidiinae). *Journal of Economic Entomology*, *98*:704–708.
Liu, Q., Zhou, J., Zhang, C. *et al.* 2019. Co-occurrence of thelytokous and bisexual *Trichogramma dendrolimi* Matsumura (Hymenoptera: Trichogrammatidae) in a natural population. *Scientific Report*, *9*: 17480. 10.1038/s41598-019-53992-8
Lopez-Vaamonde, C., Godfray, H.C.J., West, S.A., Hansson§, C. & Cook, J.M. (2005). The evolution of host use and unusual reproductive strategies in *Achrysocharoides* parasitoid wasps. *Journal of Evolutionary Biology*, *18*: 1029–1104.
Louis, C., Pintureau, B. & Chapelle, L. (1993). Recherches sur l'origine de l'unisexualite: la thermotherapie elimine a la fois rickettsies et parthenogenese thelytoque chez un *Trichogramma* (Hymenoptera: Trichogrammatidae). *Comptes Rendus de l'Academie des Sciences, Paris, Serie III*, *316*: 27–33.
Mackauer, M., Sequeira, R. & Otto, M. (1997). Growth and development in parasitoid wasps: Adaptation to variable host resources. In K. Dettner *et al.* (Eds.), *Vertical food web interactions. Evolutionary patterns and driving forces*. Berlin, Germany: Springer, pp. 191– 203.
Mackauer, M. & Sequeira, R. (1993). Patterns of development in insect parasites. In N.E. Beckage *et al.* (Eds.), *Parasites and pathogens of insects*. New York, NY, USA: Academic Press, pp. 1–20.
Manzano, M.R., van Lenteren, J.C., Cardona, C. & Drost, Y.C. (2000). Developmental Time, Sex Ratio, and Longevity of *Amitus fuscipennis* MacGown & Nebeker (Hymenoptera: Platygasteridae) on Greenhouse Whitefly. *Biological Control*, *18*(2): 94–100.
Marcotte, M., Delisle, J. & McNeil, J.N. (2007). Effects of different male remating intervals on reproductive success of *Choristoneura rosaceana* males and females. *Journal of Insect Physiology*, *53*: 139–145.
Matalon, Y., Katzir, N., Gottlieb, Y., Portnoy, V. & Zchori-Fein, E. (2007). *Cardinium* in *Plagiomerus diaspidis* (Hymenoptera: Encyrtidae). *Journal of Invertebrate Pathology*, *96*: 106–108.
Matalon, Y., Katzir, N., Gottlieb, Y., Portnoy, V. & Zchori-Fein, E. (2007). *Cardinium* in *Plagiomerus diaspidis* (Hymenoptera: Encyrtidae). *Journal of Invertebrate Pathology*, *96*(2): 106–108.
Maynard Smith, J. (1978). *The evolution of sex*. UK.: Cambridge University Press.
Meyer, J.M. & Hoy, M.A. (2007). *Wolbachia*-Associated thelytoky in *Diaphorencyrtus aligarhensis* (Hymenoptera: Encyrtidae), a parasitoid of asian citrus psyllid. *Florida Entomologist*, *90*: 776–779.
Mochiah, M.B., Ngi-Song, A.J., Overholt, W.A. & Stouthamer, R. (2002). Wolbachia infection in *Cotesia sesamiae* (Hymenoptera: Braconidae) causes cytoplasmic incompatibility: Implications for biological control. *Biological Control*, *25*(1): 74–80.
Mock, D.W. & Parker, G.A. (1997). *The evolution of sibling rivalry*. Oxford: Oxford University Press.
Monti, M.M., Nugnes, F., Gualtieri, L., Gebiola, M. & Bernardo, U. (2016). No evidence of parthenogenesis-inducing bacteria involved in *Thripoctenus javae* thelytoky: An unusual finding in Chalcidoidea. *Entomologia Experimentalis et Applicata*, *160*(3): 292–301.
Murthy, K.S., Venkatesan, T., Jalali, S.K. & Ramya, S.L. (2015). Reproductive alterations by *Wolbachia* in Braconid *Cotesia vestalis* (Haliday). In A.K. Chakravarthy (Ed.), *New horizons in insect science: Towards sustainable pest management*, Springer, New Delhi. pp. 347–351.
Nagaraja, H. (1987). Recent advances in biosystematics of *Trichogramma* and *Trichogrammatoidea* (Hymenoptera, Trichogrammatidae). *Proceedings of Indian National Science Academy (Anim. Sci.)*, *96*(5): 469–477.
Nagaraja, H. & Nagarkatti, S. (1973). A key to some new world species of Trichogramma (Hym., Trichogrammatidae), with descriptions of four new species. *Proceedings of Entomological Society of Washington*, *75*: 288–297.
Nagarkatti, S. & Nagaraja, H. (1978). Experimental comparison of laboratory reared vs. wild-type *Trichogramma confusum* (Hym.: Trichogrammatidae) I. Fertility, fecundity and longevity. *BioControl*, *23*: 129–136.
Navarro, L., Maldonado, C., & Benavides, P. (2019). Molecular evidence for *Wolbachia* infection in *Prorops nasuta* (Hymenoptera: Bethylidae). A Parasitoid wasp of the coffee berry borer. *Comite Nacional*, 70(2): 45–54.
Normark, B.B. & Kirkendall, L.R. (2009). Parthenogenesis in insects and mites. In V.H. Resh & R.T. Carde (Eds.), *Encyclopedia of insects*, 2nd edn. Berkeley, CA, USA: Elsevier Inc., pp. 753–757.

Nunney, L. & Luck, R.F. (1988). Factors influencing optimum sex ratio in a structured population. *Theoretical Population Biology*, *33*:1–30.
Nunney, L. (1985). Female-biassed sex ratios: Individual or group selection? *Evolution*, *39*: 349–361.
Nyabuga, F.N., Outreman, Y., Simon, J.-C., Heckel, D.G. & Weissner, W.W. (2010). Effects of pea aphid secondary endosymbionts on aphid resistance and development of aphid parasitoid *Aphidius ervi*: A correlative study. *Entomologia Experimentalis et Applicata*, *136*: 243–253.
Ode, P.J. & Hardy, I.C.W. (2008). Parasitoid sex ratios and biological control. In E. Wajnberg, C. Bernstein & J.J.M. van Alphen (Eds.), *Behavioural ecology of insect parasitoids*: *From theoretical approaches to field applications*. Oxford: Blackwell Publishing, pp. 253–291.
Pandey, R.R. & Johnson, M.W. (2005). Effects of cool storage on *Anagyrus ananatis* Gahan (Hymenoptera: Encyrtidae). *Biological Control*, *35*: 9–16.
Pannebakker, B.A., Pijnacker, L.P., Zwaan, B.J. & Beukeboom, L.W. (2004). Cytology of *Wolbachia*-induced parthenogenesis in *Leptopilinia clavipes* (Hymenoptera: Figitidae). *Genome*, *47*: 299–303.
Perlman, S.J., Kelly, S.E., Zchori-Fein, E. & Hunter, M.S. (2006). Cytoplasmic incompatibility and multiple symbiont infection in ash whitefly parasitoid, *Encarsia inaron*. *Biological Control*, *54*: 961–968.
Pexton, J.J. & Mayhew, P.J. (2005). Clutch size adjustment, information use and evolution of gregarious development in parasitoid wasps. *Behavioral Ecology and Sociobiology*, *58*(1): 99–110.
Pijls, W.A.M., van Steenbergen, H.J. & van Alphen, J.J.M. (1996). Asexuality cured: relations and differences between sexual and asexual *Apoanagyrus diversicomis*. *Heredity 76*: 506–513.
Pinto, J.D. 1998. Systematics of North American species of *Trichogramma* (Hymenoptera: Trichogrammatidae). *Memoirs of Entomological Society of Washington*, *22*: 287 pp.
Pintureau, B., Grenier, S., Boleat, B., Lassabliere, F., Heddi, A. & Khatchadourian, C. 2000. Dynamics of *Wolbachia* populations in transfected lines of *Trichogramma*. *Journal of Invertebrate Pathology*, *76*: 20–25.
Pintureau, B., Gerding, M. & Cisternas, E. (1999). Description of three new species of Trichogrammatidae (Hymenoptera) from Chile. *Canadian Entomology*, *131*: 53–63.
Poirié, M., Colinet, D. & Gatti, J.L. (2014). Insights into function and evolution of parasitoid wasp venoms. *Current Opinion in Insect Science*, *6*: 52–60. doi: 10.1016/j.cois.2014.10.004.
Prabhulinga, T., Jalali, S.K., Kumar, K.P. & Doddabasappa, B. (2016). Biological characteristics of Arrhenotokous and Thelytokous *Trichogrammapretiosum* Riley. In *Arthropod Diversity and Conservation in Tropics and Sub-Tropics*. pp. 327–333, Springer, Singapore.
Pratissoli, D. & Fornazier, M.J. (1999). Ocorrência deSingapore *Trichogramma acacioi* Brun, Moraes & Soares (Hym.: Trichogrammatidae) em ovos de Nipteria panacea Thierry-Mieg (Lep.: Geometridae), um geometrídeo desfolhador do abacateiro. *Anais da Sociedade Entomológica do Brasil*, *28*: 347–349.
Price, P.W. (1975). Reproductive strategies of parasitoids. In P.W. Price (Ed.), *Evolutionary strategies of parasitic insects and mites*. Boston, MA, USA: Springer, pp. 87–111.
Querino, R.B. & Zucchi, R.A. (2003). New species of *Trichogramma* Westwood (Hymaneoptera: Trichogrammatidae) associated with Lepidopterous eggs in Brazil. *Zootaxa*, *163*: 1–10.
Quezada, J.R., DeBach, P. & Rosen, D. (1973). Biological and taxonomic studies of *Signophora borinquensis,* new species, (Hym: Signiphoridae), a primary parasite of diaspidine scales. *Hilgardia*, *41*: 543–604.
Ramirez-Romero, R., Sivinski, J., Copeland, C.S. & Aluja, M. (2011). Are individuals from thelytokous and arrhenotokous populations equally adept as biocontrol agents? Orientation and host searching behavior of a fruit fly parasitoid. *BioControl*, *57*(3): 427–440.
de Réaumur, R.A.F. (1738). *Mémoires pour servir à l'histoire des Insectes*. Vol. 4. Paris, France: De l'Imprimerie Royale, 636 p.
Reumer, B.M., van Alphen, J.J.M. & Kraaijeveld, K. (2013). Population genetics of Wolbachia-infected, parthenogenetic and uninfected, sexual populations of *Tetrastichus coeruleus* (Hymenoptera: Eulophidae). *Molecular Ecology 22*: 4433–4444.
Roopa, G.R., Jalali, S.K., Nagaraja, H. & Lalitha, Y. (2009). Comparison of biological parameters among arrhenotokous and thelytokous forms of *Trichogramma* (Hymenoptera: Trichogrammatidae). *Pest Management in Horticultural Ecosystems*, *15*(1): 28–32.
Rossler, Y. & DeBach, P. (1973). Genetic variability in a thelytokous form of *Aphytis mytilaspidis* (Le Baron) (Hymenoptera: Aphelinidae). *Hilgardia*, *42*: 149–175

Rousset F., Bouchon, D., Pintureau, B., Juchault, P. & Solignac, M. (1992). *Wolbachia* endosymbionts responsible for various alterations of sexuality in arthropods. *Proceedings of Royal Society B*, *250*: 91–98.
Rundle, B.J., Thomson, L.J. & Hoffmann, A.A. (2004). Effects of cold storage on field and laboratory performance of *Trichogramma carverae* (Hymenoptera: Trichogrammatidae) and response of three *Trichogramma* spp. (*T. carverae*, *T.* nr. *brassicae*, and *T. funiculatum*) to cold. *Journal of Ecological Entomology*, *97*: 213–221.
Sarmiento, M.C.E. (1993). Una nueva especie de *Trichogramma* (Hymenoptera: Trichogrammatidae) de los Andes de Colombia. *The Revista Colombiana de Entomología*, *19*: 3–5.
Sazama, E.J., Ouellette, S.P. & Wesner, J.S. (2019). Bacterial endosymbionts are common among, but not necessarily within, insect species. *Environmental Entomology*, *48*: 127–133.
Schilthuizen, M. & Stouthamer, R. (1997). Horizontal transmission of parthenogenesis inducing microbes in *Trichogramma* wasps. *Proceedings of Royal Society of London B*, *264*: 361–366.
Schlinger, E.I. & Hall, J.C. (1959). A synopsis of biologies of three imported parasites of spotted alfalfa aphid. *Journal of Economic Entomology*, *52*: 154–157.
Schuler, H., Kern, P., Arthofer, W., Vogt, H., Fischer, M., Stauffer, C. & Riegler, M. (2016). *Wolbachia* in parasitoids attacking native European and introduced eastern cherry fruit flies in Europe. *Environmental Entomology*, *45*(6): 1424–1431.
Segoli, M., Harari, A.R., Rosenheim, J.A., Bouskila, A. & Keasar, T. (2010). The evolution of polyembryony in parasitoid wasps. *Journal of Evolutionary Biology*, *23*(9): 1807–1819.
Segoli, M., Stouthamer, R., Stouthamer, C.M., Rugman-Jones, P. & Rosenheim, J.A. (2013). The effect of *Wolbachia* on lifetime reproductive success of its insect host in field. *Journal of Evolutionary Biology*, *26*(12): 2716–2720.
Semiatizki, A., Weiss, B., Bagim, S., Rohkin-Shalom, S., Kaltenpoth, M. & Chiel, E. (2020). Effects, interactions, and localization of *Rickettsia* and *Wolbachia* in house fly parasitoid, *Spalangia endius*. *Microbial Ecology*, 10.1007/s00248-020-01520-x.
Sgrò, C.M., Terblanche, J.S. & Hoffmann, A.A. (2016). What can plasticity contribute to insect responses to climate change? *Annual Review of Entomology*, *61*: 433–451.
Shuker, D.M., Pen, I., Duncan, A.B., Reece, S.E. & West, S.A. (2005). Sex ratio under asymmetrical local mate competition: Theory and a test with parasitoid wasps. *The American naturalist*, *166*: 301– 316.
Shuker, D.M., Sykes, E.M., Browning, L.E., Beukeboom, L.W. & West, S.A. (2006). Male influence on sex allocation in parasitoid wasp *Nasonia vitripennis*. *Behavioral Ecology and Sociobiology*, *59*: 829–835.
Silva, I.M.M.S., van Meer, M.M.M., Roskam, M.M., Hoogenboom A., Gort, G. & Stouthamer, R. (2000). Biological control potential of *Wolbachia*-infected versus uninfected wasps: Laboratory and greenhouse evaluation of *Trichogramma cordubensis* and *T. deion* strains. *Biocontrol Science and Technology*, *10*: 223–238.
Silva, R.B.Q. (2002). Taxonomia do gênero *Trichogramma* Westwood 1833 (Hymenoptera: Trichogrammatidae) na América do Sul. *Tese de Doutorado, Escola Superior de Agricultura Luiz de Queiroz-USP*. Piracicaba. 214 pp.Doctoral dissertation, Universidade de São Paulo, Brazil.
Silveira, L.C.P., Souza, I.L., Tomazella, V.B., Mendez, H.A.G. (2019). Parasitoid insects. In B. Souza, L. Vázquez & R. Marucci (Eds.), *Natural enemies of insect pests in neotropical agroecosystems*. Springer, Cham, pp. 97–109.
Simonato, M., Pilati, M., Magnoux, E., Courtin, C., Sauné, L., Rousselet, J., Battistib, A., Auger-Rozenberg, A.N. & Kerdelhué, C. (2019). A population genetic study of egg parasitoid *Baryscapus servadeii* reveals large scale automictic parthenogenesis and almost fixed homozygosity. *Biological Control*, *139*: 104097.
Skinner, S.W. (1985). Son-killer: A third extrachromosomal factor affecting sex ratio in parasitoid wasp, *Nasonia (=Mormoniella) vitripennis*. *Genetics*, *109*: 745–759.
Slobodchikoff, C.N. & Daly, H.V. (1971). Systematic and evolutionary implications of parthenogenesis in Hymenoptera. *Animal Zoology*, *11*: 273–282.
Sorakina, A.P. (1987) Biological and morphological substantiation of specific distinctness of *Trichogramma telengai* sp. n. *Entomological Review*, *66*: 20–34.
Speicher, B.R. (1937). Oogenesis in a thelytokous wasp *Nemeritis canescens* (Grav.). *Journal of Morphology*, *61*: 453–471.
Speicher, K.G. (1934). Impaternate females in *Habrobracon*. *The Biological Bulletin*, *67*: 277–293.

Srinatha, H.S., Jalali, S.K., Sriram, S. & Chakravarthy, A.K. (2015). Isolation of microbes associated with field-collected populations of egg parasitoid, *Trichogramma chilonis* capable of enhancing biotic fitness. *Biocontrol Science and Technology*, *25*(7): 789–802.

Stearns, S.C. (1992). *Evolution of life histories*. Oxford: Oxford University Press.

Steiner, S. & Ruther, J. (2009). How important is sex for females of a haplodiploid species under local mate competition? *Behavioral Ecology*, *20*: 570–574.

Stolk, C. & Stouthamer, R. (1996). Influence of a cytoplasmic incompatibility-inducing *Wolbachia* on fitness of parasitoid wasp *Nasonia vitripennis*. *Proceedings of sections Experimental and Applied Entomology of Netherlands Entomological Society*, *7*: 33–37.

Stouthamer, R. & Kazmer, D.J. (1994). Cytogenetics of microbe associated parthenogenesis and its consequences for gene flow in *Trichogramma* wasps. *Heredity*, *73*: 317–327.

Stouthamer, R., Breeuwer, J.A.J., Luck, R.F. & Werren, J.H. (1993). Molecular identification of microorganisms associated with parthenogenesis. *Nature*, *361*: 66–68.

Stouthamer, R., Liiko, S. & Mak, F. (1994). Influence of parthenogenesis *Wolbachia* on host fitness. *Norwegian Journal of Agricultural Sciences*, *16*: 117–122.

Stouthamer, R., Luck, R.F. & Hamilton, W.D. (1990a). Antibiotics cause parthenogenetic *Trichogramma* to revert to sex. *Proceedings of National Academy of Sciences of USA*, *87*: 2424–2427.

Stouthamer, R., Pinto, J.D., Platner, G.R. & Luck, R.F. (1990b). Taxonomic status of thelytokous forms of *Trichogramma*. *Annals of Entomological Society of America*, *83*: 475–481.

Strand, M.R. (2000). Developmental traits and life-history evolution in parasitoids. In M.E. Hochberg & A.R. Ives (Eds.), *Parasitoid population biology*. Princeton University, p. 139–162.

Tagami, Y., Doi, M., Sugiyama, K., Tatara, A. & Saito, T. (2006). Survey of leafminers and their parasitoids to find endosymbionts for improvement of biological control. *Biological Control*, *38*: 210–216.

Tagami, Y., Miura, K. & Stouthamer, R. (2001). How does infection with parthenogenesis-inducing *Wolbachia* reduce fitness of *Trichogramma*? *Journal of Invertebrate Pathology*, *78*(4): 267–271.

Taylor, G.P., Coghlin, P.C., Floate, K.D. & Perlman, S.J. (2011). The host range of male-killing symbiont *Arsenophonus nasoniae* in filth fly parasitoids. *Journal of Invertebrate Pathology*, *106*(3): 371–379.

Thornhill, R. & Alcock, J. (1983). *The evolution of insect mating systems*. Cambridge, MA, USA: Harvard University Press.

Tougeron, K., Brodeur, J., Le Lann, C. & Van Baaren, J. (2020). How climate change affects seasonal ecology of insect parasitoids. *Ecological Entomology*, *45*: 167–181.

Turlings, T.C.L., Wäckers, F.L., Vet, L.E.M., Lewis, W.J. & Tumlinson, J.H. (1993). Learning of Host-finding cues by hymenopterous parasitoids. In D.R. Papaj & A.C. Lewis (Eds.), *Insect learning*. Boston, MA, USA: Springer.

Ueno, T. & Tanaka, T. (1994). Comparative biology of six polyphagous solitary pupal endoparasitoids (Hymenoptera: Ichneumonidae): Differential host suitability and sex allocation. *Annals of Entomological Society of America*, *87*: 592–598.

Ueno, T. & Tanaka, T. (1997). Comparison between primary and secondary sex ratios in parasitoid wasps using a method for observing chromosomes. *Entomologia Experimentalis et Applicata*, *82*: 105–108.

Urano, T. & Hijii, N. (1995). Resource utilization and sex allocation in response to host size in two ectoparasitoid wasps on subcortical beetles. *Entomolgia Experimentalis et Applicata*,*74*: 23–35.

van Dijken, M.J. & Waage, J.K. (1987). Self and conspecific superparasitism by egg parasitoid *Trichogramma evanecsens*. *Entomologia Experimentalis et Applicata*, *43*: 183–192.

van Meer, M.M.M., Witteweldt, J. & Stouthamer, R. (1999). Phylogeny of arthropod endosymbionts *Wolbachia* based on *wsp* gene. *Insect Molecular Biology*, *8*: 399–408.

van Meer, M.M.M. (1999). *Phylogeny and host-symbiont interactions of thelytoky inducing Wolbachia in Hymenoptera*. PhD Thesis. Wageningen: Wageningen Agricultural University, p. 118.

Vasquez, C.J., Stouthamer, R., Jeong, G. & Morse, J.G. (2011). Discovery of a CI-inducing Wolbachia and its associated fitness costs in biological control agent *Aphytis melinus* DeBach (Hymenoptera: Aphelinidae). *Biological Control*, *58*(3): 192–198.

Vavre, F., Girin, C. & Bouletreau, M. (1999). Phylogenetic status of a fecundity-enhancing *Wolbachia* that does not induce thelytoky in *Trichogramma*. *Insect Molecular Biology*, *8*: 67–72.

Vet, L.E.M., Datema, A., Janssen, A. & Snellen, H. (1994). Clutch size in a larval pupal endoparasitoid consequences for fitness. *Journal of Animal Ecology*, *63*: 807–815.

Visser, M.E. & Rosenheim, J.A. (1998). The influence of competition between foragers on clutch size decisions in insect parasitoids. *Biological Control, 11*: 169–174.

Visser, M.E. (1994). The importance of being large relationship between size and fitness in females of parasitoid *Aphaereta minuta* (Hymenoptera: Braconidae). *Journal of Animal Ecology, 63*: 963–978.

Waage, J.K. (1982). Sex ratio and population dynamics of natural enemies – some possible interactions. *Annals of Applied Biology, 101*: 159–164.

Waage, J.K. & Ng, S.M. (1984). The reproductive strategy of a parasitic wasp. I. Optimal progeny and sex allocation in *Trichogramma evanescens. Journal of Animal Ecology, 53*: 401–415.

Walter, G.H. (1983). Divergent male ontogenies in Aphelinidae (Hymenoptera: Chalcidoidea): A simplified classification and a suggested evolutionary sequence. *Biological Journal of Linnean Society, 19*: 63–82.

Wang, Z. & Smith, S.M. (1996). Phenotypic differences between thelytokous and arrhenotokous *Trichogramma minutum* from *Zeiraphera Canadensis. Entomologia Experimentalis et Applicata 78*: 315–323.

Weeks, A.R., Marec, F. & Breeuwer, J.A.J. (2001). A mite species that consist entirely of haploid females. *Science, 292*: 2479–2482.

Weinert, L.A. (2015). The diversity and phylogeny of *Rickettsia*. In S. Morand, B.R. Krasnov & D. Littlewood (Eds.), *Parasite diversity and diversification. Evolutionary Ecology Meets Phylogenetics.* Cambridge: Cambridge University Press.

Werren, J.H., Baldo, L. & Clark, M.E. (2008). *Wolbachia*: Master manipulators of invertebrate biology. *Nature Reviews Microbiology, 6*: 741–751.

Werren, J.H., Zhang, W. & Guo, L.R. (1995). Evolution and phylogeny of *Wolbachia* bacteria: Reproductive parasites of arthropods. *Proceedings of Royal Society of London, 5261*: 55–63.

Wetherington, M.T., Jennings, D.E., Shrewsbury, P.M. & Duan, J.J. (2017). Climate variation alters synchrony of host-parasitoid interactions. *Ecology and Evolution, 7*(20): 8578–8587. doi:10.1002/ece3.3384

White, J.A., Kelly, S.E., Cockburn, S.N., Perlman, S.J. & Hunter, M.S. (2011). Endosymbiont costs and benefits in a parasitoid infected with both *Wolbachia* and *Cardinium. Heredity, 106*(4): 585–591.

White, M.J.D. (1973). *Animal cytology and evolution*, 3rd edn. Cambridge: Cambridge University Press, p. 468.

Whu, M. & Valdivieso, L. (1999). Distribución y comportamiento de ocho especies de *Trichogramma* y *Trichogrammatoidea* (Hymenoptera: Trichogrammatidae) en el Perú. *Revista Peruana de Entolomogía, 41*: 61–68.

Wilson, F. & Woolcock, L.T. (1960). Temperature determination of sex in a parthenogenetic parasite, *Ooencyrtus submetallicus* (Howard) (Hymenoptera: Encyrtidae). *Australian Journal of Zoology, 8*: 153–169.

Wilson, K. & Lessells, C.M. (1994). Evolution of clutch size in insects. I. A review of static optimality models. *Journal of Evolutionary Biology, 7*: 339–363.

Wu, L.H., Hoffmann, A.A. & Thomson, L.J. (2016). Taiwanese *Trichogramma* of Asian corn borer: morphology, ITS-2 rDNA characterization, and natural *Wolbachia* infection. *Journal of Insect Science, 16*(1): 1–7.

Ye, Z., Vollhardt, I.M.G., Parth, N., Rubbmark, O., Traugott, M. & Tylianakis, J. (2018). Facultative bacterial endosymbionts shape parasitoid food webs in natural host populations: A correlative analysis. *Journal of Animal Ecology, 87*: 1440–1451.

Zchori-Fein E., Roush, R. & Rosen D. (1998). Distribution of parthenogenesis-inducing symbionts in ovaries and eggs of *Aphytis* (Hymenoptera: Aphelinidae). *Current Microbiology, 36*: 1–8.

Zchori-Fein, E., Faktor, O., Zeidan, M., Gottlieb, Y., Czosnek, H. & Rosen, D. (1995). Parthenogenesis-inducing microorganisms in *Aphytis. Insect Molecular Biology, 4*: 173–178.

Zchori-Fein, E., Perlman, S.J., Kelly, S.E., Katzir, N. & Hunter, M.S. (2004). Characterization of a Bacteroidetes symbiont in *Encarsia* wasp (Hymenoptera: Aphelinidae): Proposal of Candidatus *Cardiniumhertigii. International Journal of Systematic and Evolutionary Microbiology, 54*: 961–968.

Zchori-Fein, E., Rosen, D. & Roush, R.T. (1994). Microorganisms associated with thelytoky in *Aphytis lignanensis. International Journal of Insect Morphology and Embryology, 23*: 169–172.

Zchori-Fein, E., Roush, R.T. & Hunter, M.S. (1992). Male production induced by antibiotic treatment in *Encarsia formosa,* an asexual species. *Experientia, 48*: 102–105.

Zucchi, R.A. (1988). New species of *Trichogramma* (Hymenoptera: Trichogrammatidae) associated with sugar cane borer *Diatraea saccharalis* (F.) (Lepidoptera: Pyralidade) in Brazil. *Colloques de l'INRA*, *43*: 133–140.

Zucchi, R.A. & Monteiro, R.C. (1997). O gênero *Trichogramma* na América do Sul. In J.R.P. Parra & R.A. Zucchi (Eds.), *Trichogramma e o controle biolólgico aplicado*. Piracicaba: FEALQ, pp. 41–66.

Zucchi, R.A. (1985). Taxonomia de espécies de *Trichogramma* (Hymenoptera: Trichogrammatidae) associados a algumas pragas (Lepidoptera) no Brasil. Tese (Livre- Docência). Piracicaba: Escola Superior de Agricultura Luiz de Queiroz-USP, p. 77.

Zucchi, R.A., Parra, J.R.P. & Silveira, N.S. (1991). *Trichogramma* species associated with some lepidopterous pests in Brazil. *Colloques de l'INRA*, *53*: 131–134.

Zug, R. & Hammerstein, P. (2012). Still a host of hosts for *Wolbachia*: Analysis of recent data suggests that 40% of terrestrial arthropod species are infected. *PLoS One*, *7*: e38544.

Zuparko, R.L. (2008) Parasitic Hymenoptera (Parasitica). In J.H. Frank *et al.* (Eds.), *Encyclopedia of entomology*. Dordrecht: Springer, pp. 2730–2736.

15

Oviposition Strategies

Desh Deepak Chaudhary[1], Bhupendra Kumar[2], and Omkar[3]
[1]*Indira Gandhi National tribal University, Amarkantak, Madhya Pradesh, India*
[2]*Banaras Hindu University, Varanasi, Uttar Pradesh, India*
[3]*University of Lucknow, Lucknow, Uttar Pradesh, India*

CONTENTS

15.1 Introduction

Insects are the most successful organisms on earth with multiple supporting factors that enhance their adaptability under constantly changing environmental conditions (Khaliq *et al.*, 2014). A wide range of food, shelter, body structure, body size, and an organized life style of living are responsible for their success on earth (Hasan *et al.*, 2013). Oviposition or act of laying eggs is an important part of their life

DOI: 10.1201/9781003043195-15

cycle (Birch, 1953). Females initiate a species-specific search for best oviposition spot, considering sites with less enemies and abundant food resources (Hamed & Khattak, 1985).

Insect oviposition behaviour is central to investigations of population dynamics, life-history evolution, plant-insect and host-parasitoid interactions, and biological control. Investigations of proximal causes of variations in insect feeding and oviposition behaviour have focused on responses to environmental cues (Van Alphen & Visser, 1990) and role of learning, and genetic variation (Futuyma & Peterson, 1985). However, internal physiological processes modulating insect oviposition behaviour are rarely evaluated extensively (Thompson & Pellmyr, 1991).

Insects choose oviposition sites that enhance fitness of immature stages (Gmeinbauer & Crailsheim, 1993). Oviposition site selection and oviposition by females differs significantly with insect orders (Hamed & Khattak, 1985). While phytophagous insects may prefer varying plant surfaces or insides of plants (Minkenberg *et al.*, 1992), in predatory insects oviposition is influenced by visual and olfactory cues (Obata, 1997), colour of prey (Harmon *et al.*, 1998), volatiles released by prey infested plants (Han & Chen, 2002), honeydew of prey (Carter & Dixon, 1984), intensity of smell of prey (Hatano *et al.*, 2008), alarm pheromones of prey (Francis *et al.*, 2004), quantity and quality of prey (Sarmento *et al.*, 2007; Singh *et al.*, 2016), and age of prey colony (Hemptinne *et al.*, 2000). In oviparous species, oviposition-site choice is a female's assessment of potential nest sites and selection of a particular site. In some insect orders, accessory reproductive gland secretions produce substances for attaching eggs with each other and to substrate. Many insects cover their eggs with fecal matter, hairs or scales, and secretions from their glands that provide a physical barrier to natural enemies. Adults of moth, *Spodoptera frugiperda*, take up poisonous larval hairs and cover their eggs with them. In order Neuroptera, eggs of Chrysopidae and Mantispidae are placed onto tip of a stalk to protect from predation and cannibalism (Hilker & Meiners, 2002). Eggs of European beewolf, *Philanthus triangulum* (Hymenoptera: Crabronidae) emit a strong antifungal agent, a gaseous radical nitric oxide, which protects them and beewolf's provisions (paralysed honeybees) against fungi (Strohm *et al.*, 2019).

Egg protection by parental care can reach unusual forms. Giant water bugs (Hemiptera: Belostomatidae) lay eggs on male's wings (Tallamy, 2001). Females of some dragonflies and damselflies drop their eggs into water, or lay them on or inside plant tissues near water level (Dijkstra & Lewington, 2018). Some egg parasitoids like wasp, *Trichogramma evanescens,* lay their eggs below the eye of *P. brassicae* females (order Lepidoptera) that transport them to oviposition sites (Clausen, 1978). Numerous hymenopteran parasitoids are specialized to parasitize ootheca of cockroaches (Blattodea) and deposit their eggs (Lebeck, 1991). Shield bugs, *Platynopiellus septumdecemmaculati* (Heteroptera: Acanthosomatidae), deposit their eggs with minimal contact onto plant surface; this oviposition mode might be a strategy to prevent induction of plant cues attracting parasitoids (Tallamy, 2001). Coreid golden egg bug, *Phyllomorpha laciniata* (Heteroptera: Coreidae), is a specialist of *Paronynchia argentea* plant species but deposits its eggs on other conspecifics, acting as 'mobile nests'.

A female's decision about where to lay her eggs can have serious consequences for her own reproductive fitness, as oviposition site significantly affects embryo survival, juvenile performance, offspring phenotype, and potentially survival of ovipositing female. Thus, choice for oviposition-site is a life-history trait of critical importance (Resetarits, 1996).

In the present chapter, we have discussed oviposition strategies in insects across different natures and habitat, viz. phytophagous insects, parasitoids, aquatic insects, and predatory insects. We anticipate that the present chapter not only helps in understanding insect oviposition strategies but also evolutionary significance of oviposition site selection and nature of egg production on embryo survival, juvenile performance, offspring phenotype, and potentially survival of ovipositing female.

15.2 Oviposition-site Selection

15.2.1 Maximizing Embryo Survival

Oviposition sites that minimize predation risk and maintain a microclimate suitable for embryo development are preferentially selected. However, selection is not so simple but complex and not straightforward.

Females ovipositing in terrestrial (Higashiura, 1989) and aquatic habitats (Petranka & Fakhoury, 1991) prefer predator-free oviposition sites. Previous studies have reported reduced egg survivorship with increased density of eggs on a host plant (Mitchell, 1975); early hatched larvae may feed on neighbouring conspecific eggs (Williams & Gilbert, 1981). Thus, plants bearing conspecific eggs are often avoided.

15.2.2 Maximizing Maternal Survival

Maximizing maternal survival is another determining factor while selecting oviposition sites, especially in iteroparous species, where lifetime reproductive success of a female depends on her survival across periodic breeding events. *Fecundity-survival hypothesis* (Refsnider & Janzen, 2010) discusses that females have to choose between increasing their fecundity despite concurrent increase in mortality risk to themselves and reducing their own mortality risk at a cost to their offspring.

The *preference–performance hypothesis* states that females should prefer to oviposit on hosts with highest nutritional quality for offspring, which would promote short development time, higher adult biomass, and increased maternal fecundity (Pöykkö, 2006). However, oviposition-site choice by many phytophagous insects fails to support this hypothesis. In pierid butterfly, *Anthocharis cardamines*, oviposition occurs on host plants that provide poor larval; nutrition and not on plant on which larvae have highest performance (Courtney, 1981). Similarly, grass miner, *Chromatomyia nigra*, optimizes number of eggs laid per host plant rather than optimizing quality of host plant (Scheirs *et al.*, 2000). Thus, rather than optimizing habitat quality for offspring, such species optimizes maternal fitness and survival. In above two examples, while females of *A. cardamines* fed extensively on nectar of plant selected for oviposition (Courtney, 1981), grass miners optimized their own fecundity over performance of their offspring (Scheirs *et al.*, 2000).

Such suboptimal selection of oviposition sites may also result from time limitation rather than egg limitation (Rosenheim *et al.*, 2008). Therefore, sometimes suitable host plants may be chosen for oviposition simply to maximize maternal fecundity, even though such host plants are not ideal for offspring performance.

15.2.3 Proximity to Suitable Habitat for Offspring

Eggs and larval stages of insects are most often vulnerable to predation, desiccation, and/or starvation, thus suitable oviposition habitat is critical. Studies on many insects support preference–performance hypothesis, which suggests that oviposition sites that provide suitable habitat with highest nutritional benefits to larvae are selected (Pöykkö, 2006). Larvae that developed on such habitats grew larger (Rausher, 1983), had shorter development time (Vacek *et al.*, 1985), higher digestive efficiency (Sadeghi & Gilbert, 1999), and sequestered defensive chemicals (Thompson & Pellmyr, 1991). Certain ant species allow larvae to move among several host plant species to achieve a mixed diet, minimizing intraspecific competition and also allowing sufficient time to complete development before host plant dehisces (Rausher, 1983; Wiklund & Friberg, 2009). Mosquito females avoid ovipositing in pools that may dry (Blaustein & Margalit, 1996) and contain predators of their larvae (Spencer *et al.*, 2002) and/or competitor species (Petranka & Fakhoury, 1991).

15.2.4 Maintaining Natal Philopatry

In some insect taxa, oviposition-site choice may be an artifact of natal philopatry with females returning to nest at same location where they themselves hatched during their development. If oviposition-site choice is heritable, females that oviposit at sites that produce high-quality offspring would pass on this oviposition site choice to their daughters, thereby maintaining a lineage that produces high-quality offspring resulting from oviposition-site choice. Selection favours philopatry when dispersal from natal habitat entails high costs. Damselfly, *Enallagma* spp., that are unable to determine if a potential oviposition site contains predators continue to oviposit in habitat in which they themselves hatched (McPeek, 1989).

15.2.5 *Indirect Oviposition-Site Choice through Mate Choice*

In insects where males construct nests and/or tend eggs, quality of oviposition site is correlated with quality of males. Therefore, offspring from eggs deposited in high-quality oviposition sites are more likely to be sired by males with high-quality phenotypes; oviposition site selection is indirectly affected by mate choice (Kelly, 2008).

15.3 Evolutionary Ecology and Oviposition Strategies

Most insect larvae are small and relatively immobile and ovipositing females single-handedly make choices of larval nutrition. In addition, to choosing a safe and suitable habitat, females must maximize their search efficiency and host encounter rates, and they must decide on how long they should spend searching for best host before accepting alternatives. Ovipositing females' survival till full complement of eggs is laid is also important. In short, process of choosing oviposition sites is a daunting task for insects.

One of the earliest non-cladistic attempts at a broad comparative study of host use was Ehrlich and Raven's (1964) on coevolution between relatively specialized butterfly–plant associations. Plants' secondary metabolites, and escape and radiation of associated lineages, might thus have coevolved. Factors such as habitat choice, plant growth form (Beccaloni & Symons, 2000), predation (Keese, 1997) and diapause pattern (Carey, 1994) may also play a role.

Modest modifications in oviposition preference within a geographic mosaic of populations resulted in a preference hierarchy pattern of host use with ranking and degree of specialization (Thompson & Pellmyr, 1991; Sadeghi & Gilbert, 1999). These studies indicate that oviposition preference is a relatively flexible trait allowing for rapid evolution of host associations constrained only by ability to feed and survive on plants. With enough knowledge we should be able to make predictions about future colonizations and host shifts, which could be of high importance for pest management.

15.4 Oviposition Strategies in Aquatic Insects

Measuring oviposition strategies is highly challenging for aquatic insects with complex life cycles and highly mobile dispersal stages (Peckarsky *et al.*, 2000). Since larval survival and growth are primary needs for selection of oviposition sites, specialized oviposition behaviour can be found in many aquatic taxa, including Coleoptera (Binckley & Resetarits, 2008); Diptera (Baba & Takaoka, 1989, 1991); Ephemeroptera (Encalada & Peckarsky, 2006, 2007); and Trichoptera (Reich & Downes, 2003a, b).

Some aquatic insects lay their eggs on or within plants, others do not attach their eggs to any particular object, while some attach their eggs to inanimate objects (Lancaster *et al.*, 2010a, b; Macqueen & Downes, 2015). Some aquatic insects broadcast their fertilized eggs into water bodies from air which are dispersed far and wide via currents to locations where they may eventually attach themselves; making studies very difficult.

15.5 Oviposition Strategies in Phytophagous Insects

Phytophagous insects generally avoid laying eggs on unsuitable oviposition sites (Renwick & Chew, 1994; Gripenberg *et al.*, 2010). A variety of indicators are used to assess suitability of oviposition sites. In *Euura lasiolepis*, a shoot-galling sawfly, shorter shoots that are likely to fall off are avoided (Craig *et al.*, 1989). Plant secondary chemical compounds (Wennström *et al.*, 2010), conspecific host-marking pheromones (Huth & Pellmyr, 1999), fungal infection on sites (Biere & Honders, 2006), and age of plant parts (Heard, 1995) are likely cues for selection of oviposition sites.

15.5.1 Factors Involved in Oviposition Site Selection by Phytophagous Insects

15.5.1.1 Specialist and Generalist Insect Species

While some insect species are specialists, i.e. they lay eggs on particular plants, others are generalists and lay eggs on a number of plant species. In generalist species, females exhibit preference hierarchies among a range of host plants, which may change with experience.

Specialization can be favoured if trade-offs lead to individuals being able to utilize a single host plant species more efficiently than multiple plant species (Futuyma, 1983; Bernays & Wcislo, 1994). However, preference hierarchies will be favoured if they reflect quality and temporal abundance of different host plant species in insect habitat (Ward, 1987; Courtney *et al.*, 1989). Generalist species may learn more quickly about habitat or availability of food resources at oviposition sites than specialist species (Papaj & Prokopy, 1989).

15.5.1.2 Plant Features, Repellents, and Chemical Defense of Host Plants

Host plants have different mechanisms for their defense, like hairs, trichomes, and chemicals, like 4-methoxy-3-indolylacetonitrile, 3-indolylacetonitrile, and (E)-ß-caryophyllene (Lucas-Barbosa *et al.*, 2011). Some plants produce flavonoid and phenolic compounds, oils and fatty acids (Traynier *et al.*, 1994), insecticidal peptides or proteins, (Conti *et al.*, 2008) (Table 15.1), lectins (Sulzenbacher *et al.*, 2010), and non-protein amino acids (Huang *et al.*, 2011) in plants are detrimental to phytophagous insects.

Nutritional quality of host plant is also positively correlated with oviposition site-selection (Barros & Zucoloto, 1999). While soluble carbohydrate contents of host plants have positive effects on oviposition; they adversely affect larval duration (Zhang *et al.*, 2011). Density of natural predators and parasitoids is also a deciding factor (Kaasik *et al.*, 2014).

Plant growth (Bone *et al.*, 2009), shape, and size of leaves (Damman & Feeny, 1988), plant odour (Hattori, 1988), plant/flower colour (Jayaprakasha *et al.*, 1997), trichome density (Molina-Montenegro *et al.*, 2006), and hypertrophic trichomes (Conti *et al.*, 2008; Karioti *et al.*, 2011) also affect oviposition in many phytophagous insects as does temperature (Lindblad & Sigvald, 1999).

15.6 Oviposition Strategies in Parasitoids

Insects that are parasitic during their immature stages and free living as adults are termed as protelean parasites (Askew, 1971) or parasitoids (Reuter, 1913). Parasitoids make four kinds of oviposition decisions, viz. (i) mode of searching for habitat; (ii) which host to attack, including problem of superparasitism; (iii) clutch or brood size per host, and (iv) sex ratio to produce (Charnov & Skinner, 1985).

TABLE 15.1

Plant-derived chemicals suppressing oviposition in insects (Khaliq *et al.*, 2015)

S.N.	Chemical Nature	Name of Plant	Insect Species	References
1	Volatile	*Pinus sylvestris*	*Diprion pini*	Pschorn-Walcher (1982)
3	Extract	*Citrullus colocynthis*	*Bactrocera* spp.	Rehman *et al.* (2009)
4	Terpenoid volatiles	*Pinus sylvestris*	*Chrysonotomia ruforum*	Hilker *et al.* (2002a)
5	CO_2 concentration	Flowering plant	*Pieris rapae*	Langan *et al.* (2001)
7	Hypersensitivity response	*Brassica nigra*	*Pieris* spp.	Hilker *et al.* (2002b)
10	Extract	*Veleriana jatamansi*	*Bactocera zonata*	Rehman *et al.* (2009)
11	Neoplasms	*Pisum sativum*	*Bruchid* species	Doss *et al.* (1995)

15.6.1 Factors Affecting Oviposition Strategies in Parasitoids

15.6.1.1 Habitat Choice

The problem of habitat choice is one of identifying chain of stimuli that bring female into close proximity with a patch of hosts (Doutt, 1964; Vinson, 1976). According to Cook and Hubbard (1977), several parasitoids simultaneously search in an environment with multiple patches that differ in their host densities. Within each patch, females find and oviposit on hosts according to *Random Parasite Equation* (Rogers, 1972). This model predicts that most profitable patches (i.e., those with greatest host density) should be exploited first, with others of lower rank added in as value of exploited patches declines to equal them. General habitat preference or selection of areas where searching is concentrated may be influenced by temperature, humidity, light intensity, wind, and food sources, as well as flying and crawling habits of parasitoids (Leius, 1967; Shahjahan, 1974).

15.6.1.2 Host Selection

Upon emergence, adult female parasitoid is found often in an alien habitat, is removed from a host population and must locate a suitable host in order to propagate. Post habitat selection, host is systematically searched. Doutt (1964) divided process resulting in successful parasitism into four steps, viz. (i) host habitat location, (ii) host location, (iii) host acceptance, and (iv) host suitability. A fifth step, host regulation, has also been added by Vinson (1975). Host plant may also influence host selection primarily by interference. One such way is through interference (Flanders, 1937; Arthur, 1962) such as secretion of a sticky exudate from trichomes of tobacco, which deters oviposition (Rabb & Bradley, 1968).

Volatile chemicals derived from host's food, organisms associated with host, host itself, or a combination of these factors are responsible for directing parasitoid to a host habitat (Table 15.2). Chemicals deposited during host oviposition stimulate searching by egg parasitoids (Lewis *et al.*, 1972; Vinson, 1975) as do host pheromones (Mitchell & Mau, 1971). Host odours (kairomones) appear to orient parasitoid only from a short distance (Schmidt, 1974) and are analogous to short-range factors released from injured plant tissues (allomones (Vinson, 1975).

Carton (1974) found that *Pimpla instigator* was attracted initially by odour from a distance to an area containing hosts, then a cylindrical shape with a degree of relief increased acceptance. Size of host also influences selection (Duodu & Davis, 1974). Age, independent of size, also influences host acceptance (Lewis & Redlinger, 1969; Schmidt, 1974).

TABLE 15.2

Compounds involved in host selection process of insect parasitoids (Vinson, 1976)

S.N.	Family	Parasitoids	Host	Source	Chemical	References
1	Braconidae	*Diaeretiella rapae*	*Myzus persicae*	Host plant	Allylisothiocyanate	Read *et al.* (1970)
2	Tachinidae	*Cyzenis albieans*	*Operophtera brumata*	Damaged host plant	Sucrose and fructose	Hassell (1968)
3	Braconidae	*Microplifis croceipes*	*Heliothis zea*	Frass	13-methyl hentriacontane	Jones *et al.* (1971)
4	Trichogrammatidae	*Trichogramma evanescens*	Eggs of wide range of Lepidoptera hosts	Female	Tricosane	Jones *et al.* (1973)
5	Braconidae	*Cardiochiles nigriceps*	*Heliothis virescens*	Moth scale	Several methyl compounds	Vinson (1975)
8	Ichneumonidae	*Itoplectis conquisitor*	*Galleria mellonella*	Hemolymph	Serine, arginine, Leucine, MgCl	Arthur (1966)

Host movement or vibrations are also important stimuli for release of ovipositor probing and acceptance of concealed hosts (Ryan & Rudinsky, 1962; Baier, 1963). Smith (1952) reported that motionless hosts were palpated by female *Microtonus vittatae* and were attacked only when movement was induced.

Parasitoids are attracted to certain colours or wavelengths (Hollingsworth *et al.*, 1970). Takahashi and Pimentel (1967) found that black pupae of house fly were preferred over brown pupae. Arthur (1966) found that *Itoplectis conquisitor* could be conditioned to attack its host larvae in blue-coloured tubes.

To prevent super or hyperparasitism, parasitoids leave chemical markers, viz. *spoor factors* (Flanders, 1951), trail odours (Price, 1970), *search-deterrent substances* (Matthews, 1974), *deterrent pheromones* (Greany & Oatman, 1972), and *host-marking pheromones* (Vinson, 1972).

15.6.1.3 Clutch Size

Offspring survival declines with increasing clutch size, due to less availability of food per offspring (Lack, 1947), and those that emerge will be smaller, which in turn may affect adult fitness, altering both life span and egg production (Charnov *et al.*, 1981). Charnov & Krebs (1974) suggested that if yearly survival rates of adults decreased with increasing clutch size, selection would favour smaller clutches than predicted, because production of smaller clutches would increase chances of surviving to produce subsequent ones.

Parasitoids clutch size may be smaller than normal for either or both of two reasons: (i) trade-off between maternal survival and egg production rate, or (ii) trade-off between time and clutch size, as mediated through oviposition time. Together, or singly, they may favour a smaller than normal clutch per host. Selection has probably acted to match egg production rate to general scarcity of hosts, as shown by correlation of ovariole number versus host abundance discovered by Price (1973, 1974). Since these correlations hold for species that lay one egg per host, they suggest that selection has made egg production rate as small as possible to reasonably meet expected times between hosts.

15.6.1.4 Sex Ratio

Sex ratio in parasitoids may be affected by host size (when single egg is laid per host) or local mate competition. In former case, since small hosts produce small adults and vice versa, natural selection favours a strategy of producing sons in small hosts and daughters in large hosts (Charnov, 1979; Bull, 1981).

The *'Local Mate Competition'* (LMC) hypothesis (Hamilton, 1967) suggests that a typical LMC wasp lays many eggs in a single host and mating takes place on or near host before daughters disperse to lay their own eggs. If hosts are themselves not clumped, all females are mated by males emerging from same host. If single mother oviposits in a host, there is strict brother-sister mating. If more females oviposit in a host, matings may be between members of several broods. If host itself is clumped, presence of several ovipositing females on a clump may likewise lead to mating among members of several broods. A natural selection approach to sex ratio evolution under these conditions shows that if there is strict sib-mating, mostly daughters would be produced. When mating takes place between broods of several females, favoured sex ratio approaches one-half males (Hamilton, 1967; Charnov *et al.*, 1981).

15.7 Oviposition Strategies in Predaceous Insects

The preference of predaceous insects for specific plant species (Ballal & Singh, 1999) or parts of plants (Richards & Harper, 1970; Richards & Schmidt, 1996) for oviposition indicates that chemical cues might be involved in their selection (Table 15.3). For mind bug, *Mucrolophus culiginosus*, it has been demonstrated that more eggs are laid on substrates treated with plant extracts as compared to moist substrates (Constant *et al.*, 1996). Syrphids, chrysopids, and coccinellids are stimulated to oviposit using honeydew (Sadeghi & Gilbert, 2000) and aphid density (Scholz & Poehling, 2000).

TABLE 15.3

Chemicals used as oviposition cues by insect predators (Steidle and van Loon, 2002)

S.N.	Species (Family)	Prey	Cues or their Origin	References
2	*Metasyrphus corolla* (Syrphidae)	Specialist on *Aphis fabae*	Aphid extract (Aqueous)	Shonouda (1996)
			Aphid odour, alkalines (C23 to C26, C28)	Shonouda *et al.* (1998)
3	*Chrysopa carnea* (Chrysopidae)	Generalist	Tryptophan	Van Emden and Hagen (1976)
			Attracted by indol acetaldehyde and other breakdown products of tryptophan	Van Emden and Hagen (1976)
4	*Coccinella septempundata* (Coccinellidae)	Generalist also on aphid	Aphid odour honeydew	Evans and Dixon (1986)
5	*Macrolophus caliginosus* (Miridae)	Generalist	Leaf extract (ethanol) of *Inula viscosa*	Constant *et al.* (1996)
6	*Iphiseius degenerans* (Phytoseiidae)	Generalist	Water soluble cues from conspecific and heterospecific eggs	Faraji *et al.* (2000)
7	*Pherbellia cinerella* (Sciomyzidae)	Terrestrial helicid snails	Fresh snail faeces	Coupland (1996)

In *Chrysopa carnea*, breakdown product of amino acid tryptophan, especially indole acetaldehyde, has been identified as key stimulant in honeydew (Van Emden & Hagen, 1976). A mixture of long chain hydrocarbons (C23–C28) of limited volatility (Shonouda *et al.*, 1998) and salicyladehyde (Rank & Smiley, 1994) are indicator cues for oviposition. For aphidophagous ladybird beetles, aphid patch quality is an important component during oviposition site selection (Kindlmann & Dixon, 1993, 1999). Perception of prey density even in actual absence of prey (Singh *et al.*, 2016) stimulates oviposition.

Patches with conspecific tracks are avoided by ladybird beetles (Mishra *et al.*, 2012). Aphid colonies in early phase of development are selected for oviposition (Hemptinne *et al.*, 1992). Kindlmann and Dixon (1993, 1999) propose that there is in development of aphid patches a time, 'egg window', when it is most advantageous for ladybirds to lay their eggs. This stresses importance of age of an aphid colony in determining its suitability for oviposition. Since ladybird larvae and aphid colonies have similar development times, theory stipulates that ladybirds should lay a limited number of eggs early in development of aphid colonies. If eggs are laid later aphid colony might disperse and disappear before ladybird larvae complete their development. As a result, risk of cannibalism will increase greatly due to absence of food (Agarwala & Dixon, 1992).

15.8 Egg Deposition Patterns

15.8.1 Egg Clustering

Oviposition in clusters or egg clumping, egg pooling, aggregation of eggs, and batch of eggs is a common phenomenon in various insect taxa (Agarwala & Dixon, 1993). It is defined as a group of eggs produced by one or more females in which each egg touches rest of group through contact with at least one other egg.

Explanations for evolution of egg clustering are three pronged. The first group of hypotheses suggests egg clustering occurs because of reduced search for oviposition sites saves energy, incurs lower mortality risk, and/or reduces chances of failure (Stamp, 1980).

The second school of thought attributes egg clustering to increased fitness of larvae because of predator saturation (Lawrence, 1990), communal defence against predators and parasitoids (Stamp, 1980), an additive effect of warning colouration (Stamp, 1980), increased efficiency of thermoregulation

(Porter, 1982; Seymour, 1974), and/or increased efficiency of resource use (Le Masurier, 1994; Clark & Faeth, 1997). However, larval co-operation is likely to have evolved secondarily to egg clustering (Courtney, 1984).

The third group of hypotheses explains evolution of egg clustering through fitness effects on eggs. Clark and Faeth (1998) show that hatching success of eggs was greater for larger clusters with eggs in multiple layers. Another effect of clustering on egg survival is via protection against predators (Stamp, 1980; Agarwala & Dixon, 1993). Fisher (1999) suggested that, having fed on a distasteful egg, a nauseous predator would avoid feeding on nearby siblings of prey, favouring evolution of aggregation with distastefulness and aposematism via kin selection. Thus, egg clustering is expected to have preceded evolution of aposematic colouration rather than reverse (Courtney, 1984). A last effect of clustering on egg survival is that egg clusters may protect eggs in centre of cluster from predation (Agarwala & Dixon, 1993).

15.8.2 Concealed Eggs

The majority of unwinged insects deposit their eggs into soil, leaf litter, or under stones. Their eggs are also among smallest, which makes them less suitable for development of egg parasitoids (Church *et al.*, 2019). Although such insects that bury their eggs and/or egg masses appear to escape from egg parasitoids by this strategy, a huge percentage in this category are still susceptible to parasitism. Many genera successfully parasitize concealed eggs of Orthoptera, Coleoptera, Hemiptera, and Diptera (Noyes, 2019).

15.8.3 Role of Bright Egg Colouration in Defense

Several moths of Pyralidae and Olethreutidae families have transparent shells that make them look like they have same colour as background (Peterson, 1962). Many insect eggs on plant leaves have intrinsic green or yellowish-green colouration, matching substrate thereby reducing visibility (Kawooya *et al.*, 1985; Kang *et al.*, 1995). Egg mimicry in several families of Lepidoptera (e.g. Notodontidae, Lymantriidae, Saturniidae, etc.) involves making unparasitized eggs look like eggs from which parasitoids have emerged (Hinton, 1981).

15.8.4 Role of Egg Surface Chemicals in Defense

Chemical barriers (sticky fluids with repellent chemicals), alarm pheromones, chemical weapons, and defence secretions are used to defend eggs against predators and pathogens. Ovipositing female uses secretions from oviduct and accessory reproductive glands (ARGs), which other than supporting egg to glide through oviduct to outside, attachment, and protection of eggs, also protect eggs against predators, parasitoids, and pathogens (Hilker & Meiners, 2002). Alder leaf beetle, *Agela sticaalni* (Coleoptera: Chrysomelidae), has its eggs covered with spumaline and it protects eggs by sticking on mouthparts or legs of natural enemies (Hilker & Meiners, 2002). In some insects, ARGs also produce cement for attaching eggs to each other and to substrate. In some cases, female may protect eggs with natural products, which are biosynthesized *de novo*, or are of dietary origin (Hilker & Meiners, 2002).

Many insects cover eggs with faecal matter, hairs, or scales, secretions from glands, providing a physical barrier to natural enemies (Kenis *et al.*, 2019). Eggs of Chrysopidae and Mantispidae (both Neuroptera) are placed onto tip of a stalk, which serves as protection against predation and cannibalism (Hilker & Meiners, 2002). Egg stalk of chrysopid, *Ceraeochysa smithi*, is coated by a protective secretion that repels ants (Johnson & Bin, 1982). Autogenously produced defensive components within eggs that are of anti-microbial activity are known from many Coleopteran species, e.g. cantharidin produced by blister beetles (Meloidae), anthraquinones produced by leaf beetles (Chrysomelidae) (Hilker & Meiners, 2002; Pankewitz *et al.*, 2007), and coccinelline, adaline, and harmonine present in eggs of ladybird beetles, *Coccinella septempunctata*, *Adalia* spp., and *Harmonia axyridis*, respectively (Hodek *et al.*, 2012). Eggs of several ladybird species produce alkaloids (Hilker & Meiners, 2002). Yet, a number of parasitoids are known from eggs of herbivorous ladybird species (Riddick *et al.*, 2009). However, effect of many defensive components of extrinsic origin (host plant) within insect eggs on egg parasitoids are yet to be studied (Dettner, 2015).

15.9 Conclusions

Oviposition is a major event in life of female insects. The behavioural processes involved in this strategy are habitat choice, host finding, host acceptance, and choice of oviposition sites. However, causes and consequences of these behavioural processes associated with oviposition strategies are exceedingly complex. Different insects, viz. phytophagous insects, aquatic insects, parasitoids, and predatory insects have different strategies for oviposition according to their habitat types. Among them, phytophagous insects and parasitoids have complex processes involved in their oviposition strategies that include additional steps, like governing host defense mechanisms. Oviposition strategies also include egg deposition pattern, like single egg or concealed egg deposition, or depositing eggs in clutches. While eggs have different protective mechanisms like cementing their surfaces with accessory reproductive gland secretions or toxic nature of egg surface chemicals, and repulsive nature of odour and taste of such surface chemicals; still eggs display survival strategies, like aposematic colouration, camouflage, and mimicry. Although a lot many ecological and evolutionary studies have been conducted emphasizing oviposition strategies in different insects, still many more are yet to be conducted specifically emphasizing molecular mechanisms of such behaviour in both primitive and advanced insect orders. Studies assessing ecological significance of defensive components of host plants within eggs of phytophagous insects need more rigorous experimentation. In addition, role of learning in governing oviposition strategies in different insect taxa is yet to be elucidated.

REFERENCES

Agarwala, B.K. & Dixon, A.F. (1992). Laboratory study of cannibalism and interspecific predation in ladybirds. *Ecological Entomology*, *17*(4): 303–309.

Agarwala, B.K. & Dixon, A.F.G. (1993). Why do ladybirds lay eggs in clusters? *Functional Ecology*, *7*(5), 541–548.

Arthur, A.P. (1962). Influence of host tree on abundance of *Itoplectisconquisitor* (Say), a polyphagous parasite of the European shoot moth *Rhyacionia buoliana* (Schiff). *Candian Entomologist*, *94*: 337–347.

Arthur, A.P. (1966). Associative learning in *Itoplectis conquisitor* (Say)(Hymenoptera: Ichneumonidae). *The Canadian Entomologist*, *98*(2): 213–223.

Askew, R. (1971). *Parasitic insects*. New York: American Elsevier, p. 820.

Baba, M. & Takaoka, H. (1989). Oviposition sites and the number of larval instars of two mountain blackfly species, *Simulium japonicum* and *S. rufibasis*: Diptera: Simuliidae. *Medical Entomology and Zoology*, *40*(4): 307–313.

Baba, M. & Takaoka, H. (1991). Oviposition habits of a univoltine blackfly, *Prosimulium kiotoense* (Diptera: Simuliidae), in Kyushu, Japan. *Medical and Veterinary Entomology*, *5*(3): 351–358.

Encalada, A.C. & Peckarsky, B.L. (2006). Selective oviposition of the mayfly *Baetis bicaudatus*. *Oecologia*, *148*(3): 526–537.

Baier, M. (1963). Zurbiologie und gradologie der sattelmücke Wagner (Diptera, Cecidomyiidae). *ZeitschriftfürangewandteEntomologie*, *53*(1-4): 217–273.

Ballal, C.R. & Singh, S.P. (1999). Host plant-mediated orientational and ovipositional behavior of three species of chrysopids (Neuroptera: Chrysopidae). *Biological Control*, *16*(1): 47–53.

Barros, H.C.H. & Zucoloto, F.S. (1999). Performance and host preference of *Asciamonuste* (Lepidoptera, Pieridae). *Journal of Insect Physiology*, *45*(1): 7–14.

Beccaloni, G.W. & Symons, F.B. (2000). Variation of butterfly diet breadth in relation to host-plant predictability: Results from two faunas. *Oikos*, *90*(1): 50–66.

Bernays, E.A. & Wcislo, W.T. (1994). Sensory capabilities, information processing, and resource specialization. *The Quarterly Review of Biology*, *69*(2): 187–204.

Biere, A. & Honders, S.C. (2006). Coping with third parties in a nursery pollination mutualism: *Hadenabicruris* avoids oviposition on pathogen-infected, less rewarding *Silene latifolia*. *New Phytologist*, *169*(4): 719–727.
Binckley, C.A. & Resetarits Jr, W.J. (2008). Oviposition behavior partitions aquatic landscapes along predation and nutrient gradients. *Behavioral Ecology*, *19*(3): 552–557.
Birch L.C. (1953). Experimental background to the study of the distribution and abundance of insects-l: The influence of temperature, moisture and food on the innate capacity for increase of three grain-beetles. *Ecology*, *34*: 698–711.
Blaustein, L. & Margalit, J. (1996). Priority effects in temporary pools: Nature and outcome of mosquito larva-toad tadpole interactions depend on order of entrance. *Journal of Animal Ecology*, *65*(1), 77–84.
Bone, N.J., Thomson, L.J., Ridland, P.M., Cole, P. & Hoffmann, A.A. (2009). Cover crops in Victorian apple orchards: Effects on production, natural enemies and pests across a season. *Crop Protection*, *28*(8): 675–683.
Bull, J.J. (1981). Sex ratio evolution when fitness varies. *Heredity*, *46*(1): 9–26.
Carey, D.B. (1994). Diapause and the host plant affiliations of lycaenid butterflies. *Oikos*, *69*(2), 259–266.
Carter, M.C. & Dixon, A.F.G. (1984). Honeydew: An arrestant stimulus for coccinellids. *Ecological Entomology*, *9*(4): 383–387.
Carton, Y. (1974). Biologie de pimpla instigator (Ichneumonidae: Pimplinae) III. Analyseexpérimentale du processus de reconnaissance de l'hôte-chrysalide. *Entomologia Experimentaliset Applicata*, *17*(2): 265–278.
Charnov, E.L. (1979). The genetical evolution of patterns of sexuality: Darwinian fitness. *The American Naturalist*, *113*(4): 465–480.
Charnov, E.L. & Krebs, J.R. (1974). On clutch-size and fitness. *Ibis*, *116*(2): 217–219.
Charnov, E.L. & Skinner, S.W. (1985). Complementary approaches to the understanding of parasitoid oviposition decisions. *Environmental Entomology*, *14*(4): 383–391.
Charnov, E.L., Los-Den Hartogh, R.L., Jones, W.T. & Van Den Assem, J. (1981). Sex ratio evolution in a variable environment. *Nature*, *289*(5793): 27–33.
Church, S.H., Donoughe, S., de Medeiros, B.A.S. & Extavour, C.G. (2019). Insect egg size and shape evolve with ecology but not developmental rate. *Nature*, *571*: 58–62.
Clark, B.R. & Faeth, S.H. (1997). The consequences of larval aggregation in the butterfly *Chlosyne lacinia*. *Ecological Entomology*, *22*: 408.
Clark, B.R. & Faeth, S.H. (1998). The evolution of egg clustering in butterflies: A test of the egg desiccation hypothesis. *Evolutionary Ecology*, *12*(5): 543–552.
Clausen, C.P. (1978). Introduced parasites and predators of arthropod pests and weeds: A world review. *USDA Agric Handbook*, *480*: 1–551.
Constant, B., Grenier, S. & Bonnot, G. (1996). Artificial substrate for egg laying and embryonic development by the predatory Bug (Heteroptera: Miridae). *Biological Control*, *7*(2): 140–147.
Conti, E., Zadra, C., Salerno, G., Leombruni, B., Volpe, D., Frati, F., Marucchini, C. & Bin, F. (2008). Changes in the volatile profile of *Brassica oleracea* due to feeding and oviposition by *Murgantia histrionica* (Heteroptera: Pentatomidae). *European Journal of Entomology*, *105*(5), 839–847.
Cook, R.M. & Hubbard, S.F. (1977). Adaptive searching strategies in insect parasites. *The Journal of Animal Ecology*, *46*(1), 115–125.
Coupland, J.B. (1996). Influence of snail feces and mucus on oviposition and larval behavior of *Pherbellia cinerella* (Diptera: Sciomyzidae). *Journal of Chemical Ecology*, *22*, 183–189.
Courtney, S.P. (1981). Coevolution of pierid butterflies and their cruciferous foodplants. *Oecologia*, *51*(1): 91–96.
Courtney, S.P. (1984). The evolution of egg clustering by butterflies and other insects. *The American Naturalist*, *123*(2): 276–281.
Courtney, S.P., Chen, G.K. & Gardner, A. (1989). A general model for individual host selection. *Oikos*, *55*(1), 55–65.
Craig, T.P., Itami, J.K. & Price, P.W. (1989). A strong relationship between oviposition preference and larval performance in a shoot-galling sawfly. *Ecology*, *70*(6): 1691–1699.
Damman, H. & Feeny, P. (1988). Mechanisms and consequences of selective oviposition by the zebra swallowtail butterfly. *Animal Behaviour*, *36*(2): 563–573.

Dettner, K. (2015). Toxins, defensive compounds and drugs from insects. *Insect Molecular Biology and Ecology*, 39–93.
Dijkstra K.D. & Lewington R. (2018). *Field guide to the dragonflies of Britain and Europe*. Dorset, UK: British Wildlife Publishing.
Doutt, R.L. (1964). Biological characteristics of entomophagous adults. *In Biological Control of Insect Pests and Weeds* (P. DeBach. ed.) Chapman and Hall, London, 145–167.
Duodu, Y.A. & Davis, D.W. (1974). Selection of alfalfa weevil larval instars by, and mortality due to, the parasite *Bathyplectes curculionis* (Thomson). *Environmental Entomology*, *3*(3): 549–552.
Ehrlich, P.R. & Raven, P.H. (1964). Butterflies and plants: A study in coevolution. *Evolution*, 586–608.
Encalada, A.C. & Peckarsky, B.L. (2007). A comparative study of the costs of alternative mayfly oviposition behaviors. *Behavioral Ecology and Sociobiology*, *61*(9): 1437–1448.
Evans, E.W., & Dixon, A.F.G. (1986). Cues for oviposition by ladybird beetles (Coccinelidae): Response to aphids. *The Journal of Animal Ecology*, *55*, 1027–1034.
Faraji, F., Janssen, A., Van Rijn, P.C.J., & Sabelis, M. (2000). Kin recognition by the predatory mite *Iphiseius degenerans*: Discrimination among own, conspecific, and heterospecific eggs. *Ecological Entomology*, *25*, 147–155.
Fisher, R.A. (1999). *The genetical theory of natural selection: A complete variorum edition*. Oxford University Press, Oxford, England.
Flanders, S.E. (1937). Habitat selection by Trichogramma. *Annals of the Entomological Society of America*, *30*(2): 208–210.
Flanders, S.E. (1951). Mass cultures of California red scale and its golden chalcid parasites. *Hilgardia*, *21*: 1–41.
Francis, F., Lognay, G. & Haubruge, E. (2004). Olfactory responses to aphid and host plant volatile releases: (E)-β-farnesene an effective kairomone for the predator *Adalia bipunctata*. *Journal of Chemical Ecology*, *30*(4): 741–755.
Futuyma, D.J. (1983). Selective factors in the evolution of host choice by phytophagous insects. *In Herbivorous Insects: Host Seeking Behavior and Mechanisms, ed.* SAmahd, Academic, New York, 227-244.
Futuyma, D.J. & Peterson, S.C. (1985). Genetic variation in the use of resources by insects. *Annual Review of Entomology*, *30*(1): 217–238.
Gmeinbauer, R. & Crailsheim, K. (1993). Glucose utilization during flight of honeybee (*Apis mellifera*) workers, drones and queens. *Journal of Insect Physiology*, *39*(11): 959–967.
Greany, P.D. & Oatman, E.R. (1972). Analysis of host discrimination in the parasite *Orgilus lepidus* (Hymenoptera: Braconidae). *Annals of the Entomological Society of America*, *65*(2): 377–383.
Gripenberg, S., Mayhew, P.J., Parnell, M. & Roslin, T. (2010). A meta-analysis of preference–performance relationships in phytophagous insects. *Ecology Letters*, *13*(3): 383–393.
Hamed, M. & Khattak, S.U. (1985). Red flour beetle: Development and losses in various stored food stuffs. *Sarhad Journal of Agriculture*, 1, 97-101.
Hamilton, W.D. (1967). Extraordinary sex ratios. *Science*, *156*(3774): 477–488.
Han, B.Y. & Chen, Z.M. (2002). Composition of the volatiles from intact and mechanically pierced tea aphid– tea shoot complexes and their attraction to natural enemies of the tea aphid. *Journal of Agricultural and Food Chemistry*, *50*(9): 2571–2575.
Harmon, J.P., Losey, J.E. & Ives, A.R. (1998). The role of vision and color in the close proximity foraging behavior of four coccinellid species. *Oecologia*, *115*(1-2): 287–292.
Hasan, M., Sagheer, M., Khaliq, A., Khan, F.Z.A., Gul, H.T., Ahmad, K., Manzoor, S.A., Yasir, M. & Javed M, M.N. (2013). Assessment of relative resistance in advanced rice genotypes in response to variation in abiotic factors and development of *Tribolium castaneum* (Herbst) (Coleoptera: Tenebrionidae). *International Journal of Biosciences*, *3*: 33–38.
Hatano, E., Kunert, G., Bartram, S., Boland, W., Gershenzon, J., & Weisser, W.W. (2008). Do aphid colonies amplify their emission of alarm phermomone?, Journal of chemical ecology, *34*(9): 1149–1152.
Hattori, M. (1988). Host-plant factors responsible for oviposition behaviour in the limabean pod borer, *Etiella zinckenella* Treitschke. *Journal of Insect Physiology*, *34*(3): 191–196.
Heard, T.A. (1995). Oviposition preferences and larval performance of a flower-feeding weevil, *Coelocephalapion aculeatum*, in relation to host development. *Entomologia Experimentalis et Applicata*, *76*(2): 195–201.

Hemptinne, J.L., Dixon, A.F.G. & Coffin, J. (1992). Attack strategy of ladybird beetles (Coccinellidae): Factors shaping their numerical response. *Oecologia*, *90*(2): 238–245.
Hemptinne, J.L., Doumbia, M. & Dixon, A.F.G. (2000). Assessment of patch quality by ladybirds: Role of aphid and plant phenology. *Journal of Insect Behavior*, *13*(3): 353–359.
Higashiura, Y. (1989). Survival of eggs in the gypsy moth *Lymantria dispar*. II. Oviposition site selection in changing environments. *The Journal of Animal Ecology*, *58*(2), 413–426.
Hilker M. & Meiners T. (2002). *Chemoecology of insect eggs and egg deposition*. Berlin, Germany: Blackwell Publishing.
Hinton, H.E. (1981). *Biology of insect eggs. Volume I, Volume II, Volume III*. Oxford: Pergammon Press, The Boulevard, Langford Lane, Kidlington.
Hodek, I., Honek, A. & Van Emden, H.F. (Eds.). (2012). *Ecology and behaviour of the ladybird beetles (Coccinellidae)*. John Wiley & Sons, Hoboken, New Jersey.
Hollingsworth, J.P., HartstackJr, A.W. & Lingren, P.D. (1970). The spectral response of *Campoletis perdistinctus*. *Journal of Economic Entomology*, *63*(6): 1758–1761.
Huang, T., Jander, G. & de Vos, M. (2011). Non-protein amino acids in plant defense against insect herbivores: Representative cases and opportunities for further functional analysis. *Phytochemistry*, *72*(13): 1531–1537.
Huth, C.J. & Pellmyr, O. (1999). Yucca moth oviposition and pollination behavior is affected by past flower visitors: Evidence for a host-marking pheromone. *Oecologia*, *119*(4): 593–599.
Jayaprakasha, G.K., Singh, R.P., Pereira, J. & Sakariah, K.K. (1997). Limonoids from *Citrus reticulata* and their moult inhibiting activity in mosquito *Culex quinquefasciatus* larvae. *Phytochemistry*, *44*(5): 843–846.
Johnson, N.F. & Bin, F.E. (1982). Species of Telenomus (Hym.,Scelionidae), parasitoids of stalked eggs of Neuroptera (Chrysopidae et berothidae). *Redia*, 65: 189-206.
Kaasik, R., Kovács, G., Kaart, T., Metspalu, L., Williams, I.H. & Veromann, E. (2014). *Meligethesaeneus* oviposition preferences, larval parasitism rate and species composition of parasitoids on *Brassica nigra*, *Raphanus sativus* and *Eruca sativa* compared with on *Brassica napus*. *Biological Control*, *69*: 65–71.
Kang, Y., Kulakosky, P.C., Van Antwerpen, R. & Law, J.H. (1995). Sequestration of insecticyanin, a blue hemolymph protein, into the egg of the hawkmoth *Manducasexta*. Evidence for receptor-mediated endocytosis. *Insect Biochemistry and Molecular Biology*, *25*(4): 503–510.
Karioti, A., Tooulakou, G., Bilia, A.R., Psaras, G.K., Karabourniotis, G. & Skaltsa, H. (2011). Erinea formation on *Quercus ilex* leaves: Anatomical, physiological and chemical responses of leaf trichomes against mite attack. *Phytochemistry*, *72*(2-3): 230–237.
Kawooya, J.K., Keim, P.S., Law, J.H., Riley, C.T., Ryan, R.O. & Shapiro, J.P. (1985). Why are green caterpillars green? In: Hedin, P.A. (Ed.), Bioregulators for Pest Control. ACS Symposium series 276, Washington, DC, 511–521.
Keese, M.C. (1997). Does escape to enemy-free space explain host specialization in two closely related leaf-feeding beetles (Coleoptera: Chrysomelidae)?. *Oecologia*, *112*(1): 81–86.
Kelly, C.D. (2008). The interrelationships between resource-holding potential, resource-value and reproductive success in territorial males: How much variation can we explain?. *Behavioral Ecology and Sociobiology*, *62*(6): 855–871.
Kenis, M., Du Plessis, H., Van den Berg, J., Ba, M.N., Goergen, G., Kwadjo, K.E. & Offord, L. (2019). *Telenomusremus*, a candidate parasitoid for the biological control of *Spodoptera frugiperda* in Africa, is already present on the continent. *Insects*, *10*(4): 92.
Khaliq, A., Javed, M., Sohail, M. & Sagheer, M. (2014). Environmental effects on insects and their population dynamics. *Journal of Entomology and Zoology Studies*, *2*(2): 1–7.
Khaliq, A., Ahmad, M.H. & Anas, M. (2015). Behavioral ecology of oviposition in insects-A dumpy review. Sky Journal of Agricultural Research, *4*(1): 1–7.
Kindlmann, P. & Dixon, A.F. (1993). Optimal foraging in ladybird beetles (Coleoptera: Coccinellidae) and its consequences for their use in biological control. *European Journal of Entomology*, *90*(4): 443–450.
Kindlmann, P. & Dixon, A.F. (1999). Generation time ratios determinants of prey abundance in insect predator–prey interactions. *Biological Control*, *16*(2): 133–138.
Lack, D. (1947). The significance of clutch-size. *Ibis*, *89*(2): 302–352.
Lancaster, J., Downes, B.J. & Arnold, A. (2010a). Oviposition site selectivity of some stream-dwelling caddisflies. *Hydrobiologia*, *652*(1): 165–178.

Lancaster, J., Downes, B.J., & Arnold, A. (2010b). Environmental constraints on oviposition limit egg supply of a stream insect at multiple scales. *Oecologia, 163*, 373–384.
Lawrence, W.S. (1990). The effects of group size and host species on development and survivorship of a gregarious caterpillar *Halisidotacaryae* (Lepidoptera: Arctiidae). *Ecological Entomology, 15*(1): 53–62.
Le Masurier, A.D. (1994). Costs and benefits of egg clustering in *Pieris brassicae. Journal of Animal Ecology*, 677–685.
Lebeck L.M. (1991). A review of the Hymenopterous natural enemies of cockroaches with emphasis on biological control. *Entomophaga, 36*: 335–352.
Leius, K. (1967). Influence of wild flowers on parasitism of tent caterpillar and codling moth. *The Canadian Entomologist, 99*(4): 444–446.
Lewis, W.J. & Redlinger, L.M. (1969). Suitability of eggs of the almond moth, *Cadracautella*, of various ages for parasitism by *Trichogramma evanescens. Annals of the Entomological Society of America, 62*(6): 1482–1484.
Lewis, W.J., Jones, R.L. & Sparks, A.N. (1972). A host-seeking stimulant for the egg parasite *Trichogramma evanescens*: Its source and a demonstration of its laboratory and field activity. *Annals of the Entomological Society of America, 65*(5): 1087–1089.
Lindblad, M. & Sigvald, R. (1999). Frit fly infestation of oats in relation to growth stage and weather conditions at oviposition. *Crop Protection, 18*(8): 517–521.
Lucas-Barbosa, D., van Loon, J.J. & Dicke, M. (2011). The effects of herbivore-induced plant volatiles on interactions between plants and flower-visiting insects. *Phytochemistry, 72*(13): 1647–1654.
Macqueen, A. & Downes, B.J. (2015). Large-scale manipulation of oviposition substrata affects egg supply to populations of some stream-dwelling caddisflies. *Freshwater Biology, 60*(4): 802–812.
Matthews, R.W. (1974). Biology of Braconidae. *Annual Review of Entomology, 19*(1): 15–32.
McPeek, M.A. (1989). Differential dispersal tendencies among *Enallagma* damselflies (Odonata) inhabiting different habitats. *Oikos*, 187–195.
Minkenberg, O.P., Tatar, M. & Rosenheim, J.A. (1992). Egg load as a major source of variability in insect foraging and oviposition behavior. *Oikos, 65*(1), 134–142.
Mishra, G., Singh, N., & Shahid, M. (2012). Effect of presence and semiochemicals of conspecific stages on oviposition by ladybirds (Coleoptera: Coccinellidae). *European Journal of Entomology, 109*, 363–371.
Mitchell, R. (1975). The evolution of oviposition tactics in the bean weevil, *Callosobruchus maculatus* (F.). *Ecology, 56*(3): 696–702.
Mitchell, W.C. & Mau, R.F. (1971). Response of the female southern green stink bug and its parasite, *Trichopoda pennipes*, to male stink bug pheromones. *Journal of Economic Entomology, 64*(4): 856–859.
Molina-Montenegro, M.A., Ávila, P., Hurtado, R., Valdivia, A.I. & Gianoli, E. (2006). Leaf trichome density may explain herbivory patterns of *Actinote* spp. (Lepidoptera: Acraeidae) on *Liabum mandonii* (Asteraceae) in a montane humid forest (Nor Yungas, Bolivia). *Actaoecologica, 30*(2): 147–150.
Noyes J.S. (2019). *Universal Chalcidoidea database*. World Wide Web Electronic Publication.
Obata, S. (1997). The influence of aphids on the behaviour of adults of the ladybird beetle, *Harmonia axyridis* (Col.: Coccinellidae). *Entomophaga, 42*(1–2): 103–106.
Pankewitz, F., Zöllmer, A., Gräser, Y. & Hilker, M. (2007). Anthraquinones as defensive compounds in eggs of Galerucini leaf beetles: Biosynthesis by the beetles?. *Archives of Insect Biochemistry and Physiology: Published in Collaboration with the Entomological Society of America, 66*(2): 98–108.
Papaj, D.R. & Prokopy, R.J. (1989). Ecological and evolutionary aspects of learning in phytophagous insects. *Annual Review of Entomology, 34*(1): 315–350.
Peckarsky, B.L., Taylor, B.W. & Caudill, C.C. (2000). Hydrologic and behavioral constraints on oviposition of stream insects: Implications for adult dispersal. *Oecologia, 125*(2): 186–200.
Peterson, A. (1962). Some eggs of insects that change color during incubation. *The Florida Entomologist, 45*(2): 81–87.
Petranka, J.W. & Fakhoury, K. (1991). Evidence of a chemically-mediated avoidance response of ovipositing insects to blue-gills and green frog tadpoles. *Copeia, 1991*(1): 234–239.
Porter, K. (1982). Basking behaviour in larvae of the butterfly *Euphydry asaurinia. Oikos, 38*(3), 308–312.
Pöykkö, H. (2006). Females and larvae of a geometrid moth, *Cleorodes lichenaria*, prefer a lichen host that assures shortest larval period. *Environmental Entomology, 35*(6): 1669–1676.

Price, P.W. (1970). Trail odors: Recognition by insects parasitic on cocoons. *Science, 170*(3957): 546–547.

Price, P.W. (1973). Reproductive strategies in parasitoid wasps. *The American Naturalist, 107*(957): 684–693.

Price, P.W. (1974). Strategies for egg production. *Evolution, 28*(1), 76–84.

Pschorn-Walcher, H. (1982). Symphyta. In: Forstscha¨dlinge Europas. Vol. 4 (Ed. by Schwenke, W.), pp. 66e103. Hamburg: Parey.

Rabb, R.L. & Bradley, J.R. (1968). The influence of host plants on parasitism of eggs of the tobacco hornworm. *Journal of Economic Entomology, 61*(5), 1249–1252.

Rank, N.E. & Smiley, J.T. (1994). Host-plant effects on *Parasyrphusmelanderi* (Diptera: Syrphidae) feeding on a willow leaf beetle *Chrysomela aeneicollis* (Coleoptera: Chrysomelidae). *Ecological Entomology, 19*(1): 31–38.

Rausher, M.D. (1983). Ecology of host-selection behavior in phytophagous insects. *Variable Plants and Herbivores in Natural and Managed Systems*, 223–257.

Refsnider, J.M. & Janzen, F.J. (2010). Putting eggs in one basket: Ecological and evolutionary hypotheses for variation in oviposition-site choice. *Annual Review of Ecology, Evolution, and Systematics, 41*: 39–57.

Reich, P. & Downes, B.J. (2003a). The distribution of aquatic invertebrate egg masses in relation to physical characteristics of oviposition sites at two Victorian upland streams. *Freshwater Biology, 48*(9): 1497–1513.

Reich, P. & Downes, B.J. (2003b). Experimental evidence for physical cues involved in oviposition site selection of lotic hydrobiosid caddis flies. *Oecologia, 136*(3): 465–475.

Renwick, J.A.A. & Chew, F.S. (1994). Oviposition behavior in Lepidoptera. *Annual Review of Entomology, 39*(1): 377–400.

Resetarits Jr, W.J. (1996). Oviposition site choice and life history evolution. *American Zoologist, 36*(2): 205–215.

Reuter, O.M. (1913). *Lebensgewohnheitenund Instinkte der Insekten*. Berlin: Friedlander. 488 pp.

Richards, L.A. & Harper, A.M. (1970). Oviposition by *Nubisalternutus* (Hemiptera: Nabiidae) in alfalfa. *Can. Entomol. 110*: 1359–1362.

Richards, P.C. & Schmidt, J.M. (1996). The suitability of some natural and artificial substrates as oviposition sites for the insidious flower bug *Orius insidiosus*, *Entomologia Experimentalis et Applicata, 80*(2): 325–333.

Riddick, E.W., Cottrell, T.E. & Kidd, K.A. (2009). Natural enemies of the Coccinellidae: parasites, pathogens, and parasitoids. *Biological Control, 51*(2): 306–312.

Rogers, D. (1972). Random search and insect population models. *The Journal of Animal Ecology. 41*(2), 369–383.

Rosenheim, J.A., Jepsen, S.J., Matthews, C.E., Smith, D.S. & Rosenheim, M.R. (2008). Time limitation, egg limitation, the cost of oviposition, and lifetime reproduction by an insect in nature. *The American Naturalist, 172*(4): 486–496.

Ryan, R.B. & Rudinsky, J.A. (1962). Biology and habits of the Douglas-fir beetle parasite, *Coeloides brunneri* Viereck (Hymenoptera: Braconidae), in Western Oregon. *The Canadian Entomologist, 94*(7): 748–763.

Sadeghi, H. & Gilbert, F. (1999). Individual variation in oviposition preference, and its interaction with larval performance in an insect predator. *Oecologia, 118*(4): 405–411.

Sadeghi, H. & Gilbert, F. (2000). Oviposition preferences of aphidophagous hoverflies. *Ecological Entomology, 25*(1): 91–100.

Sarmento, R.A., Venzon, M., Pallini, A., Oliveira, E.E. & Janssen, A. (2007). Use of odours by *Cycloneda sanguinea* to assess patch quality. *Entomologia Experimentalis et Applicata, 124*(3): 313–318.

Scheirs, J., Bruyn, L.D. & Verhagen, R. (2000). Optimization of adult performance determines host choice in a grass miner. *Proceedings of the Royal Society of London. Series B: Biological Sciences, 267*(1457): 2065–2069.

Schmidt, G.T. (1974). Host-acceptance behavior of *Campoletis sonorensis* toward *Heliothiszea*. *Annals of the Entomological Society of America, 67*(6): 835–844.

Scholz, D. & Poehling, H.M. (2000). Oviposition site selection of *Episyrphus balteatus*. *Entomologia Experimentalis et Applicata, 94*(2): 149–158.

Seymour, R.S. (1974). Convective and evaporative cooling in sawfly larvae. *Journal of Insect Physiology, 20*(12): 2447–2457.

Shahjahan, M. (1974). Erigeron flowers as a food and attractive odor source for *Peristenus pseudopallipes*, a braconid parasitoid of the tarnished plant bug. *Environmental Entomology, 3*(1): 69–72.

Shonouda, M.L. (1996). Crude aqueous-extract (kairomone) from *Aphis fabae* Scop.(Hom., Aphidae) and its effect on the behaviour of the predator *Metasyrphus corollae* Fabr.(Dipt.,Syrphidae) female. *Journal of Applied Entomology*, *120*(1-5): 489–492.

Shonouda, M.L., Bombosch, S., Shalaby, A.M. & Osman, S.I. (1998). Biological and chemical characterization of a kairomone excreted by the bean aphids, *Aphis fabae* Scop.(Hom., Aphididae) and its effect on the predator *Metasyrphus corollae* Fabr. I. Isolation, identification and bioassay of aphid-kairomone. *Journal of Applied Entomology*, *122*(1-5): 15–23.

Singh, S., Mishra, G. & Omkar (2016). Perceived prey quantity modulates oviposition in the ladybird *Menochilus sexmaculatus*. *Journal of Ethology*, *34*(1): 59–64.

Smith, O.J. (1952). Biology and behaviour of *Microtonus vittatae* Muesebeck. *Univ. Calif. Berkeley Publ. Entomoi*, 9: 315-344.

Spencer, M., Blaustein, L. & Cohen, J.E. (2002). Oviposition habitat selection by mosquitoes (*Culiseta longiareolata*) and consequences for population size. *Ecology*, *83*(3): 669–679.

Stamp, N.E. (1980). Egg deposition patterns in butterflies: Why do some species cluster their eggs rather than deposit them singly? *The American Naturalist*, *115*(3): 367–380.

Steidle, J.L., & van Loon, J.J. (2002). Chemoecology of parasitoid and predator oviposition behaviour. *Chemoecology of insect eggs and egg deposition*, 219–317.

Strohm, E., Herzner, G., Ruther, J., Kaltenpoth, M. & Engl, T. (2019). Nitric oxide radicals are emitted by wasp eggs to kill mold fungi. *Elife*, *8*: e43718.

Sulzenbacher, G., Roig-Zamboni, V., Peumans, W.J., Rougé, P., Van Damme, E.J. & Bourne, Y. (2010). Crystal structure of the GalNAc/Gal-specific agglutinin from the phytopathogenic ascomycete reveals novel adaptation of a β-trefoil domain. *Journal of Molecular Biology*, *400*(4): 715–723.

Takahashi, F. & Pimentel, D. (1967). Wasp preference for black-, brown-, and hybrid-type pupae of the house fly. *Annals of the Entomological Society of America*, *60*(3): 623–625.

Tallamy, D.W. (2001). Evolution of exclusive paternal care in Arthropods. *Annual Review of Entomology*, *46*: 139–165.

Thompson, J.N. & Pellmyr, O. (1991). Evolution of oviposition behavior and host preference in Lepidoptera. *Annual Review of Entomology*, *36*(1): 65–89.

Traynier, R.M.M., Schumacher, R.K. & Lau, D.M. (1994). Oviposition site selection by *Tineola bisselliella*, *Tinea spp* (Lepidoptera: Tineidae) and *Anthrenus flavipes* (Coleoptera: Dermestidae). *Journal of Stored Products Research*, *30*(4): 321–329.

Vacek, D.C., East, P.D., Barker, J.S.F. & Soliman, M.H. (1985). Feeding and oviposition preferences of *Drosophila buzzatii* for microbial species isolated from its natural environment. *Biological Journal of the Linnean Society*, *24*(2): 175–187.

Van Alphen, J.J. & Visser, M.E. (1990). Superparasitism as an adaptive strategy for insect parasitoids. *Annual Review of Entomology*, *35*(1): 59–79.

Van Emden, H.F. & Hagen, K.S. (1976). Olfactory reactions of the green lacewing, *Chrysopacarnea*, to tryptophan and certain breakdown products. *Environmental Entomology*, *5*(3): 469–473.

Vinson, S.B. (1972). Factors involved in successful attack on *Heliothis virescens* by the parasitoid *Cardiochiles nigriceps*. *Journal of Invertebrate Pathology*, *20*(1): 118–123.

Vinson, S.B. (1975). Source of material in the tobacco budworm which initiates host-searching by the egg-larval parasitoid, *Chelonustexanus*. *Annals of the Entomological Society of America*, *68*(2): 381–384.

Vinson, S.B. (1976). Host selection by insect parasitoids. *Annual Review of Entomology*, *21*(1): 109–133.

Ward, S.A. (1987). Optimal habitat selection in time-limited dispersers. *The American Naturalist*, *129*(4): 568–579.

Wennström, A., Hjulström, L.N., Hjältén, J. & Julkunen-Tiitto, R. (2010). Mother really knows best: Host choice of adult phytophagous insect females reflects a within-host variation in suitability as larval food. *Chemoecology*, *20*(1): 35–42.

Wiklund, C. & Friberg, M. (2009). The evolutionary ecology of generalization: Among-year variation in host plant use and offspring survival in a butterfly. *Ecology*, *90*(12): 3406–3417.

Williams, K.S. & Gilbert, L.E. (1981). Insects as selective agents on plant vegetative morphology: Egg mimicry reduces egg laying by butterflies. *Science*, *212*(4493): 467–469.

Zhang, B., Huai, L.I.U., Helen, H.S. & Wang, J.J. (2011). Effect of host plants on development, fecundity and enzyme activity of *Spodoptera exigua* (Hübner) (Lepidoptera: Noctuidae). *Agricultural Sciences in China*, *10*(8): 1232–1240.

16

Trophic Eggs

Richard Evans and Michael J. Grodowitz
USDA-ARS National Biological Control Laboratory, 59 Lee Road, Stoneville, MS38776

CONTENTS

16.1 Introduction

Trophic eggs have been reported from numerous species of invertebrates and vertebrates and have been proposed to be an alternative to egg cannibalism. It is considered a type of maternal care, and a source of nutrition for siblings and/or nest mates in both sub-social and eusocial species. Depending on the species, trophic eggs have been described as morphologically different from viable eggs in shape and size, internally lacking micropyles and embryonic development. Trophic eggs have been variously defined as inviable, degenerate in form and/or fed to colony members (Holldobler & Wilson 1990). Trophic eggs are produced by the ovaries, similar to fertile eggs, but are not viable and are normally consumed (Crespi 1992). Trophic eggs have also been called 'nurse eggs'; in this case they do not develop and are consumed by offspring (Perry & Roitberg 2006). Since these definitions were published, the term 'trophic egg' has been expanded to include uses other than consumption, i.e., a type of extended maternal care that provides protection from predation and parasitism. This chapter reviews the literature concerning the development of trophic eggs across a wide spectrum of different organisms, mainly concentrating on insect species. While previous authors have attempted to define trophic egg production in the context of the organisms or groups of organisms under study, trophic egg production is much broader than previously noted.

16.2 Trophic Eggs and Maternal Care in Sub-Social Species

In many species, maternal care ends after oviposition, but some species have extended maternal care to include host food provisioning, supplying trophic eggs for consumption as well as egg and larvae protection from predation. In the semelparous Hemipteran, Family Acanthosomatidae, *Elasmucha grisea* (L.), and the Family Pentatomidae, *Ramosiana insignis* (Blanchard), females provide extended maternal care by selecting plant oviposition sites with lower predation risks and actively guarding the eggs and offspring from predation (Mappes & Kaitala 1995, Lopez-Ortega & Williams 2018). There are no reports of trophic egg production in either species. In a seminivorous Hemipteran, family Cydnidae, *Adomerus triguttulus* (Motschulsky) females provision host seeds and provide trophic eggs

DOI: 10.1201/9781003043195-16

for the young to feed on (Nakahira & Kudo 2008). In another Cydnidae, *Canthophorus niveimarginatus* (Scott), females guard the clutch, provision seeds and produce pre-hatch and post-hatch trophic eggs for the young to eat (Filippi *et al.*, 2009). In the Family Parastrachiidae, another Hemipteran, *Parastrachia japonensis* (Scott) must vertically transmit the obligate endosymbiont *Candidatus Benitsuchiphilustojoi* to her offspring. Just prior to hatching, the female spreads a white mucus from her anus containing the symbiont on trophic eggs, which are consumed by the nymphs (Hosokawa *et al.,* 2012). In Araneae and the Family Amaurobiidae, the species *Amaurobius ferox* (Walkckenaer) females exhibit maternal care in the form of prey provisioning, regurgitating fluids extracted from prey, allowing the offspring to consume her haemolymph, providing trophic eggs and then ultimately sacrificing herself to matriphagy (Kim & Roland 2000). In Orthoptera and the Family Gryllidae, the species *Anurogryllus muticus* (Saussure), mothers dig and provision a brood chamber (Figure 16.1), provide inviable miniature trophic eggs which are eaten by the nymphs, guard the chamber and upon death are eaten by the offspring (West & Alexander 1963). In the Order Coleoptera, Family Passalidae, *Cylindrocaulus patalis* (Lewis) third instar larvae have been observed soliciting trophic eggs from adult females (Ento *et al.,* 2008). In the Family Coccinellidae, the aphidophagous species *Harmonia axyridis* (Pallas) is reported to produce trophic eggs (Noriyuki *et al.,* 2012) which are consumed by larvae (Perry & Roitberg 2005). In the Family Chrysomelidae, another species uses trophic eggs as a protective barrier for the viable eggs. *Mimosestes amicus* (Horn) females cover viable eggs with a layer of eggs that protect the lower from parasitization by the hymenopteran parasitoid in the Family Trichogrammatidae, *Uscana simifumipennis* (Girault) (Deas & Hunter 2012). In the Order Neuroptera and the family Ascalaphidae, the females possess dimorphic ovaries (Figure 16.2) adapted to produce highly differentiated eggs called repagula (New 1971). The females of *Ululodes mexicanus* (McLachlan) oviposit viable eggs on host plants near the apical end of branches and then encircle the branch proximal with rod-like barriers called repagula (Figure 16.3). The repagula or 'little barriers' protect the eggs and larvae from predation (Henry 1972). In these cases, trophic eggs are a considered a type of maternal care. The evolution of trophic eggs is most likely an adaptation to reduce cannibalism of viable eggs (Crespi 1992). If offspring

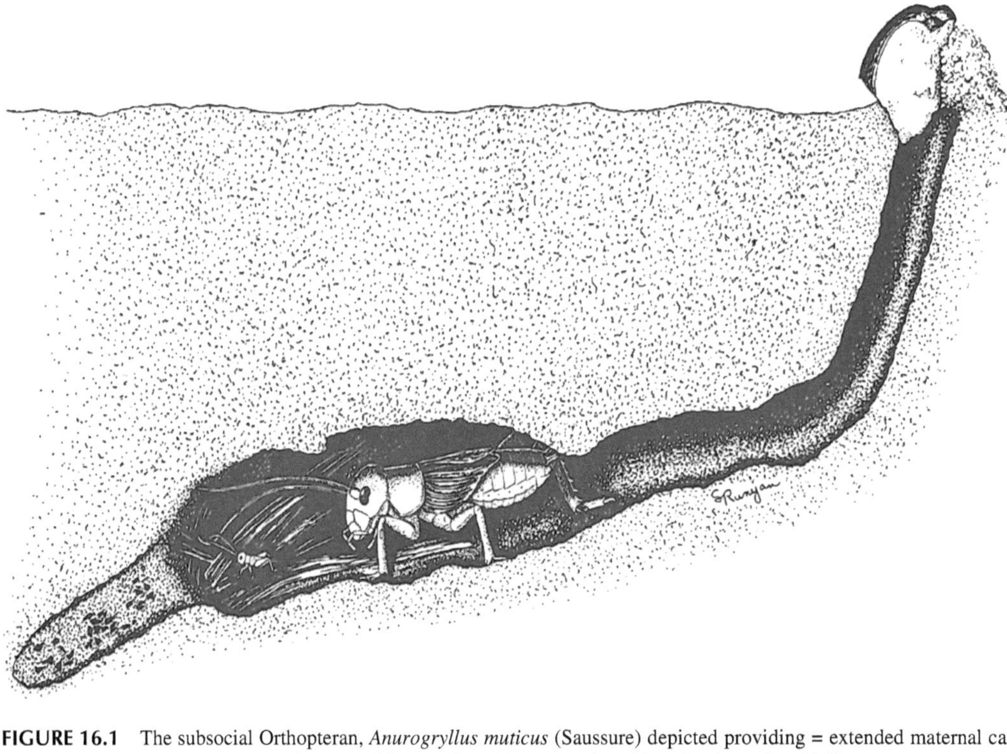

FIGURE 16.1 The subsocial Orthopteran, *Anurogryllus muticus* (Saussure) depicted providing = extended maternal care with burrow, protection of young, food provision, and trophic eggs (after West & Alexander, 1963).

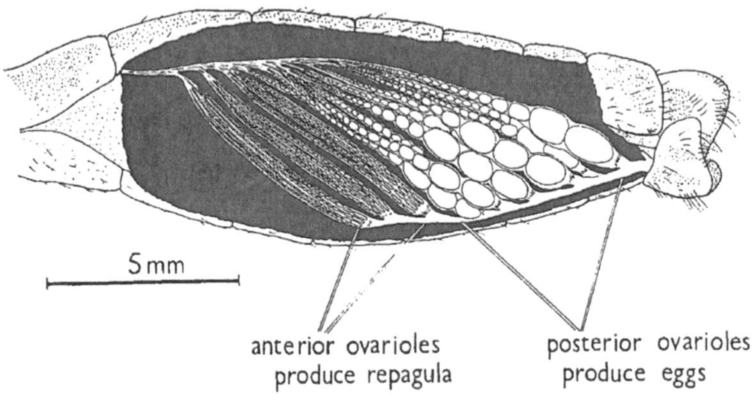

FIGURE 16.2 Dimorphic ovaries of an Ascalaphidae for producing reproductive eggs and repagula (after New, 1971).

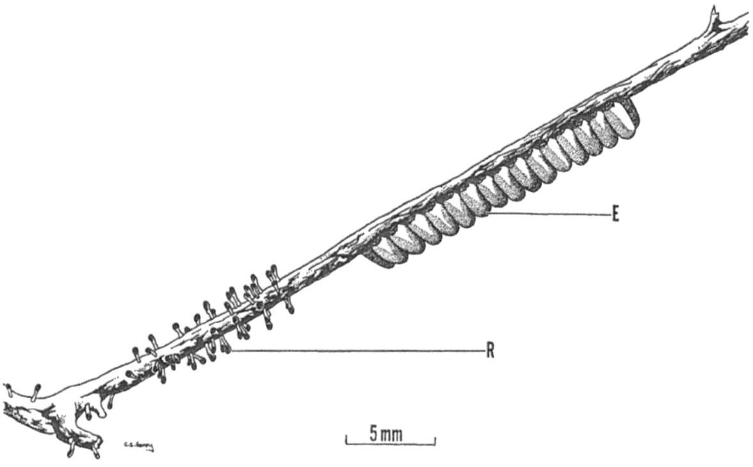

FIGURE 16.3 Repagula (R) proximal to apical reproductive eggs (E) of *Ululodes mexicanus* (after Henry, 1972).

are engaging in sibling oophagy, it may benefit the mother to produce unviable trophic eggs for consumption (Crespi 1992).

16.3 Trophic Eggs and Cannibalism in Sub-Social Species

The evolution of trophic eggs is theorized to reduce cannibalism of reproductive eggs (Crespi 1992). Cannibalism is widespread in the animal kingdom among invertebrates and vertebrates. Cannibalism occurs in various forms: 1) conspecific cannibalism: the consumption of the same species, 2) sibling cannibalism: the consumption of a sibling, or 3) filial cannibalism: the consumption of young by one or both parents. One theory supporting cannibalism resulted in the evolution of trophic eggs (Crespi 1992). Adelphophagy or intrauterine cannibalism, a type of sibling cannibalism, occurs in some ovoviviparous species of sharks. This can be traced back to the Lower Carboniferous period in a fossil of the now extinct Holocephalan fish species, *Delphyodontos dacriformes*, from Montana, USA (Lund 1980). The extant, Condrichtheys in the Order Lamniformes and Family Odontaspididae, species *Carcharias taurus* (Raphinesque) and family Allopiidae, species *Alopias superciliosus* (Lowe) both exhibit cannibalistic oviphagous behaviour.

Filial cannibalism occurs when parents kill and consume their own young (Bartlett, 1987). Coleopterans in the family Silphidae in the Genus *Nicrophorus* exhibit maternal and/or parental care.

Nicrophorous males search for small vertebrate carcasses after dark using extremely sensitive chemoreceptors in the antennae. When a suitable size carcass is found the male produces pheromones to attract a mate and they proceed to bury the carcass, strip away fur or feathers, and lay a clutch of eggs nearby. After hatching the larvae feed on the carcass. In the species *Nicrophorus orbicollis* (Say), the female does not adjust the size of the clutch to corpse size and if the carcass is too small to rear all offspring both parents are known to commit filial cannibalism (Bartlett, 1987). In other species, the production of trophic eggs should reduce sibling intrauterine cannibalism and conspecific oophagy (Crespi, 1992). In the vertebrate Class Teleostei and the Order Suluriformes and family Bagridae, the catfish species *Bagrus meridionalis* (Gunther) is reported to produce trophic eggs (McKaye, 1986). Some amphibians are reported to produce trophic eggs as in the Family Leptodactylidae and species *Leptodactylus labyrinthicus* (Spix). The genus *Leptodactylus* is reported to produce eggs for consumption embedded inside foam nests inside burrows in the ground (Prado *et al.*, 2005). In the species *L. fallax* (Muller) it has been reported that tadpoles displayed obligatory oophagy, feeding only on trophic eggs (Prado *et al.*, 2005). The young of the vampire tree frog, *Rhacophorus vampyrus* (Rowley, Le, Thi & Hoanghave) evolved a specialized oral disc armed with a pair of hooks on the lower jaw which are reported to be an adaptation for oophagy (Vassilieva *et al.*, 2013).

16.4 Trophic Eggs and Maternal Care in Eusocial Species

The term eusocial is derived from the Greek loosely meaning, 'to associate', and is best represented by species of hexapods in the Orders Hymenoptera and Blattodea by certain traits. These traits include cooperation in caring for offspring, a division of reproduction, labor, and overlapping generations or life stages that contribute to the colony's success (Wilson 1971). Eusocial insects exhibit a high degree of maternal care and utilize trophic eggs in such a manner not reported in most hexapod arthropods. In the Family Apidae there is a group of mostly tropical stingless bees in the Subfamily Apinae and Tribe Meliponini. Studies have shown the queen of some species in this group may have a stimulatory effect on ovarian development as well as fertile and trophic egg production (Cruz-Landim, 2000). It is reported that two female castes occur in this subfamily, one fertile and the other sterile. Ovarian dimorphism is reported in queens and workers of these highly eusocial bees (Cruz-Landim, 2000). *Apis mellifera* (L.) exhibits expressed dimorphism in number of ovarioles per ovary (Cruz-Landim 2000). In the species of sting-less bee *Tetragonisca angustula* (Illiger), eggs of workers are placed inside the walls of brood cells and normally consumed by the queen and therefore considered trophic eggs (Koedam *et al.*, 1996). This oviposition behaviour is also recorded as aggregations of workers in 'rosette' formations around brood cells, with a production of 35% mature chorionated eggs (Koedam *et al.*, 1996). Other Apidae in the genus *Melipona* produce trophic eggs that are reported to be morphologically different than viable eggs. In *Melipona asilvai* (Moure) queens and reproductive workers produce eggs that are longer, and the anterior region is wider than trophic eggs produced by workers (Aparecido Pereira *et al.*, 2006). In *M. scutellaris* (Latreille) and *M. compressipes* (Fabricius), trophic eggs are shorter than eggs produced by the queen but showed no significant anterior difference (Aparecido Pereira *et al.*, 2006). In *M. bicolor* (Lepeletier), a facultatively polygynous species, only reproductive workers consumed reproductive eggs from either queen or worker, but the queens consumed trophic eggs (Koedam *et al.*, 2007). In *Plebeia remota* (Holmberg) workers offer trophic eggs to the queen while retreating backwards and laying an egg on the comb which is consumed by the queen or other workers (Van Benthem *et al.*, 1995). Hymenopterans in the family Formicidae exhibit a high level of eusocial behavior and in his book, The *Insect Societies*, E. O. Wilson (1971) regarded them as the premier social insects because of their wide distribution and abundance (Wilson 1971). In a primitive monandrous or polydomous crytobiotic ant, *Prionopelta amabilis* (Borgmeier), it is reported that workers have well-developed ovaries and oviposit trophic eggs which are fed to the queen (Holldobler and Wilson 1986). Virgin workers of the ponerine species *Gnamptogenys menadensis* (Mayr) lay trophic eggs with different micropyles and membrane ultrastructure, which are fed to larvae, and workers can lay reproductive haploid eggs (Gobin *et al.*, 1999). In another ponerine ant, *Pachycondyla apicalis* (Smith), queens have an influence on the production of reproductive and trophic eggs. The workers of this species lay trophic eggs which are fed to the queen and reproductive haploid eggs in the queen's absence. It is believed that the queen produces pheromones that signal her presence which influences worker egg production (Dietemann and Peeters 2000). In *Paraponera clavata* (Fabricius), the first workers forage and produce trophic eggs which are fed to larvae (Peeters 2017). In the sub-family Ectatomminae, workers of *Ectatomma tuberculatum* (Olivier) produce trophic eggs which are fed to queen and larvae (Azevedo *et al.*, 2011). Similar to reports of queen influence on trophic egg production by workers in *P. apicalis*, it has been reported that *Myrmecia gulosa* (Fabricius) (Figure 16.4) reproductive and non-reproductive workers are influenced by queen cuticular hydrocarbons (Dietemann *et al.*, 2003). In the species *Acanthomyrmex ferox* (Emery), queens produce both reproductive and trophic eggs but workers only produce trophic eggs; both are used to redistribute nutrients (Gobin and Ito 2000). A reproductive division of labor occurs in *Crematogaster minutissima smithi* (Creighton) where a caste of large workers with different morphologies and number of ovarioles from the queen are specialized to produce unfertilized eggs eaten by the queen and larvae. In queenless colonies, these eggs may develop into males (Heinze *et al.*, 1995). In *Basiceros manni* (Brown & Kempf), the youngest workers have the most developed ovaries and are reported to be associated with the production of trophic eggs for consumption (Wilson & Holldobler 1986). In a highly invasive species of fire ants, *Solenopsis invicta* (Buren), the virgin queens are reported to oviposit two types of eggs, reproductive and trophic. These eggs are eaten by the queen and then fed to larvae by trophallaxis (Cassill 2002). This species is known to establish by claustral colony founding. Alate virgin queens leave the mound on nuptial flights to mate. After mating the queens land, shed their wings and dig a chamber in the ground. Until they can rear the first workers they must live on the reabsorbed wing muscle tissue, stored body reserves and trophic eggs (Cassill 2002). In *Aphaenogster smythiesii japonica* (Forel) workers with functional ovaries lay trophic eggs when the queen is present but produce viable eggs in her absence and may be attacked by other workers and believed to be a type of reproductive control (Iwanishi *et al.*, 2003). In the presence of the queen, *Aphaenogaster cockerelli* (Andre) workers produce trophic eggs and, since trophallaxis does not appear to occur, this may be the major source of nutrition for the queen (Holldobler and Carlin 1989). In *Myrmica rubra* (L.) workers produce two types of eggs

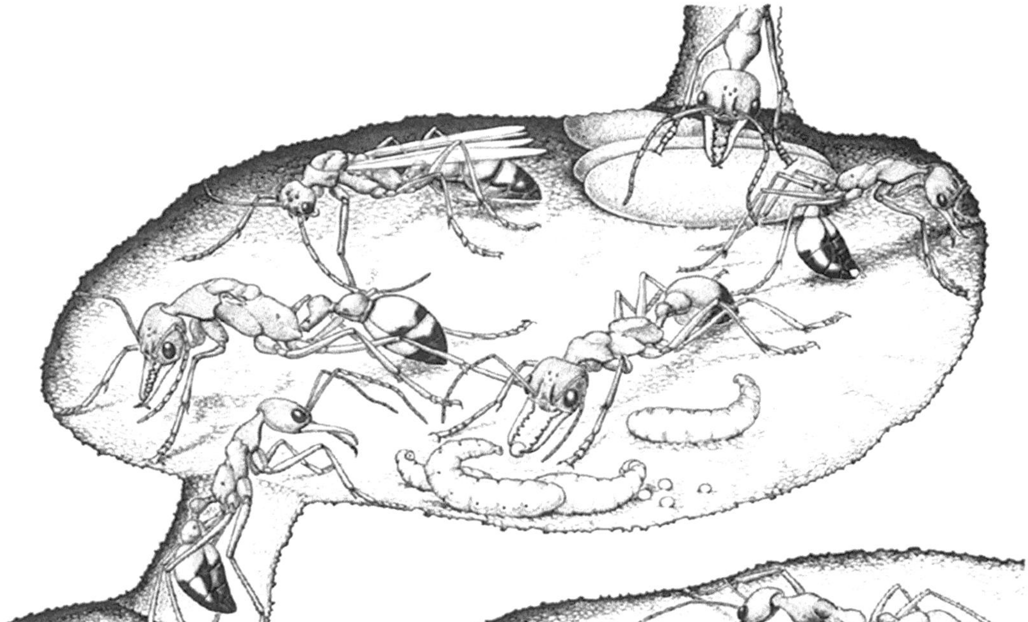

FIGURE 16.4 *Myrmecia gulosa* queen and alate male on left, on right a worker laying a trophic egg and another worker feeding a trophic egg to larvae (after Wilson, 1971).

depending on the presence of the queen. In the absence of the queen, workers produce large reproductive D-shaped eggs, which develop into male larvae. In the presence of the queen, workers produce O-shaped trophic eggs which are fed to larvae and the queen (Brian & Rigby 1978). In *Myrmica ruginodes* (Nylander), *M. schencki* (Viereck) and *M. sulcinodis* (Nylander), queens lay both trophic and reproductive eggs with or without workers (Wardlaw and Elmes 1995). In *Acromyrmex octospinosus* (Reich) and *A. octospinosus echinatior* (Forel) workers produce trophic eggs as a source of food for larvae, similar in size to the queen's eggs but with the amount of yolk reduced (Dijkstra *et al.*, 2005). Leafcutter female ants possess paired telotrophic ovaries that may each contain several hundred ovarioles. In *Atta cephalotes* (L.) and *A. sexdens* (L.), small to medium workers form a retinue around the queen and lay high numbers of trophic eggs for her consumption (Dijkstra *et al.*, 2005). Leafcutter ants are known to cultivate a fungus as a source of nutrition for the colony. A newly mated queen will transfer a small portion of the fungus (inoculum) in her mouth to start a new colony. The queen will then lay reproductive eggs that will hatch into larvae and trophic eggs for her and her larvae to consume. In *Atta texana* (Buckley) the queen lays a third type of egg (Figure 16.5) that is trophic for fungal nutrient consumption (Fang and Mueller 2019). Some species of Myrmicinae do not redistribute nutrient resources via trophallaxis but produce trophic eggs for this purpose. In *Pogonomyrmex barbatus* (Smith), queens produce trophic eggs during colony founding. Both trophic and reproductive eggs may be consumed (Volny, Greene & Gordon 2006). *P. badius* (Latreille) lack trophallaxis and contain a polymorphic worker class able to produce haploid eggs without a queen and assumed to produce trophic eggs for consumption (Smith *et al.*, 2007). In the Dolichoderinae, *Technomyrmex albipes* (Smith) creates polydomous colonies where winged reproductives disperse and inbred wingless reproductives increase colonies via budding. Trophallaxis does not occur and

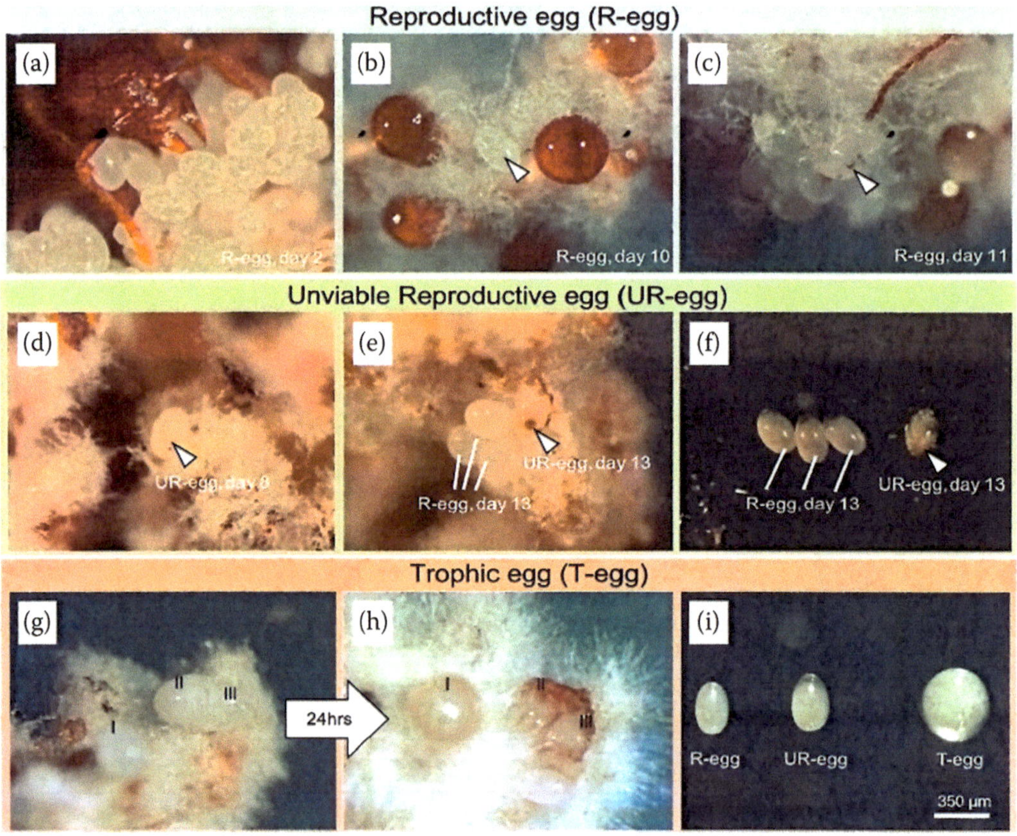

FIGURE 16.5 Differences between reproductive and trophic eggs for consumption by brood and fungal garden in *Atta texana* (after Fang & Mueller, 2019).

female dealates, intercasts and workers appear to produce trophic eggs (Yamauchi *et al.*, 1991). In the Formicinae *Oecophylla longinoda* (Latreille) and *O. smaragdina* (Fabricius), major workers are attracted to the queen which produces a chemical stimulus inducing regurgitation or trophic egg production while suppressing production of viable eggs (Holldobler & Wilson 1983). In *Anoplolepis gracilipes* (Smith) physogastric workers produce trophic eggs for queen and larvae consumption and haploid eggs which develop into males (Lee *et al.*, 2017).

The adaptive and evolutionary significance of trophic egg production has not been studied to any great extent. Probably the first author to address such concerns is Crespi (1992) where he suggested that the development of trophic eggs arose from cannibalism by larva of both viable eggs and those that failed to develop properly. He put forth that the transition from a competitive stance to one of cooperation provides evidence of the evolutionary development of the use of trophic eggs. He explains that under the two forms of sociality (i.e., subsociality and eusociality) that the dependence of the immature on the parents as well as between the queen, workers and offspring most likely aided in trophic egg development. The other important publication that attempts to define experimental procedures to document the evolutionary development of trophic eggs is Perry and Roitberg (2006). In this publication, the authors develop a framework in which to document the adaptive and evolutionary significance of trophic egg use including the quantity of eggs produced, specialized cellular development, the cost associated with trophic egg production in terms of nutritional composition, their distribution, and the plasticity involved in laying eggs such as adjusting numbers based on prevailing environmental conditions. These authors go further and suggest testing the trophic egg production hypothesis using comparative predictions and byproducts associated with trophic egg production. More research is warranted.

16.5 Conclusions

Trophic eggs have been shown to be ovarian derived structures produced for reproduction of male offspring, reduction of sibling cannibalism, egg protection from predation and parasitism, nutrient redistribution, a nutrient source for cluastral founding, a surrogate for trophallaxis and an alternative for filial cannibalism. In this review we discuss over 100 cases of trophic egg production in both vertebrates and invertebrates with a special emphasis on insect species. Trophic egg production has been noted in several vertebrate species including amphibians, catfish and sharks and found in a wide variety of insect species from the Orders Orthoptera, Hemiptera, Neuroptera, Coleoptera, and Hymenoptera. As noted previously, the production of trophic eggs and associated uses vary widely and express a diversity that has not often been captured in previous definitions. Undoubtedly, with additional research and observation the development and use of trophic eggs will be demonstrated in other insect species.

For a detail listing of all references used in this chapter with corresponding summaries see Table 16.1.

Acknowledgements

We would like to thank Drs. Nathan Harms and Jian Chen for critical review of this manuscript. Mention of trade names or commercial products in this publication is solely for the purpose of providing specific information and does not imply recommendation or endorsement by the U. S. Department of Agriculture.

TABLE 16.1

Literature examined with supplemental information including family, subfamily, and year

Family	Subfamily	Species	Citation
			Perry and Roitberg (2006)
			Perry (2004)
			Perry and Roitberg (2005)
			Bell and Bohm (1975)
			Hölldobler and Wilson (1990)
			Wilson (1971)
			Crespi (1992)
		Delphyodontos dacriformes (extinct)	Lund (1980)
Odontaspididae		***Carcharias taurus (Odonta spistaurus)***	Wourms (1977); Springer (1948); Gilmore *et al.* (1983)
Alopiidae		***Alopias superciliosus (Alopias profundus)***	Gruber and Compagn (1981)
Bagridae		***Bagrus meridionalis***	McKaye (1986)
Leptodactylidae		***Leptodactylus labyrinthicus***	Prado *et al.* (2005)
Leptodactylidae		***Leptodactylus fallax***	Prado *et al.* (2005)
Rhacophoridae	Rhacophorinae	***Rhacophorus vampyrus***	Vassilieva *et al.* (2013)
Rhacophoridae	Rhacophorinae	***Kurixalus eiffingeri***	Kishimoto and Hayashi (2017)
Dendrobatidae	Dendrobatinae	***Oophaga pumilio***	Dugas *et al.* (2013)
Spionidae		***Boccardia proboscidea (Polydora proboscidae)***	Smith and Gibson (1999)
Buccinidae		***Lirabuccinum dirum (Searlesia dira)***	Rivest (1983)
Helicidae		***Arianta arbustorum***	
Agelenidae		***Coelotes terrestris***	
Thomisidae		***Diaea ergandros***	Unglaub *et al.* (2013)
Amaurobiidae		***Amaurobius ferox***	Kim and Roland (2000)
Gryllidae	Gryllinae	***Anurogryllus muticus***	West and Alexander (1963)
Archipsocidae	Archipsocinae	***Archipsocus***	
Cydnidae		***Admerus triguttulus***	(Kudo *et al.* (2006); Kudo and Nakahira (2004, 2005, 2008)
Cydnidae		***Canthophorus niveimarginatus***	Filippi *et al.* (2009); Baba *et al.* (2011)
Parastrachiidae		***Parastrachia japonensis***	Hosokawa *et al.* (2012); Hironaka *et al.* (2005); Filippi *et al.* (2012)
Acanthosomatidae		***Elasmucha grisea***	Tallamy and Brown (1999)
Pentatomidae		***Ramosiana insignis***	Lopez-Ortega and Williams (2018)
Passalidae		***Cylindrocaulus patalis***	Ento *et al.* (2008)
Coccinellidae	Coccinellinae	***Harmonia axyridis***	Perry and Roitberg (2005); Noriyuki *et al.* (2012)
Coccinellidae	Coccinellinae	***Harmonia yedoensis***	Noriyuki *et al.* (2012); Noriyuki (2017); Noriyuki *et al.* (2016)
Coccinellidae		***Coccinella 11-pungtata***	Hawkes and Marriner (1927)
Coccinellidae		***Sospita oblongoguttata***	Noriyuki (2017)
Silphidae		***Nicrophorous orbicollis***	Bartlett (1987)
Chrysomelidae	Bruchinae	***Mimosestes amicus***	Deas and Hunter (2012)
Ascalaphidae	Ascalaphinae	***Ululodes*sp.**	McClendon (1902); New (1971)
Ascalaphidae	Haplogleniinae	***Ascalobyas (Byas)* sp.**	New (1971)
Ascalaphidae	Haplogleniinae	***Amoea (Episperches)*sp.**	New (1971)

TABLE 16.1 (Continued)

Literature examined with supplemental information including family, subfamily, and year

Family	Subfamily	Species	Citation
Ascalaphidae	Ascalaphinae	***Ululodes mexicanus (Ululode smexicana)***	Henry (1972); Henry (1977); Aldrich and Zhang (2016)
Ascalaphidae	Haplogleniinae	***Ascaloptynx appendiculata (Ascaloptynx appendiculatus)***	Henry (1972); Henry (1978)
Apidae	Apinae	***Tetragonisca angustula (Trigona angustula)***	Koedam *et al.* (1996)
Apidae	Apinae	***Apis mellifera***	Cruz-Landim (2000)
Apidae	Apinae	***Melipona quadrifasciata (Melipona quadrifasciata anthidiodes)***	Cruz-Landim (2000)
Apidae	Apinae	***Scaptotrigona postica***	Cruz-Landim (2000)
Apidae	Apinae	***Leurotrigona muelleri***	Cruz-Landim (2000)
Apidae	Apinae	***Frieseomelitta silvestrii (Frieseomellita silvestri)***	Cruz-Landim (2000)
Apidae	Apinae	***Trigona spinipes***	Cruz-Landim (2000)
Apidae	Apinae	***Melipona scutellaris***	Aparecido Pereira *et al.* (2006)
Apidae	Apinae	***Melipona compressipes (Melipona compressipes fasciculata)***	Aparecido Pereira *et al.* (2006)
Apidae	Apinae	***Melipona asilvai***	Aparecido Pereira *et al.* (2006)
Apidae	Apinae	***Melipona bicolor***	Koedam *et al.* (2007)
Apidae	Apinae	***Plebeia remota***	Van Benthem *et al.* (1995)
Formicidae	Amblyoponinae	***Prionopelta amabilis***	Hölldobler and Wilson (1986)
Formicidae	Ponerinae	***Neoponera apicalis (Pachycondyla apicalis)***	Dietemann and Peeters (2000)
Formicidae	Ponerinae	***Gnamptogenys menadensis***	Gobin *et al.* (1998); Gobin *et al.* (1999)
Formicidae	Ponerinae	***Rhytidoponera*sp.**	Peeters (1987)
Formicidae	Ponerinae	***Paraponera clavata***	Peeters (2017)
Formicidae	Ectatomminae	***Ectatomma tuberculatum***	Azevedo *et al.* (2011, 2016)
Formicidae	Myrmicinae	***Myrmecia forceps***	Freeland (1958)
Formicidae	Myrmicinae	***Myrmecia gulosa***	Dietemann *et al.* (2003)
Formicidae	Myrmicinae	***Acanthomyrmex ferox***	Gobin and Ito (2000)
Formicidae	Myrmicinae	***Crematogaster minutissimasmithi* (subspecies)**	Heinze *et al.* (1995); Heinze *et al.* (1999)
Formicidae	Myrmicinae	***Basiceros manni***	Wilson and Hölldobler (1986)
Formicidae	Myrmicinae	***Solenopsis invicta***	Cassill (2002)
Formicidae	Myrmicinae	***Aphaenogaster smythiesii japonica* (subspecies)**	Iwanishi *et al.* (2003)
Formicidae	Myrmicinae	***Aphaenogaster cockerelli (Novomessor cockerelli)***	Hölldobler and Carlin (1989)
Formicidae	Myrmicinae	***Myrmica rubra***	Brian and Rigby (1978)
Formicidae	Myrmicinae	***Myrmica rubra***	Wardlaw and Elmes (1995)
Formicidae	Myrmicinae	***Myrmica ruginodis***	Wardlaw and Elmes (1995)
Formicidae	Myrmicinae	***Myrmica schencki***	Wardlaw and Elmes (1995)
Formicidae	Myrmicinae	***Myrmica sulcinodis***	Wardlaw and Elmes (1995)
Formicidae	Myrmicinae	***Acromyrmexocto spinosus***	Djkstra *et al.* (2005)
Formicidae	Myrmicinae	***Acromyrmexocto spinosus echinatior* (subspecies)**	Djkstra *et al.* (2005)
Formicidae	Myrmicinae	***Atta cephalotes***	Djkstra *et al.* (2005)

(Continued)

TABLE 16.1 (Continued)

Literature examined with supplemental information including family, subfamily, and year

Family	Subfamily	Species	Citation
Formicidae	Myrmicinae	***Atta sexdens***	Djkstra *et al.* (2005);
Formicidae	Myrmicinae	***Atta columbica***	
Formicidae	Myrmicinae	***Atta texana***	Fang and Mueller (2019)
Formicidae	Myrmicinae	***Pogonomyrmex*sp.**	Cahan *et al.* (2011)
Formicidae	Myrmicinae	***Pogonomyrmex barbatus***	Corona *et al.* (2013); Volny *et al.* (2006)
Formicidae	Myrmicinae	***Pogonomyrmex badius***	Smith *et al.* (2007)
Formicidae	Dolichoderinae	***Technomyrmex albipes***	Yamauchi *et al.* (1991)
Formicidae	Formicinae	***Oecophylla longinoda***	Hölldobler and Wilson (1983)
Formicidae	Formicinae	***Oecophyllas maragdina***	Hölldobler and Wilson (1983)
Formicidae	Formicinae	***Anoplolepis gracilipes***	Lee *et al.* (2017); Schuh and Slater (1995)

REFERENCES

Aldrich, J.R. and Zhang, Q.H., 2016. Chemical ecology of Neuroptera. *Annual Review of Entomology*, *61*, pp. 197–218.

Aparecido Pereira, R., Manfrini Morais, M., Domingos Gioli, L., Santos Nascimento, F., Rossi, M.A. and Rolandi Bego, L., 2006. Comparative morphology of reproductive and trophic eggs in *Melipona* bees (Apidae, Meliponini). *Brazilian Journal of Morphological Sciences*, *23*(3–4), pp. 349–354.

Azevedo, D.O., de Paula, S.O., Zanuncio, J.C., Martinez, L.C. and Serrão, J.E., 2016. Juvenile hormone downregulates vitellogenin production in *Ectatomma tuberculatum* (Hymenoptera: Formicidae) sterile workers. *Journal of Experimental Biology*, *219*(1), pp. 103–108.

Azevedo, D.O., Zanuncio, J.C., Delabie, J.H.C. and Serrao, J.E., 2011. Temporal variation of vitellogenin synthesis in *Ectatomma tuberculatum* (Formicidae: Ectatomminae) workers. *Journal of Insect Physiology*, *57*(7), pp. 972–977.

Baba, N., Hironaka, M., Hosokawa, T., Mukai, H., Nomakuchi, S. and Ueno, T., 2011. Trophic eggs compensate for poor offspring feeding capacity in a subsocial burrower bug. *Biology Letters*, *7*(2), pp. 194–196.

Bartlett, J., 1987. Filial cannibalism in burying beetles. *Behavioral Ecology and Sociobiology*, *21*(3), pp. 179–183.

Bell, W.J. and Bohm, M.K., 1975. Oosorption in insects. *Biological Reviews*, *50*(4), pp. 373–396.

Brian, M.V. and Rigby, C., 1978. The trophic eggs of *Myrmica rubra* L. *Insectes Sociaux*, *25*(1), pp. 89–110.

Cahan, S.H., Graves, C.J. and Brent, C.S., 2011. Intergenerational effect of juvenile hormone on offspring in *Pogonomyrmex* harvester ants. *Journal of Comparative Physiology B*, *181*(8), pp. 991–999.

Cassill, D., 2002. Brood care strategies by newly mated monogyne *Solenopsis invicta* (Hymenoptera: Formicidae) queens during colony founding. *Annals of Entomological Society of America*, *95*(2), pp. 208–212.

Corona, M., Libbrecht, R., Wurm, Y., Riba-Grognuz, O., Studer, R.A. & Keller, L. (2013). Vitellogenin underwent subfunctionalization to acquire caste and behavioral specific expression in harvester ant *Pogonomyrmex barbatus*. *PLoS Genetics*, *9*(8), pp. e1003730.

Crespi, B.J., 1992. Cannibalism and trophic eggs in subsocial and eusocial insects. In Elgar, M.A. and B.J. Crespi (Eds.), *Cannibalism. Ecology and Evolution among Diverse Taxa.* Oxford Science Publications. pp. 172–212.

Cruz-Landim, C.D., 2000. Ovarian development in Meliponine bees (Hymenoptera: Apidae): Effect of queen presence and food on worker ovary development and egg production. *Genetics and Molecular Biology*, *23*(1), pp. 83–88.

Deas, J.B. and Hunter, M.S., 2012. Mothers modify eggs into shields to protect offspring from parasitism. *Proceedings of Royal Society of London B: Biological Sciences*, *279*(1730), pp. 847–853.
Dietemann, V. and Peeters, C., 2000. Queen influence on shift from trophic to reproductive eggs laid by workers of ponerine ant*Pachycondyla apicalis*. *Insectes Sociaux*, *47*(3), pp. 223–228.
Dietemann, V., Peeters, C., Liebig, J., Thivet, V. and Hölldobler, B., 2003. Cuticular hydrocarbons mediate discrimination of reproductives and nonreproductives in ant *Myrmecia gulosa*. *Proceedings of National Academy of Sciences of USA*, *100*(18), pp. 10341–10346.
Dijkstra, M.B., Nash, D.R. and Boomsma, J.J., 2005. Self-restraint and sterility in workers of *Acromyrmex* and *Atta* leafcutter ants. *Insectes Sociaux*, *52*(1), pp. 67–76.
Dugas, M.B., Yeager, J. and Richards-Zawacki, C.L., 2013. Carotenoid supplementation enhances reproductive success in captive strawberry poison frogs *(Oophagapumilio)*. *Zoo Biology*, *32*(6), pp. 655–658.
Ento, K., Araya, K. and Kudo, S.I., 2008. Trophic egg provisioning in a passalid beetle (Coleoptera). *European Journal of Entomology*, *105*(1), p. 99.
Fang, C.C. and Mueller, U.G., 2019. A new type of egg produced by foundress queens of *Atta texana* (Attini, Formicidae). *Insectes Sociaux*, *66*(4), pp. 601–609.
Filippi, L., Baba, N., Inadomi, K., Yanagi, T., Hironaka, M. and Nomakuchi, S., 2009. Pre-and post-hatch trophic egg production in subsocial burrower bug,*Canthophorusniveimarginatus* (Heteroptera: Cydnidae). *Naturwissenschaften*, *96*(2), pp. 201–211.
Filippi, L., Hironaka, M. and Nomakuchi, S., 2012. Individual variation in trophic egg production: Evidence for maternal manipulation in response to resource and competition levels. *Ethology*, *118*(5), pp. 503–510.
Freeland, J., 1958. Biological and social patterns in Australian bulldog ants of genus *Myrmecia*. *Australian Journal of Zoology*, *6*(1), pp. 1–18.
Gilmore, R.G., Dodrill, J. and Linley, P.A., 1983. The sand tiger shark, *Odontaspis taurus* (Rafinesque). *Fishery Bulletin*, *81*(2), p. 201.
Gobin, B. and Ito, F., 2000. Queens and major workers of *Acanthomyrmex ferox* redistribute nutrients with trophic eggs. *Naturwissenschaften*, *87*(7), pp. 323–326.
Gobin, B., Billen, J., & Peeters, C. (1999). Policing behaviour towards virgin egg layers in a polygynous ponerine ant. *Animal Behaviour*, *58*(5), pp. 1117–1122.
Gobin, B., Peeters, C. and Billen, J., 1998. Production of trophic eggs by virgin workers in ponerine ant *Gnamptogenys menadensis*. *Physiological Entomology*, *23*(4), pp. 329–336.
Gordh, G. and Headrick, D. Eds., 2001. *A Dictionary of Entomology*. CABI, UK, pp. 1032.
Gruber, S.H. and Compagn, L.J.V., 1981. The bigeye thresher, *Alopias superciliosus*. *Fishery Bulletin*, *79*(4), p. 617.
Hawkes, O.M. and Marriner, T.F., 1927. A preliminary account of life-history of *Coccinella 11-pungtata* (L.). *Transactions of Royal Entomological Society of London*, *75*(1), pp. 47–52.
Heinze, J., Cover, S.P. and Hölldobler, B., 1995. Neither worker, nor queen: An ant caste specialized in production of unfertilized eggs. *Psyche: A Journal of Entomology*, *102*(3–4), pp. 173–185.
Heinze, J., Foitzik, S., Oberstadt, B., Rueppell, O. and Hoelldobler, B., 1999. A female caste specialized for production of unfertilized eggs in ant *Crematogaster smithi*. *Naturwissenschaften*, *86*(2), pp. 93–95.
Henry, C.S., 1972. Eggs and rapagula of *Ululodes* and *Ascaloptynx* (Neuroptera: Ascalaphidae): A comparative study. *Psyche: A Journal of Entomology*, *79*(1–2), pp. 1–22.
Henry, C.S., 1977. The behavior and life histories of two North American ascalaphids. *Annals of Entomological Society of America*, *70*(2), pp. 179–195.
Henry, C.S., 1978. An evolutionary and geographical overview of Repagula (abortive eggs) in Ascalaphidae (Neuroptera). *Proceedings Entomological Society of Washington*, *80*(1), pp. 75–86
Hironaka, M., Nomakuchi, S., Iwakuma, S. and Filippi, L., 2005. Trophic egg production in a subsocial shield bug, *Parastrachia japonensis* Scott (Heteroptera: Parastrachiidae), and its functional value. *Ethology*, *111*(12), pp. 1089–1102.
Hölldobler, B. and Carlin, N.F., 1989. Colony founding, queen control and worker reproduction in ant *Aphaenogaster* (= *Novomessor) cockerelli* (Hymenoptera: Formicidae). *Psyche: A Journal of Entomology*, *96*(3–4), pp. 131–151.
Hölldobler, B. and Wilson, E.O., 1983. Queen control in colonies of weaver ants (Hymenoptera: Formicidae). *Annals of Entomological Society of America*, *76*(2), pp. 235–238.

Hölldobler, B. and Wilson, E.O., 1986. Ecology and Behavior of primitive cryptobiotic ant *Prionopelta amabilis* (Hymenoptera: Formicidae). *Insectes Sociaux*, *33*(1), pp. 45–58.
Hölldobler, B. and Wilson, E.O., 1990. *The Ants*. Harvard University Press, Cambridge, MA, p. 732.
Hosokawa, T., Hironaka, M., Mukai, H., Inadomi, K., Suzuki, N. and Fukatsu, T., 2012. Mothers never miss moment: A fine-tuned mechanism for vertical symbiont transmission in a subsocial insect. *Animal Behaviour*, *83*(1), pp. 293–300.
Iwanishi, S., Hasegawa, E. and Ohkawara, K., 2003. Worker oviposition and policing behaviour in myrmicine and *Aphaenogaster smythiesi japonica* Forel. *Animal Behaviour*, *66*(3), pp. 513–519.
Kim, K.W. and Roland, C., 2000. Trophic egg laying in spider, *Amaurobius ferox*: Mother–offspring interactions and functional value. *Behavioural Processes*, *50*(1), pp. 31–42.
Kishimoto, K. and Hayashi, F., 2017. The complete embryonic and larval stages of oophagous frog *Kurixalus eiffingeri* (Rhacophoridae). *Current Herpetology*, *36*(1), pp. 37–45.
Koedam, D., Aponte, O.I.C. and Imperatriz-Fonseca, V.L., 2007. Egg laying and oophagy by reproductive workers in polygynous stingless bee *Melipona bicolor* (Hymenoptera, Meliponini). *Apidologie*, *38*(1), pp. 55–66.
Koedam, D., Velthausz, P.V., Krift, T.V., Dohmen, M.R. and Sommeijer, M.J., 1996. Morphology of reproductive and trophic eggs and their controlled release by workers in *Trigona (Tetragonisca) angustula*Illiger (Apidae, Meliponinae). *Physiological Entomology*, *21*(4), pp. 289–296.
Kudo, S.I. and Nakahira, T., 2004. Effects of trophic-eggs on offspring performance and rivalry in a sub-social bug. *Oikos*, *107*(1), pp. 28–35.
Kudo, S.I. and Nakahira, T., 2005. Trophic-egg production in a subsocial bug: Adaptive plasticity in response to resource conditions. *Oikos*, *111*(3), pp. 459–464.
Kudo, S.I., Nakahira, T. and Saito, Y., 2006. Morphology of trophic eggs and ovarian dynamics in subsocial bug *Adomerus triguttulus* (Heteroptera: Cydnidae). *Canadian Journal of Zoology*, *84*(5), pp. 723–728.
Lee, C.C., Nakao, H., Tseng, S.P., Hsu, H.W., Lin, G.L., Tay, J.W., Billen, J., Ito, F., Lee, C.Y. and Lin, C.C., 2017. Worker reproduction of invasive yellow crazy ant *Anoplolepis gracilipes*. *Frontiers in Zoology*, *14*(1), p. 24.
Lopez-Ortega, M. and Williams, T., 2018. Natural enemy defense, provisioning and oviposition site selection as maternal strategies to enhance offspring survival in a sub-social bug. *PLoS One*, *13*(4), e0195665.
Lund, R., 1980. Viviparity and intrauterine feeding in a new holocephalan fish from lower Carboniferous of Montana. *Science*, *209*(4457), pp. 697–699.
McClendon, J.F., 1902. The life history of *Ulula hyalina* Latreille. *The American Naturalist*, *36*(426), pp. 421–429.
McKaye, K.R., 1986. Trophic eggs and parental foraging for young by catfish *Bagrus meridionalis* of Lake Malawi, Africa. *Oecologia*, 69(3), pp. 367–369.
Nakahira, T. and Kudo, S. I., 2008. Maternal care in burrower bug *Adomerus triguttulus*: Defensive behavior. *Journal of Insect Behavior*, *21*(4), pp. 306–316.
New, T.R., 1971. Ovariolar dimorphism and repagula formation in some South American Ascalaphidae (Neuroptera). *Journal of Entomology Series A, General Entomology*, *46*(1), pp. 73–77.
Noriyuki, S., 2017. Functional convergence and phenotypic divergence in two specialist species of pine-associated ladybirds. *Evolutionary Ecology*, *31*(6), pp. 885–898.
Noriyuki, S., Kawatsu, K. and Osawa, N., 2012. Factors promoting maternal trophic egg provisioning in non-eusocial animals. *Population Ecology*, *54*(3), pp. 455–465.
Noriyuki, S., Suzuki-Ohno, Y. and Takakura, K.I., 2016. Variation of clutch size and trophic egg proportion in a ladybird with and without male-killing bacterial infection. *Evolutionary Ecology*, *30*(6), pp. 1081–1095.
Peeters, C., 2017. Independent colony foundation in *Paraponera clavata* (Hymenoptera, Formicidae): First workers lay trophic eggs to feed queen's larvae. *Sociobiology*, *64*(4), pp. 417–422.
Peeters, C.P., 1987. The reproductive division of labour in queenless ponerine and *Rhytidoponera* sp. 12. *Insectes Sociaux*, *34* (2), pp. 75–86.
Perry, J.C. and Roitberg, B.D., 2006. Trophic egg laying: Hypotheses and tests. *Oikos*, *112*(3), pp. 706–714.
Perry, J.C. and Roitberg, B.D., 2005. Games among cannibals: Competition to cannibalize and parent-offspring conflict lead to increased sibling cannibalism. *Journal of Evolutionary Biology*, *18*(6), pp. 1523–1533.

Perry, J.C. and Roitberg, B.D., 2005. Ladybird mothers mitigate offspring starvation risk by laying trophic eggs. *Behavioral Ecology and Sociobiology*, *58*(6), pp. 578–586.

Perry, J.C., 2004. *The Behavioral Ecology of Trophic Egg-laying*. Doctoral dissertation, Science: Department of Biological Sciences.

Prado, C., Toledo, L.F., Zina, J. and Haddad, C.F., 2005. Trophic eggs in foam nests of *Leptodactyluslabyrinthicus* (Anura, Leptodactylidae): An experimental approach. *The Herpetological Journal*, *15*(4), pp. 279–284.

Rivest, B.R., 1983. Development and influence of nurse egg allotment on hatching size in *Searlesia dira* (Reeve, 1846) (Prosobranchia: Buccinidae). *Journal of Experimental Marine Biology and Ecology*, *69*(3), pp. 217–241.

Schuh, R.T. and Slater, J.A., 1995. *True Bugs of World (Hemiptera: Heteroptera): Classification and Natural History*. Cornell University press.

Smith, C.R., Schoenick, C., Anderson, K.E., Gadau, J. and Suarez, A.V., 2007. Potential and realized reproduction by different worker castes in queen-less and queen-right colonies of *Pogonomyrmex badius*. *Insectes Sociaux*, *54*(3), pp. 260–267.

Smith, H.L. and Gibson, G.D., 1999. Nurse egg origin in polychaete *Boccardia proboscidea* (Spionidae). *Invertebrate Reproduction & Development*, *35*(3), pp. 177–185.

Springer, S., 1948. Oviphagous embryos of sand shark, *Carcharias taurus*. *Copeia*, *1948*(3), pp. 153–157.

Tallamy, D.W. and Brown, W.P., 1999. Semelparity and evolution of maternal care in insects. *Animal Behaviour*, *57*, pp. 727–730.

Unglaub, B., Ruch, J., Herberstein, M.E. and Schneider, J.M., 2013. Hunted hunters? Effect of group size on predation risk and growth in Australian subsocial crab spider *Diaea ergandros*. *Behavioral Ecology and Sociobiology*, *67*(5), pp. 785–794.

Van Benthem, F.D.J., Imperatriz-Fonseca, V.L. and Velthuis, H.H.W., 1995. Biology of stingless bee *Plebeia remota* (Holmberg): Observations and evolutionary implications. *Insectes Sociaux*, *42*(1), pp. 71–87.

Vassilieva, A.B., Galoyan, E.A. and Poyarkov Jr, N.A., 2013. *Rhacophorus vampyrus* (Anura: Rhacophoridae) reproductive biology: A new type of oophagous tadpole in Asian Treefrogs. *Journal of Herpetology*, *47*(4), pp. 607–614.

Volny, V.P., Greene, M.J. and Gordon, D.M., 2006. Brood production and lineage discrimination in red harvester ant (*Pogonomyrmex barbatus*). *Ecology*, *87*(9), pp. 2194–2200.

Wardlaw, J.C. and Elmes, G.W., 1995. Trophic eggs laid by fertile *Myrmica* queens (Hymenoptera: Formicidae). *Insectes Sociaux*, *42*(3), pp. 303–308.

West, M. J. and Alexander, R. D., 1963. Subsocial behavior in a burrowing cricket *Anurogryllus muticus*(De Geer) Orthoptera: Gryllidae. *The Ohio Journal of Science*, *63*(1), pp. 19–24.

Wilson, E.O. and Hölldobler, B., 1986. Ecology and behavior of neotropical cryptobiotic ant*Basiceros manni* (Hymenoptera: Formicidae: Basicerotini). *InsectesSociaux*, *33*(1), pp. 70–84.

Wilson, E.O., 1971. The insect societies. *The Insect Societies* ,Belknap Press, Harvard, p. 548.

Wourms, J.P., 1977. Reproduction and development in chondrichthyan fishes. *American Zoologist*, *17*(2), pp. 379–410.

Yamauchi, K., Furukawa, T., Kinomura, K., Takamine, H. and Tsuji, K. (1991). Secondary polygyny by inbred wingless sexuals in dolichoderine ant *Technomyrmex albipes*. *Behavioral Ecology and Sociobiology*, *29*(5), pp. 313–319.

17

Parental Care

Joël Meunier[1], Maximilian Körner[2], and Jos Kramer[3]
[1]*Institut de Recherche sur la Biologie de l'Insecte (IRBI), UMR CNRS 7261, University of Tours, Tours, France*
[2]*Institute of Evolutionary Animal Ecology, University of Bayreuth, Bayreuth, Germany*
[3]*Department of Quantitative Biomedicine, University of Zurich, Winterthurerstrasse 190, 8057 Zurich, Switzerland*

CONTENTS

17.1 Introduction

Among the multiple strategies that males and females can adopt to ensure the proper development of their offspring, an important one is the expression of parental care. Following Smiseth *et al.* (2012), parent care is defined as 'any parental trait that enhances the fitness of a parent's offspring, and that is likely to have originated and/or is currently maintained for this function'. This broad definition of parental care encompasses behavioural (e.g. egg grooming) as well as non-behavioural (e.g. provisioning of gametes) traits that may be expressed both before and after oviposition, but excludes traits whose bearers are individuals other than the parents (e.g. adult siblings). In nature, parental care can take a broad diversity of forms, ranging from egg provisioning and nest construction over brood attendance to food provisioning (Clutton-Brock, 1991; Costa, 2006; Smiseth *et al.*, 2012; Wong *et al.*,

DOI: 10.1201/9781003043195-17

2013). The diverse forms of care can be provided by either the mother, the father or both parents; they can last from only few minutes to several years; and can range from mildly beneficial to essential for offspring survival (Klug *et al.*, 2012; Trumbo, 2012; Kramer & Meunier, 2019).

Regardless of its form, parental care often has a profound impacts on the fitness of both offspring and parents. On the one hand, parental care is beneficial to offspring, as it can improve their quality and increase their survival by neutralizing environmental hazards (Alonso-Alvarez & Velando, 2012; Klug & Bonsall, 2014). On the other hand, investing into parental care often comes with costs to parents. This is because its expression negatively impacts the condition and survival of the tending parents (due to an increased energy loss or elevated risk of predation), and thus reduces their lifetime reproductive success (Trivers, 1972; Alonso-Alvarez & Velando, 2012). Moreover, parental care can entail consequential costs to both parents and offspring that may arise from evolutionary conflicts of interest among the family members over the expression of parental care (Parker *et al.*, 2002; Royle *et al.*, 2004). Because of relatedness asymmetries among the family members, offspring often behave more selfishly than their parents desire, both by trying to manipulate their parents into increasing their parental investment (parent-offspring conflict; Trivers 1974; Kilner & Hinde 2012; Kölliker *et al.*, 2015), and by trying to monopolize limited parental resources at the expense of their (current or future) siblings (sibling rivalry; Mock & Parker 1997; Roulin & Dreiss 2012). For the same reason, one parent might often try to reduce its parental effort at the other parent's expense (sexual conflict; Trivers 1972; Lessells 2012). Investigating the conditions that allow parents to gain sufficient benefits to offset these costs of parental care (Hamilton, 1964; Smiseth *et al.*, 2012) has thus long been considered of key importance to improve our general understanding of its evolution as a reproductive strategy in animals (Clutton-Brock, 1991; Gross, 2005; Kramer & Meunier, 2019).

In the vast majority of cases, examples of parental care describe pre-ovipositional traits, with less than 5% of known genera exhibiting post-ovipositional care behaviours (Table 17.1). For instance, post-ovipositional parental care can be found in 78.9% and 21.7% of the Strepsiptera and Blattodea genera (not taking into account eusocial species), respectively, whereas it has never been reported in seven insect orders (Grylloblattidae, Mantophasmatodea, Mecoptera, Neuroptera, Phasmatodea, Siphonaptera and Zoraptera; Table 17.1; Machado & Trumbo 2018). These numbers, however, need to be taken with caution. Insect orders encompass very different numbers of genera and species, and the biology of a vast majority of these species remains poorly known (Foottit & Adler, 2018). This notwithstanding, the most common forms of post-ovipositional parental care in insects are internal egg carrying, as well as egg and offspring attendance (Machado & Trumbo, 2018).

In this book chapter, we aim to give an introduction to parental care as a reproductive strategy in insects by providing a broad overview of the different forms of parental care that can be found in insects. We will detail these different traits in the chronological order in which they take effect on offspring during their development (Figure 17.1). We begin with parental care before oviposition, when parents may select a site to lay their eggs or build a nest to protect the future clutch. Such pre-ovipositional care also includes females adjusting the quantity and nature of substances provisioned to the eggs to enhance the fitness of the resulting offspring. We then move to forms of parental care occurring between oviposition and egg hatching, when parents may guard and/or brood their eggs. We subsequently present forms of post-hatching parental care, when parents may tend, brood or food-provision their juveniles until they reach nutritional independence, and even later. Finally, we discuss the possibility for 'unconventional' forms of parental care and how it calls for further experimental studies exploring their occurrence in insects. Note that even if eusocial insects such as ants, termites and some bees and wasps exhibit extended family groups with multiple forms of care towards eggs and/or juveniles (Wilson, 1971; Kramer & Meunier, 2019), most of these forms of care are provided by adult offspring (workers). Such alloparental care does not fall under the definition of parental care introduced above. We therefore did not include care provided by individuals other than the parents in this chapter, and refer readers interested in the nature of family interactions in eusocial insects to the existing comprehensive reviews of this topic (Wilson, 1971; Bourke & Franks, 1995; Choe & Crespi, 1997). The few forms of care provided by the parents (mostly queens) of eusocial species to young offspring, i.e. parental care sensu stricto, are typically limited to egg provisioning and some forms of nest construction and egg care during colony foundation and fall within the framework outlined below.

TABLE 17.1

Taxonomic distribution of post-ovipositional parental care in insects

Order Name	Common Name	Known Species Number	Number of Genera	Number of Genera with Known Forms of Parental Care[a] (%)		Most Represented Form(s) of Parental Care among Genera[a]
Archaeognatha	Jumping bristletails	500	52	1	1.9%	Egg and offspring attendance
Blattodea	Cockroaches, termites	7,570	775	168[b]	21.7%	Internal egg carrying
Coleoptera	Beetles	390,000	4,722	331	7.0%	Egg and offspring attendance
Dermaptera	Earwigs	2,000	203	28	13.8%	Egg and offspring attendance
Diptera	Flies	160,000	3,175	676	21.3%	Internal egg carrying
Embioptera	Web spinners	450	90	6	6.7%	Offspring attendance
Ephemeroptera	Mayflies	3,328	441	3	0.7%	Internal egg carrying
Grylloblattodae	Ice and rock crawlers	32	5	0	0.0%	–
Hemiptera	True bugs	106,970	6,185	599	9.7%	Internal and external egg carrying/ guarding
Hymenoptera	Bees, ants, wasps, sawflies	115,000	1,579	85[b]	5.4%	Egg and offspring attendance
Lepidoptera	Moths, butterflies	174,250	16,298	23	0.1%	Internal egg carrying
Mantodea	Mantids	2,400	426	4	0.9%	Offspring attendance
Mantophasmatodea	Heelwalkers, gladiators	19	13	0	0.0%	–
Mecoptera	Scorpionflies	737	38	0	0.0%	–
Neuropteridae	Lacewings, owlflies, antlions	6,434	671	0	0.0%	–
Odonata	Dragonflies, damselflies	5,680	642	1	0.2%	Internal egg carrying
Orthoptera	Grasshoppers, crickets, katydids	28,000	2,687	6	0.2%	Egg and offspring attendance
Phasmatodea	Stick insects	3,000	473	0	0.0%	–
Phthiraptera	Lice	5,316	304	1	0.3%	Internal egg carrying
Plecoptera	Stoneflies	3,497	286	10	3.5%	Internal egg carrying
Psocoptera	Book and bark lice	5,500	576	3	0.5%	Egg attendance
Siphonaptera	Fleas	2,185	246	0	0.0%	–
Strepsiptera	Twisted wing parasitoids	630	57	45	78.9%	Internal egg carrying
Thysanoptera	Thrips	6,339	842	42[b]	5.0%	Egg and offspring attendance
Trichoptera	Caddisflies	14,500	380	1	0.3%	Internal egg carrying
Zoraptera	Zorapterans	50	1	0	0.0%	–
		TOTAL	41167	2033	4.9%	

[a] More details on forms of post-ovipositional care per genera can be found in Machado and Trumbo (2018).
[b] This number does not include eusocial species.
Data gathered from multiple sources (Wong *et al.*, 2013; Alcock 1975; Foottit & Adler 2018; Machado *et al.*, 2018).

(caption on next page)

17.2 Pre-oviposition Care

17.2.1 Site Selection for Oviposition

Females can influence the survival and phenotype of their offspring by selecting where to lay their eggs, a process called oviposition site selection (Refsnider & Janzen, 2010). Across species and taxa, oviposition site selection can be defined as a form of parental care if it (1) maximizes embryo survival, (2) induces adaptive modifications of offspring phenotype, and/or (3) offers proximity to suitable habitat for the resulting offspring. First, oviposition site selection can improve the chances of offspring survival by limiting the risk of predators finding the eggs, of the eggs drying out or of conspecifics competing with each other after the eggs have hatched. In the bean weevil *Callosobruchus maculatus*, for instance, females avoid laying eggs on beans that already contain conspecific eggs, and thus prevent high levels of within-bean competition that would otherwise increase the development time and mortality of their offspring (Mitchell, 1975). Conversely in the damselfly *Lestes macrostigmata*, females prefer to lay their eggs on plant species growing in the deeper parts of temporary ponds to minimize the risk of egg desiccation at the end of the drought season (Lambret *et al.*, 2018). Second, oviposition site selection may lead to adaptive modifications of offspring phenotypes. An example of such modification has been reported in the monarch butterfly *Danaus plexippus*, where mothers infected by the protozoan parasite *Ophryocystis elektroscirrha* lay their eggs on specific food plants (Milkweed species). The resulting offspring therefore only have access to this plant as a food source, and feeding on this plant has been shown to modify offspring physiology in a way that improves their resistance against this parasite (Lefèvre *et al.*, 2012). Finally, oviposition site selection can favour the proximity of future offspring to a suitable habitat such as one providing direct access to nutrients for growth and development. This is the case in many phytophagous insects, such as the lichenivorous moth *Cleorodes lichenaria* (Pöykkö, 2006). In this species, females prefer host lichens that ensure the shortest developmental period at the expense of higher growth rates at the beginning of the larval period. Similarly, females of the mosquitoes *Culiseta longiareolata* and *Anopheles punctipennis* avoid laying eggs in pools containing predators or competitors of their future larvae (Petranka & Fakhoury, 1991; Spencer *et al.*, 2002; Silberbush & Blaustein, 2011). Finally, this benefit may also involve the production of eggs on or within living animals to allow their future carnivorous offspring to feed on these hosts. Parasitoid wasps are textbook examples of this phenomenon. In these species, females typically evolved an ovipositor (up to 10 cm long, such as in *Megarhyssa macrurus*) to pierce the skin of the host and lay their eggs into its body, or to pierce wood pieces or fruits skins to reach hosts hidden within (Heatwole & Davis, 1965; Vinson, 1976; Elias *et al.*, 2018).

Note that oviposition-site selection may also exclusively favour the survival of the egg-laying mothers and does therefore not necessarily reflect a form of parental care. This is the case in the orange tip butterfly *Anthocharis cardamines*. In this species, females feed extensively on the nectar of the plant they select for oviposition, whereas this plant only has a poor nutritional value – compared to other neighbouring plants – for their larvae (Courtney, 1981). Similarly in the water strider *Aquarius paludum insularis*, females avoid ovipositing in sites where the backswimmer N*otonecta triguttata*, a predator that only attacks adults, is present (Hirayama & Kasuya, 2013).

FIGURE 17.1 Examples of different forms of parental care in insects. (a) Site selection for oviposition in *Callosobruchus maculatus;* egg laying on seed (Photo: J. Ritchey). (b) Nest building in the mason bee *Osmia* spp., nest of flower petals in form of tube held with scissors (Photo: A. Mallewick). (c) Egg provisioning in the house cricket *Acheta domesticus*. In this species, mothers adjust the quantity of hormones transferred to their eggs to better fit to their future environment (Photo: S. Pincebourde). (d) Egg attendance in the European earwig *Forficula auricularia*. A female grooms her eggs to remove fungal spores present on their surface (Photo: J. Meunier). (e) Egg brooding in the golden egg bug *Phyllomorpha laciniate*. A male carries conspecific eggs on its back (Photo: D. Carrasco). (f) Offspring attendance in the shield bug *Cumare pallida*. A male and a female guard their nymphs (Photo: G Monteith). (g) Offspring brooding in the shield bug *Peltocopta crassiventris*. A female carrying its nymphs under her abdomen (Photo: J. Wright). (h) Food provisioning in the burying beetle *Nicrophorus vespilloides*. A female regurgitates pre-digested carrion to a begging larva (Photo: P.T. Smiseth).

17.2.2 Nest Building

Nests typically serve as a protection against environmental hazards, either of an abiotic source such as (too high or too low) temperature and humidity, or of a biotic source such as predators and pathogenic microbes (Fialho *et al.*, 2018). For instance, in the beewolf *Philanthus triangulum*, females protect their offspring against microbial pathogens by incorporating elements with antipathogenic properties into the nest. To this end, females cultivate *Streptomyces* bacteria in their antenna and apply them to the brood cell prior to oviposition. The bacteria are then taken up by the larva and integrated into the walls of the cocoon where they offer protection against fungal infection and thus enhance larva survival (Kaltenpoth *et al.*, 2005). This protection results from the production of nine antibiotic substances by the bacteria (Kroiss *et al.*, 2010).

A broad diversity of materials can be used by insects to build their nest, such as silk in web spinners and some bark lice, mud and wax in wasps and bees, and modified plant tissues in the form of galls in some thrips (Machado & Trumbo, 2018). In other species, females directly use unmodified plant parts as nest material. For instance, female mason bees *Osmia avosetta* smear the inside of their underground nest with petals from surrounding flowers, probably to safeguard the nest from flooding, as well as to help retaining moisture and thus avoid desiccation (Sedivy *et al.*, 2010). Specifically, the *O. avosetta* female first digs a main tunnel, then brings in large pieces of petals to line the chamber, next carries in fine moist soil to plaster the inner surface of this lining and finally imports new petals to completely line the cell cavity (Sedivy *et al.*, 2010). She then lays a single egg in this chamber, provisions it with a sticky mixture of pollen and nectar and seals the chamber. The use of petals as nest material can also be found in other osmiine bees, such as *Wainia atrorufa*, *W. gessorum* and *W. sexsignata*, where mothers nest in empty snail shells and use petals to partition brood cells within shells (Müller *et al.*, 2018).

Parentally-built nests can exhibit a broad diversity of architectures. The nest can be simple and take the form of a small chamber or tunnel, such as in the necrophagous beetle *Coprophanaeus bellicosus* (Fialho *et al.*, 2018) and the European earwig *Forficula auricularia* (Meunier *et al.*, 2012; Diehl & Meunier, 2018). It can also take the form of a silken gallery spun from parental silk glands, such as in the webspinner *Anisembia texana* (Choe, 1994). Conversely, parents can build sophisticated nests, composed of a several chambers dedicated to different functions. In the short-tailed cricket *Anurogryllus muticus*, for instance, females built nests with a brood chamber to maintain the eggs and larvae, and a defecation chamber to store feces (West & Alexander, 1963). Note that in some species of ants with independent colony foundation - i.e. where queens found new colonies by themselves - the foundations of the future colony are typically build by the mother (queen) and take the form of a simple nest with only few chambers (Hölldobler & Wilson, 1990). The architecture of the colony then becomes more and more complex when the first adult offspring (workers) take the lead of colony construction.

17.2.3 Egg Provisioning

Another pre-ovipositional, widespread form of parental care consists in transferring hormones, energy, nutrients, or other substances into and/or onto the eggs to enhance the fitness of the resulting offspring (Smiseth *et al.*, 2012). Interestingly, many females can vary the nature and amount of the substances they transfer, and such adaptive plasticity in egg provisioning typically facilitates egg development, and helps to protect the eggs against environmental risks such as desiccation, drowning, predation and parasitism (Zeh *et al.*, 1989). In insects, multiple examples of adaptive egg provisioning can be found and reflect responses to future social constraints, food availability and temperature. The level of maternal hormone deposition to eggs reflects social constraints, for instance, in the house cricket *Acheta domesticus*. Mothers of this species provision their eggs with lower doses of ecdysteroid hormones when they are reared under high (as compared to low) social density. This has been suggested to be an adaptive strategy, as it allows mothers to program the production of slow-growing hatchlings (which is often associated with a reduced metabolic budget) in environments with high levels of resource competition, and therefore to enhance the likelihood of offspring survival (Crocker & Hunter, 2018). On the other hand, egg provisioning in the seed beetle *Stator limbatus* is driven by food availability.

Here, mothers adaptively change their investment into egg size depending on the type of host plant on which their eggs are laid. Females produce large but few eggs when laying on *Cercidium floridum*, where larvae need to be large and strong enough to penetrate the tough seed coat to feed. By contrast, they produce small but numerous eggs when laying on *Acacia greggii*, as the resulting small larvae can easily penetrate the softer seed coat of this host plant (Fox *et al.*, 1997). Finally, expected future temperatures drive egg provisioning in the butterfly *Bicyclus anynana*. In this species, Fischer and colleagues (2003) demonstrated that it pays off for females to produce larger but fewer eggs at a lower temperature, but more and smaller eggs at a higher temperature.

Egg provisioning may also involve defensive chemicals that are either produced by the mothers themselves or obtained from food via sequestration (Blum & Hilker, 2002). In the European earwig *F. auricularia*, mothers regularly apply self-produced hydrocarbons onto the surface of their eggs, probably to reduce the risk of mould germination and growth (Boos *et al.*, 2014). Females of the chrysomelid beetle *Cryptophalus hypochaeridis* also coat their eggs with a mixture of secretions and faecal materials, which together form a hard defensive structure protecting the eggs against predators (Ang *et al.*, 2008). Performing such an egg coating is costly to *C. hypochaeridis* mothers, and equivalent to producing half an extra egg (Ang *et al.*, 2008).

It is noteworthy that the adaptive effects of egg provisioning may not always be apparent, as they can be masked or altered when other forms of parental care are expressed later during offspring development. In the burying beetle *Nicrophorus vespilloides*, for instance, egg size has a positive effect on offspring body mass in absence of post-hatching care, whereas this effect is absent in presence of post-hatching care (Monteith *et al.*, 2012). By contrast, in the European earwig *F. auricularia*, it is the amount of pre-hatching care that masks the effects of egg size: the mean egg mass is positively associated with offspring number only in clutches receiving high levels (and not in clutches receiving low levels) of pre-hatching care (Koch & Meunier, 2014).

17.3 Pre-hatching Care

17.3.1 Egg Attendance/Guarding

After egg deposition, parents can remain in the vicinity to physically protect the eggs against biotic and abiotic threats, a phenomenon called egg attendance or egg guarding. Egg attendance is the most common form of post-ovipositional pre-hatching parental care in insects. It has been reported in almost all Dermaptera, as well as many Embiidina, Hemiptera, Membracidae and a few Diptera (Lounibos & Machado-Allison, 1983; Tallamy & Wood, 1986; Edgerly, 1997; Matzke & Lass, 2005). Egg attendance can provide multiple benefits to the eggs, offering protection against predators, pathogens, and social parasites. The protection against predators can be found in the nymphalid butterfly *Hypolimnas anomala*, where females straddle their eggs to exclude predatory ants (Nafus & Schreiner, 1988). This behaviour, however, does not deter all types of predator, as it appears to be non-effective against parasitoid wasps (Nafus & Schreiner, 1988). In the coreid bug *Physomeris grossipes*, egg guarding females do not deter predators with body movements, but by discharging chemicals from anal glands (Tallamy & Wood, 1986). Egg guarding may also protect against conspecific predators, such as in the maritime earwig *Anisolabis maritima*, where maternal egg guarding reduces the risk of egg cannibalism by conspecifics (Miller *et al.*, 2011). Egg guarding may also be associated with protection against pathogens. This has been reported in two earwig species, *F. auricularia* and *Labidura riparia*, where mothers frequently groom their eggs to remove fungal spores that otherwise grow and ultimately kill the entire clutch (Shepard *et al.*, 1973; Boos *et al.*, 2014).

17.3.2 Egg Brooding

After egg laying, insect parents can carry their eggs and provide care to these eggs during carrying, a phenomenon called egg brooding. Note that egg brooding has a different definition in birds, where it is defined as the act of sitting on eggs to incubate them. Insect egg brooding can be found in several insect orders, such as Coleoptera and Hemiptera (Machado & Trumbo, 2018). In the water bug *Abedus herbeti*, for instance, females lay their eggs on the males' back. The males then take over egg brooding and ensure that the eggs receive a proper level of oxygenation (Smith, 1997). Similarly in the golden egg bug *Phyllomorpha laciniata*, females can lay their eggs on plants, but mainly use conspecific males and females as egg-laying substrates. This surprising habit comes with no apparent costs to the recipient individual, whereas it generally improves egg survival when parasitoids are present in the environment (Carrasco & Kaitala, 2009). Interestingly, mothers may also evolve specific morphological traits for egg brooding. In the lobster cockroach *Nauphoeta cinera*, for instance, eggs are extruded, rotated by 90 degrees, and then retracted into a specialized brood sac, where development proceeds until the larvae are ready to hatch (Moore *et al.*, 2007). Similarly, females of the florida woods cockroach *Eurycotis floridana* and the brown-banded cockroach *Supella longipalpa* briefly carry their eggs in an extruded ootheca as they travel (Tallamy & Wood, 1986).

17.4 Post-hatching and Post-birth Care

17.4.1 Offspring Attendance

Once the eggs have hatched, parents can remain with and provide care to their offspring either at a fixed location or by escorting them. This offspring attendance is a taxonomically widespread phenomenon in insects and can last from a few hours, such as in the cockroaches *Blaberus craniifer*, *Gromphadorhina protentosa* and *Thanatophyllum akinetum*, to several days or weeks, such as in burying beetles and earwigs, over multiple years, as in the Australian Wood-feeding cockroach *Panesthia cribata* (Lamb, 1976; Scott & Traniello, 1990; Bell *et al.*, 2007; Wong & Kölliker, 2012). Offspring attendance typically enhances offspring protection against predators and improves the ability of newly hatched juveniles to defend against microbes and pathogens. The protection of juveniles against predation can be found in the lacebug *Gargaphia solani*. In this species, mothers aggressively protect their newly hatched offspring against predatory coccinellid adults by showing offensive movements towards the predator and/or expressing wing fanning (Tallamy & Denno, 1981). Similarly in the earwig *A. maritima*, mothers have been shown to fiercely defend their brood of nymphs with their forceps (Suzuki, 2010). Parental defences against microbes have been well studied in the burying beetle *N. vespilloides*. In this species, parents apply antimicrobial anal and oral secretions to the carcass used for breeding to limit the growth of microbes that could infect the larvae or compete with them over food access (Rozen *et al.*, 2008; Arce *et al.*, 2012). Mothers of the European earwig *F. auricularia* also line their nest with feces exhibiting antimicrobial properties (Diehl *et al.*, 2015).

17.4.2 Offspring Brooding

Similar to egg brooding, offspring brooding consists of carrying juveniles after egg hatching – a phenomenon also called 'nymphal phoresy'. Just like egg brooding, offspring brooding is a rare phenomenon in insects. It has been reported in the aquatic cockroach *Phlebonotus pallens*, where mothers carry their nymphs under their wing covers (Bell *et al.*, 2007). Offspring brooding is also well documented in the peculiar pentatomoid family Phloeidae, which includes the Australian shield bugs *Peltocopta crassiventris* and *Cumare pallida*. In these species, adults have a flattened body whereby all their covering body parts are remarkably expanded into flattened lobes under which mothers typically carry their first instar nymphs (Monteith, 2006).

17.4.3 Food Provisioning to Offspring

Insect parents exhibit multiple ways of providing food to their newly hatched offspring: the production of trophic eggs, the production of milk-like substances and other feeding secretions, the regurgitation of food, as well as the progressive and non-progressive (i.e. mass) provisioning of food items.

17.4.3.1 Trophic Egg Consumption

Trophic eggs (also called 'nurse eggs') are ovariole-produced structures that often exhibit morphological or biochemically specialized phenotypes (Perry & Roitberg, 2006). The production of trophic eggs can be found from solitary to social insects (Perry & Roitberg, 2006), and exhibits great temporal variation during a species' life cycle, ranging from a production synchronized with that of fertilized eggs, to a delayed production during egg development or even long after first eggs have hatched (Kudo & Nakahira, 2005; Filippi *et al.,* 2008). This latter option is quite unusual but can be found in the Passalid beetle *Cylindrocaulus patalis*. In this species, third instar larvae trigger the production of a trophic egg by their mothers through a series of behavioural solicitations in the form of head-rubbing and stridulation (Ento *et al.*, 2008).

Because they are typically non-fertile, the nutritional resources contained in trophic eggs are fully available for offspring (Crespi, 1992; Perry & Roitberg, 2006; Filippi *et al.*, 2008). Having access to and consuming trophic eggs may thus offer crucial nutritional benefits to newly hatched juveniles. These benefits have, for instance, been demonstrated in the Asian cydnid *Canthophorus niveimarginatus*, where the experimental removal of trophic eggs came with a significant decrease in the nymphs' body weight and survival rate (Filippi *et al.*, 2008; Baba *et al.*, 2011). Similarly the consumption of trophic eggs improved larval nutritional condition and the subsequent expression of anti-predator responses in the beetle *Leptinotarsa decemlineata* (Tigreros *et al.*, 2017). The benefits of trophic egg consumption are not only associated with the eggs' nutritional value, as they can also protect viable eggs from predators (Kudo & Nakahira, 2004) or meditate the vertical transfer of essential symbionts to juveniles. Such a transfer occurs in the stinkbug *Parastrachia japonensis*, where mothers cover trophic eggs with symbiont-containing white mucus so that their juveniles can obtain symbionts while consuming these trophic eggs (Hosokawa *et al.*, 2012). This use of trophic eggs to transfer symbionts to offspring resembles the use of symbiont capsules (which look like small eggs) in the Japanese common plastaspid stinkbug *Megacopta punctatissima* (Fukatsu & Hosokawa, 2002). However, it is important to note that the role of trophic eggs in the transfer of symbionts is not universal among insects (Baba *et al.*, 2011).

The evolution of trophic eggs might be favoured if they allow parents to provide offspring with resources that would be hard or impossible to access for newly hatched individuals (Crespi, 1992). In line with this hypothesis, mothers of the Asian ladybird *Harmonia axyridis* increase the number of trophic eggs when the amount of resources available for their future offspring decreases in the environment (Perry & Roitberg, 2005). Similarly, newly hatched nymphs of the burrower bug *C. niveimarginatus* are still unable to feed on the tough vessels of seeds, their primary food source, and therefore require trophic eggs to obtain their first energetic resources (Baba *et al.*, 2011). Alternatively, the production of trophic eggs may also help mothers to increase their investment into offspring quality by provisioning each of their young with more resources than can be contained in one egg (Crespi, 1992). For this reason, it has been suggested that the evolution of trophic egg production should be favoured in species where egg size is subject to strong mechanical and/or physiological constraints, e.g. due to the size of the ovipositor (Mock & Parker, 1997; Noriyuki *et al.*, 2012). Such a constraint would prevent females from producing larger eggs and would thus only allow females to increase their investment into offspring quality by offering them additional eggs as food sources (Mock & Parker, 1997). This hypothesis, however, received only little attention in the literature and has no clear empirical support (Noriyuki *et al.*, 2012). Overall, this suggests that maternal adaptations to environmental changes rather than morphological constraints on egg size are sufficient to explain the evolution of trophic egg-laying.

Not all egg consumption targets trophic eggs (Alexander, 1974), since eating fertilized eggs may be beneficial to individual offspring in the light of sibling rivalry, (Elgar & Crespi, 1992; Mock & Parker, 1997) or queen-worker competition over male production in eusocial insects (Ratnieks & Visscher, 1989; Ratnieks *et al.*, 2006). In family-living species, the difference of genetic relatedness among offspring is expected to generate competition between siblings to monopolise parental resources (Trivers, 1974; Mock & Parker, 1997), so that the consumption of sibling eggs immediately after hatching could be a solid and efficient method to win such a competition. The consumption of fertile eggs has been reported in several insect species (Frechette & Coderre, 2000; López-Ortega & Williams,

2018), making it difficult to understand what is the main driver of egg cannibalism. In other species such as in the cricket *Velarifictorus asperses*, however, nymphs have been shown to discriminate between fertile and non-fertile eggs and to selectively consume the latter (Zeng *et al.*, 2017).

17.4.3.2 Milk and Other Feeding Secretions

While feeding offspring with milk is a well-known example of maternal care in mammals, it only occurs in a few insects exhibiting adenotrophic vivipary (Ostrovsky *et al.*, 2016). Contrary to mammals, however, this milk feeding typically occurs within the mother's body and involves poorly developed larvae or embryos. The larva typically feeds orally from secretions (milk) produced by accessory glands (milk glands) consisting of tubular organs opening into the mother's uterus. Milk feeding is a characteristic of several Diptera families, such as the Glossinidae (tsetse flies), Hippoboscidae (louse or wallaby flies, keds), Nycteribiidae and the Streblidae (bat flies)(Meier *et al.*, 1999). It has also been reported in the Pacific beetle cockroach *Diploptera punctata*, which is the only known species of cockroach where females possess milk glands (Stay & Coop, 1974; Bell *et al.*, 2007; Tobe *et al.*, 2013). Finally, a recent study suggests that milk-feeding could occur in the viviparous earwig *Hemimerus talpoides* (Dermaptera). In this species, the transformed follicular/ovarian cells of mothers and the invaginated dorsal vessel of embryos indeed converge to form a cephalic vesicle analogous to a placenta (Biliński *et al.*, 2018). Just like with mammals, insect milk typically contains a large proportion of proteins as well as some symbionts that can be essential to food digestion by offspring (Tobe *et al.*, 2013; Nováková *et al.*, 2015).

Insect mothers can also feed juveniles outside of their body with non-milk-like secretions. This is the case in several cockroaches, such as *Trichoblatta sericea, Thorax porcellana* or *Gromphadorhina Oxyhaloinae* (Nalepa & Bell, 1997). In these species, mothers produce feeding secretions either ventrally, under their tegmina, in their abdominal tip, or in their mouthparts. The food source provided to the juveniles can either consist of hemolymph, sternal and tergal exudates, or secretion from the brood sac (Nalepa & Bell, 1997). In non-viviparous earwig species such as *F. auricularia* and *L. riparia*, young nymphs were also observed actively gnawing at the pleural region of their mother's abdomen (Meunier J, pers. obs.)(Shepard *et al.*, 1973). The function of this behaviour and whether it mediates the transfer of feeding substances from earwig parents to offspring remains, however, unknown.

17.4.3.3 Food Regurgitation

Food regurgitation (also called stomodeal trophallaxis) typically occurs through mouth-to-mouth contact between parents and offspring. This process allows parents to soften food, add and transfer digestive enzymes, as well as to transfer symbionts to their offspring (Trumbo, 2012). Known cases of parental food regurgitation are scarce in non-eusocial insects. One clear example can be found in the European earwig *F. auricularia*, a species where mothers and nymphs frequently show mouth-to-mouth contact (Mas & Kölliker, 2011). Making use of the relative transparency of newborn earwig nymphs, Staerkle & Kölliker (2008) demonstrated that earwig nymphs become green if they are tended by mothers previously fed with green-colored food. In another earwig species, the striped earwig *L. riparia*, Shepard and colleagues (1973) used radioactive labelling of food, and came to the same conclusion: nymphs showed radioactivity after they had been tended by mothers previously fed with radioactive food. Other well-known examples of food regurgitation in insects are present in burying beetles, such as *Nicrophorus orbicollis*, *N. vespilloides* and *N. quadripuncatus* (Milne & Milne, 1976; Eggert *et al.*, 1998; Suzuki, 2015). In these species, adults have been suggested to regurgitate pre-digested carrion to the young larvae on the carcass. Contrary to earwigs, however, the claim that food regurgitation occurs in burying beetles mostly relies on behavioural observations of mouth-to-mouth contacts between parents and larvae, as well as on studies investigating the effects of starvation on the intensity of larval begging behaviours (e.g. Smiseth & Moore, 2002, 2004; Suzuki, 2015) (but see Capodeanu-Nägler *et al.*, 2018). In the same way, food regurgitation has been suggested to occur in the woodroach genus *Salganea* based on the frequent mouth-to-mouth contacts between mothers and offspring (Nalepa *et al.*, 2008).

17.4.3.4 Provisioning of Small and Large Food Items

Juveniles of many insect species are capable of processing food items but are rarely able to obtain these items by themselves, e.g. due to poor foraging capabilities. Parents can provide care to circumvent this issue in the form of collecting and bringing back food items to or close to the nest. This is the case in the burrower bug *Sehirus cinctus*, where mothers provide *Lamium purpureum* nutlets to newly hatched nymphs until they reach their second developmental instar (Sites & McPherson, 1982). Similarly, *Embia ramburi* females provide pieces of masticated food to their nymphs (Ledoux, 1958), and females of the short-tailed cricket *A. muticus* provide nymphs with bits of retrieved food (West & Alexander, 1963). In the earwig *A. maritima*, mothers also carry small food items with their mouth to the nest (Suzuki, 2010) and adapt the amount of food to the number and need of their nymphs (Suzuki, 2018).

Parents can also provision their offspring with larger food items that simultaneously serve as nesting material, a phenomenon called mass provisioning. A textbook example can be found in burying beetles of the genus *Nicrophorus*. Here, parents provide the carcass of a small vertebrate to their larvae, on which they subsequently feed either directly, or indirectly through parental regurgitations (Scott, 1998). Once *Nicrophorus* females find a carcass of a small vertebrate, they oviposit in its vicinity and start preparing the carcass (alone, or together with the male) by removing the hair or feathers and then creating an opening in the integument of the carcass before or shortly after egg hatching to favour its consumption by the larvae (Eggert *et al.*, 1998; Schrader *et al.*, 2015; Trumbo, 2017). They then apply oral and anal antimicrobial secretions to its surface to help preserving the carcass and allow its use as a food source until the end of larvae development (Suzuki, 2001; Rozen *et al.*, 2008). Another example of brood mass provisioning is found in the dung beetles. In 'tunneller' dung beetles, such as the ones found in the genus *Onthophagus*, females pack portions of dung into the end of tunnels excavated beneath the dung pad, and then seal the tunnel after depositing a single egg in each tunnel (Hunt & Simmons, 1997, 2002). Conversely, in 'roller' dung beetles, such as the ones found in the family Scarabaeinae, females (sometimes together with males) extract a portion of the dung pad, lay one egg on the top of it, and then roll it to form a brood ball that will be tended until hatching (Sato, 1997). Notwithstanding the method, the portion of dung contained in the brood ball varies in size between females (Hunt & Simmons, 2002) and represents the only food source that is available to the larvae during development.

Finally, an extreme form of parental food provisioning occurs when parents provision their offspring with their bodies. This rare phenomenon has been reported in the hump earwig, *Anechura harmandi*, where offspring completely consume their mother at the end of post-hatching care (Suzuki *et al.*, 2005). Matriphagy also occurs in the beetle *Micromalthus debilis*, the sole species of the family Micromalthidae. In this insect with a very peculiar life-cycle, female offspring feed on fungus-infested rotting wood, whereas male offspring exclusively consume their mother (Pollock & Normark, 2002; Normark, 2013). Why obligatory matriphagy has only evolved in males and why it is maintained in this species remains yet unclear (Pollock & Normark, 2002; Normark, 2013).

17.4.4 Care after Offspring Nutritional Independence

Providing care to offspring after they have reached nutritional independence is a very unusual phenomenon in insects (Clutton-Brock, 1991). It can nevertheless be found in the European earwig *F. auricularia* and the burying beetle *N. vespilloides*. In the European earwig, juveniles attain mobility and can forage for themselves only few days after egg hatching (Wong & Kölliker, 2012). However, earwig mothers remain with their nymphs up to several weeks after their emergence (Lamb, 1976), probably to enhance offspring protection against predators and/or pathogens (Diehl *et al.*, 2015). In the burying beetle *N. vespilloides*, larvae become nutritionally independent about three days after emergence, but mothers remain with the larvae a couple of additional days to defend them against predators and conspecific intruders (Smiseth *et al.*, 2003). Laboratory experiments of orphaning in this burying beetle reveal that reduced post-hatching care reduces offspring survival rate and mass (Eggert *et al.*, 1998; Capodeanu-Nägler *et al.*, 2016). Finally, in the wood cockroach *Cryptocercus punctulatus* nymphs are nutritionally independent after they reach the third developmental instar, which occurs within several months, but parents remain with their offspring for more than three years (Nalepa, 1990). During this family life, mothers typically provide their offspring with hindgut fluids containing nutrients and hindgut symbionts – two components that are surprisingly not required to ensure offspring survival after their third instar (Nalepa, 1984, 1990).

17.4.5 Unconventional Forms of Parental Care

In addition to the 'classical' forms of parental care detailed above, it has recently been suggested that other 'unconventional' parental behaviours could function as parental care. In particular, Davenport *et al.* (2019) hypothesised that offspring abandonment and filial cannibalism could be considered (and evolve) as forms of parental care, even in the absence of other classical forms of care. This hypothesis relies on the idea that when within-clutch density negatively affects offspring survival – an effect reported in many species (see in Davenport *et al.*, 2019) – offspring abandonment and filial cannibalism can allow parents to actively reduce the density of their clutch and thus the overall mortality of their young. This benefit would allow classifying these two behaviours as forms of parental care. This hypothesis was supported by their modelling approach (Davenport *et al.*, 2019), particularly when several specific conditions were met: when within-clutch density effects are relatively high, when abandoning or eating some young indeed reduces such density effects, when offspring require care and when eggs are laid at higher than optimal densities. However, this quite limited set of conditions suggests that offspring abandonment and filial cannibalism can only be considered forms of parental care in a very limited number of species and calls for empirical works testing the validity of this interpretation in insects.

17.5 Conclusions

Insects represent a prime illustration of the diversity of parental care that can be found in animals. Across species, families and genera, forms of parental care range from the simple provisioning of eggs with nutrients and hormones, over the regurgitation of food to older juveniles, to fierce protection of eggs and newly hatched offspring against intruders. This unique diversity makes insects excellent biological models for improving our general understanding of the evolution, diversification and underlying physiological and genetic mechanisms shaping family life in nature.

Acknowledgments

JM would like to thank the Fyssen Foundation and the French National Research Agency (ANR; Project MicroSoc) for financial support during the writing of this book chapter. JK was supported by the German Science Foundation (DFG; KR 5017/2-1).

REFERENCES

Alcock, J. (1975). Territorial behaviour by males of *Philanthus multimaculatus* (Hymenoptera: Sphecidae) with a review of territoriality in male sphecids. *Animal Behaviour 23*: 889–895.

Alexander, R.D. (1974). The evolution of social behavior. *Annual Review of Ecology and Systematics, 5*: 325–383.

Alonso-Alvarez, C., and Velando, A. (2012). Benefits and costs of parental care. In *The evolution of parental care*, eds. N.J. Royle, P.T. Smiseth, and M. Kölliker. Oxford: Oxford University Press, pp. 40–61.

Ament, S.A., Velarde, R.A., Kolodkin, M.H., Moyse, D., & Robinson, G.E. (2011). Neuropeptide Y-like signalling and nutritionally mediated gene expression and behaviour in the honey bee. *Insect Molecular Biology*, 20, 335–345.

Amsalem, E., Malka, O., Grozinger, C., and Hefetz, A. (2014). Exploring role of juvenile hormone and vitellogenin in reproduction and social behavior in bumble bees. *BMC Evolutionary Biology 14*: 1–13.
Anderson, R.S. (1982). Burying beetle larvae: Nearctic nicrophorus and oriental *Ptomascopus morio* (Silphidae). *Systematic Entomology, 7*, 249–264.
Ang, T.Z., O'Luanaigh, C., Rands, S.A., Balmford, A., and Manica, A. (2008). Quantifying costs and benefits of protective egg coating in a Chrysomelid beetle. *Ecological Entomology, 33*, 484–487.
Arce, A.N., Johnston, P.R., Smiseth, P.T., and Rozen, D.E. (2012). Mechanisms and fitness effects of antibacterial defences in a carrion beetle. *Journal of Evolutionary Biology, 25*, 930–937.
Arrese, E.L., and Soulages, J.L. (2010). Insect fat body: Energy, metabolism, and regulation. *Annual Review of Entomology, 55*, 207–225.
Baba, N., Hironaka, M., Hosokawa, T., Mukai, H., Nomakuchi, S., *et al.* (2011). Trophic eggs compensate for poor offspring feeding capacity in a subsocial burrower bug. *Biology Letters, 7*, 194–196.
Balshine, S., and Trumbo, S.T. (2012). Parental care in invertebrates. In *The evolution of parental care*, eds. N.J. Royle, P.T. Smiseth, and M. Kölliker. Oxford: Oxford University Press, pp. 3–51.
Bell, W.J., Roth, L.M., and Nalepa, C.A. (2007). *Cockroaches: Ecology, behavior, and natural history*,. Baltimore: The Johns Hopkins University Press.
Ben-Shahar, Y., Dudek, N.L., and Robinson, G.E. (2004). Phenotypic deconstruction reveals involvement of manganese transporter malvolio in honey bee division of labor. *Journal of Experimental Biology, 207*, 3281–3288.
Benowitz, K.M., McKinney, E.C., Roy-Zokan, E.M., Cunningham, C.B., and Moore, A.J. (2017). The role of lipid metabolism during parental care in two species of burying beetle (Nicrophorus spp.). *Animal Behaviour, 129*, 143–149.
Bilinski, S.M., Jaglarz, M.K., Halajian, A., and Tworzydlo, W. (2018). Unusual morphological adaptations and processes associated with viviparity in an epizoic dermapteran. *PLoS One, 13*, 1–10.
Blum, M.S., and Hilker, M. (2002). Chemical protection of insect eggs. In *Chemo-ecology of insects eggs and egg deposition*, eds. M. Hilker and T. Meiners. Oxford: Blackwell Publishing Inc, pp. 61–90.
Bonsall, M.B., and Klug, H. (2011). The evolution of parental care in stochastic environments. *Journal of Evolutionary Biology, 24*, 645–655.
Boos, S., Meunier, J., Pichon, S., and Kölliker, M. (2014). Maternal care provides antifungal protection to eggs in European earwig. *Behavioral Ecology, 25*, 754–761.
Boulay, R., Hooper-Bui, L.M., and Woodring, J. (2001). Oviposition and oogenesis in virgin fire ant females *Solenopsis invicta* are associated with a high level of dopamine in brain. *Physiological Entomology, 26*, 294–299.
Capodeanu-Nägler, A., Keppner, E.M., Vogel, H., Ayasse, M., Eggert, A.-K., *et al.* (2016). From facultative to obligatory parental care: Interspecific variation in offspring dependency on post-hatching care in burying beetles. *Scientific Reports, 6*, 29323.
Capodeanu-Nägler, A., Eggert, A.-K., Vogel, H., Sakaluk, S.K., and Steiger, S. (2018). Species divergence in offspring begging and parental provisioning is linked to nutritional dependency. *Behavioral Ecology*, 42–50.
Capodeanu-Nägler, A., Prang, M.A., Trumbo, S.T., Vogel, H., Eggert, A.-K., *et al.* (2018). Offspring dependence on parental care and role of parental transfer of oral fluids in burying beetles. *Frontiers in Zoology, 15*, 1–12.
Carrasco, D., and Kaitala, A. (2009). Egg-laying tactic in *Phyllomorpha laciniata* in presence of parasitoids. *Entomologia Experimentalis et Applicata, 131*, 300–307.
Chaboo, C.S., Frieiro-Costa, F.A., Gómez-Zurita, J., and Westerduijn, R. (2014). Origins and diversification of subsociality in leaf beetles (Coleoptera: Chrysomelidae: Cassidinae: Chrysomelinae). *Journal of Natural History, 48*, 2325–2367.
Choe, J.C. (1994). Communal nesting and subsociality in a webspinner, *Anisembia texana* (Insecta: Embiidina: Anisembiidae). *Animal Behaviour, 47*, 971–973.
Choe, J.C., and Crespi, B.J. (1997). *The evolution of social behaviour in insects and arachnids*,. Cambridge: Cambridge University Press.
Clutton-Brock, T.H. (1991). *The evolution of parental care*,. Princeton, NJ: Princeton University Press.

Courtney, S.P. (1981). Coevolution of pierid butterflies and their cruciferous foodplants - III. *Anthocharis cardamines* (L.) survival, development and oviposition on different hostplants. *Oecologia*, *51*, 91–96.
Costa, J.T. (2006). The other insect societies.Cambridge, Massachusetts & London, England, MA: Harvard University Press.
Crespi, B.J. (1992). Cannibalism and trophic eggs in sub-social and eusocial insects. In *Cannibalism: ecology and evolution among diverse taxa*, eds. M.A. Elgar and B.J. Crespi. Oxford: Oxford University Press, pp. 176–213.
Crocker, K.C., and Hunter, M.D. (2018). Social density, but not sex ratio, drives ecdysteroid hormone provisioning to eggs by female house crickets (*Acheta domesticus*). *Ecology and Evolution*, *8*, 10257–10265.
Cunningham, C.B., Badgett, M.J., Meagher, R.B., Orlando, R., and Moore, A.J. (2017). Ethological principles predict neuropeptides co-opted to influence parenting. *Nature Communications*, *8*, 30–36.
Cunningham, C.B., Ji, L., Wiberg, R.A.W., Shelton, J., McKinney, E.C., *et al.* (2015). The genome and methylome of a beetle with complex social behavior, *Nicrophorus vespilloides* (Coleoptera: Silphidae). *Genome Biology and Evolution*, *7*, 3383–3396.
Cunningham, C.B., VanDenHeuvel, K., Khana, D.B., McKinney, E.C., and Moore, A.J. (2016). The role of neuropeptide F in a transition to parental care. *Biology Letters*, *12*, 20160158.
Davenport, M.E., Bonsall, M.B., and Klug, H. (2019). Unconventional care: Offspring abandonment and filial cannibalism can function as forms of parental care. *Frontiers in Ecology and Evolution*, *7*, 1–11.
Diehl, J.M., Körner, M., Pietsch, M., and Meunier, J. (2015). Feces production as a form of social immunity in an insect with facultative maternal care. *BMC Evolutionary Biology*, *15*, 15, 40.
Diehl, J.M., and Meunier, J. (2018). Surrounding pathogens shape maternal egg care but not egg production in European earwig. *Behavioral Ecology*, *29*, 128–136.
Dierick, H.A., and Greenspan, R.J. (2007). Serotonin and neuropeptide F have opposite modulatory effects on fly aggression. *Nature Genetics*, *39*, 678–682.
Elgar, M.A., and Crespi, B.J. (1992). *Cannibalism: Ecology and evolution among diverse taxa*,. Oxford: Oxford University Press.
Elias, L. G., Kjellberg, F., Farache, F.H.A., Almeida, E.A.B., Rasplus, J.-Y., Cruaud, A., Peng, Y.-Q., Yang, D.-R., and Pereira, R.A.S. (2018). Ovipositor morphology correlates with life history evolution in agaonid fig wasps. *Acta Oecologica*, 90, 109–116.
Eggert, A.-K., Reinking, M., and Müller, J.K. (1998). Parental care improves offspring survival and growth in burying beetles. *Animal Behaviour*, 55, 97–107.
Engel, K.C., Stökl, J., Schweizer, R., Vogel, H., Ayasse, M., *et al.* (2016). A hormone-related female anti-aphrodisiac signals temporary infertility and causes sexual abstinence to synchronize parental care. *Nature Communications*, *7*, 11035.
Ento, K., Araya, K., and Kudo, S. (2008). Trophic egg provisioning in a passalid beetle (Coleoptera). *European Journal of Entomology*, *105*, 99–104.
Fellowes, M.D.E. (1998). Do non-social insects get (kin) recognition they deserve? *Ecological Entomology*, *23*, 223–227.
Fréchette, B., and Coderre, D. (2000). Oviposition strategy of the green lacewing *Chrysoperla rufilabris* (Neuroptera: Chrysopidae) in response to extraguild prey availability. *European Journal of Entomology*, 97, 507–510 10.14411/eje.2000.078.
Fialho, V.S., Rodrigues, V.B., and Elliot, S.L. (2018). Nesting strategies and disease risk in necrophagous beetles. *Ecology and Evolution*, *8*, 3296–3310.
Filippi, L., Baba, N., Inadomi, K., Yanagi, T., Hironaka, M., *et al.* (2008). Pre- and post-hatch trophic egg production in subsocial burrower bug, *Canthophorus niveimarginatus* (Heteroptera: Cydnidae). *Naturwissenschaften*, *96*, 201–211.
Fischer, E.K., and O'Connell, L.A. (2017). Modification of feeding circuits in evolution of social behavior. *The Journal of Experimental Biology*, *220*, 92–102.
Fischer, K., Brakefield, P.M., and Zwaan, B.J. (2003). Plasticity in butterfly egg size: Why larger offspring at lower temperatures? *Ecology*, *84*, 3138–3147.
Foottit, R.G., and Adler, P.H. Eds. (2018). *Insect Biodiversity: science and society - Volume II*,. Wiley-Blackwell.

Fox, C.W., Thakar, M.S., and Mousseau, T.A. (1997). Egg size plasticity in a seed beetle: An adaptive maternal effect. *The American Naturalist, 149*, 149–163.
Fukatsu, T., and Hosokawa, T. (2002). Capsule-transmitted gut symbiotic bacterium of the Japanese common plataspid stinkbug, *Megacopta punctatissima. Applied and Environmental Microbiology*, 68, 389–39610. 1128/aem.68.1.389-396.2002.
Gilbert, J.D.J., and Manica, A. (2015). The evolution of parental care in insects: A test of current hypotheses. *Evolution, 69*, 1255–1270.
Gilbert, J.D.J., Thomas, L.K., and Manica, A. (2010). Quantifying benefits and costs of parental care in assassin bugs. *Ecological Entomology, 35*, 639–651.
Gillespie, J.P., Burnett, C., and Charnley, A.K. (2000). The immune response of desert locust *Schistocerca gregaria* during mycosis of entomopathogenic fungus, *Metarhizium anisopliae* var acridum. *Journal of Insect Physiology, 46*, 429–437.
Gross, M.R. (2005). The evolution of parental care. *The Quarterly Review of Biology, 80*, 37–45.
Hamilton, W.D.W.D. (1964). The genetical evolution of social behaviour. I. *Journal of Theoretical Biology, 7*, 1–16.
Harshman, L.G., and Zera, A.J. (2007). The cost of reproduction: Devil in details. *Trends in Ecology & Evolution, 22*, 80–86.
Heatwole, H., and Davis, D. M. (1965). Ecology of three sympatric species of parasitic insects of the Genus *Megarhyssa* (Hymenoptera: Ichneumonidae). *Ecology*, 46, 140–15010.2307/1935265.
Henshaw, J.M., Fromhage, L., and Jones, A.G. (2019). Sex roles and evolution of parental care specialization. *Proceedings of Royal Society B: Biological Sciences, 286*, 20191312.
Hirayama, H., and Kasuya, E. (2013). Effect of adult females' predation risk on oviposition site selection in a water strider. *Entomologia Experimentalis et Applicata, 149*, 250–255.
Hölldobler, B., and Wilson, E.O. (1990). *The ants*,. Harvard: Harvard University Press.
Hosokawa, T., Hironaka, M., Mukai, H., Inadomi, K., Suzuki, N., *et al.* (2012). Mothers never miss moment: A fine-tuned mechanism for vertical symbiont transmission in a subsocial insect. *Animal Behaviour, 83*, 293–300.
Hunt, J., and Simmons, L.W. (1997). Patterns of fluctuating asymmetry in beetle horns: An experimental examination of honest signalling hypothesis. *Behavioral Ecology and Sociobiology, 41*, 109–114.
Hunt, J., and Simmons, L.W. (2002). The genetics of maternal care: Direct and indirect genetic effects on phenotype in dung beetle *Onthophagus taurus. Proceedings of National Academy of Sciences of USA, 99*, 6828–6832.
Kaltenpoth, M., Göttler, W., Herzner, G., and Strohm, E. (2005). Symbiotic bacteria protect wasp larvae from fungal infestation. *Current Biology, 15*, 475–479.
Klug, H., and Bonsall, M.B. (2010). Life history and evolution of parental care. *Evolution, 64*, 823–835.
Klug, H., and Bonsall, M.B. (2014). What are the benefits of parental care? The importance of parental effects on developmental rate. *Ecology and Evolution, 4*, 2330–2351. 10.1002/ece3.1083.
Klug, H., Alonso, S., Bonsall, M.B., and Alonzo, S.H. (2012). Theoretical foundations of parental care. In *The Evolution of Parental Care*, eds. N.J. Royle, P.T. Smiseth, and M. Kölliker. Oxford, UK: Oxford University Press, pp. 21–39.
Klug, H., and Bonsall, M.B. (2019). Coevolution influences evolution of filial cannibalism, offspring abandonment and parental care. *Proceedings of Royal Society B: Biological Sciences, 286*, 20191419.
Klug, H., Bonsall, M.B., and Alonzo, S.H. (2013). Sex differences in life history drive evolutionary transitions among maternal, paternal, and bi-parental care. *Ecology and Evolution, 3*, 792–806.
Koch, L.K., and Meunier, J. (2014). Mother and offspring fitness in an insect with maternal care: Phenotypic trade-offs between egg number, egg mass and egg care. *BMC Evolutionary Biology, 14*, 125.
Kokko, H., and Jennions, M.D. (2008). Parental investment, sexual selection and sex ratios. *Journal of Evolutionary Biology, 21*, 919–948.
Kokko, H., and Jennions, M.D. (2012). *Sex differences in parental care in the evolution of parental care*, eds. N.J. Royle, P.T. Smiseth, and M. Kölliker. Oxford, UK: Oxford University Press, pp. 101–116.
Kölliker, M. (2007). Benefits and costs of earwig (*Forficula auricularia*) family life. *Behavioral Ecology and Sociobiology, 61*, 1489–1497.

Kölliker, M., Boos, S., Wong, J.W.Y., Röllin, L., Stucki, D., *et al.* (2015). Parent-offspring conflict and genetic trade-offs shaping parental investment. *Nature Communications*, *6*, 6850.

Kölliker, M., Royle, N.J., and Smiseth, P.T. (2012). Parent-offspring coadaptation. In *The evolution of parental care*,. Oxford: Oxford University Press, pp. 285–303.

Kramer, J., Körner, M., Diehl, J.M., Scheiner, C., Yüksel-Dadak, A., *et al.* (2017). When earwig mothers do not care to share: Parent-offspring competition and evolution of family life. *Functional Ecology*, *31*, 2098–2107.

Kramer, J., and Meunier, J. (2016). Maternal condition determines offspring behavior toward family members in European earwig. *Behavioral Ecology*, *27*, 494–500.

Kramer, J., and Meunier, J. (2019). The other facets of family life and their role in evolution of animal sociality. *Biological Reviews*, *94*, 199–215.

Kramer, J., Thesing, J., and Meunier, J. (2015). Negative association between parental care and sibling cooperation in earwigs: A new perspective on early evolution of family life? *Journal of Evolutionary Biology*, *28*, 1299–1308.

Kroiss, J., Kaltenpoth, M., Schneider, B., Schwinger, M.-G., Hertweck, C., *et al.* (2010). Symbiotic streptomycetes provide antibiotic combination prophylaxis for wasp offspring. *Nature Chemical Biology*, *6*, 261–263.

Kudo, S., and Nakahira, T. (2005). Trophic-egg production in a subsocial bug: Adaptive plasticity in response to resource conditions. *Oikos*, *111*, 459–464.

Kvarnemo, C. (2006). Evolution and maintenance of male care: Is increased paternity a neglected benefit of care? *Behavioral Ecology*, *17*, 144–148.

Lamb, R.J. (1976). Dispersal by nesting earwigs, *Forficula auricularia* (Dermaptera: Forficulidae). *Canadian Journal of Entomology*, *108*, 213–216.

Lambret, P., Rutter, I., Grillas, P., and Stoks, R. (2018). Oviposition plant choice maximizes offspring fitness in an aquatic predatory insect. *Hydrobiologia*, *823*, 1–12.

Ledoux, A. (1958). Biologie et comportement de l'Embioptère *Monotylota ramburi* Rims.-Kors. *Annales des sciences naturelles Zoologie*, *20*, 515–523.

Lefèvre, T., Chiang, A., Kelavkar, M., Li, H., Li, J., *et al.* (2012). Behavioural resistance against a protozoan parasite in monarch butterfly. *The Journal of Animal Ecology*, *81*, 70–79.

Lessells, C.M. (2012). *Sexual conflict in the evolution of parental care*, eds. N.J. Royle, P.T. Smiseth, and M. Kölliker. Oxford: Oxford University Press, pp. 150–170.

Lin, C.-P., Danforth, B.N., and Wood, T.K. (2004). Molecular phylogenetics and evolution of maternal care in membracine treehoppers. *Systematic Biology*, *53*, 400–421.

Lion, S., and van Baalen, M. (2007). From infanticide to parental care: Why spatial structure can help adults be good parents. *The American Naturalist*, *170*, E26–E46.

López-Ortega, M., and Williams, T. (2018). Natural enemy defense, provisioning and oviposition site selection as maternal strategies to enhance offspring survival in a sub-social bug. *PLoS One*, 13, e019566510.1371/journal.pone.0195665.

Machado, G., and Trumbo, S.T. (2018). *Parental care in Insect behavior: From mechanisms to ecological and evolutionary consequences*, eds. A. Cordoba-Aguilar, D. Gonzalez-Tokman, and I. Gonzalez-Santoyo. Oxford: Oxford University Press, pp. 203–218.

Mas, F., and Kölliker, M. (2011). Differential effects of offspring condition-dependent signals on maternal care regulation in European earwig. *Behavioral Ecology and Sociobiology*, *65*, 341–349.

Matzke, D.A.M., and Lass, K.-D. (2005). Reproductive biology and nymphal development in basal earwig Tagalina papua (Insecta: Dermaptera: Pygidicranidae), with a comparison of brood care in dermaptera and embioptera. *Entomologische Adhandlungen*, *62*, 99–116.

Maynard Smith, J., Burian, R., Kauffman, S., Alberch, P., Cambell, J., *et al.* (1985). Developmental Constraints and Evolution. *The Quarterly Review of Biology*, *60*, 265–287.

Mehlferber, E.C., Benowitz, K.M., Roy-Zokan, E.M., McKinney, E.C., Cunningham, C.B., *et al.* (2017). Duplication and sub/neofunctionalization of Malvolio, an insect homolog of Nramp, in subsocial beetle *Nicrophorus vespilloides*. *G3: Genes, Genomes, Genetics*, *7*, 3393–3403.

Meier, R., Kotrba, M., and Ferrar, P. (1999). Ovoviviparity and viviparity in Diptera. *Biological Reviews*, *74*, 199–258.

Meunier, J. (2015). Social immunity and evolution of group living in insects. *Philosophical Transactions of Royal Society B: Biological Sciences, 370*, 20140102.
Meunier, J., and Kölliker, M. (2012). When it is costly to have a caring mother: Food limitation erases benefits of parental care in earwigs. *Biology Letters, 8*, 547–550.
Meunier, J., Wong, J.W.Y., Gómez, Y., Kuttler, S., Röllin, L., *et al.* (2012). One clutch or two clutches? Fitness correlates of coexisting alternative female life-histories in European earwig. *Evolutionary Ecology, 26*, 669–682.
Miller, J.S., Rudolph, L., and Zink, A.G. (2011). Maternal nest defense reduces egg cannibalism by conspecific females in maritime earwig *Anisolabis maritima*. *Behavioral Ecology and Sociobiology, 65*, 1873–1879.
Mitchell, R. (1975). The evolution of oviposition tactics in bean weevil, *Callosobruchus maculatus*. *Ecology, 56*, 696–702.
Mock, D.W., and Parker, G.A. (1997). *The evolution of sibling rivalry*,. Oxford: Oxford University Press.
Monteith, G.B. (2006). Maternal care in Australian oncomerine shield bugs (Insecta, Heteroptera, Tessaratomidae). *Denisia, 19*, 1135–1152.
Monteith, K.M., Andrews, C., and Smiseth, P.T. (2012). Post-hatching parental care masks effects of egg size on offspring fitness: A removal experiment on burying beetles. *Journal of Evolutionary Biology, 25*, 1815–1822.
Moore, A.J., and Benowitz, K.M. (2019). From phenotype to genotype: Precursor hypothesis predicts genetic influences that facilitates transitions in social behavior. *Current Opinion in Insect Science, 34*, 91-96.
Moore, P.J., Harris, W.E., and Moore, A.J. (2007). The cost of keeping eggs fresh: Quantitative genetic varitation in females that mate late relative to sexual maturation. *The American Naturalist, 169*, 311.
Müller, A., Praz, C., and Dorchin, A. (2018). Biology of Palaearctic Wainia bees of subgenus *Caposmia* including a short review on snail shell nesting in osmiine bees (Hymenoptera, Megachilidae). *Journal of Hymenoptera Research, 65*, 61–89.
Nafus, D.M., and Schreiner, I.H. (1988). Parental care in a tropical nymphalid butterfly *Hypolimnas anomala*. *Animal Behaviour, 36*, 1425–1431.
Naghdi, M., Maestro, J., Belles, X., and Bandani, A. (2016). Transduction of vitellogenic signal of juvenile hormone by Methoprene-tolerant in cockroach *Blattella germanica* L. (Dictyoptera, Blattellidae). *Arthropods, 5*, 130–137.
Nalepa, C.A. (1990). Early development of nymphs and establishment of hindgut symbiosis in *Cryptocercus punctulatus* (Dictyoptera: Cryptocercidae). *Annals of Entomological Society of America, 83*, 786–789.
Nalepa, C.A., and Bell, W.J. (1997). *Postovulation parental investment and parental care in cockraches in The evolution of social behavior in insects and arachnids*, eds. J.C. Choe and B.J. Crespi. Cambridge: Cambridge University Press, pp. 26–51.
Nalepa, C.A., Maekawa, K., Shimada, K., Saito, Y., Arellano, C., *et al.* (2008). Altricial development in subsocial wood-feeding cockroaches. *Zoological Science, 25*, 1190–1198.
Nelson, C.M., Ihle, K.E., Fondrk, M.K., Page, R.E., and Amdam, G.V. (2007). The gene vitellogenin has multiple coordinating effects on social organization. *PLoS Biology, 5*, 0673–0677.
Noriyuki, S., Kawatsu, K., and Osawa, N. (2012). Factors promoting maternal trophic egg provisioning in non-eusocial animals. *Population Ecology, 54*, 455–465.
Normark, B.B. (2013). Micromalthus debilis. *Current Biology, 23*, R430–R431.
Nováková, E., Husník, F., Šochová, E., and Hypša, V. (2015). *Arsenophonus* and *Sodalis* symbionts in louse flies: An analogy to *Wigglesworthia* and *Sodalis* system in tsetse flies. *Applied and Environmental Microbiology, 81*, 6189–6199.
Nowak, M.A., Tarnita, C.E., and Wilson, E.O. (2010). The evolution of eusociality. *Nature, 466*, 1057–1062.
Ostrovsky, A.N., Lidgard, S., Gordon, D.P., Schwaha, T., Genikhovich, G., *et al.* (2016). Matrotrophy and placentation in invertebrates: A new paradigm. *Biological Reviews, 91*, 673–711.
Palmer, W.J., Duarte, A., Schrader, M., Day, J.P., Kilner, R., *et al.* (2016). A gene associated with social immunity in burying beetle *Nicrophorus vespilloides*. *Proceedings of Royal Society B: Biological Sciences, 283*, 20152733.
Parker, D.J., Cunningham, C.B., Walling, C. a, Stamper, C.E., Head, M.L., *et al.* (2015). Transcriptomes of parents identify parenting strategies and sexual conflict in a subsocial beetle. *Nature Communications, 6*, 8449.

Parker, G.A., Royle, N.J., and Hartley, I.R. (2002). Intrafamilial conflict and parental investment: A synthesis. *Philosophical Transactions of Royal Society of London B: Biological sciences, 357*, 295–307.

Perry, J.C., and Roitberg, B.D. (2005). Ladybird mothers mitigate offspring starvation risk by laying trophic eggs. *Behavioral Ecology and Sociobiology, 58*, 578–586.

Perry, J.C., and Roitberg, B.D. (2006). Trophic egg laying: Hypotheses and tests. *Oikos, 112*, 706–714.

Pilakouta, N., Richardson, J., and Smiseth, P.T. (2016). If you eat, I eat: Resolution of sexual conflict over consumption from a shared resource. *Animal Behaviour, 111*, 175–180.

Pöykkö, H. (2006). Females and larvae of a geometrid moth, *Cleorodes lichenaria*, prefer a lichen host that assures shortest larval period. *Environmental Entomology, 35*, 1669–1676.

Pollock, D.A., and Normark, B.B. (2002). The life cycle of *Micromalthus debilis* LeConte (1878) (Coleoptera: Archostemata: Micromalthidae): historical review and evolutionary perspective. *Journal of Zoological Systematics and Evolutionary Research, 40*, 105–11210.1046/j.1439-0469.2002.00183.x.

Ratnieks, F.L.W., Foster, K.R., and Wenseleers, T. (2006). Conflict resolution in insect societies. *Annual Review of Entomology, 51*, 581–608.

Roulin, A. , and Dreiss, A.N. (2012). Sibling competition and cooperation over parental care. In The evolution of parental care,. Oxford: Oxford University Press, p. pp. 133-149.

Ratnieks, F.L.W., and Visscher, P.K. (1989). Worker policing in honeybee. *Nature, 342*, 796–797.

Reavey, C.E., Beare, L., and Cotter, S.C. (2014). Parental care influences social immunity in burying beetle larvae. *Ecological Entomology, 39*, 395–398.

Refsnider, J.M., and Janzen, F.J. (2010). Putting eggs in one basket: Ecological and evolutionary hypotheses for variation in oviposition-site choice. *Annual Review of Ecology, Evolution, and Systematics, 41*, 39–57.

Rehan, S.M., and Toth, A.L. (2015). Climbing social ladder: Molecular evolution of sociality. *Trends in Ecology & Evolution, 30*, 426–433.

Roff, D.A. (2002). *Life history evolution,*. Sunderland, MA: Sinauer.

Rossiter, M.C., Cox-Foster, D.L., and Briggs, M.A. (1993). Initiation of maternal effects in Lymantria dispar: Genetic and ecological components of egg provisioning. *Journal of Evolutionary Biology, 6*, 577–589.

Roulin, A., and Dreiss, A. N. (2012). Sibling competition and cooperation over parental care. In *The Evolution of Parental Care*, eds. N. J. Royle, P. T. Smiseth, and M. Kölliker. Oxford: Oxford University Press, pp. 133–149.

Roy-Zokan, E.M., Cunningham, C.B., Hebb, L.E., McKinney, E.C., and Moore, A.J. (2015). Vitellogenin and vitellogenin receptor gene expression is associated with male and female parenting in a subsocial insect. *Proceedings of Royal Society B: Biological Sciences, 282*, 20150787.

Royle, N.J., Alonzo, S.H., and Moore, A.J. (2016). Co-evolution, conflict and complexity: What have we learned about evolution of parental care behaviours? *Current Opinion in Behavioral Sciences, 12*, 30–36.

Royle, N.J., Hartley, I.R., and Parker, G.A. (2002). Begging for control: When are offspring solicitation behaviours honest? *Trends in Ecology & Evolution, 17*, 434–440.

Royle, N.J., Hartley, I.R., and Parker, G.A. (2004). Parental investment and family dynamics: Interactions between theory and empirical tests. *Population Ecology, 46*, 231–241.

Royle, N.J., Smiseth, P.T., and Kölliker, M. (2012). *The evolution of parental care,*. Oxford Uni. Oxford: Oxford University Press.

Rozen, D. E., Engelmoer, D. J. P., and Smiseth, P. T. (2008). Antimicrobial strategies in burying beetles breeding on carrion. *Proceedings of the National Academy of Sciences*, 105, 17890–1789510.1073/pnas.0805403105.

Sabatier, L., Jouanguy, E., Dostert, C., Zachary, D., Dimarcq, J.L., *et al.* (2003). Pherokine-2 and -3: Two *Drosophila* molecules related to pheromone/odor-binding proteins induced by viral and bacterial infections. *European Journal of Biochemistry, 270*, 3398–3407.

Sato, H. (1997). Two nesting behaviours and life history of a subsocial African dungrolling beetle, *Scarabaeus catenatus* (Coleoptera: Scarabaeidae). *Journal of Natural History, 31*, 457–469.

Scott, M.P., and Traniello, J.F.A. (1990). Behavioural and ecological correlates of male and female parental care and reproductive success in burying beetles (*Nicrophorus* spp.). *Animal Behaviour, 39*, 274–283.

Schrader, M., Jarrett, B.J.M., andKilner, R.M. (2015). Parental care masks a density-dependent shift from cooperation to competition among burying beetle larvae. *Evolution*, 69, 1077–108410.1111/evo.12615.
Sedivy, C., Ascher, J.S., Özbek, H., Müller, A., Praz, C., *et al.* (2010). Nests, petal usage, floral preferences, and immatures of *Osmia (Ozbekosmia) avosetta* (Megachilidae: Megachilinae: Osmiini), including biological comparisons with other Osmiine bees. *American Museum Novitates*, *3680*, 1–22.
Silberbush, A., and Blaustein, L. (2011). Mosquito females quantify risk of predation to their progeny when selecting an oviposition site. *Functional Ecology*, *25*, 1091–1095.
Sites, R.W., and McPherson, J.E. (1982). Life history and laboratory rearing of *Sehirus cinctus cinctus* (Hemiptera: Cydnidae), with descriptions of immature stages. *Annals of Entomological Society of America*, *75*, 210–215.
Smiseth, P.T., and Moore, A.J. (2002). Does resource availability affect offspring begging and parental provisioning in a partially begging species? *Animal Behaviour*, *63*, 577–585. 10.1006/anbe.2001.1944.
Smiseth, P. T., andMoore, A. J. (2004). Signalling of hunger when offspring forage by both begging and self-feeding. *Animal Behaviour*, *67*, 1083–108810.1016/j.anbehav.2003.10.012.
Smiseth, P.T., Darwell, C.T., and Moore, A.J. (2003). Partial begging: An empirical model for early evolution of offspring signalling. *Proceedings of Royal Society B: Biological Sciences*, *270*, 1773–1777.
Smiseth, P.T., Dawson, C., Varley, E., and Moore, A.J. (2005). How do caring parents respond to mate loss? Differential response by males and females. *Animal Behaviour*, *69*, 551–559.
Smiseth, P.T., Kölliker, M., and Royle, N.J. (2012). What is parental care? In *The evolution of parental care*, Oxford: Oxford University Press, pp. 1–17.
Smiseth, P.T., Lennox, L., and Moore, A.J. (2007). Interaction between parental care and sibling competition: Parents enhance offspring growth and exacerbate sibling competition. *Evolution*, *61*, 2331–2339.
Smiseth, P.T., and Royle, N.J. (2018). The resolution of conflict in families. *Current Opinion in Insect Science*, *28*, 8–12.
Smith, R.L. (1997). Evolution of paternal care in giant water bugs (Heteroptera: Belostomatidae. In *The evolution of social behavior in insects and arachnids*, eds. J.C. Choe and B.J. Crespi. Cambridge: Cambridge University Press, pp. 116–149.
Socias-Martínez, L., and Kappeler, P.M. (2019). Catalyzing transitions to sociality: Ecology builds on parental care. *Frontiers in Ecology and Evolution*, *7*, 1–18.
Søvik, E., LaMora, A., Seehra, G., Barron, A.B., Duncan, J.G., *et al.* (2017). *Drosophila* divalent metal ion transporter Malvolio is required in dopaminergic neurons for feeding decisions. *Genes, Brain and Behavior*, *16*, 506–514.
Staerkle, M., and Kölliker, M. (2008). Maternal food regurgitation to nymphs in earwigs (*Forficula auricularia*). *Ethology*, *114*, 844–850.
Stay, B., and Coop, A. C. (1974). 'Milk' secretion for embryogenesis in a viviparous cockroach. *Tissue and Cell*, *6*, 669–69310.1016/0040-8166(74)90009-3.
Suzuki, S. (2001). Suppression of dungal development on carcasses by burying beetle *Nicrophorus quadripunctatus* (Coleoptera: Silphidae). *Entomological Science*, *4*, 403–405.
Suzuki, S. (2010). Progressive provisioning by females of earwig, *Anisolabis maritima*, increases survival rate of young. *Journal of Insect Science*, *10*, 184.
Suzuki, S. (2015). Recognition of sex of parents by young: Effects of hunger and biparental care on begging behavior of *Nicrophorus quadripuncatus* larvae. *Journal of Insect Behavior*, *28*, 338–344.
Suzuki, S. (2018). Provisioning control during maternal care by earwig *Anisolabis maritima* (Dermaptera: Anisolabididae): Do mothers adjust provisioning according to offspring need? *Entomological Science*, *21*, 428–432.
Suzuki, S., Kitamura, M., and Matsubayashi, K. (2005). Matriphagy in hump earwig, *Anechura harmandi* (Dermaptera: Forficulidae), increases survival rates of offspring. *Journal of Ethology*, *23*, 211–213.
Tallamy, D.W. (1984). Insect parental care. *BioScience*, *34*, 20–24.
Tallamy, D.W. (2001). Evolution of exclusive paternal care in arthropods. *Annual Reviews of Entomology*, *46*, 139–165.
Tallamy, D.W., and Denno, R.F. (1981). Maternal care in *Gargaphia solani* (Hemiptera: Tingidae). *Animal Behaviour*, *29*, 771–778.

Tallamy, D.W., and Schaefer, C. (1997). Maternal care in Hemiptera: Ancestry, alternatives, and current adaptive value. In *The evolution of social behaviour in insects and arachnids*, ed. B.J. Crespi. Cambridge: Cambridge University Press, pp. 94–115.

Tallamy, D.W., and Wood, T.K. (1986). Convergence patterns in subsocial insects. *Annual Review of Entomology*, *31*, 369–390.

Thompson, A., Vo, D., Comfort, C., and Zakon, H.H. (2014). Expression evolution facilitated convergent neofunctionalization of a sodium channel gene. *Molecular Biology and Evolution*, *31*, 1941–1955.

Tigreros, N., Norris, R.H., Wang, E.H., and Thaler, J.S. (2017). Maternally induced intraclutch cannibalism: An adaptive response to predation risk? *Ecology Letters*, *20*, 487–494.

Tobe, S.S., Huang, J., Stay, B., Hult, E.F., and Marchal, E. (2013). *Diploptera punctata* as a model for studying endocrinology of arthropod reproduction and development. *General and Comparative Endocrinology*, *188*, 85–93.

Trivers, R.L. (1972). Parental investment and sexual selection. In *Sexual selection and descent of man*,. Chicago, IL: B. Campbell, pp. 136–179.

Trivers, R.L. (1974). Parent-offspring conflict. *American Zoologist*, *14*, 249–264.

Trumbo, S.T. (1996) Parental care in invertebrates. *Advances in Study of Behavior*, *25*, 3–51.

Trumbo, S.T. (2012) Patterns of parental care in invertebrates. In *The evolution of parental care*, eds. N.J. Royle, P.T. Smiseth and M. Kölliker. Oxford and New-York: Oxford University Press, pp. 81–100.

Trumbo, S.T. (2017). Feeding upon and preserving a carcass: Function of prehatch parental care in a burying beetle. *Animal Behaviour*, *130*, 241–249.

Trumbo, S.T. (2018). Juvenile hormone and parental care in subsocial insects: Implications for role of juvenile hormone in evolution of sociality. *Current Opinion in Insect Science*, *28*, 13–18.

Trumbo, S.T. (2019). The physiology of insect families: A door to study of social evolution in *Advances in Insect Physiology*,. Elsevier Ltd., pp. 203–250.

Uller, T. (2012). Parental effects in development and evolution. In *The Evolution of Parental Care*, eds. N.J. Royle, P.T. Smiseth, and M. Kölliker. Oxford: Oxford University Press, p.247–266.

Van Meyel, S., Körner, M., and Meunier, J. (2018). Social immunity: Why we should study its nature, evolution and functions across all social systems. *Current Opinion in Insect Science*, *28*, 1–7.

Vinson, S. B. (1976). Host selection by insect parasitoids. *Annual Review of Entomology*, 21, 109–133 10.1146/annurev.en.21.010176.000545.

Walsh, J.T., Signorotti, L., Linksvayer, T.A., and d'Ettorre, P. (2018). Phenotypic correlation between queen and worker brood care supports role of maternal care in evolution of eusociality. *Ecology and Evolution*, *8*, 10409–10415.

West-Eberhard, M.J. (1996). Wasp societies as microcosms for study of development and evolution. In *Natural History and Evolution of Paper-Wasps*, eds. S. Turillazzi and M.J. West-Eberhard. Oxford and New York: Oxford University Press, pp. 290–317.

West, M.J., and Alexander, R.D. (1963). Sub-social behavior in a burrowing cricket, *Anurogryllus muticus* (De Geer). *The Ohio Journal of Science*, *63*, 19–24.

West, S.A., Pen, I., and Griffin, A.S. (2002). Cooperation and competition between relatives. *Science*, *296*, 72–75.

Wise, R.A. (2004). Dopamine, learning and motivation. *Nature Reviews Neuroscience*, *5*, 483–494.

Wong, J.W.Y., and Kölliker, M. (2012). The effect of female condition on maternal care in European earwig. *Ethology*, *118*, 450–459.

Wong, J.W.Y., Meunier, J., and Kölliker, M. (2013). The evolution of parental care in insects: the roles of ecology, life history and the social environment. *Ecological Entomology*, 38, 123–137.

Wu, M., Walser, J.-C., Sun, L., and Kölliker, M. (2020). The genetic mechanism of selfishness and altruism in parent-offspring coadaptation. *Science Advances*, *6*, 1–11.

Wyatt, T.D. (1986). How a subsocial intertidal beetle, *Bledius spectabilis*, prevents flooding and anoxia in its burrow. *Behavioral Ecology and Sociobiology*, *19*, 323–331.

Zeh, D.W., and Smith, R.L. (1985). Paternal investment by terrestrial arthropods. *Integrative and Comparative Biology*, *25*, 785–805.

Zeh, D.W., Zeh, J.A., and Smith, R.L. (1989). Ovipositors, amnions and eggshell architecture in diversification of terrestrial arthropods. *The Quarterly Review of Biology*, *64*, 147–168.

Zera, A.J., and Zhao, Z. (2003). Life-history evolution and microevolution of intermediary metabolism: Activities of lipid-metabolizing enzymes in life-history morphs of a wing-dimorphic cricket. *Evolution*, *57*, 586–596.

Ziadie, M.A., Ebot-Ojong, F., McKinney, E.C., and Moore, A.J. (2019). Evolution of personal and social immunity in context of parental care. *The American Naturalist*, *193*, 296–308.

Zeng, Y., Zhou, F.-H., Kang, W.-N., and Zhu, D.-H. (2017). Availability of unfertilised eggs increases the fitness of nymphal crickets (Gryllidae). *Ecological Entomology*, *42*, 500–50510.1111/een.12411.

Index